TRENDS IN THE THEORY AND
PRACTICE OF NON-LINEAR ANALYSIS

NORTH-HOLLAND MATHEMATICS STUDIES 110

NORTH-HOLLAND – AMSTERDAM • NEW YORK • OXFORD

TRENDS IN THE THEORY AND PRACTICE OF NON-LINEAR ANALYSIS

Proceedings of the VIth International Conference on Trends in the Theory and Practice of Non-Linear Analysis held at The University of Texas at Arlington, June 18-22, 1984

Edited by:

V. LAKSHMIKANTHAM
The University of Texas at Arlington
Arlington
Texas
U.S.A.

1985

NORTH-HOLLAND – AMSTERDAM • NEW YORK • OXFORD

ISBN: 0 444 87704 5

Publishers:

ELSEVIER SCIENCE PUBLISHERS B.V.
P.O. Box 1991
1000 BZ Amsterdam
The Netherlands

Sole distributors for the U.S.A. and Canada:

ELSEVIER SCIENCE PUBLISHING COMPANY, INC.
52 Vanderbilt Avenue
New York, N.Y. 10017
U.S.A.

Library of Congress Cataloging in Publication Data

International Conference on Trends in the Theory and
Practice of Non-linear Analysis (6th : 1984 :
University of Texasat Arlington)
Trends in the theory and practice of non-linear
analysis.

(North-Holland mathematics studies ; 110)
Bibliography: p.
1. Mathematical analysis--Congresses. 2. Nonlinear
theories--Congresses. I. Lakshmikantham, V. II. Title.
III. Series.
QA299.6.I57 1984 515 84-28759
ISBN 0-444-87704-5 (U.S.)

PRINTED IN THE NETHERLANDS

PREFACE

An International Conference on *Trends in Theory and Practice of Nonlinear Analysis* was held at The University of Texas at Arlington during June 18-22, 1984. This conference was sponsored by the U. S. Army Research Office and The University of Texas at Arlington. This was the sixth in a series of conferences organized by The University of Texas at Arlington. It is a pleasure to acknowledge the financial support received from the various sponsoring agencies which made the conference possible.

The present volume consists of the proceedings of this sixth conference. It includes papers that were delivered as invited survey talks and research reports as well as contributed papers. There were well over seventy talks and twelve countries were represented.

The aim of the conference was to feature recent trends in theory and practice of nonlinear analysis. The contents of this conference are very broad including theory as well as applications. The works, in some cases, reflected collaborative efforts between mathematicians and other scientists and engineers. Indeed there are talks concerned with optimal control and variational methods which deal with problems that are either deterministic or stochastic in nature. Several papers in qualitative behavior of solutions of nonlinear evolution equations as well as partial differential equations including current status of Hamilton-Jacobi equations, are presented. Some of the partial differential equations as well as Volterra equations and ordinary differential equations represented models in the life and natural sciences. In related papers numerical techniques are developed to analyze the models mentioned above as well as others. There is also a group of mathematical scientists concerned with dynamical systems; particularly, stability theory, bifurcation analysis, chaos, and pattern formation. There are a few papers on delay differential equations, particularly for boundary value problems.

One technique found in many papers includes asymptotics and perturbation methods. This is especially evident in those papers dealing with the qualitative behavior of solutions of these nonlinear equations.

In summary, this book attempts to put together the work of a wide range of mathematicians, scientists, and engineers. The problems are both theoretical as well as computational, deterministic as well as stochastic, and the models include differential equations, with and without delay, as well as integral equations in such fields as biology, chemistry, and physics.

I wish to express my special thanks to my colleagues Professors C. Corduneanu, S. Bernfeld, J. Bolen and G. Ladde for helping me in planning and organizing the conference and to my secretaries Mrs. Gloria Brown and Ms. Marian Pruiett, for assisting me during the conference. I am extremely thankful to Mrs. Pruiett for typing partially and finalizing the proceedings of the conference. My immense thanks are due to publishers North-Holland Publishing Company, Inc. for their cooperation. Finally, my appreciation is extended to Lisa Rathert, Debbie Brantley and Patricia Jackson for their typing assistance.

V. Lakshmikantham

TABLE OF CONTENTS

Preface v

On the Solutions of Third Order Non-Linear Boundary Value Problems
A.R. AFTABIZADEH and J. WIENER 1

Recent Results on Multipoint Boundary Value Problems for Higher Order Differential Equations
R.P. AGARWAL 7

A Theory of Nonlinear Evolution Equations
M. ALTMAN 17

Asymptotic Analysis of a Functional-Integral Equation Related to Cell Population Kinetics
O. ARINO and M. KIMMEL 27

On the B-Convergence of the θ-Method Over Infinite Time for Time Stepping for Evolution Equations
O. AXELSSON 33

Filter Stability in Flows
P.N. BAJAJ 41

Containment of Solutions to Strongly Coupled Parabolic Systems
P.W. BATES 45

Developments in Fixed Point Theory for Nonexpansive Mappings
L.P. BELLUCE and W.A. KIRK 55

Hopf Bifurcation with a General Transversality Condition
S.R. BERNFELD 63

Stability Regions of Certain Linear Second Order Periodic Differential Equations
S.R. BERNFELD and M. PANDIAN 67

Well-Posedness of Functional Differential Equations with Nonatomic D Operators
J.A. BURNS, T.L. HERDMAN and J. TURI 71

Periodicity in Linear Volterra Equations
T.A. BURTON 79

A Nonlinear Diffusion Epidemic System with Boundary Feedback
V. CAPASSO and L. MADDALENA 85

Applications of Admissibility for Volterra Integral Equations
H. CASSAGO, JR. 91

Multiple Solutions for a Dirichlet Problem with Jumping Nonlinearities
A. CASTRO and R. SHIVAJI 97

Some Estimates for a System of Multiple Reactions
J. CHANDRA and P. DAVIS 103

Qualitative Problems for Some Hyperbolic Equations
C. CORDUNEANU and H. POORKARIMI 107

Viscosity Solutions of Hamilton-Jacobi Equations in Banach Spaces
M.G. CRANDALL and P.-L. LIONS 115

Maximal Regularity for Abstract Differential Equations and Applications to the Existence of Periodic Solutions
G. DA PRATO 121

A Finite Element Error Estimate for Regularized Compressible Flow
H. DINH and G.F. CAREY 127

Some Results on Non-Resonant Non-Linear Delay Differential Equations
L.D. DRAGER and W. LAYTON 131

Asymptotics of Numerical Methods for Nonlinear Evolution Equations
L.D. DRAGER, W. LAYTON and R.M.M. MATTHEIJ 137

Exponential Dichotomy of Nonlinear Systems of Ordinary Differential Equations
S. ELAYDI and O. HAJEK 145

A Flow Associated with a Semiflow
S. ELAYDI and S.K. KAUL 155

A Method of Finding Critical Points of Nonlinear Functionals
A. EYDELAND 161

Asymptotic Behavior of Nonlinear Functional Evolution Equations in Fading Memory Spaces
W.E. FITZGIBBON 167

Friendly Spaces for Functional Differential Equations with Infinite Delay
J.R. HADDOCK 173

The First Boundary Value Problem for Nonlinear Diffusion
C.J. HOLLAND and J.G. BERRYMAN 183

An Asymptotic Analysis of a Reaction-Diffusion System
F.A. HOWES 187

On a Nonlinear Hyperbolic Integrodifferential Equation with a Singular Kernel
W.J. HRUSA and M. RENARDY 193

Boundary Trajectories of Generalized Control Systems
B. KAŠKOSZ 201

Isolation of the Zeros of a Complex Polynomial by Exploring Function Structure Uniqueness of the Solution Set Established
D.A. KLIP 207

Distal, Equicontinuous, Zero Characteristic, and Recurrent Dynamical Systems
R.A. KNIGHT 217

On the Asymptotic Behavior of Solutions to Nonlinear Volterra Equations
K. KOBAYASI 223

Random Difference Inequalities
G.S. LADDE and M. SAMBANDHAM 231

Singularly Perturbed Stochastic Differential Systems
G.S. LADDE and O. SIRISAENGTAKSIN 241

Systems of First Order Partial Differential Equations and Monotone Iterative Technique
G.S. LADDE and A.S. VATSALA 249

Improved a Posteriori Error Bounds for Quasilinear Boundary-Value Problems by the Method of Pseudolinear Equations
J.E. LAVERY 257

Periodic or Unbounded Solutions for a Class of Three-Dimensional ODE Systems with Biological Applications
D.S. LEVINE 263

The Phenomenon of Quenching: A Survey
H.A. LEVINE 275

Interpolation Between Spaces of Continuous Functions
A. LUNARDI 287

Constructive Existence of Solution for Negative Exponent Generalized Emden-Fowler Nodal Problems
C.D. LUNING and W.L. PERRY 295

Existence Results for the Inverse Problem of the Volume Potential
C. MADERNA, C.D. PAGANI and S. SALSA 301

Variational Problems Governed by a Multi-Valued Evolution Equation
T. MARUYAMA 305

On the Stability of Equilibrium for Periodic Mechanical Systems
V. MOAURO 309

Remarks on Time Discretization of Contraction Semigroups
O. NEVANLINNA 315

Low-Dimensional Behavior of the Pattern Formation Cahn-Hilliard Equation
B. NICOLAENKO and B. SCHEURER 323

Stochastic Continuity and Random Differential Inequalities
J.J. NIETO 337

Information Processing in Vertebrate Retina
M.N. OGUZTORELI, T.M. CAELLI and G. STEIL 345

Numerical Solution of Quasilinear Boundary Value Problems
M.C. PANDIAN 357

Numerical Methods for Simultaneously Approaching Roots of Polynomials
L. PASQUINI and D. TRIGIANTE 363

The Parallel Sum of Generalized Gradients
G.B. PASSTY and R. TORREJÓN 371

An Exact, Direct, Formal Integral (DFI) Approach to Differential Equations
F.R. PAYNE 377

A Quasi-Autonomous Second-Order Differential Inclusion
E.I. POFFALD and S. REICH 387

On Systems with Transfer Functions Related to the Riemann Zeta Function
V.M. POPOV 393

Limit Sets in Infinite Horizon Optimal Control Systems
E.O. ROXIN 401

Exchange of Stability and Bifurcation for Periodic Differential Systems
L. SALVADORI 409

On Invariant Transformations of a Class of Evolution Equations
K. SEITZ 417

Cauchy Problem for Hyper-Parabolic Partial Differential Equations
R.E. SHOWALTER 421

Existence, Uniqueness, and Galerkin Approximations for Semilinear Periodically Forced Wave Equations at Resonance
M.W. SMILEY 427

Hopf Bifurcation for Periodic Systems
F. VISENTIN 435

Remarks on Some Strongly Nonlinear Degenerate Sturm-Liouville Eigenvalue Problems
P.A. VUILLERMOT 441

Point Data Boundary Value Problems for Functional Differential Equations
J. WIENER and A.R. AFTABIZADEH 445

On the Use of Iterative Methods with Supercomputers for Solving Partial Differential Equations
D.M. YOUNG and D.R. KINCAID 455

A Non-Linear Result About Almost-Periodic Solutions of Abstract Differential Equations
S. ZAIDMAN 467

A Mathematical Model for the Study of the Motion of a Mixture of Two Viscous Incompressible Fluids
A. ZARETTI 469

A Survey of the Oscillation of Solutions to First Order Differential Equations with Deviating Arguments
B.G. ZHANG 475

Author Address List 485

Trends in the Theory and Practice of Non-Linear Analysis
V. Lakshmikantham (Editor)

ON THE SOLUTIONS OF THIRD ORDER NON-LINEAR BOUNDARY VALUE PROBLEMS*

A.R. Aftabizadeh, Joseph Wiener

Department of Mathematics
Pan American University
Edinburg, Texas 78539
U.S.A.

Existence and uniqueness of solutions of third order non-linear boundary value problems are discussed. Two different methods have been used, (i) contraction mapping principle and (ii) successive approximations. It is found that the results obtained by method(ii) are better than those given by (i) and by [1].

1. INTRODUCTION

This paper has been inspired by R.P. Agarwal's paper [1]. Here, we are concerned with existence and uniqueness results for non-linear third order differential equation

(1.1) $$y''' = f(x,y,y',y'')$$

when the boundary conditions are defined at two points. It will always be assumed that the function $f(x,y,y',y'')$ is continuous in the interior of its domain. In Section 2, some existence theorems are given. We also prove uniqueness of solutions by using contraction mapping. In Section 3, by using successive approximation method, we obtain better bounds, given in Section 2 and given by [1]. To do this, we transform equation (1.1) into a second order integro-differential equation. Then we apply known results for second order boundary value problems, Schauder's fixed point theorem, contraction mapping and successive approximation, to obtain existence and uniqueness results for third order boundary value problems.

2. EXISTENCE AND UNIQUENESS

In this section, we prove a number of theorems, for the existence and uniqueness of solutions of (1.1).

Theorem 2.1: For some positive real numbers M,N, and L consider the set $\Omega = \{(x,y,u,v):\ a \le x \le b,\ |y| \le 2M,\ |u| \le 2N,\ |v| \le 2L\}$. Let $|f| \le Q$ on Ω. If

(2.1) $$b - a \le \min \left\{ \frac{2M - |A|}{2N},\ \left(\frac{8N}{Q}\right)^{1/2},\ \frac{2L}{Q} \right\}$$

and

(2.2) $$|B| + |C-B| \le N, \quad |C-B| \le (b-a)L,$$

then the boundary value problem

*Research partially supported by U. S. Army Research Grant No. DAAG29-84-G-0034

(2.3)
$$y''' = f(x,y,y',y'')$$
$$y(a) = A, y'(a) = B, y'(b) = C$$

has a solution on [a,b].

Proof. Let $y'(x) = u(x)$. Since $y(a) = A$, then one has

(2.4)
$$y(x) = A + \int_a^x u(s)ds,$$

and

(2.5)
$$u'' = f(x,A + \int_a^x u(s)ds,u,u')$$
$$u(a) = B, \ u(b) = C.$$

The existence of solutions of (2.3) is assured by showing the existence of solutions for (2.5). The set $B[a,b] = \{u(x) \in C^{(1)}[a,b]: \|u\|_0 \leq 2N, \ \|u'\|_0 \leq 2L\}$ is a closed convex subset of the Banach space $C^{(1)}[a,b]$. The mapping $T: C^{(1)}[a,b] \to C^{(1)}[a,b]$ defined by

(2.6)
$$(Tu)(x) = B + (C-B)\frac{x-a}{b-a} + \int_a^b G(x,t)f(t,A + \int_a^t u(s)ds,u,u')dt$$

where $G(x,t)$ is the Green's function for the B.V.P.

$$u'' = 0, \ u(a) = u(b) = 0, \text{ i.e.}$$
$$G(x,t) = \frac{-1}{b-a}\begin{cases}(b-x)(t-a), & a \leq t \leq x \leq b\\ (b-t)(x-a), & a \leq x \leq t \leq b,\end{cases}$$

is completely continuous. For $u \in B[a,b]$, we have, using (2.6), (2.1), and (2.2),

$$|(Tu)(x)| \leq |B| + |C-B| + \frac{1}{8}Q(b-a)^2 \leq 2N,$$

and

$$|(Tu)'(x)| \leq \frac{1}{b-a}|C-B| + \frac{1}{2}Q(b-a) \leq 2L,$$

This means T maps $B[a,b]$ into itself. It then follows from the Schauder fixed-point theorem that T has a fixed point in $B[a,b]$ which is a solution of (2.5). Proof is complete.

Theorem 2.2. For some positive real numbers M,N, and L consider the set $\Omega = \{(x,y,u,v): a \leq x \leq b, \ |y| \leq 2M, \ |u| \leq 2N, \ |v| \leq 2L\}$. Let $|f| \leq Q$ on Ω. If

(2.7)
$$b - a \leq \min\left\{\frac{2M-|A|}{2N}, \left(\frac{2N}{Q}\right)^{1/2}, \frac{L}{Q}\right\},$$

and

(2.8)
$$|B| + (b-a)|C| \leq N, \quad |C| \leq L,$$

then the boundary value problem

(2.9)
$$y''' = f(x,y,y',y'')$$
$$y(a) = A, \ y'(a) = B, \ y''(b) = C$$

has a solution on [a,b].

There is an analogous result for the boundary value problem

$$(2.10) \qquad y''' = f(x,y,y',y'')$$
$$y(a) = A, \quad y''(a) = B, \quad y'(b) = C.$$

Remark 2.1. The conditions of Theorems 2.1 and 2.2 are different from the conditions of Theorems 2.2 and 2.3 of [1].

Now, let us assume that $f(x,y,y',y'')$ is Lipschitzian. Then we obtain

Theorem 2.3. Let $f(x,y,y',y'')$ be continuous and

$$(2.11) \qquad |f(x,y,u,v) - f(x,\bar{y},\bar{u},\bar{v})| \le L_1|y-\bar{y}| + L_2|u-\bar{u}| + L_3|v-\bar{v}|,$$

for (x,y,u,v), $(x,\bar{y},\bar{u},\bar{v}) \in [a,b] \times R^3$. Then, if

$$(2.12) \qquad \frac{1}{12}(b-a)^3 L_1 + \frac{1}{8}(b-a)^2 L_2 + \frac{1}{2}(b-a)L_3 < 1$$

the boundary value problem

$$(2.13) \qquad y''' = f(x,y,y',y'')$$
$$y(a) = A, \quad y'(a) = B, \quad y'(b) = C$$

has a unique solution.

Proof. Let the space S consist of continuously differentiable functions on $[a,b]$ with the norm

$$||u|| = [\tfrac{2}{3}(b-a)L_1 + L_2]\max|u(x)| + L_3\max|u'(x)|, \quad a \le x \le b.$$

Define the operator $T: C^{(1)}[a,b] \to C^{(1)}[a,b]$ by (2.6). We shall show that T is contracting operator. For u and $v \in S$ we have, using (2.11),

$$|(Tu)(x)-(Tv)(x)| \le \int_a^b |G(x,t)|[L_1(t-a)|u(t)-v(t)| + L_2|u(t)-v(t)| + L_3|u'(t)-v'(t)|]dt,$$

or

$$|(Tu)(x) - (Tv)(x)| \le \frac{(b-a)^2}{8}\left[\left(\frac{8\sqrt{3}}{27}(b-a)L_1 + L_2\right)\max|u-v| + L_3\max|u' - v'|\right],$$

and

$$|(Tu)'(x) - (Tv)'(x)| \le \frac{(b-a)}{2}\left[\left(\frac{2}{3}(b-a)L_1 + L_2\right)\max|u-v| + L_3\max|u'-v'|\right],$$

or

$$|(Tu)(x) - (Tv)(x)| \le \frac{1}{8}(b-a)^2\,\|u-v\|, \quad |(Tu)'(x) - (Tu)'(x)| \le \frac{1}{2}(b-a)\,\|u-v\|.$$

Therefore

$$\|Tu - Tv\| \le \left[\frac{1}{12}(b-a)^3 L_1 + \frac{1}{8}(b-a)^2 L_2 + \frac{1}{2}(b-a)L_3\right]\|u-v\|.$$

This inequality and the contraction mapping principle yield the theorem.

Theorem 2.4. Suppose f satisfies (2.11) with

$$(2.14) \qquad \frac{1}{3}L_1(b-a)^3 + \frac{1}{2}L_2(b-a)^2 + L_3(b-a) < 1,$$

Then the boundary value problem (2.9) has a unique solution.

There is an analogous result for the boundary value problem (2.10).

Remark 2.2. The conditions of Theorems 2.3 and 2.4 are the same as the conditions of Theorems 2.5 and 2.6 of [1].

3. AN IMPROVED BOUND FOR THE EXISTENCE AND UNIQUENESS

In Section 2, we employed contraction mapping principle and we obtained conditions on the length of the interval [a,b] in terms of Lipschitz constants L_1, L_2, and $L_3 > 0$. Here we shall use the successive approximations method given by R.P. Agarwal [1], to find the length of the interval. The result obtained indicate that the solution exists over a length of interval greater than that obtained in section 2, and also greater than that obtained in [1].

Theorem 3.1. Let $f(x,y,y',y'')$ be continuous and Lipschitzian on $[a,b] \times R^3$. If

$$\frac{1}{20}(b-a)^3L_1 + \frac{5}{48}(b-a)^2L_2 + \frac{1}{3}(b-a)L_3 = \alpha < 1, \tag{3.1}$$

then the boundary value problem (2.13) has a unique solution.

Proof. Define the successive approximations

$$\begin{aligned} u_o(x) &= \ell(x) = B + (C-B)\frac{x-a}{b-a} \\ u_{n+1}(x) &= \ell(x) + \int_a^b G(x,t)f(t,A+\int_a^t u_n(s)ds, u_n(t), u_n'(t))dt. \end{aligned} \tag{3.2}$$

Then we have

$$|u_1(x) - u_0(x)| \leq \frac{1}{2}Q(x-a)(b-x), \quad a \leq x \leq b, \tag{3.3}$$

$$|u_1'(x) - u_0'(x)| \leq \frac{1}{2(b-a)}Q[(x-a)^2+(b-x)^2], \quad a \leq x \leq b. \tag{3.4}$$

Now, using the Lipschitz condition we have

(3.5)
$$|u_2(x)-u_1(x)| \leq \int_a^b |G(x,t)|[L_1(t-a)|u_1(t)-u_0(t)| + L_2|u_1(t)-u_0(t)| + L_3|u_1'(t)-u_0'(t)|]dt,$$

and

(3.6)
$$|u_2'(x)-u_1'(x)| \leq \int_a^b |G_x(x,t)|[L_1(t-a)|u_1(t)-u_0(t)| + L_2|u_1(t)-u_0(t)| + L_3|u_1'(t)-u_0'(t)|]dt.$$

We need the following estimates:

$$\int_a^b |G(x,t)||u_1(t) - u_o(t)|(t-a)dt \leq \frac{1}{2}Q(b-x)(x-a)(\frac{1}{20}(b-a)^3) \tag{3.7}$$

$$\int_a^b |G(x,t)||u_1(t) - u_0(t)|dt \leq \frac{1}{2}Q(b-x)(x-a)(\frac{5}{48}(b-a)^2), \tag{3.8}$$

$$\int_a^b |G(x,t)||u_1'(t) - u_0'(t)|dt \leq \frac{1}{2}Q(b-x)(x-a)(\frac{1}{3}(b-a)), \tag{3.9}$$

$$\int_a^b |G_x(x,t)||u_1(t) - u_0(t)|(t-a)dt \leq \frac{1}{2(b-a)}Q[(x-a)^2 + (b-x)^2](\frac{1}{20}(b-a)^3), \tag{3.10}$$

$$(3.11)\quad \int_a^b |G_x(x,t)||u_1(t)-u_0(t)|dt \le \frac{1}{2(b-a)} Q[(x-a)^2 + (b-x)^2] \left(\frac{5}{48}(b-a)^2\right),$$

$$(3.12)\quad \int_a^b |G_x(x,t)||u_1'(t)-u_0'(t)|dt \le \frac{1}{2(b-a)} Q[(x-a)^2 + (b-x)^2] \left(\frac{1}{3}(b-a)\right).$$

Using (3.7) - (3.9) in (3.5) and (3.10) - (3.12) in (3.6) and condition (3.1) we have

$$(3.13)\quad |u_2(x) - u_1(x)| \le \alpha \cdot \frac{1}{2} Q(b-x)(x-a),$$

and

$$(3.14)\quad |u_2'(x) - u_1'(x)| \le \alpha \cdot \frac{1}{2(b-a)} Q[(x-a)^2 + (b-x)^2].$$

By induction and using the above estimates, we obtain

$$(3.15)\quad |u_{n+1}(x) - u_n(x)| \le \alpha^n \cdot \frac{1}{2} Q(b-x)(x-a), \quad a \le x \le b,$$

$$(3.16)\quad |u_{n+1}'(x) - u_n'(x)| \le \alpha^n \cdot \frac{1}{2(b-a)} Q[(x-a)^2 + (b-x)^2], \quad a \le x \le b.$$

Since $\alpha < 1$, estimates (3.15) and (3.16) ensure that the boundary value problem (2.5) has a solution. This implies (2.13) has a solution.

Now, we shall prove that the solution is unique, or in other words the boundary value problem (2.5) has a unique solution. Let us assume that there are two solutions, $u(x)$ and $v(x)$, then

(3.17)

$$|u(x)-v(x)| \le \int_a^b |G(x,t)|[L_1(t-a)|u(t)-v(t)| + L_2|u(t)-v(t)|L_3|u'(t)-v'(t)|]dt.$$

Let

$$\|u-v\| = \max[L_1(b-a)|u(x)-v(x)| + L_2|u(x)-v(x)| + L_3|u'(x)-v'(x)|], a \le x \le b,$$

then

$$|u(x)-v(x)| \le \frac{1}{2}(b-x)(x-a)\ \|u-v\|, \quad a \le x \le b,$$

$$|u'(x)-v'(x)| \le \frac{1}{2(b-a)} [(x-a)^2 + (b-x)^2]\ \|u-v\|, \quad a \le x \le b.$$

From (3.17) and above results, we find

$$|u(x)-v(x)| \le \alpha \cdot \frac{1}{2}(b-x)(x-a)\ \|u-v\|, \quad a \le x \le b,$$

and

$$|u'(x)-v'(x)| \leq \alpha \cdot \frac{1}{2(b-a)} [(b-x)^2 + (x-a)^2] \, ||u-v|| , \quad a \leq x \leq b.$$

Continuing this way, we have

$$|u(x)-v(x)| \leq \alpha^n \cdot \frac{1}{2}(b-x)(x-a) \, ||u-v|| , \quad a \leq x \leq b,$$

and

$$|u'(x)-v'(x)| \leq \alpha^n \cdot \frac{1}{2(b-a)} [(x-a)^2 + (b-x)^2] \, ||u-v|| , \quad a \leq x \leq b.$$

Since $\alpha < 1$, result follows immediately.

Theorem 3.2. Let $f(x,y,y',y'')$ be continuous and Lipschitzian on $[a,b] \times R^3$. Then if

$$\frac{3}{10} (b-a)^3 L_1 + \frac{5}{12} (b-a)^2 L_2 + \frac{1}{2}(b-a) L_3 < 1, \tag{3.18}$$

the boundary value problem (2.9) has a unique solution.

Proof is similar to that of Theorem 3.1.

Remark 3.1. The condition (3.1) of Theorem 3.1 is better than the condition of Theorem 3.2 of [1] which is

$$\frac{7}{120} (b-a)^3 L_1 + \frac{\sqrt{3}-1}{4\sqrt{3}} (b-a)^2 L_2 + \frac{1}{3}(b-a) L_3 < 1,$$

and condition (3.18) of Theorem 3.2 is better than the one of Theorem 3.3 [1] that is

$$\frac{1}{3}(b-a)^3 L_1 + \frac{1}{2}(b-a)^2 L_2 + \frac{1}{2}(b-a) L_3 < 1.$$

REFERENCES

[1] Agarwal, R.P., Non-Linear two point boundary value problems. Indian J. of P. and Applied Math. Vol. 4, Nos. Gandlo (1973), pp. 757-769.

Trends in the Theory and Practice of Non-Linear Analysis
V. Lakshmikantham (Editor)

RECENT RESULTS ON MULTIPOINT BOUNDARY VALUE PROBLEMS FOR HIGHER ORDER DIFFERENTIAL EQUATIONS

Ravi P. Agarwal

Department of Mathematics
National University of Singapore
Kent Ridge, Singapore 0511

1. INTRODUCTION

The purpose of this talk is to present several results which have recently appeared or will be appearing soon for the following multipoint boundary value problem

$$x^{(n)} = f(t,\underset{\sim}{x}(t)) \tag{1.1}$$

$$x(a_i) = A_{1,i},\ x'(a_i) = A_{2,i},\ldots,x^{(k_i)}(a_i) = A_{k_i+1,i};\ 1 \le i \le r\ (\ge 2)$$
$$a < a_1 < a_2 < \ldots < a_r < b,\ 0 \le k_i,\ \sum_{i=1}^{n} k_i + r = n \tag{1.2}$$

where $\underset{\sim}{x}(t)$ stands for $(x(t),x'(t),\ldots,x^{(q)}(t))$, $0 \le q \le n-1$. The function f is assumed to be continuous on $(a,b) \times R^{q+1}$. For a fixed $2 \le r \le n$ but arbitrary $k_i \ge 0$ and a_i, $1 \le i \le r$ such that $a < a_1 < a_2 < \ldots < a_r < b$, the boundary value problem (1.1), (1.2) will be called r point problem.

2. PROBLEM OF BEST CONSTANTS IN HERMITE INTERPOLATION

Theorem 1. [Agarwal 1] Let $x(t) \in C^{(n)}[a_1,a_r]$ satisfying

$$x(a_i) = x'(a_i) = \ldots = x^{(k_i)}(a_i) = 0,\ 1 \le i \le r(\ge 2) \tag{2.1}$$

$$a_1 < a_2 < \ldots < a_r,\ 0 \le k_i,\ \sum_{i=1}^{r} k_i + r = n.$$

Then,

$$|x^{(k)}(t)| \le C_{n,k}\, m(a_r-a_1)^{n-k},\ 0 \le k \le n-1$$

where $m = \max_{a_1\le t\le a_r} |x^{(n)}(t)|$, and

$$C_{n,k} = \frac{1}{(n-k)!}\,\frac{(n-\alpha-1)^{n-\alpha-1}(\alpha-k+1)^{\alpha-k+1}}{(n-k)^{n-k}},\ 0 \le k \le \alpha$$

$$C_{n,\alpha+k} = \frac{k}{(n-\alpha)(n-\alpha-k)!},\ 1 \le k \le n-\alpha-1$$

$\alpha = \min(k_1,k_r)$.

The constants $C_{n,k}$ are the best possible if $\alpha = 0$; $k = 0$, $\alpha = 0$ and for $k \neq 0$, $\alpha \neq 0$ it is an open problem.

3. EXISTENCE AND UNIQUENESS

Theorem 2. [Agarwal 2] Let $K_i > 0$, $0 \leq i \leq q$ be given real numbers and D_o be a closed and bounded subset of R^{q+1}

$$D_o = \{(u_o,u_1,\ldots,u_q) : |u_i| \leq 2K_i,\ 0 \leq i \leq q\}$$

and $|f| \leq Q$ on $[a_1,a_r] \times D_o$. Further, we assume that

$$\max_{a_1 \leq t \leq a_r} |P_{n-1}^{(i)}(t)| \leq K_i, \quad (a_r-a_1) \leq (K_i/QC_{n,i})^{1/n-i};\ 0 \leq i \leq q$$

where $P_{n-1}(t)$ is the unique polynomial of degree (n-1) satisfying (1.2). Then, the boundary value problem (1.1), (1.2) has a solution in D_o.

Corollary 3. [Agarwal 2] Suppose that the function $f(t,u_o,u_1,\ldots,u_q)$ be such that

$$|f(t,u_o,u_1,\ldots,u_q)| \leq L + \sum_{i=0}^{q} L_i|u_i|^{\alpha(i)};\ 0 \leq \alpha(i) < 1,\ 0 \leq i \leq q \tag{3.1}$$

for all $(t,u_o,u_1,\ldots,u_q) \in [a_1,a_r] \times R^{q+1}$ Then, the problem (1.1), (1.2) has a solution.

Theorem 4. [Agarwal 2] Suppose that the inequality (3.1) is satisfied with $\alpha(i) = 1$, $0 \leq i \leq q$ for all $(t,u_o,u_1,\ldots,u_q) \in [a_1,a_r] \times D_1$, where

$$D_1 = \{(u_o,u_1,\ldots,u_q) : |u_i| \leq \max_{a_1 \leq t \leq a_r} |P_{n-1}^{(i)}(t)| + C_{n,i}(a_r-a_1)^{n-i} \frac{L+\ell}{1-\theta},\ 0 \leq i \leq q\}$$

where

$$\ell = \max_{a_1 \leq t \leq a_r} \sum_{i=0}^{q} L_i|P_{n-1}^{(i)}(t)|$$

and

$$\theta = \sum_{i=0}^{q} C_{n,i}\, L_i(a_r-a_1)^{n-i} < 1.$$

Then, the boundary value problem (1.1), (1.2) has a solution in D_1.

Theorem 5. [Agarwal 3] Suppose that the boundary value problem (1.1), (2.1) has a nontrivial solution x(t) and the condition (3.1) is satisfied with $L = 0$; $\alpha(i) = 1$, $0 \leq i \leq q$ for all $(t,u_o,u_1,\ldots,u_q) \in [a_1,a_r] \times D_2$, where

$$D_2 = \{(u_o,u_1,\ldots,u_q) : |u_i| \leq C_{n,i}(a_r-a_1)^{n-i}m,\ 0 \leq i \leq q\}$$

and $m = \max_{a_1 \leq t \leq a_r} |x^{(n)}(t)|$. Then, it is necessary that $\theta \geq 1$.

Definition. The function $f(t,u_0,u_1,\ldots,u_q)$ is said to be of Lipschitz class if for all $(t,u_1,u_1,\ldots,u_q)$, $(t,v_0,v_1,\ldots,v_q) \in [a_1,a_r] \times D$, $D \subseteq R^{q+1}$; the following is satisfied

$$|f(t,u_0,u_1,\ldots,u_q) - f(t,v_0,v_1,\ldots,v_q)| \leq \sum_{i=0}^{q} L_i |u_i - v_i| \ .$$

Theorem 6. [Agarwal 2] Suppose that the function f is of Lipschitz class on $[a_1,a_r] \times D_1$ where $L = \max_{a_1 \leq t \leq a_r} |f(t,0,\ldots,0)|$. Then, the boundary value problem (1.1), (1.2) has a unique solution in D_1.

4. CONVERGENCE OF PICARD'S ITERATIVE METHOD

Definition. A function $\bar{x}(t) \in C^{(n)}[a_1,a_r]$ is called an approximate solution of (1.1), (1.2) if there exist δ and ε nonnegative constants such that

$$\max_{a_1 \leq t \leq a_r} |\bar{x}^{(n)}(t) - f(t,\underset{\sim}{\bar{x}}(t))| \leq \delta \tag{4.1}$$

and

$$\max_{a_1 \leq t \leq a_r} |P_{n-1}^{(i)}(t) - \bar{P}_{n-1}^{(i)}(t)| \leq \varepsilon C_{n,i}(a_r-a_1)^{n-i}, \quad 0 \leq i \leq q$$

where $\bar{P}_{n-1}(t)$ is the unique polynomial of degree (n-1) satisfying

$$\bar{P}_{n-1}(a_i) = \bar{x}(a_i),\ \bar{P}'_{n-1}(a_i) = \bar{x}'(a_i),\ldots,\bar{P}_{n-1}^{(k_i)}(a_i) = \bar{x}^{(k_i)}(a_i),\ 1 \leq i \leq r.$$

Inequality (4.1) means that there exists a continuous function $\eta(t)$ such that

$$\bar{x}^{(n)}(t) = f(t,\underset{\sim}{\bar{x}}(t)) + \eta(t)$$

where

$$\max_{a_1 \leq t \leq a_r} |\eta(t)| \leq \delta \ .$$

Thus, the approximate solution $\bar{x}(t)$ can be expressed as

$$\bar{x}(t) = \bar{P}_{n-1}(t) + \int_{a_1}^{a_r} g(t,s)[f(s,\underset{\sim}{\bar{x}}(s)) + \eta(s)]ds$$

where $g(t,s)$ is the Green's function for the boundary value problem : $x^{(n)} = 0$, (2.1).

In theorems 7 - 12 we consider the Banach space $B = C^{(q)}[a_1,a_r]$ and for all $x(t) \in B$

$$||x|| = \max_{0 \leq j \leq q} \left\{ \frac{C_{n,0}(a_r-a_1)^j}{C_{n,j}} \max_{a_1 \leq t \leq a_r} |x^{(j)}(t)| \right\} \ .$$

Theorem 7. [Agarwal 4] With respect to (1.1), (1.2) we assume that there exists an approximate solution $\bar{x}(t)$ and

(i) the function f is of Lipschitz class on $[a_1,a_r] \times D_2$, where

$$D_2 = \{(u_0,u_1,\dots,u_q) : |u_j - \bar{x}^{(j)}(t)| \le N \frac{C_{n,j}}{C_{n,0}(a_r-a_1)^j}, \ 0 \le j \le q\}$$

(ii) $\theta < 1$

(iii) $(1-\theta)^{-1}(\varepsilon+\delta)C_{n,0}(a_r-a_1)^n \le N.$

Then,

(1) there exists a solution $x^*(t)$ of (1.1), (1.2) in $\bar{S}(\bar{x},N_0)$

(2) $x^*(t)$ is the unique solution of (1.1), (1.2) in $\bar{S}(\bar{x},N)$

(3) the Picard's sequence $\{x_m(t)\}$ defined by

$$x_{m+1}(t) = P_{n-1}(t) + \int_{a_1}^{a_r} g(t,s)f(s,\underset{\sim}{x}_m(s))ds \tag{4.2}$$

$$x_0(t) = \bar{x}(t); \ m = 0,1,\dots$$

converges to $x^*(t)$ with

$$||x^* - x_m|| \le \theta^m N_0$$

(4) for $x_0(t) = x(t) \in \bar{S}(\bar{x},N_0)$ the iterative process (4.2) converges to $x^*(t)$

(5) any sequence $\{\bar{x}_m(t)\}$ such that $\bar{x}_m(t) \in \bar{S}(x_m,\theta^m N_0)$; $m = 0,1,\dots$ converges to $x^*(t)$

where $N_0 = (1-\theta)^{-1} ||x_1 - \bar{x}||$.

5. CONVERGENCE OF APPROXIMATE PICARD'S ITERATIVE METHOD

Theorem 8. [Agarwal 4,6] With respect to (1.1), (1.2) we assume that there exists an approximate solution $\bar{x}(t)$ and

(i) condition (i) of theorem 7

(ii) for $y_m(t)$ obtained from

$$y_{m+1}(t) = P_{n-1}(t) + \int_{a_1}^{a_r} g(t,s)f_m(s,\underset{\sim}{y}_m(s))ds \tag{5.1}$$

$$y_0(t) = x_0(t) = \bar{x}(t); \ m = 0,1,\dots$$

the following inequality is satisfied

$$\max_{a_1 \le t \le a_r} |f(t,\underset{\sim}{y}_m(t)) - f_m(t,\underset{\sim}{y}_m(t))| \le a_m \max_{a_1 \le t \le a_r} |f(t,\underset{\sim}{y}_m(t))| \tag{5.2}$$

where a_m; $m = 0,1,\dots$ are nonnegative constants and $a_m \le \Delta$

(iii) $\theta_1 = (1+\Delta)\theta < 1$

(iv) $N_1 = (1-\theta_1)^{-1}(\varepsilon+\delta+\Delta F)C_{n,0}(a_r-a_1)^n \le N$

where $F = \max_{a_1 \le t \le a_r} |f(t,\underset{\sim}{\bar{x}}(t))|$.

Then,

(1) all the conclusions (1)-(5) of theorem 7 hold
(2) the sequence $\{y_m(t)\}$ obtained from (5.1) remains in $\bar{S}(\bar{x},N_1)$
(3) the sequence $\{y_m(t)\}$ converges to $x^*(t)$ the solution of (1.1), (1.2) if and only if

$$\lim_{m\to\infty} \|y_{m+1}(t) - P_{n-1}(t) - \int_{a_1}^{a_r} g(t,s)f(s,\underset{\sim}{y}_m(s))ds\| = 0 \qquad (5.3)$$

also

$$\|x^*-y_{m+1}\| \le (1-\theta)^{-1}[\theta \|y_{m+1}-y_m\| + \Delta C_{n,0}(a_r-a_1)^n \max_{a_1\le t\le a_r} |f(t,\underset{\sim}{y}_m(t))|].$$

6. CONVERGENCE OF QUASILINEARIZATION METHOD

Theorem 9. [Agarwal 5,6] With respect to (1.1), (1.2) we assume that there exists an approximate solution $\bar{x}(t)$ and

(i) the function $f(t,u_0,u_1,\ldots,u_q)$ is continuously differentiable with respect to all u_i, $0 \le i \le q$ on $[a_1,a_r] \times D_2$
(ii) there exists L_i, $0 \le i \le q$ nonnegative constants such that for all $(t,u_0,u_1,\ldots,u_q) \in [a_1,a_r] \times D_2$, $|\frac{\partial}{\partial u_i} f(t,u_0,u_1,\ldots,u_q)| \le L_i$
(iii) $3\theta < 1$
(iv) $N_2 = (1-3\theta)^{-1}(\varepsilon+\delta)C_{n,0}(a_r-a_1)^n \le N$.

Then,
(1) the Newton's sequence $\{x_m(t)\}$ defined by

$$x_{m+1}(t) = P_{n-1}(t) + \int_{a_1}^{a_r} g(t,s)[f(s,\underset{\sim}{x}_m(s)) + \sum_{i=0}^{q} (x_{m+1}^{(i)}(s)-x_m^{(i)}(s))\frac{\partial}{\partial x_m^{(i)}(s)}f(s,\underset{\sim}{x}_m(s))]ds$$

$x_0(t) = \bar{x}(t)$; $m = 0,1,\ldots$ remains in $\bar{S}(\bar{x},N_2)$

(2) the sequence $\{x_m(t)\}$ converges to the unique solution $x^*(t)$ of the boundary value problem (1.1), (1.2)
(3) a bound on error is given by

$$\|x_m - x^*\| \le (\frac{2\theta}{1-\theta})^m(1 - \frac{2\theta}{1-\theta})^{-1} \|x_1 - \bar{x}\|$$

$$\le (\frac{2\theta}{1-\theta})^m(1 - \frac{2\theta}{1-\theta})^{-1}(1-\theta)(\varepsilon+\delta)C_{n,0}(a_r-a_1)^n.$$

Theorem 10. [Agarwal 5,6] Let the conditions of theorem 9 be satisfied. Further, let $f(t,u_0,u_1,\ldots,u_q)$ be continuously twice differentiable with respect to all u_i, $0 \le i \le q$ on $[a_1,a_r] \times D_2$, and for all $(t,u_0,u_1,\ldots,u_q) \in [a_1,a_r] \times D_2$,

$$|\frac{\partial^2}{\partial u_i \partial u_j}f(t,u_0,u_1,\ldots,u_q)| \le L_i L_j K, \; 0 \le i, j \le q.$$

Then,

$$\|x_{m+1} - x_m\| \le \alpha \|x_m - x_{m-1}\|^2 \le \frac{1}{\alpha} (\alpha \|x_1 - x_0\|)^{2^m}$$

$$\le \frac{1}{\alpha} [\tfrac{1}{2} K (\varepsilon+\delta)(\frac{\theta}{1-\theta})^2]^{2^m}$$

where $\alpha = K\theta^2/2(1-\theta)C_{n,0}(a_r-a_1)^n$.

Thus, the convergence is quadratic if $\frac{1}{2} K(\varepsilon+\delta)(\frac{\theta}{1-\theta})^2 < 1$.

7. CONVERGENCE OF APPROXIMATE QUASILINEARIZATION METHOD

Theorem 11. [Agarwal 5] With respect to (1.1), (1.2) we assume that there exists an approximate solution $\bar{x}(t)$ and

(i) conditions (i) and (ii) of theorem 9

(ii) for $y_m(t)$ obtain from

$$x_{m+1}(t) = P_{n-1}(t) + \int_{a_1}^{a_r} g(t,s)[f_m(t,\underset{\sim}{y}_m(s)) + \sum_{i=0}^{q} (y_{m+1}^{(i)}(s)-y_m^{(i)}(s)) \times \frac{\partial}{\partial y_m^{(i)}(s)} f_m(s,\underset{\sim}{y}_m(s))]ds \tag{7.1}$$

$y_0(t) = x_0(t) = \bar{x}(t)$; $m = 0,1,\ldots$

inequality (5.2) is satisfied, and $f_m(t,u_0,u_1,\ldots,u_q)$ is continuously differentiable with respect to all u_i, $0 \le i \le q$ on $[a_1,a_r] \times D_2$ and

$$\left|\frac{\partial}{\partial u_i} f_m(t,u_0,u_1,\ldots,u_q)\right| \le L_i$$

(iii) $\theta_2 = (3+\Delta)\theta < 1$

(iv) $N_3 = (1-\theta_2)^{-1}(\varepsilon+\delta+\Delta F)C_{n,0}(a_r-a_1)^n \le N$.

Then,

(1) all the conclusions (1)-(3) of theorem 9 hold

(2) the sequence $\{y_m(t)\}$ obtained from (7.1) remains in $\bar{S}(\bar{x},N_3)$

(3) the sequence $\{y_m(t)\}$ converges to $x^*(t)$ the solution of (1.1), (1.2) if and only if (5.3) holds, also

$$||x^*-y_{m+1}|| \le (1-\theta)^{-1}[2\theta\, ||y_{m+1}-y_m|| + \Delta \max_{a_1\le t\le a_r} |f(t,\underset{\sim}{y}_m(t))|].$$

Theorem 12. [Agarwal 5] Let the conditions of theorem 11 be satisfied. Further, let $f_m = f_0$ for all $m = 1,2,\ldots$ and $f_0(t,u_0,u_1,\ldots,u_q)$ be continuously twice differentiable with respect to all u_i, $0 \le i \le q$ on $[a_1,a_r] \times D_2$ and $\left|\frac{\partial^2}{\partial u_i \partial u_j} f_0(t,u_0,u_1,\ldots,u_q)\right| \le L_i\, L_j\, K$, $0 \le i, j \le q$. Then,

$$||y_{m+1} - y_m|| \le \alpha ||y_m - y_{m-1}|| \le \frac{1}{\alpha} (\alpha\, ||y_1-y_0||)^{2^m}$$

$$\le \frac{1}{\alpha} [\tfrac{1}{2} K (\varepsilon+\delta+\Delta F)(\tfrac{\theta}{1-\theta})^2]^{2^m}.$$

8. PROBLEM OF BEST LIPSCHITZ CONSTANTS

Theorem 13. [Jackson 14,15] Suppose that the function f is of Lipschitz class on $(a,b) \times R^{q+1}$. Then, the boundary value problem (1.1), (1.2) has a unique solution provided $a_r - a_1 < h$, where $h = \min\{h_k : 1 \le k \le [\frac{n}{2}]\}$ and h_k is the smallest positive number such that there is a solution $x(t)$ of the boundary value problem

$$x^{(n)} = (-1)^k[L_0 x + \sum_{j=1}^{q} L_j\, |x^{(j)}|]$$

$$x^{(i)}(0) = 0,\ 0 \le i \le n-k-1$$

$$x^{(i)}(h_k) = 0,\ 0 \le i \le k-1$$

with $x(t) > 0$ on $(0,h_k)$ or $h_k = +\infty$ if no such solution exists. This result is best possible.

Theorem 14. [Agarwal 8] Let $n = 4$, $q = 0$, $r = 2$, $k_1 = 1$, $k_2 = 1$ and f be of Lipschitz class on $[a_1,a_2] \times R$. Then, the boundary value problem (1.1), (1.2) has a unique solution provided $a_2 - a_1 < h_2$.

Theorem 15. [Agarwal 7] Let $n = 4$, $q = 0$, $r = 2$, $k_1 = 2$, $k_2 = 0$ and f be of Lipschitz class on $[a_1,a_2] \times R$. Then, the boundary value problem (1.1), (1.2) has a unique solution provided $a_2 - a_1 < h_1 = \infty$.

Theorems 14 and 15 suggest the following two problems

(1) in theorem 13 can the interval (a,b) be replaced by $[a_1,a_r]$

(2) the result is best possible for which particular r point problem.

Since, for the arbitrary a_i, $1 \le i \le r$ the limiting cases cannot be avoided, it appears that the result is best possible only for $r = 2$, $k_1 = [\frac{n}{2}] - 1$.

9. CONTINUOUS DEPENDENCE ON BOUNDARY DATA

Theorem 16. [Klaasen 17] Suppose that

(i) all solutions of (1.1) extend to (a,b)

(ii) all n point boundary value problems (1.1), (1.2) have at most one solution.

Then, the solutions of (1.1), (1.2) are continuous functions of $(a_1,a_2,\ldots,a_n;\ A_1,A_{1,2},\ldots,A_{1,n})$.

10. UNIQUENESS IMPLIES EXISTENCE

Theorem 17. [Hartman 10, Klaasen 16] Suppose that

(i) conditions (i) and (ii) of theorem 16

(ii) solutions of initial value problems for (1.1) are unique on (a,b)

(iii) if $[c,d]$ is a compact subinterval of (a,b) and if $\{x_j(t)\}$ is a sequence of solutions of (1.1) which is uniformly bounded on $[c,d]$, then there is a subsequence $\{x_{j_k}(t)\}$ such that $\{x_{j_k}^{(i)}(t)\}$ converges uniformly on $[c,d]$ for each $0 \le i \le n-1$.

Then, each n point boundary value problem (1.1), (1.2) has a solution.

For $n = 2$ the condition (iii) in this theorem is not needed was first proved in [Lasota and Opial 18], whereas for $n = 3$ in [Jackson and Schrader 12], and for $n > 3$ it remains an open problem.

Theorem 18. [Hartman 9] Suppose that

(i) condition (i) of theorem 16

(ii) condition (ii) of theorem 17

(iii) all n point boundary value problems (1.1), (1.2) have a unique solution.

Then, each r point boundary value problem (1.1), (1.2) has a unique solution.

Theorem 19. [Henderson and Jackson 11] Suppose that

(i) condition (i) of theorem 16
(ii) conditions (ii) and (iii) of theorem 17
(iii) all 2 point boundary value problems (1.1), (1.2) have solutions
(iv) all (n-1) point boundary value problems (1.1), (1.2) have at most one solution.

Then, each r point boundary value problem (1.1), (1.2) has a unique solution.

11. UNIQUENESS IMPLIES UNIQUENESS

Theorem 20. [Jackson 13] Suppose that

(i) condition (ii) of theorem 16
(ii) condition (ii) of theorem 17.

Then, each r point boundary value problem (1.1), (1.2) has at most one solution.

Theorem 21. [Henderson and Jackson 11] Suppose that

(i) condition (ii) of theorem 17
(ii) all (n-1) point boundary value problem (1.1), (1.2) have at most one solution.

Then, each r point boundary value problem (1.1), (1.2) has at most one solution with $2 \leq r \leq n-1$.

Theorem 22. [Henderson and Jackson 11] Suppose that

(i) condition (ii) of theorem 17
(ii) there is an integer k such that $2 < k-1 < k < n$ and all k point and all (k-1) point boundary value problems (1.1), (1.2) have at most one solution.

Then, each r point boundary value problem (1.1), (1.2) with $2 \leq r \leq k$ has at most one solution.

REFERENCES

[1] Agarwal, R. P., Some inequalities for a function having n zeros, Proc. Conf. "General Inequalities 3" Oberwolfach, Edited by E. F. Beckenbach and W.Walter, ISNM 64 (1983), 371-378.
[2] Agarwal, R. P., Boundary value problems for higher order integro-differential equations, Nonlinear Analysis : TMA, 7 (1983), 259-270.
[3] Agarwal R. P., Necessary conditions for the existence of solutions of multi-point boundary value problems, Bull. Inst. Math. Acad. Sinica 12 (1984), 11-16.
[4] Agarwal R. P. and Loi, S. L., On approximate Picard's iterates for multipoint boundary value problems, Nonlinear Analysis : TMA, 8 (1984), 381-391.
[5] Agarwal, R. P., Quasilinearization and approximate quasilinearization for multipoint boundary value problems, J. Math. Anal. Appl. (to appear).
[6] Agarwal R. P. and Chow Y. M., Iterative methods for a fourth order boundary value problem, J. Comp. Appl. Math. 10 (1984), 203-217.
[7] Agarwal, R. P., Best possible length estimates for nonlinear boundary value problems, Bull. Inst. Math. Acad. Sinica 9 (1981), 169-177.
[8] Agarwal R. P. and Wilson, S. J., On a fourth order boundary value problem, Utilitas Mathematica (to appear)
[9] Hartman, P., Unrestricted n-parameter families, Rend. Circ. Mat. Palermo (2) 7 (1958), 123-142.
[10] Hartman, P., On n-parameter families and interpolation problems for non-linear ordinary differential equations, Trans. Amer. Math. Soc. 154 (1971), 201-226.

[11] Henderson,J. and Jackson, L., Existence and uniqueness of solutions of k-point boundary value problems for ordinary differential equations, J. Diff. Equs. 48 (1983), 373-385.
[12] Jackson,L. and Schrader, K., Existence and uniqueness of solutions of boundary value problems for third order differential equations, J. Diff. Equs. 9 (1971), 46-54.
[13] Jackson,L., Uniqueness of solutions of boundary value problems for ordinary differential equations, SIAM J. Appl. Math. 24 (1973), 535-538.
[14] Jackson,L., Existence and uniqueness of solutions of boundary value problems for Lipschitz equations, J. Diff. Equs. 32 (1979), 76-90.
[15] Jackson, L., Boundary value problems for Lipschitz equations, "Differential Equations" Ed. Shair Ahmed et. al. Academic Press, New York, (1980), 31-50.
[16] Klaasen, G., Existence theorems for boundary value problems for nth order ordinary differential equations, Rocky Mountain J. Math. 3 (1973), 457-472.
[17] Klaasen, G., Continuous dependence for N-point boundary value problems, SIAM J. Appl. Math. 29 (1975), 99-102.
[18] Lasota A., and Opial, Z., On the existence and uniqueness of solutions of a boundary value problem for an ordinary second order differential equation, Colloq. Math. 18 (1967), 1-5.

SOME ADDITIONAL REFERENCES

[1] Beesack, P. R., On the Green's function of an N-point boundary value problem, Pacific J. Math. 12 (1962), 801-812.
[2] Bernfeld S. R. and Lakshmikantham, V., An introduction to nonlinear boundary value problems, Academic Press, New York, 1974.
[3] Bessmertnyh, G. A. and Levin, A. Ju., Some inequalities satisfied by differentiable functions of one variable, Soviet Math. Dokl. 3 (1962), 737-740.
[4] Das, K. M. and Vatsala, A. S., On Green's function of an n-point boundary value problem, Trans. Amer. Math. Soc. 182 (1973), 469-480.
[5] Gustafson, G. B., A. Green's function convergence principle, with applications to computation and norm estimates, Rocky Mount. J. Math. 6 (1976), 457-492.
[6] Jackson, L., Uniqueness and existence of solutions of boundary value problems for ordinary differential equations, "Ordinary Differential Equations" Ed. L. Weiss, Academic Press, New York, (1972), 137-149.
[7] Jackson, L., A Nagumo condition for ordinary differential equations, Proc. Amer. Math. Soc. 57 (1976), 93-96.
[8] Jackson, L., A compactness condition for solutions of ordinary differential equations, Proc. Amer. Math. Soc. 57 (1976), 89-92.
[9] Jackson, L., Boundary value problems for ordinary differential equations, "Studies in Ordinary Differential Equations" Ed. J. Hale, The Mathematical Association of America (1977), 93-127.
[10] Umamaheswarm, S., $\lambda(n,k)$-parameter families and associated convex functions, Rocky Mount. J. Math. 8 (1978), 481-502.
[11] Umamaheswarm, S., $\lambda(n,k)$-convex functions, Rocky Mount. J. Math. 8 (1978), 759-764.
[12] Umamaheswarm, S., Boundary value problems for higher order differential equations, J. Diff. Equs. 18 (1975), 188-201.

This paper is in final form and no version of it will be submitted for publication elsewhere.

Trends in the Theory and Practice of Non-Linear Analysis
V. Lakshmikantham (Editor)
Elsevier Science Publishers B.V. (North-Holland), 1985

A THEORY OF NONLINEAR EVOLUTION EQUATIONS

Mieczyslaw Altman

Department of Mathematics
Louisiana State University
Baton Rouge, Louisiana
U.S.A.

A theory of solving nonlinear evolution equations based on three different methods is presented which is independent of C_0-semigroups. From the standpoint of applications, its important feature is the possibility of solving nonlinear evolution equations via linearized evolution equations.

INTRODUCTION

Kato's [12] theory of quasilinear evolution equations, which he successfully applied to nonlinear partial differential equations of mathematical physics, was the first significant step in the nonlinear direction. His famous theorem which applies only to reflexive Banach spaces was generalized to nonreflexive ones in [2] by a different method. In 1956 Nash [14] introduced the concept of smoothing operators and in 1966 Moser [13] introduced the concept of the degree of approximate linearization. Based on these notions and what is called the Nash-Moser technique constitutes a truely fundamental development in handling of the "loss of derivatives." But so far the related theorems and their generalizations are designed for nonlinear operator equations and do not apply to abstract nonlinear evolution equations (NEE). In a series of investigations [5,7,8] an attempt was made to solve the nonlinear problem for EE. Our theory is based on three global linearization iterative methods (GLIM). Global linearization means that the equation is linearized about a vector running over a nonbounded set B whereas local linearization indicates that B is bounded. The theory also applies to nonlinear operator equations (see [3]). Thus a unified approach has been developed to both nonlinear operator and evolution equations. Although the general theory is independent of C_0-semigroups, nevertheless from the point of view of applications to nonlinear partial differential equations it is rather important that nonlinear evolution equations can be solved via linearized evolution equations. In this way one can make use of the theory of C_0-semigroups which is more powerful and advanced than the theory of nonlinear semigroups which is rather limited in scope. Some highlights of our theory will be presented below.

1. CONVEX APPROXIMATE LINEARIZATION AND GLOBAL LINEARIZATION ITERATIVE METHODS (GLIM) FOR SOLVING NONLINEAR EVOLUTION EQUATIONS

Let $Z \subset Y \subset X$ be Banach spaces with norms $\|\cdot\|_Z \geq \|\cdot\|_Y \geq \|\cdot\|_X$

$(1.A_o)$ We assume that there exist constants $C > 0$ and $\bar{s}$ with $0 < \bar{s} < 1$ such that

$$\|x\|_Y \leq C\, \|x\|_X^{1-\bar{s}}\, \|x\|_Z^{\bar{s}}. \tag{1.0}$$

Given $0 < b$, denote by $C(0, b; X)$ the Banach space of all continuous functions $x = x(t)$ defined on the interval $[0, b]$ with values in X and norm

$$\|x\|_{\infty,X} = \sup_t [\|x(t)\|_X : 0 \leq t \leq b].$$

In the same way are defined the norms $\|y\|_{\infty,Y}$ and $\|z\|_{\infty,Z}$ for Y and Z, respectively. Denote by $C^1(0,b;\ x)$ the vector space of all continuously differentiable functions from $[0,b]$ to X. Let W_0 be an open ball in Y with center x_0 in Z and radius $r_0 > 0$. Put $V_0 = V_0 \cap Z$ and let V_1 be the closure of V_0 in Y.

Let $F : [0,b] \times V_1 \to X$ be a nonlinear mapping and consider the Cauchy problem

(1.1) $Px(t) \equiv dx/dt + F(t,x) + f(t,x) = 0,\ 0 \leq t \leq b,\ x(0) = x_0,$

where $f : [0,b] \times V_1 \to X$ is also a nonlinear mapping.

Let G be the set of functions $x \in C(0,b;\ V_0(\|\cdot\|_Z)) \cap C^1(0,b;\ X)$ with $x(0) = x_0$, $\|x - x_0\|_{\infty,Y} < r_0$ and $\|x\|_{\infty,Z} < \infty$.

We assume that the mapping F is differentiable in the following sense. For each $(t,x) \in [0,b] \times G$, there exists a linear operator $F'(t,x)$ such that $\varepsilon^{-1}\|F(\cdot,x) + \varepsilon h) - F(\cdot,x) - \varepsilon F'(\cdot,x)h\|_{\infty,X} \to 0$ as $\varepsilon \to 0+$, where $h \in C(0,b;\ Z) \cap C^1(0,b;\ X)$.

We make the following assumptions. $(1.A_1^i)$ with $i = 1$. Let $\{x_n\} \subset G$ be a Cauchy sequence in $C(0,b;\ Y)$ and let $\{h_n\} \subset C(0,b;\ Y) \cap C^1(0,b;\ X)$ be bounded in $C(0,b;\ Y)$. Then $\varepsilon_n \to 0+$ implies
$\varepsilon_n^{-1}\|F(\cdot,x_n + \varepsilon_n h_n) - F(\cdot,x_n) - \varepsilon_n F'(\cdot,x_n)h_n\|_{\infty,X} \to 0$ as $n \to \infty$; or

$(1.A_1^i)$ with $i = 2$. There exists a constant $C > 0$ such that
$\|F(t,u) - F(t,v) - F'(t,u)(u - v)\|_0 \leq C_1\|u - v\|_0\|u - v\|_Y$, for all $t \in [0,b];\ u,v \in V_0$; or

$(1.A_1^i)$ with $i = 3$. There exists a constant $M_1 > 0$ such that
$\|F(\cdot,x + h) - F(\cdot,x) - F'(\cdot,x)h\|_{\infty,0} \leq M_1\|h\|_{\infty,0}^{2-\beta}\|h\|_{\infty,Z}^{\beta}$, where $x \in G$, $h \in C(0,b;\ Z) \cap C^1(0,b;\ X)$, and $0 \leq \beta < 1$.

$(1.A_2)$ The functions F,f are continuous in the following sense, $\|x_n - x\|_{\infty,Y} \to 0$ implies $\|F(\cdot,x_n) - F(\cdot,x)\|_{\infty,0} \to 0$ as $n \to \infty$, and the same is true for f. There exists a constant $q_0 > 0$ such that
$\|f(\cdot,x + \varepsilon h) - f(\cdot,x)\|_{\infty,X} \leq q_0\varepsilon\|h\|_{\infty,X}$.

$(1.A_3)$ There exists a constant $C_0 > 0$ with the following property. For $x \in G$ and g with $\|g\|_{\infty,X} < \infty$, if h is a solution of the equation

$$dh/dt + F'(t,x)h + g = 0,\ 0 \leq t \leq b,\ h(0) = 0,$$

then $\|h\|_{\infty,X} \leq bC_0\|g\|_{\infty,X}$.

$(1.A_4)$ For $x \in G$, the linearized equation

(1.2) $dz/dt + F'(t,x)z + F(t,x) - F'(t,x)x + f(t,x) = 0,\ 0 \leq t \leq b,\ z(0) = x_0$

admits approximate solutions of order (μ,ν,σ) with $0 \leq \nu < 1$ in the sense of the following

Definition 1.1. [9]. Let $\mu > 0$, $\nu \geq 0$, $\sigma \geq 0$ be given numbers. Then the linearized equation (1.2) admits approximate solutions of order (μ,ν,σ) if there exists a constant $M > 0$ which has the following property. For every $x \in G$, $K > 1$, and $Q > 1$, if $\|x\|_{\infty,Z} < K$ then there exists a residual (error) function y and a function z such that

$$\|z\|_{\infty,Z} < MQK^{\nu}$$

$$\|y\|_{\infty,X} \leq MQ^{-\mu}K^{\sigma}, \quad \text{and}$$

$$dz/dt + F'(t,x)z + F(t,x) = F'(t,x)x + f(t,x) + y = 0, \tag{1.2'}$$

$$0 \leq t \leq b, \; z(0) = x_0.$$

Now for $x \in G$, let z be a solution of the equation (1.2') and put $z = x + h$. Then obviously h is a solution of the equation

$$dh/dt + F'(t,x)h + Px + y = 0, \; 0 \leq t \leq b, \; h(0) = 0, \tag{1.3}$$

and we get

$$\|h\|_{\infty,X} \leq bC_0(\|Px\|_{\infty,X} + MQ^{-\mu}K^{\sigma}) \tag{1.4}$$

$$\|h\|_{\infty,Z} \leq \|x\|_{\infty,Z} + MQK^{\nu}. \tag{1.5}$$

GLIM-I. The following is an iterative method of contractor directions [6]. Put $x_0(t) \equiv x_0$, $t_0 = 0$, and assume that $x_0,x_1,\ldots,x_n \in G$; $t_0,t_1,\ldots,t_n$ are known and put

$$x_{n+1} = x_n + \varepsilon_n h_n \quad \text{and} \quad t_{n+1} = t_n + \varepsilon_n, \tag{1.6}$$

that is

$$x_{n+1} = (1 - \varepsilon_n)x_n + \varepsilon_n z_n,$$

which justifies the term "convex approximate linearization, where $z_n = x_n + h_n$ and h_n being a solution of (1.3). To detemine ε_n let $2\bar{q}/q < c < 1$, where $q_0 bC_0 = \bar{q} < q/2 < 1/2$, and put $\Phi(\varepsilon,h,x) = \varepsilon^{-1}\|P(x + \varepsilon h) - (1 - \varepsilon)Px\|_{\infty,X}$. If $\Phi(1,h_n,x_n) \leq q\|Px_0\|_{\infty,X}\exp(-(1 - q)t_n)$, then put $\varepsilon_n = 1$. Otherwise there exists $0 < \varepsilon < 1$ such that

$$cqp_0 \exp(-(1 - q)t_n) \leq \Phi(\varepsilon,h_n,x_n) \leq qp_0 \exp(-(1 - 1)t_n),$$

where $p_0 = \|Px_0\|_{\infty,X}$ and put $\varepsilon_n = \varepsilon$.

The method (1.6) satisfies the following induction assumptions

$$\|P(x_n + \varepsilon_n h_n) - (1 - \varepsilon_n)Px_n\|_{\infty,X} \leq \varepsilon_n q_n \|Px_n\|_{\infty,X}$$

$$\|x_n\|_{\infty,Z} < A \exp(\alpha(1 - q)t_n) = K_n$$

$$\|Px_n\|_{\infty,X} \leq \|Px_0\|_{\infty,X} \exp(-(1 - q)t_n),$$

for some constants $\alpha, A > 0$ to be determined.

<u>Theorem 1.1</u> [9]. In addition to $(1.A_0$–$1.A_4)$ with $i = 1$, suppose that $\mu(1 - \nu) - \sigma > 0$, $\alpha(1 - q) - 1 > 0$, $\alpha > [\mu(1 - \nu) - \sigma]^{-1}$ and $M(2M)^{1/\mu}(\bar{q}p_0)^{-1/\mu} < A^{1-\nu-\sigma/\mu}[\alpha(1 - q) - 1]$, and b' is such that

$$N = [(1 - q)\delta]^{-1}\exp((1 - q)\delta)C[b'C_0(1 + \bar{q})p_0]^{1-\bar{s}}[\alpha(1 - q)A]^{\bar{s}} < r_0.$$

where $\delta = 1 - (1 + \alpha)\bar{s} > 0$ and $\alpha < (1 - \bar{s})/\bar{s}$. Then (1.1) with b replaced by b' has a solution x and $\|x_n - x\|_{\infty,Y} \to 0$ as $n \to \infty$.

<u>GLIM-II.</u> The following iterative method is independent of contractor directions. Put

$$(1.7) \qquad x_0(t) \equiv x_0,\ x_{n+1} = z_n = x_n + h_n\ ,\ (x_n \in G),$$

with h_n as in (1.6) and induction assumptions

$$\|x_n\|_{\infty,Z} < Aq^{-\alpha n} = K_n \quad \text{and} \quad \|Px_n\|_{\infty,X} \leq p_0 q^n,$$

where $q < 2^{-1/\alpha}$ and $(C_1 r_0 + q_0)bC_0(1 + 2\bar{q}) < q < 1$; $\bar{q} < q$.

<u>Theorem 1.2.</u> In addition to $(1.A_0$–$1.A_4)$ with $i = 2$ and $\varepsilon = 1$ in $(1.A_2)$, suppose that

$$\beta = 1 - \bar{s}(1 + \alpha) > 0,\ N(1 - q^\beta) < r_0,\ N = C[b'C_0(1 + \bar{q})p_0]^{1-\bar{s}}[A(q^{-\alpha} - 1)]^{\bar{s}},$$

$$[\mu(1 - \nu) - \sigma]^{-1} < \alpha < (1 - \bar{s})/\bar{s}, \quad \text{and} \quad M^{1+1/\mu}(\bar{q}p_0)^{-1/\mu}(q^{-\alpha} - 2)^{-1} < A^{1-\nu-\sigma/\mu}.$$

Then the statement of Theorem 1.1 holds true.

<u>GLIM-III.</u> The following is a rapidly convergent iteration method which is based on the essential technique of Moser [13].

$(1.A_5)$ Let α, τ; $\lambda < \mu$ be such that

$$1 < (\mu - \lambda)^{-1}(1 + \alpha(1 + \lambda) + \mu) < \tau < 2 - -\alpha_0 < 2 - \alpha,$$

$$0 < 2\lambda < [\mu(1 - \alpha_0) - (1 + \alpha_0)]/(1 + \alpha_0),$$

where $0 < \alpha_0 < 1$ is such that $\mu > (1 + \alpha_0)/(1 - \alpha_0) > 1$,

$$0 < \beta < \mu\lambda(\alpha_0 - \alpha)[(1 + \alpha)(1 + \mu) + \lambda(2 + \mu)]^{-1} < 1.$$

Remark. One can put $\alpha_0 = (\mu - 1)/(\mu + 3)$ in the above. Then $0 < 2\lambda < (\mu - 1)/(\mu + 1)$. We assume $\nu = \sigma = 1 + \alpha$, and put $K_{n+1} = K_n^{\tau+\alpha}$ and define $\{x_n\}$ as in (1.7) with induction assumptions

$$\|x_n\|_{\infty,Z} < K_n \quad \text{and} \quad \|Px_n\|_{\infty,X} \leq K_n^{-\lambda}.$$

Theorem 1.3. Suppose that assumptions ($1.A_0$-$1.A_5$) are satisfied with $i = 3$ and $f \equiv 0$. Then there exists $K_0(M,\beta,\mu,\lambda,\alpha) > 1$ such that: if

$$\|Px_0\|_{\infty,X} < K_0^{-\lambda}, \ \|x_0\|_{\infty,Z} < K_0 \quad \text{and} \quad C(2C_0)^{1-\overline{s}} \sum_{n=0}^{\infty} K^{-\delta} < r_0,$$

where $\delta = (1 - \overline{s})\lambda - \overline{s}\tau > 0$ and $\overline{s} < \lambda(\lambda + 2)$, then equation (1.1) with $f = 0$ has a solution x and $\|x_n - x\|_{\infty,Y} \to 0$ as $n \to \infty$. Let us notice that the constant bC_0 in (A_3) can be replaced by C_0.

($1.A_5'$) Suppose that $0 < \lambda + 1 < (\mu + 1)/2$ and $0 < \beta < \frac{\lambda}{\lambda+1}\frac{\mu}{\mu+1}\left(1 - 2\frac{\lambda-1}{\mu+1}\right)$. Then τ is a number such that $1 < \left(1 - \frac{\lambda+1}{\mu+1}\right)^{-1} < \tau < 2$.

Theorem 1.4. Theorem 1.3 remains valid if ($1.A_5$) is replaced by ($1.A_5'$) and $\alpha = 0$.

2. Smoothing operators combined with elliptic regularization. The degree $\overline{k}$ of elliptic regularization. The choice of Moser's degree μ of approximate linearization.

Let $\{X_j\}$ with $0 \leq j \leq p$ be a scale of Banach spaces with increasing norms such that $i < j$ implies $X_j \subset X_i$ and $\|\cdot\|_j \geq \|\cdot\|_i$ and let $0 < m_1 < m_2 < s < \overline{p} < p$.

($2.A_0$) We assume that there exists a one parameter family of linear operators S_θ, $\theta \geq 1$, (see Nash [14], Moser [13]) such that

$$\|(I - S_\theta)s\|_0 \leq C\theta^{-m_1}\|x\|_{m_1}; \ \|(I - S_\theta)x\|_{m_1} \leq C\theta^{-(m_2-m_1)}\|x\|_{m_2}$$

$$\|S_\theta x\|_p \leq C\theta^{p-m_2}\|x\|_{m_2}$$

for some constant $C > 0$, where I is the identity mapping. We also assume that $\|x\|_j \leq C\|x\|_r^{1-\lambda}\|x\|_p^\lambda$ for $j = (1-\lambda)r + \lambda p$. with $0 \leq \lambda \leq 1$.

Using the same notation as in Section 1 put $\|x\|_{\infty,j} = \sup_t[\|x(t)\|_j : 0 \leq t \leq b]$, and let $W_0 \subset X_s$ be an open ball with center $x_0 = 0$ and radius $r_0 > 0$. Put $V_0 = W_0 \cap X_p$ and let V_s be the closure of V_0 in X_s.

Let $F,f : [0,b] \times X_s \to X_0$ be two nonlinear mappings and consider the Cauchy problem

(2.1) $\quad Px(t) \equiv dx/dt + F(t,x) + f(t,x) = 0,\ 0 \leq t \leq b,\ x(0) = 0.$

Let G be the set of functions $x \in C(0,b;\ V_0(\|\cdot\|_p)) \cap C^1(0,b;\ X_0)$ with $x(0) = 0$, $\|x\|_{\infty,s} < r_0$ and $\|x\|_{\infty,p} < \infty$.

Assumptions (2.!$_1$-2.A$_3$) are the same as in Section 1, where $X = X_0$, $Z = X_p$ and $Y = X_s$ with s to be determined and $\bar{s} = s/p$.

(2.A$_4$) There exists a constant $C > 0$ such that

$$\|F'(t,x)h\|_{m_1} \leq C\|h\|_{m_2} \quad \text{and} \quad \|F'(t,x)h\|_0 \leq C\|h\|_{m_1}$$

$$\|F(t,x) - F'(t,x)x + f(t,x)\|_{m_2} \leq C\|x\|_{\bar{p}},$$

for all $(t,x) \in [0,b] \times V_0$.

There exists a linear (regularizing) operator $L = L(\eta)$ such that $\|Lz\|_0 \leq C\|z\|_{m_2}$ for some constant $C > 0$ and the modified linearized equation

(2.2) $\quad dz/dt + F'(t,x)z + \eta Lz + F(t,x) - F'(t,x)x + f(t,x) = 0,$

$0 \leq t \leq b$, $z(0) = 0$, with small $0 < |\eta| < 1$ and $(t,x) \in [0,b] \times G$ has a solution $\bar{z}$ such that

(2.3) $\quad \|\bar{z}\|_{m_2} \leq C|\eta|^{-\bar{k}}\|F(t,x) - F'(t,x)x + f(t,x)\|_{m_2}$

for some $\bar{k} \geq 0$ to be determind and $C > 0$.

For $(t,x) \in [0,b] \times G$, lte $\bar{z}$ be a solution of (2.1).

Lemma 2.1 [9]. The following holds for $(t,x) \in [0,b] \times G$ with $\|x\|_{\infty,p} < K$, $K > 1$ and $0 < \eta < 1$, $\bar{\nu} = \bar{p}/p$.

$$\|(I - S_\theta)d\bar{z}/dt\|_{\infty,0} \leq M_1(\theta^{-m_1} + \eta)\eta^{-\bar{k}}K^{\bar{\nu}}$$

$$\|F'(\cdot,x)(I - S_\theta)\bar{z}\|_{\infty,0} \leq M_2\theta^{-(m_2-m_1)}\eta^{-\bar{k}}K^{\bar{\nu}}$$

$$\|z\|_{\infty,p} = \|S_\theta\bar{z}\|_{\infty,p} \leq M_3\theta^{p-m_2}\eta^{-\bar{k}}K^{\bar{\nu}}$$

for some $M_1, M_2, M_3 > 0$ and $z = S_\theta\bar{z}$ and $0 \leq \bar{k} < 1$.

Now put $z = x + h$. Then z is an approximate solution of (1.2) of order (μ,ν,σ) with $\nu = \sigma$ (to be determined) and $\mu = m/p-m_2$ where $m = \min(m_1, m_2 - m_1)$. But this choice of μ requires stronger assumptions for L. The following lemma gives a better choice.

Lemma 2.2 [9]. For $\bar{\nu} < \nu < 1$, there exists μ with $\mu(1 - \nu) - \nu > 0$ such that for $K > 1$ and $Q > 1$ one can find $0 < \eta < 1$ and $\theta > 1$ which satisfy $\theta^{p-m_2}\eta^{-\bar{k}}K^{\bar{\nu}} < QK^{\nu}$ and $(\theta^{-m} + \eta)\eta^{-\bar{k}}K^{\bar{\nu}} < Q^{-\mu}K^{\nu}$ provided $0 \leq \bar{k} < 1 - \bar{\nu}$. If $\bar{\nu} = \bar{p}/p$, then the last inequality is equivalent to $p > \bar{p}/(1 - \bar{k})$.

Since $(2.A_4)$ implies $(1.A_4)$ it follows that Theorems 1.1-1.4 remain valid, where $z_n = x_n + h_n$, $z_n = S_\theta\bar{z}_n$, $\bar{z}_n$ being a solution of (2.2) satisfying (2.3), and $X = X_0$, $Y = Y_s$, $Z = X_p$, $\sigma = \nu$, $\bar{s} = s/p$ with s such that $[\mu(1 - \nu) - \nu]^{-1} < (1 - \bar{s})/\bar{s}$ in case of Theorems 1.1 and 1.2. In case of Theorem 1.3, $\sigma = \nu = 1 + \alpha$ with $\bar{s} < \lambda(\lambda + 2)$. But $\alpha = 0$ in Theorem 1.4.

Note that $\bar{k} = 1$ in Theorem 1.3 is admissible. In both Theorems 1.3 and 1.4, μ exceeds one and Lemma 2.2 has to be modified.

Remark 2.1. If $F' = F_1' + F_2'$, then F' in $(2.A_4)$ can be replaced by F_1' but F_2' is subject to the same conditions of f.

3. The case of $L = 0$ and $\bar{k} = 0$

This is a special case of Section 2 and Lemmas 2.1 and 2.2 are still valid as well as a modification of Lemma 2.2 which is needed for GLIM-III. Thus Theorems 2.1-2.4 can be proved if the assumptions of Section 2 are satisfied with $L = 0$ and $\bar{k} = 0$.

4. Nonlinear evolution equations via linearized evolution equations

Denote by $G(X_0)$ the set of all negative infinitesimal generators of C_0-semigroups $\{U(t)\}$ an X_0. If $-A \in G(X_0)$, then $\{U(t)\} = \{e^{-tA}\}$, $0 \leq t < \infty$, is the semigroup generated by $-A$. For $x \in G$ put $A(t,x(t)) = F'(t,x(t))$. We assume that A is a function from $[0,b] \times V_0$ into $G(X_0)$, and $A(t) = A(t,x(t))$ is stable in X_0 (see Kato [10-12], Yosida [16], Tanabe [15]), We also assume that X_{m_2} being dense in X_0 is admissible and preserves the stability of $A(t)$ and the evolution operators $U(t,s;\ x)$ generated by $A(t,x(t))$ exist and

(4.1) $$\|U(t,s;\ x)\|_{X_0} \leq M_0 \quad\text{and}\quad \|U(t,s;\ x)\|_{X_{m_2}} \leq M_2,$$

for all $x \in G$ and some $M_0, M_2 > 0$. These assumptions imply that (2.2) and (2.3) hold with $L = 0$ and $\bar{k} = 0$.

Example 1. consider the nonlinear equation

(4.2) $$u_t + F(t,u,u_x) = 0,\ 0 \leq t \leq b,\ -\infty < x < \infty.$$

For $x \in G$ with appropriate choice of s, it follows from Kato's [12] argument that the linearized equations for (4.2) satisfy (4.1) with $X_{m_2} = H^r(-\infty,\infty)$, $r \geq 2$, since $\|u\|_{\infty,s} < r_0$.

Example 2. Korteweg-de Vries nonlinear equation

(4.3) $$u_t + u_{xxx} + F(t,u,u_x) = 0,\ 0 \leq t \leq b,\ -\infty < x < \infty.$$

It follows from Kato's [12] argument and Remark 2.1 that the linearized equations for (4.3) satisfy (4.1) if $X_{m_2} = H^4(-\infty,\infty)$ with $r \geq 3$, where $u \in G$ and s is properly chosen.

Example 3. Consider the nonlinear system

(4.3) $$du/dt + F(t,x,u,u_x) = 0,\ 0 \leq t \leq b,$$

where $u = u(t,x) = u,(t,x),\dots,u_N(t,x))$, $x \in R^m$, $F = (F_1,\dots,F_N)$. Suppose that the linearized systems for (4.3) satisfy Kato's (see [12) assumptions for symmetric hyperbolic systems, then (4.1) is satisfied with x replaced by u provided that s is properly chosen.

Remark 4.1. The evolution operators $U(t,s;x)$ in (4.1) can be replaced by their approximations $U_m(t,s;x)$ generated by the appropriate step functions (see [3]).

Remark 4.2. In addition to the assumptions made above, suppose that

$$\|F(t,u) - F(t,v) - F'(t,v)(u - v)\|_0 \leqslant c\|u - v\|_{X_0} \|u - v\|_{X_s}$$

$$\|[F'(t,u) - F'(t,v)]h\|_{X_0} \leqslant c\|u - v\|_{X_0} \|h\|_{X_s}$$

for all $t \in [0,b]$; $h,u,v \in V_S$ and some $c > 0$. Then the solution of (1.1) is unique for small b. (see [3]).

References

[1] M. Altman, Contractors and Contractor Directions, Theory and Applications, Lecture Notes in Pure and App. Math. (M. Dekker, New York, 1977).

[2] M. Altman, Quasilinear evolution equations in nonreflexive Banach spaces, J. Integral Equ. 3 (1981), 153-164.

[3] M. Altman, Nonlinear evolution equations in Banach sapces, J. Integral Equ. 4 (1982), 307-322.

[4] M. Altman, Global linearization iterative methods and nonlinear partial differential equations I,II,III, to appear.

[5] M. Altman, Nonlinear evolution equations and smoothing operators in Banach spaces, J. Nonlin. Analys. 8(1984), 481-490.

[6] M. Altman, Iterative methods of contractor directions, J. NOnlin. Analys. 4 (1980), 761-722.

[7] M. Altman, Nonlinear equations of evolution and convex approximate linearization in Banach spaces I,II, J. NOnlin. Analys. 8 (1984), 457-470.

[8] M. Altman, Nonlinear equations of evolution in Banach spaces, J. Nonlin. Analys. 8 (1984), 491-499.

[9] M. Altman, Global linearization iterative mehtods for nonlinear evolution equations, to appear.

[10] T. Kato, Linear evolution equations of "hyperbolic" type, J. Fac. Sci. Univ. Tokyo, Sec. I. 17 (1970), 241-258.

[11] T. Kato, Linear evolution equations of "hyperbolic type II, J. Math Soc. Japan, 25 (1973), 648-666.

[12] T. Kato, Quasi-linear equations fo evolution with applications to partial differential equations, in Lecture Notes in Math. No. 448 (Springer-Verlag, New York, 1975), pp. 25-70.

[13] J. Moser, A rapidly convergent iteration method and non-linear partial differential equations -I, Ann. Scuola Norm. Sup. Pisa 20 (1966), 265-315.

[14] J. Nash, The embedding problem for Riemannian manifolds, Ann. Math. 63 (1956), 20-63.

[15] H. Tanabe, Equations of Evolutions, (Pitman, 1979).

[16] K. Yosida, Functional Analysis, 2nd ed. (Springer-Verlag, New York, 1968).

This paper is in final form and no version of it will be submitted for publication elsewhere.

Trends in the Theory and Practice of Non-Linear Analysis
V. Lakshmikantham (Editor)

ASYMPTOTIC ANALYSIS OF A FUNCTIONAL-INTEGRAL EQUATION RELATED TO CELL POPULATION KINETICS

Ovide Arino and Marek Kimmel

Departement de Mathematiques, Universite de Pau, Avenue Louis Sallenave, 64000 Pau, France
and Department of Pathology, Memorial Sloan-Kettering Cancer Center, 1275 York Avenue, New York, NY 10021, USA

A functional-integral equation describes a novel model of cell population, the kinetics of which is governed by unequal division of cell metabolic constituents. Elements of the semigroup theory are applied to determine the asymptotic behaviour of the solution. The semigroup is not compact, so that direct investigation of its spectral properties is necessary.

I. GENERAL PRESENTATION

1. Biological Background. We will consider the asymptotics of solutions of a certain novel model equation of cell kinetics. The novelty of the model lies in two assumptions: (1) that the main (if not only) source of the randomness of cell life length is the unequal division of RNA between daughter cells and (2) that the cell RNA level is a major regulating factor for the cell development. These assumptions are motivated mainly by experimental work of Darzynkiewicz et al. [2] (see also references in that paper). Complete derivation of our equation will be published elsewhere (Kimmel et al. [8]); here, we will provide only few necessary details. Suppose that a mother cell at division has RNA content X; then it is assumed that one of the daughter cells will obtain Y and the other X-Y units of RNA. Y is a random variable with conditional density $f(Y|X)$. Once the daughter cell obtains its share of RNA (say, Y), her lifetime is determined and equal to $T=\Psi(Y)$, where $\Psi(\cdot)$ is a decreasing function. Also, the RNA level (X') of this daughter at its division is $X'=\varphi(Y)$, where $\varphi(\cdot)$ is an increasing function. Combining these assumptions, one is able to derive equation for $m(t,x)$, where $m(t,x)\Delta t\Delta x$ is the expected number of cells with RNA levels between x and $x+\Delta x$, that divided in time interval $[t, t+\Delta t]$.

In this paper we will show that our model equation exhibits asymptotic behaviour (exponential growth of solutions) similar to that predicted by traditional models (based on branching process theory or von Förster partial differential equation) in which the unequal division mechanism is absent. This, together with other findings (see [8] for discussion), produces evidence that unequal division of RNA (or other metabolic cell constituents) contributes considerably to the observed variability of cell life lengths.

In the present section, we will state a form of the model equation slightly more general than that in [8] and we will make preliminary comments about it.

2. The Model Equation; Generalities. We will consider the following equation:

$$m(t,x) = \int_0^\infty g(x,u)\ m(t-\theta(x),u)\ du \qquad (1)$$

in which g and θ are continuous functions and

(H_g) $g \geqslant 0$; $\int_0^\infty g(x,u)\ dx > 1$; $\exists\ \phi_1, \phi_2 : R^+ \to R^+$, increasing, $\phi_1 < \phi_2$, such that supp $g(.,u) \subset [\phi_1(u), \phi_2(u)]$; moreover $\phi_i(u)/u > 1$, for $u > 0$ small; $\phi_i(u) = u$ for $u=0$ and a_i, $i=1,2$.

(H_θ) $0 < \theta_1 < \theta(x) < \theta_2$.

In the terms of model parameters, $g(x,u) = 2(\varphi^{-1}(x))' f(\varphi^{-1}(x)|u)$, $\theta(x) = \psi(\varphi^{-1}(x))$. The assumptions on g and θ are consequences of biologically motivated assumptions on φ and ψ (see [8] and [1] for details).

Remark 1. (1) is an integro-difference equation; it can be integrated by a step-by-step procedure if the solution is known on $\Delta = [-\theta_2, 0] \times R^+$. So, (1) can be treated as an evolution equation on a class of functions on Δ.

Remark 2. Support property. If supp $m(t,.) \subset [A,B]$, for $-\theta_2 \leqslant t \leqslant 0$, where $0 < A \leqslant B < \infty$, then supp $m(t,.) \subset [\min(a_1, \phi_1(A)), \max(a_2, \phi_2(B))]$ (with a_i, ϕ_i as in (H_g)), for $t \geqslant \theta_2$. So, asymptotically supp $m(t,.) \subset [a_1, a_2]$.

It is reasonable, from biological viewpoint, to assume that m has a compact support in x. We will keep this assumption.

3. Problems to Solve. The model corresponds to a free growth period of cell population. So at least idealistically, we expect to find the exponential growth (with a distribution of RNA content, due to the unequal division). This growth has to dominate any initial perturbation. These remarks lead to two mathematical problems:

1° To find "exponential steady state" (ESS), ie. m(t,x) of the special form $m(t,x) = \exp(\lambda t)\mu(x)$.

2° To prove the "exponential stability" of the ESS with the greatest exponent λ^*, ie. to show that $m(t,x) - C\exp(\lambda^* t)\mu^*(x) = o(\exp(\lambda^* t))$, as $t \to \infty$.

The first problem is associated to the following Perron-type equation

$$\mu(x) = e^{-\lambda\theta(x)} \int_0^\infty g(x,u)\mu(u)\ du \overset{def}{=} L_\lambda \mu . \qquad (2)$$

Observe that (2) is nonlinear in λ, in contrast to the classical Perron equation [9]. This problem has been treated in [8] and [1]. Since we want to emphasize, in this paper, the second problem we will only state the result for problem 1 and refer the reader to [8] and [1].

Theorem 1. Define $\Lambda = \{\lambda \in C;\ (\lambda, \mu)$ is a solution of $(2)\}$. Then Λ is discrete and its elements have a finite multiplicity. There is a unique element λ^* in Λ with the greatest real part; $\lambda^* \in R^+$; the solution space for λ^* is one dimensional, generated by a nonnegative function μ^*, with support in $[a_1, a_2]$. The one dimensional space of ν-s such that $L^*_{\lambda^*}\nu = \nu$, where L^* is the adjoint of L in $L^2(a_1, a_2)$, is generated by a function ν^*, $\nu^*(x) > 0$, $a_1 \leqslant x \leqslant a_2$.

II. EXPONENTIAL STABILITY

In this part we will show how we can answer problem 2 and prove:

Theorem 2. The ESS $\exp(\lambda^* t)\mu^*(x)$ of Theorem 1, is exponentially stable.

Direct computations by means of inequalities are inefficient as a method of proof here; they only give: $m(t,x) \leqslant C\exp(\lambda^* t)$. In fact, due to linearity, the problem is to estimate the evolution operator when restricted to a subspace complementary to that generated by $\exp(\lambda^* t)\mu^*(x)$. This involves a semigroup approach and the determination of the spectral radius of the semigroup on a subspace.

Let us denote by G(t) the semigroup associated with (1) and by A its infinitesimal generator. Consider the splitting of the spectrum into its point (σ_P), residual (σ_R) and continuous (σ_C) parts [7]. The following relations hold between the spectrum of A and that of G(t): $\sigma_P(G(t))=\exp\{t\,\sigma_P(A)\}$, $\sigma_R(G(t))\subset\exp\{t\,\sigma_R(A)\}$ [7]. In general, there is no relation between $\sigma_C(G(t))$ and $\exp\{t\,\sigma_C(A)\}$. The way out of this trouble is to prove that the continuous spectrum is empty. If the semigroup is compact (what is true eg. for the linear autonomous functional differential equations [6]) then $\sigma_C(G(t))=\emptyset$, automatically. Unfortunately, this is not the case here.

In the sequel, we will first define the semigroup G(t), then characterize its infinitesimal generator A and its spectrum $\sigma(A)$ and finally we will look directly at $\sigma_C(G(t))$ to prove that it is void. We will leave to the reader the conclusion of Theorem 2 using the representation of the spectral radius by norms [5].

1. The Semigroup.

Denote $\Delta = [-\theta_2,0]\times(A,B)$, where $0 < A \leqslant a_1 < a_2 \leqslant B < \infty$ (cf. Remark 2) and $m_t(\tau,x) = m(t+\tau,x)$, $(\tau,x)\in\Delta$. The evolution can be expressed by

$$G(t).m_0 = m_t \tag{3}$$

To find an appropriate space for G(t), we prealably note that:

Lemma 1. If $m_0 \in L^2(\Delta)$, then $t \to m(t,.)$ is in $C(R^+, L^2(A,B))$ and m is in $C([{}^0\theta_2,\infty)\times R^+)$.

In view of this lemma, we will consider G(t) on the space

$$X = \left\{ m_0 \in C([-\theta_2,0],L^2(A,B));\ m(0,x)=\int_A^B g(x,u)\,m(-\theta(x),u)\,du \right\} \tag{4}$$

Proposition 1. $G(t): X \to X$ is a strongly continuous semigroup of bounded linear operators.

2. The Infinitesimal Generator and Its Spectrum.

Lemma 2. Let A be the infinitesimal generator of G(t). Then A is given by: $(A\chi)(\tau,x)=\partial\chi(\tau,x)/\partial\tau$, $-\theta_2<\tau<0$, $x>0$, with $D(A) = \{\chi\in X:\ \partial\chi/\partial\tau \in X\}$.

To look at the spectral properties of A, we have to consider the resolvent equation

$$A\chi - \lambda\chi = \xi,\quad \xi\in X \tag{5}$$

In view of the definition of A, (5) can be integrated to

$$\chi(\tau,x) = \chi(0,x)\, e^{\lambda\tau} + \int_0^{\tau} e^{\lambda(\tau-s)} \xi(s,x)\, ds \tag{6}$$

and so (5) reduces to an equation to be verified by the "boundary value" $\chi(0,.)$:

$$\chi(0,x) = \int_0^{\infty} g(x,u)\, e^{-\lambda\theta(x)} \chi(0,u)\, du + \int_0^{\infty} g(x,u) \left[\int_0^{-\theta(x)} e^{-\lambda(\theta(x)+s)} \xi(s,u)\, ds \right] du \tag{7}$$

The first operator in the right hand side of (7) is L_λ (defined in (2)). We denote

$$S_\lambda \xi = \int_0^{\infty} g(x,u) \left[\int_0^{-\theta(x)} e^{-\lambda(\theta(x)+s)} \xi(s,u)\, ds \right] du \tag{8}$$

Lemma 3. A necessary and sufficient condition for the existence of χ, solution of (5), is that the equation

$$\chi_0 = L_\lambda \chi_0 + S_\lambda \xi$$

has at least one solution. (Then, χ is defined by (6) with $\chi(0,.) = \chi_0$.) Thus, if $(I-L_\lambda)$ is invertible, and so is $(A-\lambda I)$, then we have

$$(A-\lambda I)^{-1} \xi = e^{\lambda\tau} (I-L_\lambda)^{-1} S_\lambda \xi + \int_0^{\tau} e^{\lambda(\tau-s)} \xi(s,x)\, ds \tag{9}$$

Remark 3. Note that S_λ is a compact operator.

We are ready now to state the result on the spectrum of A:

Proposition 2. (i) $\sigma(A)$ is a pure point spectrum, (ii) $\sigma(A) = \Lambda$ (cf. Theorem 1), (iii) Range $(A-\lambda I)$ is closed; $\mathrm{Ker}(A-\lambda I) = \exp(\lambda\tau) \otimes \mathrm{Ker}(I-L_\lambda)$, (iv) for $\lambda=\lambda^*$, $Y=\mathrm{Range}(A-\lambda^* I)$ has codimension 1 and is a direct summand of $\mathrm{Ker}(A-\lambda^* I)$.

Remark 4. Proposition 2 implies in particular that $G(t)|_Y$ is a semigroup $G_Y(t)$ associated to $A|_Y$ and so Theorem 2 says that $\| G_Y(t) \| = o(\exp(\lambda^* t))$. Thus, the proof of Theorem 1 is reduced to showing that $\sigma_C(G_Y(t))=\emptyset$, which is equivalent to $\sigma_C(G(t))=\emptyset$.

3. The Continuous Spectrum of G(t).

We will treat this subject in more detail even if we cannot be more than sketchy in the overall.

Proposition 3. $\sigma_C(G(t))=\emptyset$, for t large.

Proof. First, we present the principle of the proof. As a preliminary it will be convenient to extend G(t) to $L^2(\Delta)$ (cf. Lemma 1). The resolvent equation of G(t), for $t>0$ fixed, is

$$G(t).m_0 - \lambda m_0 = y \tag{10}$$

which will be transformed into

$$K n_0 - \lambda n_0 = z \tag{11}$$

using a continuous operator $\mathcal{G}$ which sends y into z, m_0 into n_0. (10) and (11) will be equivalent through $\mathcal{G}$ and the large iterates of K will be compact.

For such transformation we will have:

$$\mathrm{Range}\,(G(t)-\lambda I) = \mathcal{G}^{-1}(\mathrm{Range}(K-\lambda I))$$

and from the continuity of $\mathcal{G}$ and the closedness of Range($K-\lambda I$), we will conclude that $G(t)-\lambda I$ has a closed range (except maybe for $\lambda=0$). This implies that $\sigma_C(G(t))=\emptyset$.

Let us now briefly consider how $\mathcal{G}$ and K are related to G(t). To define $\mathcal{G}$, we look at (10) for t small. It splits into two equations

$$m(t+\tau,x)-\lambda m(\tau,x)=y(\tau,x);\quad t+\tau<0 \qquad (10)_1$$

$$\int g(x,u)\, m(t+\tau-\theta(x),u)\, du-\lambda m(\tau,x)=y(\tau,x);\quad t+\tau\geqslant 0 \qquad (10)_2$$

We define

$$(\mathcal{G}m)(s,x)=\int g(x,u)m(s,u)\, du=n(s,x) \qquad (12)$$

Applying $\mathcal{G}$ to $(10)_1$ we obtain

$$n(t+\tau,x)-\lambda n(\tau,x)=z(\tau,x);\quad t+\tau<0 \qquad (13)_1$$

$$n(t+\tau-\theta(x),x)-\lambda m(\tau,x)=y(\tau,x);\quad t+\tau\geqslant 0 \qquad (13)_2$$

(13) together with $\mathcal{G}$ allows us to express n in terms of m. Applying $\mathcal{G}$ to $(13)_2$ we obtain

$$\int g(x,u)\, n(t+\tau-\theta(u),u)\, du-\lambda n(\tau,x)=z(\tau,x).$$

We define

$$\mathcal{K}n(\tau,x)=\int g(x,u)\, n(t+\tau-\theta(u),u)\, du \qquad (14)$$

and then

$$K\,n=\begin{cases} n(t+\tau,x); & \tau<-t,\\ \mathcal{K}\,n(\tau,x); & -t\leqslant\tau\leqslant 0.\end{cases} \qquad (15)$$

It is not difficult to see that $\mathcal{K}$ is compact (at least, on the range of $\mathcal{G}$). Now, we consider the iterates of K, eg.:

$$K^2n=\begin{cases} n(2t+\tau,x); & \tau<-2t,\\ \mathcal{K}\,n(t+\tau,x); & -2t\leqslant\tau<-t,\\ \mathcal{K}\,(Kn)(\tau,x); & -t\leqslant\tau\leqslant 0,\end{cases}$$

etc. ; we see that for $j>(\theta_2/t)$: $K^j=\mathcal{K}\circ$(bounded operator), so that K^j is compact. This concludes the proof of Proposition 3.

III. COMMENTS

Analysis, analogous to that of our paper has been recently carried out for a certain population dynamics equation by O. Diekman et al. [3],[4]. In [4], the authors consider an age-dependent model: they meet the same difficulty that we did, in looking for exponential stability. But instead proving that σ_C is void, they compute the radius of the essential spectrum [11], computation of which seems to be more difficult in our case.

Generally, population models involving any age-dependent effects lead to functional, functional-differential or functional-integral equations (see [1], [3], [4], for references). A very interesting example of a functional-integral iteration was published recently by Lasota and Mackey [10].

REFERENCES

[1] Arino O., Kimmel M., Asymptotic Analysis of a Cell Cycle Model Based on Unequal Division.(Submitted to SIAM Journal on Applied Mathematics, 1984.)

[2] Darzynkiewicz Z., Crissman H., Traganos F., Steinkamp J., Cell Heterogeneity During the Cell Cycle. Journal of Cell Physiology, 113 (1982) 465-474.

[3] Diekmann O., Metz J.A.J., Kooijman S.A.L.M., Heijmans H.J.A.M., Continuum Population Dynamics with an Application to Daphnia Magna. Nieuw Archies voor Wiskunde, 2 (1984) 82-109.

[4] Diekmann O., Heijmans H.J.A.M., Thieme H.R., On the Stability of the Cell Size Distribution. Journal of Mathematical Biology (in press).

[5] Dunford N., Schwartz J.T., Linear Operators. Part I (Wiley, New York, 1957).

[6] Hale J., Theory of Functional Differential Equations (Springer, Berlin, 1977).

[7] Hille E., Functional Analysis and Semigroups (American Mathematical Society Colloquium Publications, vol. 31, New York, 1948).

[8] Kimmel M., Darzynkiewicz Z., Arino O., Traganos F., Analysis of a Cell Cycle Model Based on Unequal Division of Metabolic Constituents to Daughter Cells During Cytokinesis. Journal of Theoretical Biology (to appear).

[9] Krein M.G., Rutman M.E., Linear Operators Leaving Invariant the Cone in the Banach Space (Translations of the AMS, series 1, volume 10: Functional Analysis and Measure Theory, 1962, pp. 199-327).

[10] Lasota A., Mackey M.C., Globally Asymptotic Properties of Proliferating Cell Populations. Journal of Mathematical Biology, 19 (1984) 43-62.

[11] Nussbaum R.D., The Radius of the Essential Spectrum. Duke Mathematical Journal, 38 (1970) 473-478.

The final (detailed) version of this paper will be submitted for publication elsewhere.

Trends in the Theory and Practice of Non-Linear Analysis
V. Lakshmikantham (Editor)
Elsevier Science Publishers B.V. (North-Holland), 1985

On the B-Convergence of the θ-Method Over Infinite Time for Time Stepping for Evolution Equations

O. Axelsson
Department of Mathematics
Catholic University, Nijmegen, The Netherlands*

Presented at the VIth International Conference on Nonlinear Analysis, The University of Texas at Arlington, June 18-22, 1984.

ABSTRACT

In Frank et al (5), the importance of B-convergence, i.e. convergence independent of stiffness of the problem, for numerical methods for evolution equations was pointed out. In [3] and [5] it was proven that the implicit Euler method is B-convergent.

In the present paper we extend this result to the class of θ-methods and show further that for monotone problems and $0 \le \theta \le \theta_0$, $\theta_0 \simeq \frac{1}{2} - \zeta\tau$, ζ a certain positive constant and τ the time-step, the error estimate is valid for all times. We also prove convergence for problems with little regularity of the solution. For earlier presentations of related results, see [1] and [2].

1. INTRODUCTION

Consider the evolution equation

$$\dot{u} + F(t,u) = 0,\ t > 0,\quad u(0) = u_0 \in V, \tag{1.1}$$

where $\dot{u} = \frac{du}{dt}$ and $F(\tau,\cdot) = V \to V$, V a Hilbert space with inner product $(\cdot,\cdot)$ and norm $\|v\| = (v,v)^{\frac{1}{2}}$. (The results we state in this paper are also valid in the case $F(\tau,\cdot) : V \to V'$, where V' is the dual space to V, for details, see [1] and [2] and the references cited therein.) We shall assume that $\frac{\partial F}{\partial \tau}$ is bounded, $\sup_u \| \frac{\partial F}{\partial \tau}(\tau,u) \| \le C$, $t > 0$ where in practice, C is not a "very large" number. (If we match the time-steps in the numerical methods appropriately, we may let $\frac{\partial F}{\partial \tau}$, or even F, have discontinuities at certain (finite number of) points, but for simplicity of presentation we exclude this case.)

The simplest method for numerical time-stepping for (1.1) is the Euler (forward) method,

---

* The research reported in this paper was in part supported by the North Atlantic Treaty Organization, Brussels, through Grant No. 648/83.

(1.2) $v(t+\tau) = v(t) - \tau F(t,v(t)),\ t = 0,\tau,2\tau,\ldots$

where v is the corresponding approximation to u. (It is only for notational simplicity that we assume that the time step τ is constant.)

Let $e(t) = u(t) - v(t)$ be the error function. Classical error estimates, such as in [6], make use of the two-sided Lipschitzconstant,

(1.3) $L = \sup\{\|F(t,u) - F(t,v)\| / \|u - v\|\},\ \tau > 0,\ u,v \in V_0 \subset V$

where V_0 contains all functions in a sufficiently large tube about the solution u. In the analysis of the Euler forward method we have to assume that F is two-sided Lipschitz bounded, i.e. that $L < \infty$, but for the implicit methods to be considered later, we need only a one-sided Lipschitzbound (for more details, see [2]).

From (1.1) it follows

(1.4) $u(t+\tau) = u(t) - \tau \int_0^1 F(t + \tau s,\ u(t+\tau s))ds$

and from (1.2) and (1.4) we get

$$e(t+\tau) = e(t) - \tau\{F(t,u(t)) - F(t,v(t))\} + \tau R(t,u)\ ,$$

where

$$R(t,u) \equiv \int_0^1 [F(t,u(t)) - F(t + \tau s,\ u(t+\tau s))]ds$$

$$= \int_0^1 [\dot{u}(t+\tau s) - \dot{u}(t)]ds$$

is the (normalized) local truncation error.

By (1.3) it follows

$$\|e(t+\tau)\| \le (1+\tau L)\|e(t)\| + \tau\|R(t,u)\|\ , \qquad t = 0,\tau,2\tau,\ldots$$

or, by recursion,

$$\|e(t)\| \le (1+\tau L)^{t/\tau}\|e(0)\| + \tau \sum_{j=1}^{t/\tau} (1+\tau L)^{j-1}\|R(t-j\tau,u)\|$$

$$\le e^{tL}\|e(0)\| + \frac{1}{L}(e^{tL}-1) \max_{t\ge 0} \|R(t,u)\|\ , \qquad t = \tau,2\tau,\ldots$$

Notice that the initial errors may grow as $\exp(tL)$. In case $v(0) = u_0$ we have however $e(0) = 0$ (but in practice $e(0) \ne 0$, because of round-off errors) and then

(1.5) $\|e(t)\| \le \frac{1}{L}(e^{tL}-1) \max_{t\ge 0} \|R(t,u)\|$.

Typically (under the assumption of a bounded second derivative of u) we have $\|R(t,u)\| \le C\tau$, where C depends only on the smoothness of the solution, and not on the Lipschitz constant L. However, in most problems of practical interest, L is large, so even for moderately large values of t, the truncation error is amplified by a large factor $\sim L^{-1}\exp(tL)$. This is in particular true for so-called stiff problems, where L is very large, in which case the bound (1.5) (and the method (1.2), even for very small time-steps satisfying $\tau L \ll 1$) is useless. This is in fact true for all explicit time-stepping methods.

We shall now consider monotone problems, i.e., problems (1.1) where F satisfies

$$(1.6) \qquad (F(t,u) - F(t,v),\ u-v) \geq \rho(t)\|u-v\|^2 \quad \forall\, u,v \in V,\ t > 0$$

and $\rho : (0,\infty) \to R^+$, i.e., $\rho(t) \geq 0$, $t > 0$.
For such problems we easily derive the following stability bound:

$$\|u(t)-w(t)\| \leq \exp\left(\int_0^t -\rho(s)ds\right)\|u(0)-w(0)\| \leq \|u(0)-w(0)\|,\ t > 0,$$

where u, w are solutions of (1.1) corresponding to different initial values, u(0) and w(0), respectively.

We now face the following problems:

(i) Can we find a numerical time-stepping method for which a similar stability bound is valid?

(ii) Can we derive discretization error estimates without a "nasty" large (exponentially growing) stiffness factor, such as the factor in (1.5)?

The answer to these problems is affirmative as was pointed out in [3] and [5]. Consider namely the "backward" or implicit Euler method

$$(1.7) \qquad v(t+\tau) + \tau F(t+\tau,\ v(t+\tau)) = v(t), \qquad t = 0,\tau,2\tau,\ldots$$

In practice, we will never calculate the solution $v(t+\tau)$ of the non-linear equation (1.7) (which exists and is unique) exactly. To take care of this we could add a perturbation term to (1.7). For notational simplicity we assume however that $v(t+\tau)$ is calculated exactly. It follows from (1.1) and (1.7),

$$(1.8) \qquad e(t+\tau) + \tau[F(t+\tau,\ u(t+\tau)) - F(t+\tau,\ v(t+\tau))]$$
$$= e(t) + \tau R(t,u)$$

where $R(t,u) = \int_0^1 [\dot{u}(t+\tau s) - \dot{u}(t+\tau)]ds$.

Multiplying (1.8) by $e(t+\tau)$, taking the inner product and using standard inequalities, it follows from (1.6) that

$$(1+\tau\rho_0)\|e(t+\tau)\|^2 \leq \|e(t)\|^2 + \tau\rho_0^{-1}\|R(t,u)\|^2,$$

assuming at first that $\rho(t) \geq \rho_0 > 0$, $t > 0$. Hence by recursion,

$$\|e(t)\|^2 \le (1+\tau\rho_0)^{-t/\tau}\|e(0)\|^2 + \rho_0^{-2}\sup_{t>0}\|R(t,u)\|^2.$$

We notice that in this case the influence of the initial error decays (exponentially) with increasing time, which answers question (i) (the method is said to be asymptotically stable). We also notice that the constants in this estimate are independent of stiffness. In case $v(0) = u_0$, we have if the second derivative of u is uniformly bounded,

$$(1.9)\qquad \|e(t)\| \le \rho_0^{-1}\max_{t\ge 0}\|R(t,u)\| \le C\tau, \qquad t \ge 0$$

This answers question (ii). A numerical method which satisfies these properties has been called B-convergent, see [5]. The method found is however only first order accurate.

We notice that (1.9) is valid for all times. In case $\rho_0 = 0$ one can prove

$$\|e(t)\| \le 2t\max_{t>0}\|R(t,u)\| \qquad \text{(see [2])},$$

i.e., only a linear growth with time.

Notice that in case of a linear problem, stability of a numerical method for monotone problems is equivalent to so-called A-stability, i.e., the method is stable for all problems, $\dot{u} + Au = f(t)$, for which the real parts of the eigenvalues of A are positive. By a classical result by G. Dahlquist, among the class of linear multistep methods an A-stable method can not be of order of accuracy higher than two. For an extension of A-stability to nonlinear problems, see [4]. Little is known about existence of B-convergent numerical methods (see [5]). It is the purpose of this report to discuss an extension of the above result for the Euler backward method to the class of θ-methods and to prove that there exists θ-methods of up to second order of accuracy which are B-convergent and with error estimates valid for all times. The results found complement some of the results in [2].

2. STABILITY OF THE θ-METHOD (IMPLICIT FORM)

We shall consider the implicit (also called one-leg) form of the θ-method

$$(2.1)\qquad v(t+\tau) + \tau F(\bar{t},\bar{v}(t)) = v(t), \qquad t = 0,\tau,2\tau,\ldots ,$$

$v(0) = v_0$ where $\bar{t} = \theta t + (1-\theta)(t+\tau) = t + (1-\theta)\tau$
and $\bar{v}(t) = \theta v(t) + (1-\theta)v(t+\tau)$, $\quad 0 \le \theta \le 1$.

For $\theta = 0$ and 1 we get the Euler backward and Euler forward methods, respectively. We assume that F is monotone, i.e., satisfies (1.6). Then it will follow that the nonlinear algebraic equation (2.1) has a unique solution in V, if $0 \le \theta \le \frac{1}{2}$. From (1.1) follows

(2.2) $$u(t+\tau) + \tau F(\bar{t},\bar{u}(t)) = u(t) + \tau R_\theta(t,u),$$

where the (normalized) truncation error

(2.3) $$R_\theta(t,u) = F(\bar{t},\bar{u}(t)) + \tau^{-1}[u(t+\tau) - u(t)].$$

For the discretization error $e(t) = u(t) - v(t)$ we get from (2.1) and (2.2)

(2.4) $$e(t+\tau) - e(t) + \tau[F(\bar{t},\bar{u}(t)) - F(\bar{t},\bar{v}(t))] = \tau R_\theta(t,u).$$

Multiplying (2.4) by $\bar{e}(t)$ and using (1.6) we get

$$(e(t+\tau) - e(t), \bar{e}(t)) + \tau\rho_0\|\bar{e}(t)\|^2 \le \tau(R_\theta,\bar{e}(t))$$

An elementary calculation (see [1]) shows that

$$(e(t+\tau) - e(t), \bar{e}(t)) = \tfrac{1}{2}[\|e(t+\tau)\|^2 + (1-2\theta)\|e(t+\tau) - e(t)\|^2 - \|e(t)\|^2]$$

and

$$\|\bar{e}(t)\|^2 = (1-\theta)\|e(t+\tau)\|^2 + \theta\|e(t)\|^2 - (1-\theta)\theta\|e(t+\tau) - e(t)\|^2.$$

Assume at first that $\rho_0 > 0$. Then by use of standard inequalities, such as the arithmetic-geometric mean inequality, $\tau(R_\theta,\bar{e}(t)) \le \frac{1}{2}\tau\,\rho_0^{-1}\|R_\theta\|^2 + \frac{1}{2}\tau\rho_0\|\bar{e}(t)\|^2$, we find $[1 + (1-\theta)\tau\rho_0]\|e(t+\tau)\|^2 + [1 - 2\theta - (1-\theta)\theta\tau\rho_0]\|e(t+\tau) - e(t)\|^2 \le (1-\theta\tau\rho_0)\|e(t)\|^2 + \tau\rho_0^{-1}\|R_\theta(t,u)\|^2$.

Let θ_0 be the largest number ≤ 1 for which $1 - 2\theta - (1-\theta)\theta\tau\rho_0 \ge 0$. We find

(2.5) $$\theta_0 = [1 + \tfrac{1}{2}\tau\rho_0 + (1 + (\tfrac{1}{2}\tau\rho_0)^2)^{\frac{1}{2}}]^{-1},$$

i.e., $\theta_0 = \frac{1}{2} - |O(\tau)|$, $\tau \to 0$. Then for $0 \le \theta \le \theta_0$ we have

$$\|e(t)\|^2 \le q^{t/\tau}\|e(0)\|^2 + \tau \sum_{j=0}^{(t/\tau)-1} q^{(t/\tau)-j-1}[1 + (1-\theta)\tau\rho_0]^{-1}\rho_0^{-1}\|R_\theta(\tau j,u)\|^2$$

$$\le q^{t/\tau}\|e(0)\|^2 + \rho_0^{-2}\max_{t\ge 0}\|R_\theta(t,u)\|^2,$$

where $q = (1-\theta\tau\rho_0)/[1 + (1-\theta)\tau\rho_0]$. Since $\theta < \frac{1}{2}$, we have $|q| < 1$. Hence the θ-method is unconditionally stable (independent of the stiffness and of τ) and if $e(0) = 0$, then

(2.6) $$\|e(t)\| \le \rho_0^{-1}\max_{t\ge 0}\|R_\theta(t,u)\|, \quad 0 \le \theta \le \theta_0,$$

which generalizes (1.9) to the class of θ-methods.

In the case $\rho_0 = 0$ one finds that

$$\|e(t)\| \le \|e(0)\| + t2(1-\theta)\max_{t\ge 0}\|R_\theta(t,u)\|, \qquad 0 \le \theta \le \tfrac{1}{2} \quad \text{(see [2])}.$$

3. TRUNCATION ERROR

It remains to consider the truncation error R_θ defined by (2.3). To this end we shall at first assume only that $\dot{u}$ is Höldercontinuous, i.e.,

$$(3.1)\qquad \|\dot{u}(t_1) - \dot{u}(t_2)\| \le C_1|t_1-t_2|^\alpha, \quad 0 < \alpha \le 1 \quad \forall\, t_1, t_2$$

such that $t < t_1, t_2 < t+\tau$, $t = 0, \tau, 2\tau, \ldots$

Further we let $C_0 = \sup\limits_{t\ge 0, u\in V_0} \|\frac{\partial F}{\partial t}(t,u)\|$, where V_0 contains all functions in a tube with radius $\|u(t) - v(t)\|$ about the solution. (In practice we also assume that C_1 and C_0 are small relative to the stiffness constant L.)

Since u is continuous, there exists a point $t_\theta \in (t,t+\tau)$ such that

$$(3.2)\qquad \bar{u}(t) = \theta u(t) + (1-\theta)u(t+\tau) = u(t_\theta),$$

and because $\dot{u}$ is Höldercontinuous we find that

$$(3.3)\qquad t_\theta = \bar{t} + O(\tau^{1+\alpha}).$$

Further, we have

$$(3.4)\qquad F(\bar{t},\bar{u}(t)) = F(t_\theta,\bar{u}(t)) + \frac{\partial F}{\partial t}(\tilde{t},\bar{u}(t))(\bar{t}-t_\theta),$$

where $\tilde{t} \in \text{int}(t_\theta,\bar{t})$. By (3.2),

$$F(t_\theta,\bar{u}(t)) = F(t_\theta,u(t_\theta)) = -\dot{u}(t_\theta)$$

and it follows by (2.3), (3.3), (3.4) that

$$R_\theta(t,u) = \int_0^1 [\dot{u}(t+\tau s) - \dot{u}(t_\theta)]ds + \frac{\partial F}{\partial t}(\bar{t}-t_\theta)$$

or by (3.1),

$$(3.5)\qquad \|R_\theta(t,u)\| \le C_1 \int_0^1 |t+\tau s-t_\theta|^\alpha ds + C_0|O(\tau^{1+\alpha})|$$

$$\le C_1\tau^\alpha\Big[\int_0^{1-\theta} (1-\theta-s)^\alpha ds + \int_{1-\theta}^{1} (s-1+\theta)^\alpha ds\Big](1+O(\tau)) + |O(\tau^{1+\alpha})|$$

$$= \frac{1}{\alpha} C_1\tau^\alpha[(1-\theta)^{\alpha+1} + \theta^{\alpha+1}] + |O(\tau^{1+\alpha})| = O(\tau^\alpha), \qquad \tau \to 0.$$

Assume now that $\ddot{u} = \frac{d^2u}{dt^2}$ is Höldercontinuous with exponent β,

$$(3.6)\qquad \|\ddot{u}(t_1) - \ddot{u}(t_2)\| \le C_2|t_1 - t_2|^\beta, \qquad 0 < \beta \le 1$$

$\forall t_1, t_2$ such that $t < t_1, t_2 < t+\tau$, $t = 0, \tau, 2\tau, \ldots$

Then $t_\theta = \bar{t} + O(\tau^2)$ and by Taylor expansion it follows that

$$R_\theta(t,u) = \tau^{-1}[u(t+\tau) - u(t)] - \dot{u}(t+\tfrac{1}{2}\tau) + \dot{u}(t+\tfrac{1}{2}\tau) - \dot{u}(t_\theta) + \frac{\partial F}{\partial t}(\bar{t}-t_\theta)$$

or

$$\| R_\theta(t,u) \| = O(\tau^{1+\beta})C_2 + \tau|\theta-\tfrac{1}{2}|\,\| \ddot{u}(t_1) \| + O(\tau^2), \quad t < t_1 < t+\tau .$$

If $\beta = 1$ we get

$$\| R_\theta(t,u) \| = \frac{1}{24}\tau^2 \frac{d^3u}{dt^3}(t_2) + |\theta-\frac{1}{2}|\tau C\| \ddot{u}(t_1) \| + O(\tau^2), \quad t < t_2 < t+\tau .$$

4. CONCLUSION

Combining the results of Sections 2 and 3 we have

Theorem 4.1. For the θ-method (2.1) applied on (1.1) where F satisfies (1.6) (i.e., is monotone), $\frac{\partial F}{\partial t}$ is uniformly bounded and where the time derivative of the solution is Höldercontinuous (except possibly at the stepping points $\tau, 2\tau, 3\tau, \ldots$) the discretization error satisfies the following:

a) If $\rho(t) \geq \rho_0 > 0$ and $0 \leq \theta \leq \theta_0$, where θ_0 is defined by (2.5) we have if $v_0 = u_0$, $\| e(t) \| \leq \rho_0^{-1}\alpha^{-1}C_1[(1-\theta)^{\alpha+1} + \theta^{\alpha+1}]\tau^\alpha + O(\tau^{\alpha+1})$, $\tau \to 0$, $t \geq 0$ where C_1, α are the Hölder constant and exponent, respectively.

b) If $\rho_0 = 0$, $0 \leq \theta \leq \frac{1}{2}$ and $v_0 = u_0$, we have

$$\| e(t) \| \leq t2(1-\theta)C_1\alpha^{-1}[(1-\theta)^{\alpha+1} + \theta^{\alpha+1}]\tau^\alpha + O(\tau^{\alpha+1}), \quad \tau \to 0,\ t \geq 0 .$$

If the second derivative of the solution is Höldercontinuous we have

c) If $\rho(t) \geq \rho_0 > 0$, $0 \leq \theta \leq \theta_0$ and $v_0 = u_0$,

$$\| e(t) \| \leq O(\tau^{1+\beta}) + |\theta-\tfrac{1}{2}|O(\tau), \quad \tau \to 0,\ t \geq 0$$

and similarly for the case $\rho_0 = 0$, $0 \leq \theta \leq \frac{1}{2}$.

This proves convergence of the θ-method for all $t \geq 0$ and for various orders of regularity of the solution. Note that if $\theta = \frac{1}{2} - \zeta\tau^\beta$, $\zeta > 0$ (large enough) then Theorem 4.1c implies

$$\| e(t) \| = O(\tau^{1+\beta}), \qquad \tau \to 0,\ t \geq 0.$$

REFERENCES

[1] O. Axelsson, Error estimates for Galerkin methods for quasilinear parabolic and elliptic differential equations in divergence form, Numer. Math. 28, 1-14 (1977).

[2] O. Axelsson, Error estimates over infinite intervals of some discretizations of evolution equations, Report 8405, Department of Mathematics, Catholic University, Nijmegen, The Netherlands, to appear BIT.

[3] G. Dahlquist, Error analysis for a class of methods for stiff nonlinear initial value problems, Numerical Analysis (G.A. Watson, ed.), Dundee 1975, Springer-Verlag, LNM 506, 1976.

[4] G. Dahlquist, G-stability is equivalent to A-stability, BIT 18 (1978), 384-401.

[5] R. Frank, J. Schneid and C.W. Ueberhuber, The concept of B-convergence, SIAM J. Numer. Anal. 18 (1981), 753-780.

[6] P. Henrici, Discrete variable methods in ordinary differential equations, John Wiley and Sons, Inc., New York, 1962.

This paper is in final form and no version of it will be submitted for publication elsewhere.

Trends in the Theory and Practice of Non-Linear Analysis
V. Lakshmikantham (Editor)

FILTER STABILITY IN FLOWS

Prem N. Bajaj
Department of Mathematics and Statistics
Wichita State University
Wichita, Kansas 67208

Filters have been used by J. Auslander in order to have a unified approach for ε-δ and nbd definitions for Lyapunov stability. The purpose of this note is to generalize his results to rim-compact topological spaces.

1. INTRODUCTION

The role of dynamical systems in differential equations is well known [7]. In the discussion of Lyapunov stability in dynamical systems definitions using ε-δ approach or neighborhood approach are equivalent for compact sets but differ for non-compact sets. J. Auslander ([2], [3], [4]) has used filter-stability as a device to combine the two approaches; his work is in the setting of locally compact metric spaces. We plan to extend his results to rim compact [8] spaces. It is the rim-compactness, rather local compactness that matters for stability problems. Besides we work in arbitrary Hausdorff spaces instead of in metric spaces.

2. DEFINITIONS AND NOTATION

Let X be a topological space. A dynamical system (or flow) on X is a continuous map π from XxR into X satisfying indentity and group axioms. Taking R^+ and the semi-group property respectively for R and the group property, a semi-dynamical system (also called a semi-flow) is obtained. For an x in X and t in R (or R^+), $\pi(x,t)$ is denoted by xt and the set $\{xt: x \in k \subset X\}$ by kt etc.

In a dynamical system (X,π), the trajectory and positive semi-trajectory from a point x in X are denoted, respectively, by $\gamma(x)$ and $\gamma^+(x)$.

In a semi-flow (X,π), a point x is said to be a start point [5], [6], if $yt \neq x$ for each $y \in X$ and $t > 0$. For a point x which is not a start point, negative trajectory (not necessarily unique or of infinite time length) is defined in a natural way. A sub-set k of X is said to be weakly negatively invariant if each x in k is either a start point or some negative trajectory from x lies in k.

Notions of invariance and positive invariance are obvious. We'll denote the positive limit set of a point x by L(x).

A topological space is said to be rim compact [8] if each of its points has a base of nbds with compact boundaries.

To emphasize the role of rim compactness in stability problems, we digress to prove the following two theorems.

3. THEOREM

Let (X,π) be a semi-dynamical system. Let X be rim compact. Let $x \in X$. Then $L(x)$ is weakly negatively invariant and does not contain any start points.

Proof. Let $y \in L(x)$. Then there exists a net t_i in R^+, $t_i \to +\infty$ such that $xt_i \to y$.

Let $t > 0$ be arbitrary but fixed. We may take $t_i > t$ for every i.

Now consider the net $x(t_i - t)$ in X.

If some sub-net of $x(t_i - t)$, which we take the net itself, converges to y, then $x(t_i - t)\, t \to yt$ and $x(t_i - t)\, t = xt_i \to y$ imply that $yt = y$ i.e., y is not a start point.

If $x(t_i - t)$ does not have any sub-net converging to y, let U be a nbd of y, ∂U compact such that $x(t_i - t) \notin \operatorname{int} U$ for $i \geq I$ for some I. Then there exists s_i, $t_i - t < s_i < t_i$ such that $xs_i \in \partial U$. Since ∂U is compact, a sub-net of xs_i, which we take the net itself, converges to a point $z \in \partial U$. More-over $0 < t_i - s_i \leq t$ and so the net $\{t_i - s_i\}$ has a convergent sub-net; let $t_i - s_i \to s \in [0,t]$. Now $x s_i (t_i - s_i) = xt_i \to y$, $x s_i (t_i - s_i) \to zs$ so that $zs = y$. Since $y \in \operatorname{int} U$, $z \in \partial U$, s cannot be zero. Hence y is not a start point.

Weak negative invariance of $L(x)$ can easily be seen now.

4. THEOREM

Let (X,π) be a semi-dynamical system. Let X be rim compact. Let $x \in X$. If $L(x)$ is non-empty and compact, then $\overline{xR^+}$ is compact.

Proof. First notice that if U is an open set containing $L(x)$, there exists $T > 0$ such that $xt \in U$ for all $t \geq T$. To see this suppose otherwise. Let ∂U be compact. Let $y \in L(x)$. Then $xt_i \to y$ for some net t_i in R^+, $t_i \to +\infty$ For each i there exists an $s_i > t_i$ such that $x s_i \in \partial U$. Now $\{x s_i\}$ has a cluster point $z \in \partial U$, but $s_i \to +\infty$ and so, $z \in L(x)$; contradicts that $L(x) \subset U$.

Next let $\{x_i\}$ be a net in $\overline{xR^+} = xR^+ \cup L(x)$. If the net $\{x_i\}$ has a sub-net in $L(x)$, it has a convergent sub-net. So let $x_i \in xR^+$ for all i; thus $x_i = xt_i$, $t_i \in R^+$. If t_i's are bounded, then the net $\{t_i\}$, and so the net $\{xt_i\}$ has a convergent sub-net. If $\{t_i\}$ is unbounded, we may suppose that $t_i \to +\infty$. Then it is easily seen that $\{xt_i\}$ has a sub-net converging to a point $z \in L(x)$.

5. DEFINITION

The positive prolongation of an x in X in a dynamical system (X,π) is defined by $D(x) = \cap\{\gamma^+(U) : U \text{ is a nbd of } x\}$.

The manner in which rim-compactness replaces local compactness is illustrated in the following.

6. BASIC LEMMA

Let (X,π) be a dynamical system. Let X be rim-compact. Let $x \in X$. Then $D(x)$ is compact if and only if when-ever $\{x_i\}$, $\{t_i\}$ are nets, $x_i \to x$. $t_i \in R^+$, the net $\{x_i t_i\}$ has a convergent sub-net.

Proof$\Rightarrow$ Let $D(x)$ be compact. Let $\{x_i\}$ be a net in X, $x_i \to x$, and $\{t_i\}$ a net in R^+.

If any sub-net of $\{x_i t_i\}$ does not converge to a point (of D(x)), for each y in D(x), we can find an open nbd V_y of y such that ∂V_y is compact and $x_i t_i \notin V_y$ eventually. Let the open cover $\{V_y: y \in D(x)\}$ of the compact set D(x) have finite sub-cover $\{V_1, V_2, \ldots V_m\}$. Let $V = V_1 \cup V_2 \cup V_m$. Then ∂V is compact and $x_i t_i \notin V$ eventually. Since $x_i \to x \in D(x)$ we may suppose that $x_i \in V$ for all i. Then for each i, there exists an s_i such that $0 < s_i < t_i$ and $x_i s_i \in \partial V$. Due to compactness of ∂V, the net $x_i s_i$ has a sub-net converging to a point, say z in ∂V. But, then $z \in D(x)$ which contradicts that $D(x) \subset V$.

$\Leftarrow$ Let y_j be a net in D(x). For each fixed j, there exists a net $\{z_i\}$ in X, $z_i \to x$, a net $\{s_i^j\}$ in R^+ such that $z_i s_i \to y_j$. For each fixed j, let V, V_j, respectively, be nbds of x and y_j. Choose i such that $z_i \in V$ and $z_i s_i^j \in V_j$; re-label the so chosen $z_i s_i$ as $x_j t_j$. Since the net $\{x_j t_j\}$ has a convergent sub-net, so is the case for the net $\{y_j\}$ by the way above construction has been carried out.

7. DEFINITION

A filter $\mathcal{F}$ on a set X is a non-empty collection of non-empty sub-sets of X satisfying the following properties: (i) if F, G are in $\mathcal{F}$, then $F \cap G \in \mathcal{F}$. (ii) if $F \in \mathcal{F}$, $G \supset F$, then $G \in \mathcal{F}$.

The following is a filter on the set of real numbers.

$\mathcal{F} = \{F \subset R$: there exists real numbers a,b,c,d, $a < b < 0$, $1 < c < d$ such that each of the open intervals $(-\infty, a)$, (b,c) and $d, +\infty)$ is contained in F$\}$.

In this particular filter, each F is a (non-compact) nbd of [0,1].

8. DEFINITION

Let (X,π) be a dynamical system. Let M be a closed positively invariant sub-set of X. Let $\mathcal{F}$, $\mathcal{G}$ be nbd filters of M. Then M is said to be $(\mathcal{F}, \mathcal{G})$ stable if given $G \in \mathcal{G}$, there exists an $F \in \mathcal{F}$ such that $\gamma^+(F) \subset G$.

Results on filter stability are extended to (general) topological spaces, using rim-compactness in place of local compactness, as in the illustrations above.

REFERENCES

[1.] J. Auslander and P. Seibert, Prolongations and stability in dynamical systems, Ann. Inst. Fourier, Grenoble, 14 (1964), 237-268.

[2.] J. Auslander, On stability of closed sets in dynamical systems, Seminar on differential equations and dynamical systems, II, Lecture Notes in Mathematics, Vol. 144 (1969), Springer-Verlag, 1-4.

[3.] J. Auslander, Non-compact dynamical systems, Recent Advances in Topological Dynamics, Lecture Notes in Mathematics, Vol. 318 (1972), Springer-Verlag, 6-11.

[4.] J. Auslander, Filter stability in dynamical systems, Siam J. Math Anal. 8 (1977), 573-579.

[5.] P. Bajaj, Start points in semi-dynamical systems, Funkcialaj Ekvacioj, 13 (1971), 171-177.

[6.] P. Bajaj, Connectedness properties of start points in semi-dynamical systems, Funkcialaj Ekvacioj, 14 (1971), 171-175.

[7.] M. Hirsch, The dynamical system approach to differential equations, Bulletin, Amer. Math. Soc., 11 (1984), 1-64.

[8.] S. Willard, General Topology, Addison-Wesley 1970.

The Final (detailed) version of this paper will be submitted for publication elsewhere.

Trends in the Theory and Practice of Non-Linear Analysis
V. Lakshmikantham (Editor)

CONTAINMENT OF SOLUTIONS TO STRONGLY COUPLED PARABOLIC SYSTEMS

Peter W. Bates

Department of Mathematics
Texas A&M University
College Station, Texas

1. INTRODUCTION

In [12] H. Weinberger gave a "positive invariance" principle for nonlinear parabolic systems. The essence of that result can be stated as follows: Suppose there is a vector field defining a flow in $\mathbb{R}^N$ and a closed convex set, S, which is positively invariant under that flow. Then for the parabolic system obtained by adding a scalar diffusion term to the vector field, S is positively invariant, i.e., if the initial and boundary data lie in S, so does the solution.

In Weinberger's result the condition that S be positively invariant under the flow is actually stated in terms of the direction of the vector field on the boundary of S. That condition has been weakened by H. Amann [1], by J. Bebernes and K. Schmitt [7], by K. Chueh, C. Conley and J. Smoller [8], and by others, to allow the reaction term in the parabolic system to depend upon the solution and its gradient as well as the point in space-time. In [8] the case where the diffusion term is not scalar is considered and the necessary and sufficient additional requirements on S in order for it to be positively invariant, are given.

In this paper we generalize the invariance principle to allow the "invariant" set to move under some flow. Thus, although there is no positively invariant set, there may be a set moving according to an ODE which always contains the solution to the PDE as it evolves in time. For example, we have a result which basically says: Let S be closed with nonempty interior and let S(t) be the evolution of S under some vector field. If S(t) is convex for all t, then S(t) contains the solution to the corresponding parabolic system obtained by adding a scalar diffusion term to the vector field, provided the initial data lies in S and the boundary data at time t lies in S(t).

In addition to the above, comparison results, existence theorems (extending those of [7]) and applications are given. Preliminary results along these lines are given by the author in [4] and [5] and by C. Reder in [10].

The following notation will be used: Let $n, N \geq 1$ be integers and G a set in $\mathbb{R}^n \times \mathbb{R}$ then $C(G, \mathbb{R}^N)$ represents the space of continuous $\mathbb{R}^N$-valued functions, u, on G such that $\|u\| \equiv \sup\{u(x,t) : (x,t) \in G\} < \infty$. We will suppress the $\mathbb{R}^N$ and simply write C(G), in general. If $\alpha, \beta \in (0,1)$ then $C^{\alpha,\beta}(G)$ is the space of Hölder continuous functions, u(x,t), with Hölder exponent α in x and β in t, satisfying $\|u\|_{\alpha,\beta} \equiv \|u\| + \sup|u(x,t) - u(y,t)|^{\alpha}/|x-y| + \sup u(x,t) - u(x,\tau)^{\beta} / t-\tau < \infty$, where the sup's are taken over $x \neq y$, $t \neq \tau$ with (x,t), (y,t), $(x,\tau) \in G$. If j, k are nonnegative integers and $\alpha, \beta \in (0,1)$ then $C^{j+\alpha,k+\beta}(G)$ is the space of functions, u(x,t) having j derivatives in x and k derivatives in t lying in $C^{\alpha,\beta}(G)$. This is a Banach space given the appropriate norm, (see e.g., [10]). Finally, $W_q^{j,k}(G)$ represents the space of functions, $u(x,t)$ having j weak derivatives in x and k weak derivatives in t whose qth powers are integrable over G.

2. STRONG CONTAINMENT

Let D be a bounded domain in $\mathbb{R}^n$, and for $t > 0$ define $D_t = D \times (0,t)$, and $\Gamma_t = (D \times \{0\}) \cup (\partial D \times [0,t])$. Let $T > 0$ be fixed and let L be the quasilinear operator defined for $u \in C^{2,1}(\overline{D}_T, \mathbb{R}^N)$ by

$$Lu \equiv u_t - \sum_{i,j=1}^{n} a_{ij}(x,t,u,\nabla u)\partial^2 u/\partial x_i \partial x_j + \sum_{i=1}^{n} b_i(x,t,u,\nabla u)\partial u/\partial x_i$$

where a_{ij} and b_i are defined on $\overline{D}_T \times \mathbb{R}^N \times \mathbb{R}^{nN}$ and

$$0 \leq \sum a_{ij}(x,t,u,p)\xi_i \xi_j$$

for all $(x,t) \in \overline{D}_T$, $u \in \mathbb{R}^N$, $p \in \mathbb{R}^{nN}$ and $\xi = (\xi_1,\ldots,\xi_n) \in \mathbb{R}^n$. Let $f: \overline{D}_T \times \mathbb{R}^N \times \mathbb{R}^{nN} \to \mathbb{R}^N$ and $\psi: \Gamma_T \to \mathbb{R}^N$. We shall be considering solutions to the problem

$$Lu = f(x,t,u,\nabla u) \text{ in } D_T, \tag{2.1}$$

and

$$(\partial u/\partial \nu)(x,t) = 0 \text{ on } \partial D, \quad u(x,0) = \psi(x,0) \text{ on } \overline{D} \tag*{(2.2)$_N$}$$

or

$$u = \psi \text{ on } \Gamma_T, \tag*{(2.2)$_D$}$$

comparing them with solutions to a system of ODEs. Let $g: [0,T] \times \mathbb{R}^N \to \mathbb{R}^N$ be continuous and continuously differentiable in its second argument. Let S be an open set in $\mathbb{R}^N$ and suppose that for all points s_0 in a neighborhood of $\overline{S}$ the solution to

$$\dot{s} = g(t,s) \tag{2.3}$$

satisfying $s(0) = s_0$, exists for $0 \leq t \leq T$. Let $S(t)$ be the evolution of S under (2.3) and let

$$S_T = \{(t,s): s \in \partial S(t), \ 0 \leq t \leq T\}.$$

<u>Lemma 2.1</u>. Suppose that $S(t)$ is convex for $0 \leq t \leq T$. Then there is a function $n: S_T \to \mathbb{R}^N$ such that

(i) $n(t,s)$ is an outward unit normal to $S(t)$ at $s \in \partial S(t)$ and

(ii) if $s(t)$ is any solution of (2.3) with $s(0) \in \partial S$ then $n(\cdot,s(\cdot))$ is continuous on $[0,T]$.

<u>Proof</u>. let $s_0 \in \partial S$ be fixed and let M_0 be a supporting hyperplane of S at s_0. Let M be the intersection of M_0 with a neighborhood of s_0, small enough so that $M(t)$, the evolution of M under (2.3), exists for $0 \leq t \leq T$. Since M_0

is a smooth manifold with S on one side, and since the solution map associated with (2.3) is a diffeomorphism on a neighborhood of $\overline{S}$, it follows that $M(t)$ is a C^1 manifold. Also, if $s(t)$ is the solution to (2.3) with $s(0) = s_0$, then $s(t) \in M(t)$ and $S(t)$ is on one side of $M(t)$ in a neighborhood of $s(t)$. The unique unit normal, $n(t,s(t))$, to $M(t)$ at $s(t)$, outward to $S(t)$, is a normal to $S(t)$ at $s(t)$. This can be seen by considering the hyperplane orthogonal to $n(t,s(t))$ through $s(t)$ and using the convexity of $S(t)$ and the smoothness of $M(t)$. Clearly, $n(\cdot,s(\cdot))$ is continuous by the smoothness assumption on g.

Note that we are not claiming any regularity of $n(t,\cdot)$ on $\partial S(t)$ for any fixed t.

Theorem 2.2. Suppose f, g and S are as above with $S(t)$ convex for $0 \leq t \leq T$. Suppose that u is a solution of (2.1) and $(2.2)_D$ which is of class $C^{2,1}(\overline{D}_T)$ with $\psi(x,t) \in S(t)$ for $(x,t) \in \Gamma_T$. Suppose f and g satisfy

$$(2.4)\qquad \begin{cases} n(t,s) \cdot (f(x,t,s,p) - g(t,s)) < 0 \\ \text{for all } (x,t,s,p) \text{ such that } (x,t) \in D_T,\ (t,s) \in S_T \text{ and} \\ p = (p_1,\dots,p_n) \in \mathbb{R}^{nN} \text{ with } n(t,s) \cdot p_i = 0,\ 1 \leq i \leq n. \end{cases}$$

Then $u(x,t) \in S(t)$ for all $(x,t) \in \overline{D}_T$.

Proof. Suppose that u is a solution of class $C^{2,1}(D_T)$ for which the conclusion fails. Then there is a point $(x_0,t_0) \in D_T$ such that $u(x,t) \in S(t)$ for $(x,t) \in D_{t_0}$, $u(x,t_0) \in \overline{S(t_0)}$ for $x \in \overline{D}$ and $u(x_0,t_0) \in \partial S(t_0)$. Let $s(t)$ be the solution of (2.3) with $s(t_0) = u(x_0,t_0)$ and consider $w(x,t) \equiv n(t,s(t))\cdot(u(x,t) - s(t))$. Then $w(x_0,t_0) = 0$ and $w(x,t) \leq 0$ for $(x,t) \in \overline{D}_{t_0}$. Since $n(\cdot,s(\cdot))$ is continuous and $u(x_0,t_0) = s(t_0)$, it follows that $w_t(x_0,t_0)$ exists and is nonnegative. Also $w(\cdot,t_0) \in C^2$ has a maximum at $x_0 \in D$ so the hessian of w, $(\partial^2 w/\partial x_i \partial x_j)$ is nonpositive definite and symmetric at (x_0,t_0) and $(\partial w/\partial x_i)(x_0,t_0) = n(t_0,s(t_0))\cdot(\partial u/\partial x_i)(x_0,t_0) = 0$, $1 \leq i \leq n$. Hence, (2.4) holds at $(x_0, t_0, s(t_0), \nabla u(x_0,t_0))$. Also, $\sum a_{ij}\partial^2 w/\partial x_i \partial x_j = \operatorname{trace}((a_{ij})(\partial^2 w/\partial x_i \partial x_j)) \leq 0$ at this point by the assumption on (a_{ij}), the above remark on the hessian of w, and a standard result from matrix theory. Thus, from (2.1) and (2.3) we have

$$\begin{aligned} &n(t_0,s(t_0)) \cdot (f(x_0,t_0,s(t_0),\nabla u(x_0,t_0)) - g(t_0,s(t_0))) \\ &= n(t_0,s(t_0)) \cdot (Lu(x_0,t_0) - \dot{s}(t_0)) \\ &= w_t(x_0,t_0) - \sum a_{ij}(x_0,t_0,u(x_0,t_0),\nabla u(x_0,t_0))\partial^2 w(x_0,t_0)/\partial x_i \partial x_j \geq 0, \end{aligned}$$

a contradiction to (2.4). This proves the theorem. ∎

Remarks. 1. The same result holds when D is unbounded if we assume that the distance between ∂S and the range of $\psi(x,0)$ is positive.

2. If the solution is not required to satisfy (2.2) then the above proof shows that if (x_0,t_0) is as above, we must have $x_0 \in \partial D$, and $w(x,t_0) < 0$ for $x \in D$

We will use this observation to prove:

Theorem 2.3. Suppose D has the interior sphere property and L is uniformly parabolic with bounded coefficients. Suppose f is continuous, $S(t)$ is convex for $0 \leq t \leq T$ and f and g satisfy

$$(2.5) \quad \begin{cases} n(t,s) \cdot (f(x,t,s,p) - g(t,s)) < 0 \\ \text{for } (x,t) \in \overline{D}_T, \ (t,s) \in S_T \text{ and all } p \in \mathbb{R}^{nN}. \end{cases}$$

If $\psi(x,0) \in S$ for all $x \in \overline{D}$ and if u is a solution of (2.1) and (2.2) which is of class $C^{2,1}(\overline{D}_T)$, then $u(x,t) \in S(t)$ for all $(x,t) \in \overline{D}_T$.

Proof. Suppose the conclusion fails, then by Remark 2, above, there is a solution, u, and a point (x_0,t_0) with $x_0 \in \partial D$ and $0 < t_0 \leq T$ such that $u(x_0,t_0) \in \partial S(t_0)$ and $u(x,t) \in S(t)$ for $(x,t) \in D_{t_0}$. Furthermore, if s and w are as in the proof of Theorem 2.2, then $w(x,t_0) < 0$ for $x \in D$ and (2.5) holds at $(x_0, t_0, s(t_0), \nabla u(x_0,t_0))$. Thus, we have

$$(\partial w/\partial \nu)(x_0,t_0) = n(t_0,s(t_0)) \cdot (\partial u/\partial \nu)(x_0,t_0) = 0$$

by $(2.2)_N$ and

$$0 < \Big(\sum a_{ij}(x, t_0, u(x,t_0), \nabla u(x,t_0))\partial^2/\partial x_i \partial x_j +$$

$$+ \sum b_i(x, t_0, u(x,t_0), \nabla u(x,t_0))\partial/\partial x_i\Big)w$$

for x in D, close to x_0, by the continuity of f. Applying Lemma 3.4 of [9] to the linear differential operator in parentheses above, gives a contradiction to $\partial w/\partial \nu = 0$. ∎

By strengthening the requirements on the coefficients of L and the regularity of f similar results hold with only weak inequality in (2.4) and (2.5) and with initial and boundary data in $\overline{S}$ and $\overline{S(t)}$. The idea is to impose conditions upon L and f which guarantee existence, uniqueness and continuous dependence of solutions upon boundary data and perturbations of the equation. Then one finds solutions to approximations of the problem where the conditions of the above theorems hold and passes to the limit. This is done in the next section for the case where L is linear but f satisfies only mild regularity conditions.

3. EXISTENCE AND WEAK CONTAINMENT

We shall assume in this section that the domain D is bounded with boundary of class $C^{2+\alpha}$ for some $\alpha \in (0,1)$. The coefficients of L shall be assumed to satisfy: $a_{ij} = a_{ij}(x,t)$ is of class $C^{1,\alpha/2}(\overline{D}_T)$ and for some numbers λ, μ with $0 < \lambda < \mu$

$$\lambda|\xi|^2 \leq \sum a_{ij}\xi_i\xi_j \leq \mu|\xi|^2$$

for all $\xi \in \mathbb{R}^n$, and $b_i = b_i(x,t)$ is of class $C^{\alpha,\alpha/2}(\overline{D}_R)$.

Let $f: \overline{D}_T \times \mathbb{R}^N \times \mathbb{R}^{nN} \to \mathbb{R}^N$ be of class $C^{\alpha,\alpha/2,\alpha,\alpha}$ and satisfy the growth condition

$$|f(x,t,u,p)| \leq h(|u|,|p|)(1 + |p|^2)$$

$h(|u|, |p|) \to 0$ as $|p| \to \infty$ for $|u|$ bounded. Assume that ψ is the restriction to Γ of a function of class $C^{2+\alpha}(\overline{D}_T)$. The following theorems are based on similar ones in [7], where $g \equiv 0$.

<u>Theorem 3.1.</u> Let L and f be as above, g and S be as in section 2 with $S(t)$ convex for $0 \leq t \leq T$. Suppose tht $S([0,T])$ is bounded and that $\psi(x,t) \in S(t)$ for $\overline{(x,t)} \in \Gamma_T$. Suppose ψ satisfies the compatibility condition

$$\psi_t(x,0) = \sum a_{ij}(x,0)\partial^2\psi(x,0)/\partial x_i \partial x_j - \sum b_i(x,0)\partial\psi(x,0)/\partial x_i$$

for $x \in \partial D$. Suppose that f, g and S satisfy (2.4). Then problem (2.1), (2.2) has a solution $u \in C^{2,1}(\overline{D}_T)$ such that $u(x,t) \in S(t)$ for $(x,t) \in \overline{D}_T$.

<u>Proof.</u> We may assume, without loss of generality, that there is a point $s_0 \in S(t)$ for $0 \leq t \leq T$. This is because we may partition $[0,T]$ into finitely many subintervals so that $S(t)$ has the above property on each subinterval. Then solving on successive subintervals, using the final value on one subinterval as the initial data on the next (so preserving the compatibility condition imposed upon ψ) will yield a solution on $[0,T]$.

Now choose $\varepsilon > 0$ so that the ball of radius 2ε about s_0, $B(s_0,2\varepsilon)$, lies in $S(t)$ for $0 \leq t \leq T$. Let $\hat{g}$ be a function satisfying the same continuity conditions as g but which is identically zero in $[0,T] \times B(s_0,\varepsilon)$ and equal to g on the complement of $[0,T] \times B(s_0,2\varepsilon)$. Define the Nemytski operators $F,\hat{G}: C^{1,0}(\overline{D}_T) \to C(\overline{D}_T)$ by

$$(Fu)(x,t) = f(x,\ t,\ u(x,t),\ \nabla u(x,t)) \qquad \text{and}$$

$$(\hat{G}u)(x,t) = \hat{g}(t,\ u(x,t)),$$

then F and $\hat{G}$ are continuous and bounded. Also, $F,\hat{G}: C^{1+\alpha,\alpha/2} \to C^{\gamma,\beta/2}$ are continuous for some $\gamma \in (0,\alpha]$. Define the linear operator K by letting $u = Kv$ be the solution of

$$Lu = v \quad \text{in} \quad D_T, \qquad u = 0 \quad \text{on} \quad \Gamma_T.$$

We may consider K as a mapping from $C^{\alpha,\alpha/2}$ into $C_0^{2+\alpha,1+\alpha/2}$ (the subscript denoting the zero boundary condition) or as a compact mapping from C into $C^{1+\alpha,\alpha/2} \subset C^{1,0}$ by observing $K: L_q \to W_q^{2,1}$ and using the Sobolev embedding theorem for q sufficiently large (see Ladyzenskaja et al [10]). It follows that the composite maps KF, $K\hat{G}: C^{1,0} \to C^{1,0}$ are completely continuous. Let $z \in C^{2+\alpha,1+\alpha/2}$ be the solution of

$$Lz = 0 \quad \text{in} \quad D_T, \qquad z = \psi \quad \text{on} \quad \Gamma_T.$$

Suppose for some $\lambda \in [0,1]$, there is a solution $u \in C^{1,0}$ of

$$u = \lambda(KFu + z) + (1 - \lambda)(K\hat{G}u + s_0). \tag{3.1}$$

By the above mapping properties we have immediately $u \in C^{1+\alpha,\alpha/2}$ and so $Fu, \hat{G}u \in C^{\gamma,\gamma/2}$ for some $\gamma \in (0,\alpha]$. Therefore, $u \in C^{2+\gamma,1+\gamma/2}$. By the definition of K and z it follows that u is a classical solution of

$$\begin{aligned} Lu &= \lambda f(x, t, u, \nabla u) + (1 - \lambda)\hat{g}(t,u) && \text{in } D_T, \\ u &= \lambda\psi + (1 - \lambda)s_0 && \text{on } \Gamma_T. \end{aligned} \tag{3.2}$$

Now, $\lambda f + (1 - \lambda)\hat{g}$ satisfies (2.4) since $\hat{g} = g$ on S_T. Furthermore, $\lambda\psi + (1 - \lambda)s_0 \in S(t)$ on Γ_T, so Theorem 1 implies that the range of u lies in $S([0,T])$, which is assumed to be bounded. Through the regularity assumptions on the coefficients of L and the continuity and growth conditions on f, the boundedness of u implies the existence of a constant M (independent of λ) such that $|\nabla u| \leq M$ on $\overline{D}_T$ (see [10], VII. 6). The proof will be completed by using the Leray-Schauder degree in $C^{1,0}$ to show that (3.1) has a solution when $\lambda = 1$. Let $W = \{u \in C^{1,0}: u(x,t) \in S(t), |\nabla u(x,t)| < M + 1 \text{ for } (x,t) \in \overline{D}_T\}$, then W is a nonempty bounded open set in $C^{1,0}$. Define $H: [0,1] \times \overline{W} \to C^{1,0}$ by

$$H(\lambda,u) = u - \lambda(KFu + z) - (1 - \lambda)(K\hat{G}u + s_0)$$

then H is continuous and $H(\lambda,\cdot)$ is a compact perturbation of the identity in $C^{1,0}$. For each $\lambda \in [0,1]$, $H(\lambda,u) \neq 0$ for all $u \in \partial W$ for otherwise u would satisfy (3.1) and so $u(x,t) \in S(t)$ and $|\nabla u(x,t)| \leq M < M + 1$ for $(x,t) \in \overline{D}_T$, by the remarks above. Hence, H is a homotopy relative to W and by the homotopy invariance of degree

$$d(I - KF, W, z) = d(I - K\hat{G}, W, s_0). \tag{3.3}$$

Finally, $\hat{H}: [0,1] \times \overline{W} \to C^{1,0}$ defined by

$$\hat{H}(\lambda,u) = u - \lambda K\hat{G}u + s_0$$

is a completely continuous homotopy relative to W since the unique solution to

$$Lu = \lambda\hat{g}(t,u) \quad \text{in } D_T, \qquad u = s_0 \quad \text{on } \Gamma_T$$

is $u \equiv s_0 \in W$. Thus, $d(I - K\hat{G}, W, s_0) = d(I, W, s_0) = 1$. This together with (3.3) completes the proof.

Remark 3. A similar theorem may be given for the Neumann problem, (2.1) and $(2.2)_N$. The only differences being that the compatibility condition on ψ may be dropped (see [10], p. 320) and we must require the slightly stronger condition (2.5) in place of (2.4) so that Theorem 2.3 is applicable.

Now let $(2.4)_w$ and $(2.5)_w$ denote (2.4) and (2.5), respectively with strong inequality replaced by the corresponding weak inequality. The following result may be used for instance when $f \equiv g$ and $(2.4)_w$ is trivially satisfied.

Theorem 3.2. Assume that L, f, g, ψ and S are as in Theoem 3.1 except that (2.4) is replaced by $(2.4)_w$ and assume $\psi(x,t) \in \overline{S(t)}$ on Γ_T. Then (2.1) and (2.2) has a solution $u \in C^{2,1}(D\ )$ such that $u(x,t) \in \overline{S(t)}$ for $(x,t) \in \overline{D}_T$.

Proof Let $s(t)$ be a solution to (2.3) with $s(0) \in S$, let $\varepsilon > 0$ and let ρ_ε be a smooth function taking values in $[0,1]$, having support in an ε-neighborhood of S_T and such that $\rho_\varepsilon \equiv 1$ on S_T. Define $\psi_\varepsilon = (1-\varepsilon)\psi + \varepsilon s$ and $f_\varepsilon(x, t, u, p) = f(x, t, u, p) + \varepsilon\rho_\varepsilon(t,u)(s(t) - u)$. Then, by the convexity of $S(t)$, $\psi_\varepsilon(x,t) \in S(t)$ on Γ_T and f_ε satisfies (2.4). By applying Theorem 3.1 we obtain a solution u_ε of

$$u_\varepsilon = K(Fu_\varepsilon + \varepsilon R_\varepsilon u_\varepsilon) + (1 - \varepsilon)z + \varepsilon z_1$$

where $(R_\varepsilon v)(x,t) = \rho_\varepsilon(t,v(x,t))(s(t) - v(x,t))$ and z_1 is the unique solution of $Lz_1 = 0$ in D_T, $z_1(x,t) = s(t)$ on Γ_T. Furthermore, $u_\varepsilon(x,t) \in S(t)$ in $\overline{D}_T$ so $\{R_\varepsilon u_\varepsilon : \varepsilon > 0\}$ is bounded. By the compactness of K we may find a sequence $\varepsilon = \varepsilon_n \to 0$ so that the corresponding sequence of $\{u_\varepsilon\}$ converges, to u say, as $n \to \infty$. Since F is continuous one concludes that $u \in C^{1,0}$ satisfies

$$u = KFu + z.$$

As in the proof of Theorem 2, this implies that $u \in C^{2+\gamma,1+\gamma/2}$ is a solution of (2.1), (2.2) .

For the Neumann problem we also have

Theorem 3.3. Suppose that $L \equiv \partial/\partial t - \sum a_{ij}(x,t)\partial^2/\partial x_i\partial x_j + \sum b_i(x,t)\partial/\partial x_i$ is uniformly parabolic on the bounded domain $D_T \equiv D \times (0,T]$, having coefficients of class $C^{1,\alpha/2}(\overline{D}_T)$. Suppose $f(x,t,u,p) \in C^{\alpha,\alpha/2,\alpha,\alpha}(\overline{D} \times [0,T] \times \mathbb{R}^N \times \mathbb{R}^{nN})$ and satisfies the growth condition

$$|f(x,t,u,p)| \leq h(|u|, |p|)(1 + |p|^2),$$

where $h(|u|, |p|) \to 0$ as $|p| \to \infty$ for $|u|$ bounded. Let $g: [0,T] \times \mathbb{R}^N \to \mathbb{R}^N$ be of class $C^{0,1}$. Let S be bounded, open and convex and $S(t)$ be the evolution of S under $\dot{s} = g(t,s)$, $0 \leq t \leq T$. Suppose that $S(t)$ is convex and bounded for $0 \leq t \leq T$, and that $n(t,s)$ is an outward normal to $S(t)$ as given by Lemma 2.1. Suppose that $\psi \in C^{2+\alpha}(\overline{D})$ and $\psi(x) \in \overline{S}$ for all $x \in \overline{D}$. Finally suppose that

$$(2.5)_w \qquad n(t,s) \cdot (f(x,t,s,p) - g(t,s)) \leq 0$$

for $(x,t) \in \overline{D}_T$, $s \in \partial S(t)$ and $p \in \mathbb{R}^{nN}$. Then, there exists a solution, $u(x,t)$, to

$$Lu = f(x,t,u,\nabla u) \quad \text{in } D_T,$$

$$u(x,0) = \psi(x) \quad \text{in } \overline{D},$$

$$\frac{\partial u}{\partial n}(x,t) = 0 \quad \text{on } \Gamma_T, \text{ and}$$

$$u(x,t) \in \overline{S(t)} \quad \text{on } \overline{D}.$$

The proof is similar to that of Theorems 3.1 and 3.2 and will be omitted. The only difference being in the manner in which gradient bounds are obtained, but again these follow from results in [10], using the growth condition on f, together with an interpolation lemma.

Remarks. 4. From the proof one can see that the requirement that $S([0,T])$ be bounded can be dropped if one has an independent a priori bound on solutions.

5. The requirement that $g(t,s)$ be C^1 in s can be weakened assuming g to be locally Lipschitz in s if one has the normal $n(t,s)$, existing a priori and if one has existence of solutions to (2.3) on $[0,T]$.

6. Theorems 3.2 and 3.3 remain true if S is closed with empty interior. In condition $(2.5)_w$ one must replace $n(t,s)$ by all outward normals n. This extension is proved by considering the flows which are rigid motions in the directions orthogonal to the "plane" of $S(t)$.

4. EXAMPLES

The first example shows that in certain cases the previous results may be used to show that solutions to a parbolic system do not blow up infinite time if the solution to a related ODE does not do so. When blow-up does occur in the ODE, that blow-up time provides a lower bound for the blow-up time in the parabolic system.

Let D be a smooth domain in $\mathbb{R}^n$ and let $f: \overline{D} \times \mathbb{R} \times \mathbb{R}^N \to \mathbb{R}^N$ be locally Lipschitz continuous. Let L be uniformly parabolic with smooth coefficients as in Theorem 3.3 and consider the system

$$(4.1) \qquad \begin{aligned} Lu &= f(x,t,u) && \text{in } D \times (0,\infty) \\ u(x,0) &= \psi(x) && \text{in } \overline{D} \\ \frac{\partial u}{\partial n}(x,t) &= 0 && \text{on } \partial D \times (0,\infty) \end{aligned}$$

Let $r_0 > 0$ be fixed and define

$$g(t,r) = \begin{cases} \max\{f(x,t,u)\cdot u/r^2 : \ x \in \overline{D}, \quad |u| = r\} & \text{for } r \geq r_0 \\ g(t,r_0) \ \text{for } r < r_0 \end{cases}$$

and consider the system of ODEs

$$(4.2) \qquad s' = g(t, |s|)s .$$

Let $S(t)$ be the evolution of the ball of radius r_0 under (4.2), then by the radial symmetry of that system, $S(t)$ remains a ball on its domain of definition. The radius of $S(t)$ is governed by the scalar equation

$$r' = g(t,r)r. \tag{4.3}$$

Assume $g \geq 0$ (replace g by its nonnegative part) then if ψ has its range in $S(0)$, $u(x,t)$ remains in $S(t)$ since for $u \in \partial S(t)$

$$\frac{u}{|u|}\cdot\left(f(x,t,u) - g(t,|u|)u\right) = f(x,t,u)\cdot\frac{u}{|u|} - g(t,|u|)|u| \leq 0$$

by definition of g.

Thus, an estimate of blow-up time for (4.1) may be obtained from the scalar ODE (4.3). The Dirichlet problem is handled in the same way if one chooses r so that the ball of that radius contains the boundary data.

We conclude by examining the evolution of solutions to a system proposed by J. M. Lasry as a model similar to the Hodgkin-Huxley equations of nerve conduction (for other results on this system see [2], [3], and [6]).

Let ϕ be a smooth 2π-periodic function with $\phi < 0$ on $(0,\alpha_0)$, $\phi > 0$ on $(\alpha\ ,2\pi\ )$ and let $R: [0,\infty) \to R$ be smooth with $R > 0$ on $[0,a)$, $R < 0$ on (a,∞).

Let (ρ,β) be the polar coordinates of the point $(u,v) \in R^2$. Consider the parabolic system

$$\begin{aligned} u_t &= \Delta u + R(\rho)u - \phi(\beta)v \\ v_t &= \Delta v + \phi(\beta)u + R(\rho)v \qquad x \in D,\ t > 0, \end{aligned} \tag{4.4}$$

where $D^n \subset \mathbb{R}^n$ is bounded and satisfies the interior sphere property at each point of its boundary.

The functions R and ϕ provide the radial and rotational (about 0) components of the vector field. For any β_0, $\beta_1 \in (\alpha\ ,2\pi)$ (or $(0,\alpha_0)$) with $\beta_0 < \beta_1$ consider the sector $S \equiv \{(\rho,\beta): 0 < \rho,\ \beta\ < \beta_0 < \beta_1\}$ in R^2. If $S(t)$ represents the evolution of S under the ODEs

$$\begin{aligned} u' &= R(\rho)u - \phi(\beta)v \\ v' &= \phi(\beta)u + R(\rho)v \qquad t > 0 \end{aligned} \tag{4.5}$$

then it is easy to see that $S(t)$ is a sector rotating counterclockwise (clockwise) about 0 and that as $t \to \infty$, $S(t)$ approaches the positive u axis (system (4.5) decouples when written in polar coordinates).

If (u,v) solves (4.4) and satisfies Neumann boundary conditions and has initial data (u_0,v_0) contained in S, then (u,v) tends to the positive u-axis as $t \to \infty$, provided β_0 and β_1 are chosen so that $S(t)$ remains convex for $t \geq 0$. It is then easy to show that, in fact, (u,v) approaches $(a,0)$ as $t \to \infty$. This is the essence of the main result in [6].

One may obtain a similar result for the Dirichlet problem associated with (4.4) which was impossible with the techniques in [6] because of the need to avoid the origin. The following is true:

If $(\underline{u}, v)$ satisfies (4.4) with $(u(x,0), v(x,0)) = (u(x), v(s)) \in \overline{S}$ for $x \in \overline{D}$ and $u(x,t) = 0 = v(x,t)$ for $x \in \partial D$, $t \geq 0$, and if $\beta_1(t) - \beta_0(t) \leq \pi$ for $t \geq 0$ where $\beta_i(t)$ is the solution of $\beta' = \phi(\beta)$ satisfying $\beta_i(0) = \beta_i$, $i = 0,1$, then $v(x,t) \to 0$ uniformly and $u(x,t)$ approaches a nonnegative solution of $0 = \Delta u + R(u)u$ in D, $u(x) = 0$ for $x \in \partial D$, as $t \to \infty$.

REFERENCES

[1] Amann, H., Invariant sets and existence theorems for semi-linear parabolic and elliptic systems, J. Math. Anal. Appl., 65 (1978) 432-467.

[2] Barrow D. L. and Bates P. W., Bifurcation and stability of periodic traveling waves for a reaction-diffusion system, JDE, 50 (1983) 218-233.

[3] Barrow D. L. and Bates, P. W., Bifurcation of periodic traveling waves for a reaction-diffusion system, Lecture Notes in Math. no. 964, pp. 69-76 (New York, Springer-Verlag, 1982).

[4] Bates, P. W., Containment for weakly coupled parabolic systems, Houston J. Math., to appear.

[5] Bates, P. W., Existence and containment of solutions to parabolic systems, preprint, 1983.

[6] Bates, P. W. and Brown, K. J., Convergence to equilibrium in a reaction diffusion system, Nonlinear Analysis, to appear.

[7] Bebernes, J. W. and Schmitt, K., Invariant sets and the Hukuhara-Kneser property for systems of parabolic partial differential equations, Rocky Mountain J. Math., 7 (1977), 557-567.

[8] Chueh, K., Conley, C. and Smoller, J., Positively invarient regions for systems of nonlinear diffusion equations, Ind. Math. J. (1977), 373-392.

[9] Gilbarg, D. and Trudinger, N. S., Elliptic partial differential equations of second order, (New York, Springer, 1977).

[10] Ladyzenskaja, O., Solonnikov, V. and Uralceva, N., Linear and Quasilinear Equations of Parabolic Type, A. M. S. Translations of Math. Monographs 23, Providence, 1968.

[11] Reder, C., Familles de convexes invariantes et equations de diffusion-réaction, Ann. Inst. Fourier, Grenoble, 32 (1982) 71-103.

[12] Weinberger, H., Invariant sets for weakly coupled parabolic and elliptic systems, Rend. Mat. Univ. Roma (VI)8, (1975), 295-310.

This paper is in final form and no version of it will be submitted for publication elsewhere.

Trends in the Theory and Practice of Non-Linear Analysis
V. Lakshmikantham (Editor)

DEVELOPMENTS IN FIXED POINT THEORY FOR NONEXPANSIVE MAPPINGS

L. P. Belluce
Department of Mathematics
University of British Columbia
Vancouver, B.C. V6T 1W5
Canada

and

W. A. Kirk
Department of Mathematics
The University of Iowa
Iowa City, Iowa 52242
U.S.A.

Two recent methodological developments in the study of fixed point theory for nonexpansive mappings are discussed. The first involves the use of the theory of ultraproducts, and the second entails the use of approximation theorems as a tool in obtaining new existence theorems.

INTRODUCTION

Let K be a bounded closed and convex subset of a Banach space X. A mapping $T: K \to K$ is said to be <u>nonexpansive</u> if it has Lipschitz constant 1 ($\|T(u)-T(v)\| \le \|u-v\|$, $u,v \in K$). It is easily seen that in general such a mapping need not have a fixed point. There are, however, wide classes of spaces for which such sets K do have the fixed point property for nonexpansive mappings. In fact, in a more abstract sense, it is known (cf. [9]) that if M is any bounded metric space with $T: M \to M$ nonexpansive, then there is a metric space $\widetilde{M}$ (the injective hull of M) and an isometry $e: M \to \widetilde{M}$ such that the mapping $\widetilde{T}: e(M) \to e(M)$ defined by $\widetilde{T}(e(x)) = e(T(x))$ has a nonexpansive extension to all of $\widetilde{M}$, and moreover this extension always has a fixed point in $\widetilde{M}$.

Much of the study of fixed point theory for nonexpansive mappings has centered on determining precise geometric conditions on X (or K) which will always assure the existence of fixed points (see, e.g., the surveys [8,9]). Here we look more closely at two of the less routine methodologies involved.

ULTRAPRODUCTS

Let I be a nonempty set. A filter $\mathfrak{F}$ on I is a collection of nonempty subsets of I satisfying (a) $A,B \in \mathfrak{F} \Rightarrow A \cap B \in \mathfrak{F}$, and (b) $A \in \mathfrak{F}$, $B \supset A \Rightarrow B \in \mathfrak{F}$. An ultrafilter is a filter which is contained in no larger filter, and an ultrafilter is said to be free if it is not generated by a single element. Henceforth, we suppose X is a Banach space and $\mathfrak{F}$ is a free ultrafilter on I. It is easy to verify that if C is a compact Hausdorff space and if $r_i \in C$ for $i \in I$, then $\lim_{\mathfrak{F}} r_i$ always exists, i.e., there always exists a unique $r \in C$ such that if V is any neighborhood of r, then $\{i \in I : r_i \in V\} \in \mathfrak{F}$.

Let

$$\ell_\infty(X) = \{\{x_i\} : x_i \in X \ (i \in I), \sup_{i \in I} \|x_i\| < \infty\},$$

and let

$$N = \{\{x_i\} \in \ell_\infty(X) : \lim_{\mathfrak{F}} \|x_i\| = 0\}.$$

The ultraproduct $\widetilde{X}$ of X with respect to $\mathfrak{F}$ is the quotient space $\ell_\infty(X)/N$

equipped with the usual quotient norm, and it is easy to see that

$$\|\{x_i\}+N\|_{\mathfrak{F}} = \lim_{\mathfrak{F}}\|x_i\|.$$

(See, e.g., [6] for a more detailed exposition of the above facts.)

For simplicity we shall not distinguish between $\{x_i\} \in \ell_\infty(X)$ and $\{x_i\}+N \in \tilde{X}$. (Note also that X is isometric with a subspace of $\tilde{X}$ via the mapping $x \longrightarrow (x,x,\ldots)$.)

Now let $\mathfrak{F}$ be a free ultrafilter over the natural numbers $\mathbb{N}$ (i.e., let $I = \mathbb{N}$), and let K be a bounded closed and convex subset of X. If $T: K \longrightarrow K$ is nonexpansive, then by approximating T with the contraction mappings $T_n = n^{-1}x_0 + (n-1)n^{-1}T$, $n \in \mathbb{N}$, ($x_0 \in K$ fixed) one obtains a sequence $\{x_n\}$ (the fixed points of T_n) for which $\lim_{n\to\infty}\|x_n - T(x_n)\| = 0$. If

$$\tilde{K} = \{\{u_n\} \in \tilde{X} : u_n \in K \ (n \in \mathbb{N})\},$$

then the mapping $T: \tilde{K} \longrightarrow \tilde{K}$ defined by $T(\{u_n\}) = \{T(u_n)\}$ is also nonexpansive. Moreover,

$$\|\{x_n\} - \tilde{T}(\{x_n\})\|_{\mathfrak{F}} = \lim_{\mathfrak{F}}\|x_n - T(x_n)\| = 0$$

and $\{x_n\}$ is a fixed point of $\tilde{T}$.

The above observation was made by Maurey [17] as a prelude to much deeper work, and since then others (see, e.g., [4],[16]) have also found this to be a fruitful setting for obtaining deeper fixed point results in the original space X.

The ultraproduct approach we describe next appears to be new. Let K be a subset of a Banach space X, and let $\mathcal{S}$ be a semigroup of nonexpansive self-mappings of K each finite subfamily of which has a common fixed point. Let I denote the family of finite subsets of $\mathcal{S}$ and for each $\alpha \in I$, let S_α denote the semigroup generated by α. Also, let F_α be the common fixed point set of S_α, $\alpha \in I$. Set

$$J_\alpha = \{\beta \in I : S_\alpha \subset S_\beta\}$$

and let $\mathfrak{F}_1$ denote the filter generated by $\{J_\alpha : \alpha \in I\}$, i.e.,

$$\mathfrak{F}_1 = \{\gamma \in 2^I : \exists\{\alpha_1,\ldots,\alpha_n\} \in I \text{ with } \bigcap_{i=1}^{n} J_{\alpha_i} \subset \gamma\}.$$

Now let $\mathfrak{F}$ denote an ultrafilter containing $\mathfrak{F}_1$, and let $\tilde{X}$ denote the ultraproduct of X with respect to $\mathfrak{F}$. The semigroup $\mathcal{S}$ induces a semigroup $\tilde{\mathcal{S}}$ on $\tilde{K} = \{x \in \tilde{X} : x(\alpha) \in K \ (\alpha \in I)\}$ via the formula $\tilde{f}(\{x(\alpha)\}) = \{f(x(\alpha))\}$ $(f \in \mathcal{S})$, and clearly $\tilde{f}$ is nonexpansive on $\tilde{K}$. Moreover, if $\{p(\alpha)\} \in \tilde{K}$ is defined by selecting $p(\alpha) \in F_\alpha$ for each $\alpha \in I$, then we assert that $\tilde{f}(\{p(\alpha)\}) = \{p(\alpha)\}$ for each $\tilde{f} \in \tilde{\mathcal{S}}$. To see this, let $\tilde{f} \in \tilde{\mathcal{S}}$, let $\beta = \{f\}$, and let $\alpha \in J_\beta$. Then $S_\beta \subset S_\alpha$, so $f \in S_\alpha$; hence $p(\alpha) \in F_\alpha$ implies $f(p(\alpha)) = p(\alpha)$. Thus

$$J_\beta \subset \{\alpha \in I : f(p(\alpha)) = p(\alpha)\},$$

and since $J_\beta \in \mathfrak{F}$, $\{\alpha \in I : f(p(\alpha)) = p(\alpha)\} \in \mathfrak{F}$, proving $\tilde{f}(\{p(\alpha)\}) \equiv \{f(p(\alpha))\} = \{p(\alpha)\}$.

Remark. If for $\alpha \in I$, $\alpha = \{f_1,\ldots,f_n\}$, one defines $x(\alpha) = f_1 \circ \cdots \circ f_n$, where $x \in K$ is fixed, then for weakly compact K, $p = \text{weak-}\lim_{\mathfrak{F}} x(\alpha)$ always exists.

It would be interesting to find conditions on K (or $\mathcal{S}$) which imply that p is a common fixed point of $\mathcal{S}$.

<u>Lim's Theorem</u>. Recall that a Banach space has <u>normal structure</u> ([12]) if each of its bounded convex subsets H which contains more than one point contains a nondiametral point, i.e., a point y_0 for which

$$\sup\{\|y_0 - x\| : x \in H\} < \sup\{\|y - x\| : x,y \in H\}.$$

The Chebyshev radius of such a set H is defined by

$$r(H) = \inf_{y_0 \in H}\{\sup\{\|y_0 - x\| : x \in H\}\},$$

and the Chebyshev center of H is the (possibly empty) set:

$$\mathcal{C}(H) = \{y \in H : \sup\{\|y - x\| : x \in H\} = r(H)\}.$$

It is well known that if H is also weakly compact, then $\mathcal{C}(H)$ is a nonempty closed convex subset of H, and moreover if H has normal structure, then $\mathcal{C}(H)$ is a <u>proper</u> subset of H.

We now use the fact that $\tilde{\mathcal{S}}$ has a common fixed point in $\tilde{K}$ to give another proof of the following (extremely minor) variant of a theorem of Lim ([14]). We make no claim that our proof is simpler than the original—indeed the details in most respects are a combination of those found in [14] and [15]—but the method would appear to be of potential use in related contexts.

<u>Theorem</u>. Let X be a Banach space, K a weakly compact convex subset of X which has normal structure, and $\mathcal{S}$ a semigroup of nonexpansive self-mappings of K. Suppose every finite subcollection of $\mathcal{S}$ has a common fixed point in every closed convex $\mathcal{S}$-invariant subset of K. Then $\mathcal{S}$ has a common fixed point in K.

We should remark that if $\mathcal{S}$ is commutative (Lim's assumption), then our assumption on the finite subcollections of $\mathcal{S}$ always holds by Theorem 3 of [1].

A characterization of normal structure due to Landes will facilitate our proof. A bounded sequence $\{x_n\}$ in a Banach space is said to be <u>limit constant</u> if $\lim_{n\to\infty}\|x_n - x\| \equiv d > 0$ for all $x \in \mathrm{conv}\{x_1, x_2, \ldots\}$. In [13] Landes shows that a convex set K has normal structure if and only if K contains no such sequence.

<u>Proof of Theorem</u>. By a standard Zorn lemma argument, we may suppose K is minimal with respect to being nonempty, weakly compact, convex, and invariant under $\mathcal{S}$, and clearly we may also suppose K contains more than one point. Let r and W denote, respectively, the Chebyshev radius and Chebyshev center of K. Define $\mathfrak{F}$, $\tilde{\mathcal{X}}$, $\tilde{K}$ and $\tilde{\mathcal{S}}$ as in the discussion above, and let F denote the set of all points of the form $\{\{p(\alpha)\} : p(\alpha) \in F_\alpha\}$. As seen above, for each $f \in \mathcal{S}$,

$$(*) \qquad \{\alpha \in I : f(p(\alpha)) = p(\alpha)\} \in \mathfrak{F}.$$

Now fix $p(\alpha) \in F_\alpha$ for $\alpha \in I$. Let

$$H = \{x \in K : \{\alpha \in I : \|x - p(\alpha)\| \le r\} \in \mathfrak{F}\}.$$

Note that $W \subset H$; hence $H \neq \emptyset$. If $f \in \mathcal{S}$ and $x \in H$, then (*) and the fact that f is nonexpansive implies $f(x) \in H$. Since H is obviously convex, minimality of K implies $\overline{H} = K$. Thus (letting $\{x\} = (x,x,\ldots) + N \in \tilde{\mathcal{X}}$)

$$\|\{p(\alpha)\} \dot{-} \{x\}\|_{\mathfrak{F}} \le r$$

for all $x \in K$, $\{p(\alpha)\} \in F$.

Now suppose for some $\{p(\alpha)\} \in F$, $x \in K$, and $r_1 < r$,

$$\|\{p(\alpha)\} - \{x\}\|_{\mathfrak{F}} \le r_1.$$

A minimality argument like that just given implies

$$cl\{x \in K : \|\{p(\alpha)\} - \{x\}\|_{\mathfrak{F}} \le r_1\} = K.$$

Choose $\epsilon > 0$ so that $r_1 + \epsilon < r$. Then for all $x \in K$,

$$\gamma_1 = \{\alpha \in I : \|p(\alpha) - x\| \le r_1 + \epsilon\} \in \mathfrak{F};$$

thus if $p = \text{weak-}\lim_{\mathfrak{F}} p(\alpha)$ then it follows that $\|x - p\| \le r_1 + \epsilon < r$ for all $x \in K$. This contradicts the definition of r. It follows that for all $\{p(\alpha)\} \in F$ and $x \in K$,

$$\|\{p(\alpha)\} - \{x\}\|_{\mathfrak{F}} = r.$$

We now use this fact to construct a limit constant sequence. Fix $\{p(\alpha)\} \in F$, select $\alpha_1 \in I$, and set $x_1 = p(\alpha_1)$. Since $\|\{p(\alpha)\} - x_1\|_{\mathfrak{F}} = r$ there exists $\alpha_2 \in I$ such that

$$r-1 \le \|x_1 - p(\alpha_2)\| \le r+1.$$

Set $x_2 = p(\alpha_2)$. Now suppose $\{x_1, \ldots, x_n\}$ $(n \ge 2)$ have been obtained such that

$$r - 1/(n-1) \le \|x - x_n\| \le r + 1/(n-1)$$

for all $x \in \text{conv}\{x_1, \ldots, x_{n-1}\}$. Since $C = \text{conv}\{x_1, \ldots, x_n\}$ is compact and since $\|\{p(\alpha)\} - \{x\}\|_{\mathfrak{F}} = r$ for all $x \in K$, there exists $\alpha_{n+1} \in I$ such that

$$r - 1/n \le \|x - p(\alpha_{n+1})\| \le r + 1/n$$

for all $x \in C$. Taking $x_{n+1} = p(\alpha_{n+1})$, it is clear that the resulting sequence $\{x_n\}$ is limit constant if $r > 0$. Hence $r = 0$ and K is a singleton, completing the proof.

EXISTENCE FROM APPROXIMATION

In this section we outline a general method for proving the existence of fixed points for mappings defined on product spaces, and then we discuss a few of the results which have been obtained utilizing this approach.

Let X and Y be topological spaces and let P_1 and P_2 denote the respective coordinate projections of $X \times Y$ onto X and Y. Suppose $T : X \times Y \longrightarrow X \times Y$ is given, and define $f_x : Y \longrightarrow Y$ by

$$f_x(y) = P_2 \circ T(x,y).$$

Suppose $F_x = \{y \in Y : f_x(y) = y\} \neq \emptyset$ for each $y \in Y$. Let g be a selection of the mapping $x \longrightarrow F_x$. Then $P_2 \circ T(x,g(x)) = g(x)$. Define $\varphi : X \longrightarrow X$ by $\varphi(x) = P_1 \circ T(x,g(x))$. Then if $\varphi(x) = x$ for some $x \in X$, one obtains $T(x,g(x)) = (x,g(x))$, i.e., T has a fixed point in $X \times Y$.

Three things are essential to the above approach:

1) Assumptions on T and Y must be sufficient to ensure that $F_x \neq \emptyset$ for each $x \in X$.
2) A method for defining g must be found, and in particular, a method which will enable one to conclude that φ lies in a 'nice' class of mappings.
3) X must be a space which has the fixed point property for the class containing φ.

Examples. It is in Step 2 that approximation results are used. Specifically, one needs a method of identifying a particular fixed point of f_x for each $x \in X$. The simplest situation occurs when f_x is a contraction mapping. The following is due to Fora [5]; for a proof following the approach just described, see Kirk [11].

(I) Let X be a topological space which has the fixed point property for continuous mappings, suppose Y is a complete metric space, and let $T: X \times Y \longrightarrow X \times Y$ be continuous and satisfy:

(c) For each $x \in X$ there exist $\lambda(x) \in (0,1)$ and a neighborhood V of x such that for each $w \in W$ and $u,v \in Y$, $d(P_2 \circ T(w,u), P_2 \circ T(w,v)) \leq \lambda(x) d(u,v)$.

Then T has a fixed point in $X \times Y$.

The above is proved by fixing $y_0 \in Y$, defining $g(x) = \lim_{n\to\infty} f_x^n(y_0)$, and showing that the resulting function φ is continuous.

We now summarize a number of Banach space results. Explicit proofs may be found in [10], [11], and [12]. Throughout, E and F denote Banach spaces with $X \subset E$ and $Y \subset F$. For $1 \leq p < \infty$, let $(E \times F)_p$ denote the product space with usual p-norm:

$$\|(x,y)\|_p = [\|x\|^p + \|y\|^p]^{1/p}.$$

$(E \times F)_\infty$ is defined in the standard way.

(II) Let $T: X \times Y \longrightarrow X \times Y$ be nonexpansive relative to $(X \times Y)_p$ for some $1 \leq p \leq \infty$. Suppose X has the fixed point property for nonexpansive mappings, suppose Y is convex, and suppose $f_x : Y \longrightarrow Y$ is compact for each $x \in X$. Then T has a fixed point in $X \times Y$.

In the proof of the above, $g(x) = \lim_{n\to\infty} h^n(y_0)$ where $y_0 \in Y$ is fixed and $h(u) = (u + f_x(u))/2$, $u \in Y$. Convergence of the iterates of h to a fixed point of f_x is assured by a theorem of Ishikawa [7], and the resulting mapping φ is nonexpansive.

The following is proved in [12] (see also [10]) and the approximation used involves weak convergence of approximate fixed point sequences of f_x to fixed points of f_x. The precise assumption on the set Y is that it be weakly compact, convex, and have the Browder-Göhde property: $I - f$ is demiclosed for every nonexpansive $f: Y \longrightarrow Y$ (i.e., if $\{u_j\}$ in Y converges weakly to $u \in Y$ while $u_j - f(u_j) \longrightarrow v$ strongly, then $u - f(u) = v$). This property always holds if F is uniformly convex.

(III) Suppose X has the fixed point property for nonexpansive mappings, and suppose Y is weakly compact, convex, and has the Browder-Göhde property. Then $(X\times Y)_\infty$ has the fixed point property for nonexpansive mappings.

It is not known whether the above holds for $(X\times Y)_p$, $1 \le p < \infty$.

For a number of additional results obtained by utilizing the approach described at the outset of this section, we refer to [10], [11], [12].

REFERENCES

[1] Belluce, L.P. and Kirk, W.A., Fixed point theorems for families of nonexpansive mappings, Pacific J. Math. 18 (1966) 213-217.

[2] Brodskii, M.S. and Milman, D.P., On the center of a convex set, Dokl. Akad. Nauk SSSR 59 (1948) 837-840 (Russian).

[3] Bruck, R.W., A common fixed point theorem for a commuting family of nonexpansive mappings, Pacific J. Math. 53 (1974) 59-71.

[4] Elton, J., Lin, P., Odell, E. and Szarek, W., Remark on the fixed point problem for nonexpansive mappings, in: Sine, R.C. (ed.), Fixed Points and Nonexpansive Mappings, Contemporary Mathematics vol. 18 (Amer. Math. Soc., Providence, 1983).

[5] Fora, A., A fixed point theorem for product spaces, Pacific J. Math. 99 (1982) 327-335.

[6] Heinrich, S., Ultraproducts in Banach space theory, J. Reine Angew. Math. 313 (1980) 72-104.

[7] Ishikawa, S., Fixed points and iteration of nonexpansive mappings in a Banach space, Proc. Amer. Math. Soc. 59 (1976) 65-71.

[8] Kirk, W.A., Fixed point theory for nonexpansive mappings, in: Fadell, E. and Fournier, G. (eds.), Fixed Point Theory (Springer-Verlag, Berlin, Heidelberg, New York, 1981).

[9] Kirk, W.A., Fixed point theory for nonexpansive mappings II, in: Sine, R.C. (ed.), Fixed Points and Nonexpansive Mappings, Contemporary Mathematics vol. 18 (Amer. Math. Soc., Providence, 1983).

[10] Kirk, W.A., Nonexpansive mappings in product spaces, set valued mappings, and k-uniform rotundity, in: Browder, F. (ed.), Proc. 1983 AMS Summer Institute on Nonlinear Functional Analysis and Applications (Amer. Math. Soc., Providence, to appear).

[11] Kirk, W.A., Fixed point theorems in product spaces, in: Singh, S.P. (ed.), Operator Equations and Fixed Points (to appear).

[12] Kirk, W.A. and Sternfeld, Y., The fixed point property for nonexpansive mappings in certain product spaces, Houston J. Math. 10 (1984) 207-214.

[13] Landes, T., Permanence properties of normal structure, Pacific J. Math. 110 (1984) 125-143.

[14] Lim, T.C., A fixed point theorem for families of nonexpansive mappings, Pacific J. Math. 53 (1974) 487-493.

[15] Lim, T.C., Characterizations of normal structure, Proc. Amer. Math. Soc. 43 (1974) 313-319.

[16] Lin, P., Unconditional Bases and Fixed Points of Nonexpansive Mappings, (to appear).

[17] Maurey, B., Points fixe des contractions de certains faiblement compacts de L^1, in: Seminaire d'Analyse fonctionnelle 1980-81, Expose no. VIII (Ecole Polytechnique, Palaiseau, 1981).

This paper is in final form and no version of it will be submitted for publication elsewhere.

Trends in the Theory and Practice of Non-Linear Analysis
V. Lakshmikantham (Editor)

HOPF BIFURCATION WITH A GENERAL TRANSVERSALITY CONDITION

S.R. Bernfeld

Department of Mathematics
University of Texas at Arlington
Arlington, TX 76019
USA

INTRODUCTION

In a typical setting of the two dimensional version of the Hopf bifurcation theorem ([see [2] or [5] for example) we are given the system

$$(1)_\mu \qquad \begin{aligned} \dot{x} &= \alpha(\mu)x - \beta(\mu)y + X(\mu,x,y) \\ \dot{y} &= \alpha(\mu)y + \beta(\mu)x + Y(\mu,x,y) \end{aligned}$$

where $\mu \in (-\bar{\mu}\ \bar{\mu})$ for $\bar{\mu}$ sufficiently small, $\alpha(0) = 0$, $\beta(0) = 1$ and X, Y $C^\infty[(-\bar{\mu},\bar{\mu}) \times B^2(r_0), R]$, where $B^2(r_0) = \{w \in R^2: \|w\| < r_0\}$. Moreover X and Y are $0(x^2+y^2)$. Under these conditions we can prove using the transversality condition

$$(T) \qquad \alpha'(0) \neq 0$$

that there exists a function $\mu(c)$ defined for c sufficiently small, such that for each $\hat{c}$ there exists a $\hat{\mu}$ such that system $(1)_{\hat{\mu}}$ has a periodic orbit $(\hat{x}(t),\hat{y}(t))$ whose period is close to 2π and whose amplitude depends on $\hat{c}$. Moreover if $(x(t),y(t))$ is any periodic orbit of $(1)_{\hat{\mu}}$ ($\hat{\mu}$ sufficiently small) lying near the origin with period near 2π then there exists $\hat{c}$ such that $\hat{\mu} = \hat{\mu}(\hat{c})$. We describe the above by simply stating that there exists a one-parameter family of periodic orbits of $(1)_\mu$ bifurcating from the origin of $(1)_0$.

The existence of the unique function $\mu = \mu(c)$ depends heavily upon the transversality condition (T). When this condition does not hold then other cases of bifurcation phenomena may exist. To understand this we assume a generalized transversality condition; namely there exists an integer $s \geq 1$ such that

$$(GT) \qquad \alpha^j(0) = 0,\ j = 0,\ldots,\ s-1,\ \alpha^{(s)}(0) \neq 0$$

In the case that the right hand side of $(1)_\mu$ is analytic, Flockerzi [3] proved that for each μ there exists at most s periodic orbits. Kielhöfer [4] extended this result to the case of differential equations in a Hilbert space. Although Flockerzi [3] also claims his result is true in the C^∞ case we give an example showing this is not necessarily true. We also introduce a relationship between the stability behavior of the origin of $(1)_0$, the condition (GT), and the maximum number of bifurcating periodic orbits. To this end we say that the origin of $(1)_0$ is h-asymptotically stable if it is asymptotically stable under all perturbations of order greater than h and h is the smallest integer for which this is true (h must be odd).

In a joint paper with Professor Salvadori, to appear soon, we provide details of our work here as well as related results.

RESULTS

We shall always assume the conditions on $(1)_\mu$ given in the first sentence of the introduction. Define the class, $\mathcal{F}$, as the set of functions $f:R^2 \to R$ which are C^∞, flat at the origin, and for which there exists a sequence of points (x_n,y_n) such that $(x_n,y_n) \to (0,0)$ as $n \to \infty$ so that $f(x_n,y_n) = \frac{\partial f}{\partial x}(x_n,y_n) = \frac{\partial f}{\partial y}(x_n,y_n) = 0$.

Let $X(\mu,x,y) = \tilde{X}(\mu,x,y) + X_1(\mu,x,y)$ and $Y(\mu,x,y) = \tilde{Y}(\mu,x,y) + Y_1(\mu,x,y)$, where $\tilde{X}$, $\tilde{Y}$ areanalytic in (x,y) for each μ and X_1, Y_1 are flat in (x,y) at the origin for each μ. We are now able to give our results.

Theorem 1. Assume X_1, Y_1 are not in $\mathcal{F}$., for each μ. Let (GT) be satisfied for some integer $s \geq 1$. Then for each $\mu \in (-\bar{\mu},\bar{\mu})$ there exists at most s periodic solutions of $(1)_\mu$ lying near the origin with period near 2π.

Theorem 2. In addition to the hypotheses of Theorem 1 assume the origin of $(1)_\mu$ for $\mu = 0$ is h-asymptotically stable. Then there exists at most ℓ periodic orbits lying near the origin with period near 2π, for each μ, where $\ell = \min(s, \frac{h-1}{2})$.

In [5] it is proved that if the right hand side of $(1)_0$ is analytic and the origin is asymptotically stable then the origin is h-asymptotically stable for some h. Then we have

Corollary 1. Assume $X_1 \equiv Y_1 \equiv 0$ and the origin of $(1)_0$ is asymptotically stable. If (GT) is satisfied then there exists an odd integer h greater than two and at most ℓ periodic orbits with period near 2π, lying near the origin, for each μ where $\ell = \min(s, \frac{h-1}{2})$.

This result was incorrectly announced by the author in [1] where it was stated for the C^∞ case (Example 1 below is a counterexample).

The proof of our results depend on properties of the Poincaré map, the Malgrange Preparation Theorem, algebraic function theory, and Newton's diagram.

EXAMPLE AND CONCLUDING REMARKS

This example brings out the differences between (T) and (GT) as well as the role of $\mathcal{F}$ in Theorems 1 and 2. It provides a counterexample to the result in [1] and [3] in the nonanalytic case.

Example 1. Consider

$$(2)_\mu \quad \begin{aligned} \dot{x} = {} & \mu^3 x - y - 3\mu^2 x(x^2+y^2) + 3\mu x(x^2+y^2)^2 - x(x^2+y^2)^3 \\ & - x\exp\left(\frac{-1}{2(x^2+y^2)}\right) \sin^2\left(\exp\left(\frac{1}{x^2+y^2}\right)\right) \end{aligned}$$

$$\dot{y} = \mu^3 y + x - 3\mu^2 y(x^2+y^2) + 3\mu y(x^2+y^2)^2 - y(x^2+y^2)^3$$

$$- y \exp\left(-\frac{1}{2(x^2+y^2)}\right) \sin^2(\exp\left(\frac{1}{x^2+y^2}\right))$$

In polar coordinates the radial component, r, of $(2)_\mu$ is given by

$$(3)_\mu \qquad \frac{dr}{d\theta} = \mu^3 r - 3\mu^2 r^3 + 3\mu r^5 - r^7 - r\exp -\left(\frac{1}{2r^2}\right) \sin^2(\exp\left(\frac{1}{r^2}\right))$$

For $\mu = 0$ we see from $(3)_0$ that the origin of $(2)_0$ is 7 asymptotically stable, satisfies (GT) with $s = 3$ (thus $\ell = 3$) but the last term of the right hand side of $(2)_\mu$ is in F. The periodic orbits of $(2)_\mu$ are given by the zeros of the right hand side of $(3)_\mu$. Letting $c = r^2$ we obtain $0 = \mu^3 - 3\mu^2 c + 3\mu c^2 - c^3 - \exp -\left(\frac{1}{2c}\right) \sin^2(\exp\left(\frac{1}{c}\right))$ or

$$(4) \qquad \mu = c + (\exp\left(-\frac{1}{2c}\right) (\sin^2(\exp\left(\frac{1}{c}\right)))^{1/3} .$$

Notice for each c there is exactly one μ for which $(2)_\mu$ has a periodic solution. However given any number N we can show there exist $\delta(N)$ such that for any given $\bar{\mu} < \delta(N)$ there exist at least N distinct values of c for which (4) holds, that is, there exists at least N periodic orbits of $(2)_{\bar{\mu}}$ lying near the origin with period 2π.

Other examples, which will be given in the complete paper with L. Salvadori, depict other qualitative differences in the behavior of the periodic solutions of $(1)_\mu$. These depend again upon whether (T) or (GT) holds, whether X, Y are analytic or C^∞, and whether the origin of $(1)_0$ is h-asymptotically stable or not.

BIBLIOGRAPHY

[1] Bernfeld, S.R., Generalized transversality, exchange of stability and Hopf bifurcation, Conference on Differential Equations and Applications in Ecology, Epidemics, and Population Problems (Academic Press, 1981).

[2] Chow, S.N. and Hale, J.K., Methods of Bifurcation Theory (Springer-Verlag, New York, 1982)

[3] Flockerzi, D., Existence of small periodic solutions of ordinary differential equations in R^2, Arch. d. Math., 33 (1979) 263-278.

[4] Kielhöfer, H., Generalized Hopf bifurcation in Hilbert Space, Math. Meth. in Appl. Sci., 1 (1979) 498-513.

[5] Negrini, P. and Salvadori, L., Attractivity and Hopf bifurcation, J. Nonl. Anal. T.M.A., 3 (1979) 87-100.

The final (detailed) version of this paper will be submitted for publication elsewhere.

Trends in the Theory and Practice of Non-Linear Analysis
V. Lakshmikantham (Editor)

STABILITY REGIONS OF CERTAIN LINEAR SECOND ORDER PERIODIC DIFFERENTIAL EQUATIONS

S.R. Bernfeld and M. Pandian

Department of Mathematics
University of Texas at Arlington
Arlington, Texas 76019
USA

INTRODUCTION

The difficulty in finding the Floquet multipliers of Hill's equation

$$\ddot{x} + p(t)\dot{x} + q(t)x = 0 \tag{1}$$

is well known (see [1] or [2]), where $p(t)$ and $q(t)$ are 2π-periodic. Consequently the stability properties of (1) are very difficult to ascertain in general unless $p(t)$ and $q(t)$ are constant. Recall that (1) is stable if and only if all solutions are bounded and is unstable if there exists an unbounded solution. If the two multipliers ρ_1, ρ_2 of (1) are within the unit circle, $\rho_1{}^2 + \rho_2{}^2 < 1$, then (1) is stable whereas if either $|\rho_1| > 1$ or $|\rho_2| > 1$ then (1) is unstable. If $\rho_1{}^2 + \rho_2{}^2 = 1$ and both multipliers have simple elementary divisors then (1) is stable; if one of the multipliers does not have a simple elementary divisor than (1) is unstable.

Second order periodic equations have been analyzed in various contexts. For example if one wants to study the stability properties of periodic solutions of autonomous second order nonlinear equations then a useful technique is to analyze the stability properties of the corresponding variational equation which is of the form (1). Moreover in problems of mathematical physics, such as in the theory of elastic vibrations, particular cases of (1) such as the Mathieu equation ($p(t) \equiv 0$, $q(t) = a + b\cos t$) and other forms of Hill's equation are used [1].

In some problems in ecology and nonlinear chemical kinetics, for example, one is often led to (1) when certain parameters are not identically constant but rather fluctuate periodically around a constant value.

As indicated above $p(t)$ and $q(t)$ may depend on parameters. Consequently knowledge of the dependence of the multipliers on these parameters yields information on the dependence of the stability regions in the parameter space. Such regions have been depicted for the Mathieu equation [2] and other forms of Hill's equation [2]. Moreover in problems in bifurcation theory it is also important to have information on changes of the stability behavior of (1) as parameters vary, again leading to an analysis of the dependence of the multipliers on parameters.

In this preliminary note we begin with a study of the dependence of multipliers on parameters for a particular case of (1).

STABILITY REGIONS

We begin our study by expanding $p(t)$ and $q(t)$ in terms of its Fourier series and discarding all terms beyond the fundamental harmonics. We then rewrite (1) as

(2) $$\ddot{x} + (A + B\sin(t+\alpha))\dot{x} + (C + D\cos(t+\beta))x = 0,$$

where A, B, C, D, α,β, are real constants. (the case A = B = 0, $\beta = 0$ is the Mathieu equation)

We will look at the critical case A = C = 0, B = D = 1 and later study a neighborhood in parameter space of the critical case. Thus consider

(3) $$\ddot{x} + \sin(t+\alpha)\dot{x} + \cos(t+\beta)x = 0;$$

we will attempt to determine the behavior of the two multipliers $\rho_1(\alpha,\beta)$, $\rho_2(\alpha,\beta)$ on α and β. In fact we may restrict α,β to $S = [0,2\pi] \times [0,2\pi]$ by observing $(x,v) \cong (y,w)$ if and only if $x \equiv y \bmod 2\pi$ and $v \equiv w \bmod 2\pi$, is an equivalence relationship. Floquet theory tells us $\rho_1\rho_2 = 1$ and ρ_i, i = 1, 2 satisfies the characteristic equation

(4) $$x^2 - (\rho_1+\rho_2)x + 1 = 0.$$

Define $A(\alpha,\beta) = \frac{\rho_1+\rho_2}{2}$. System (3) is stable if $|A| < 1$, that is, if $\rho_2 = \bar{\rho}_1$ and $\rho_1 \neq \pm 1$; and (3) is unstable if $|A| > 1$, that is, if $0 < \rho_1 < 1 < \rho_2$. The boundary separating the stability regions in the (α,β) plane is thus given by $\rho_1 = \pm 1$. On the boundary $\rho_1 = 1$ there are periodic solutions of period 2π, and if $\rho_1 = 1$ has simple divisors for a particular (α,β) then all solutions of (3) are of period 2π for this particular (α,β). Similar statements hold when $\rho_1 = -1$ except now the solutions are of period 4π.

Numerical studies (with the assistance of B. Asner) lead to the following observations in the stability analysis of (3):

<u>Observations</u>: (a) The region of instability, $I(\alpha,\beta)$ can be written as

$$I = \bigcup_{i=1}^{\infty} W_i,$$

where each W_i is the union of a closed unbounded set U_i with a nonempty interior and a one dimensional disjoint curve C_i. Moreover $W_i \cap W_j = \emptyset$ when $i \neq j$. On C_i $\rho_1 = 1$, whereas on the boundary of U_i, $\rho_1 = -1$.

(b) The region of stability, S, can be written as

$$S = \bigcup_{i=1}^{\infty} O_i,$$

where each O_i is open, and each O_i can be written as $O_i = O_{i1} \cup O_{i2}$ such that O_{i1}, O_{i2} are each open and $\overline{O_{i1}} \cap \overline{O_{i2}} = C_i$.

(c) The range of the mapping $(\rho_1(\alpha,\beta),\rho_2(\alpha,\beta))$, which is periodic in (α,β) of period 2π, is the set

$$T = \{(\rho_1,\rho_2): \rho_1^2+\rho_2^2 = 1\} \cup \{-a \geq \rho_2 > -1 > \rho_1 \geq -b\},$$

where a is approximately .021 and b is approximately 46.54.

(d) For each constant $c \in \{[-b,-1] \cup (z: |z|=1)\}$ there exists a function $\beta_c(\alpha)$ such that $\rho_1(\alpha,\beta_c(\alpha)) \equiv c$ for all α. There exists in addition, a family $\psi_c(\alpha)$ orthogonal to the family $\beta_c(\alpha)$ such that the range of $(\rho_1(\alpha,\psi_c(\alpha))$,

$\rho_2(\alpha,\psi_c(\alpha))$ is the set T for each c.

Remarks: For c=1 we can show that $\beta_1(\alpha) = \alpha$ in (d). Indeed in this case we can integrate (3) obtaining

$$\frac{d}{dt}(x \cos(t+\alpha)) = \ddot{x}$$

leading to

$$\dot{x} = x \cos(t+\alpha) + K$$

where K is a constant. The solutions are 2π periodic if and only if K = 0. Hence $\rho_1 = \rho_2 = 1$.

In subsequent work with B. Asner details of this note will be given as well as other developments in the study of two dimensional systems depending on two or more parameters.

Acknowledgement: The authors would like to thank Professor B. Asner of the University of Dallas for both his important numerical contributions as well as for several interesting conversations.

BIBLIOGRAPHY

[1] Stoker, J.J., Nonlinear Vibrations, Interscience (New York, 1957).

[2] Yakubovich, V.A. and V.M. Starzhinskii, Linear Differential Equations with Periodic Coefficients (English translation--two volumes) (Wiley, New York, 1975).

The final (detailed) version of this paper will be submitted for publication elsewhere.

$L_p([a,b]; \mathbb{R}^n)$ for $1 \le p < +\infty$. The usual Banach space $C([a,b]; \mathbb{R}^n)$ of continuous $\mathbb{R}^n$-valued functions will be denoted by $C(a,b)$ and similarly the Sobolev space $W^{1,p}([a,b]; \mathbb{R}^n)$ will be denoted by $W^{1,p}(a,b)$. Throughout the remainder of the paper $r>0$ is a fixed real number and we shall simply write L_p, C and $W^{1,p}$ for $L_p(-r,0)$, $C(-r,0)$ and $W^{1,p}(-r,0)$, repsectively. If $x:[-r,+\infty) \to \mathbb{R}^n$, then we define $x_t:[-r,0] \to \mathbb{R}^n$ by $x_t(s) = x(t+s)$.

II. NFDEs ON PRODUCT SPACES

During the past few years considerable attention has been given to the study of semigroups generated by linear functional differential equations (see [1,2, 5-9]). In 1969 Borisovic and Turbabin [2] considered the retarded equation

$$\dot{x}(t) = Lx_t \tag{2.1}$$

with initial data

$$x(0) = \eta, \; x_0(s) = \phi(s) \;\; -r \le s < 0 \tag{2.2}$$

where $(\eta, \phi(\cdot)) \varepsilon \mathbb{R}^n \times L_p$ and L was a bounded linear operator from C into $\mathbb{R}^n$, i.e. $L \varepsilon \mathcal{B}(C, \mathbb{R}^n)$. They defined the operator

$$A(\eta, \phi(\cdot)) = (L\phi(\cdot), \; \phi(\cdot)) \tag{2.3}$$

on the domain

$$\mathcal{D}(A) = \{(\eta, \phi(\cdot)) / \phi(\cdot) \varepsilon W^{1,p}, \; \phi(0) = \eta\} \tag{2.4}$$

and noted that A was the infinitesimal generator of the C_0-semigroup $S(t)$ on $\mathbb{R}^n \times L_p$ defined by

$$S(t)(\eta, \phi(\cdot)) = (x(t), \; x_t(\cdot)) \tag{2.5}$$

where $x(\cdot)$ is the solution to (2.1)-(2.2). This result was established for a more general L by Vinter in [15] and later Delfour [8] showed that the result remained valid for any $L \varepsilon \mathcal{B}(W^{1,p}, \mathbb{R}^n)$. In fact, Delfour's result was necessary in that if A defined by (2.3)-(2.4) generated a C_0-semigroup on $\mathbb{R}^n \times L_p$, then L must be bounded as an operator from $W^{1,p}$ into $\mathbb{R}^n$. Therefore, Delfour has established the largest class of <u>retarded</u> hereditary systems defining C_0-semigroups on product spaces of the form $\mathbb{R}^n \times L_p$.

In [6] and [7] Burns, Herdman and Stech extended Delfour's results to a class of NFDEs of the form

$$\frac{d}{dt} Dx_t = Lx_t \tag{2.6}$$

with initial data

$$Dx_0(\cdot) = \eta \quad x_0(s) = \phi(s) \quad -r \le s < 0. \tag{2.7}$$

Defining the operator

$$A(\eta,\phi(\cdot)) = (L\phi(\cdot),\ \dot{\phi}(\cdot)) \tag{2.8}$$

on the domain

$$\mathcal{D}(A) = \{(\eta,\phi(\cdot))\,|\,\phi(\cdot)\varepsilon W^{1,p},\ D\phi(\cdot)=\eta\} \tag{2.9}$$

it was established in [6] that L and D must belong to $\mathcal{B}(W^{1,p}, \mathbb{R}^n)$ if A defined by (2.8)-(2.9) generates a C_0-semigroup on $\mathbb{R}^n \times L_p$. Moreover, it was shown that if $L\varepsilon\mathcal{B}(W^{1,p}, \mathbb{R}^n)$ and $\mathcal{D}\varepsilon\mathcal{B}(C, \mathbb{R}^n)$ has an atom at s=0, then A defined by (2.8)-(2.9) generates a C_0-semigroup on $\mathbb{R}^n \times L_p$. Recall that $D\varepsilon\mathcal{B}(C, \mathbb{R}^n)$ is atomic at s=0 if there is a matrix valued function μ and a non-singular matrix A_0 such that for $\phi(\cdot)\varepsilon C$

$$D\phi(\cdot) = A_0\phi(0) + \int_{-r}^{0} d\mu(s)\phi(s) \tag{2.10}$$

where $\mu(\cdot)$ is of bounded variation on [-r,0] (we shall always assume $\mu(\cdot)$ is normalized to be continuous from the right on (-r,0) and $\mu(-r)=0$) and

$$\lim_{\varepsilon\to 0^+} \int_{-\varepsilon}^{0} |d\mu(s)| = 0. \tag{2.11}$$

The sufficient conditions cited above lead to the well-posedness for a large class of hereditary systems and allows one to define generalized solutions to the NFDE (2.6)-(2.7) for L_p initial data.

Although the assumption that D belongs to $\mathcal{B}(C, \mathbb{R}^n)$ is not very restrictive, the assumption that D has an atom at s=0 seemed to rule out a number of equations, including those that occur in certain models of aeroelasticity. It was shown in [6] that the assumption that D be atomic at s=0 is not necessary. Let $0<\alpha<1$ and consider the integral equation

$$\int_{-r}^{0} |s|^{-\alpha}x(t+s)ds = \eta \qquad 0<t \tag{2.12}$$

where $x(s)=\phi(s)$, $-r\le s<0$. If L and D are defined by

$$L\phi(\cdot)\equiv 0,\quad D\phi(\cdot) = \int_{-r}^{0} |s|^{-\alpha}\phi(s)ds, \tag{2.13}$$

then (2.12) can be viewed as a neutral equation of the form (2.6)-(2.7). Note that $D\varepsilon\mathcal{B}(C, \mathbb{R})$ and D is not atomic at s=0. However, the following result may be found in [6].

THEOREM 1. Assume L and D are defined by (2.13). If $p<1/(1-\alpha)$, then A defined by (2.8)-(2.9) genreates a C_0-semigroup on $\mathbb{R}\times L_p$ defined by $S(t)(\eta,\phi(\cdot)) = (\eta,x_t(\cdot))$ where $x(\cdot)$ is the unique solution to the Abel equation (2.12). If $p\ge 1/(1-\alpha)$, then Abel's equation has a unique solution for each $(\eta,\phi(\cdot))$ in a dense subset of $\mathbb{R}\times L_p$. However, A does not generate a C_0-semigroup on $\mathbb{R}\times L_p$.

Theorem 1 above lead Kappel and Zhang [12, 17] to consider more general necessary conditions for well-posedness in C of problems of the form

$$Dx_t = D\phi \quad t>0 \tag{2.14}$$

with initial data

$$x_0(s) = \phi(s) \quad -r\le s<0 \tag{2.15}$$

where $\phi(\cdot)\varepsilon C$ and D is not atomic at s=0. For the scalar case only, they defined an operator D to be weakly atomic at s=0 if for real λ

$$\lim_{\lambda\to\infty}|\lambda\, \Delta_0(\lambda)| = \infty \tag{2.16}$$

where

$$\Delta_0(\lambda) = D(e^{\lambda\cdot}). \tag{2.17}$$

Any D operator that is atomic at s=0 is weakly atomic. However, the D operator defined by (2.13) is not atomic at s=0 but is weakly atomic at s=0. Zhang [17] proved that if (2.14)-(2.15) is well-posed in C, then it is necessary that D be weakly atomic at s=0. It was shown in [7] (see Theorem 3.2) that if (2.6)-(2.7) is well-posed in $\mathbb{R}^n\times L_p$, then (2.6)-(2.7) leads to a well-posed problem on C. It follows that if (2.6)-(2.7) is well-posed in $\mathbb{R}^n\times L_p$, then D must be weakly atomic at s=0.

For the scaler case considered in [12] and [17], the assumption that D be weakly atomic implies that for all real λ sufficiently large the operator A has λ in its resolvent. This is true because $L\phi(\cdot)\equiv 0$ and $\lambda\varepsilon\rho(A)$ if and only if

$$\Delta(\lambda) = \lambda D(e^{\lambda\cdot}I) - L(e^{\lambda\cdot}I) \tag{2.18}$$

is nonsingular (see Lemma 2.3 in [6]). However, for vector equations the problem becomes much more complex. In particular, condition (2.16) does not imply that $\rho(A)$ is not empty.

EXAMPLE 1. Let $L\phi(\cdot)\equiv 0$ and $D\varepsilon B(C,\mathbb{R}^2)$ be defined by

$$D\phi(\cdot) = \begin{bmatrix} 1 & 0 \\ 0 & 0 \end{bmatrix} \phi(0).$$

Note that $\|\lambda\Delta_0(\lambda)\| \to +\infty$ but $\Delta(\lambda) = \lambda\Delta_0(\lambda)$ is singular for all λ. Consequently, A defined by (2.8)-(2.9) can not generate a C_0-semigroup on $\mathbb{R}^n\times L_p$

In the next section we consider a special problem that occurs in aeroelasticity and establish the well-posedness of this problem. A complete treatment of more general systems (including infinite delay problems) will appear in a forthcoming paper.

III. WELL-POSEDNESS FOR AN AEROELASTIC SYSTEM

We consider the FDE

$$\frac{d}{dt}[A_0x(t)+ \int_{-r}^{0} A_1(s)x(t+s)ds] = B_0x(t)+ \int_{-r}^{0} B_1(s)x(t+s)ds \tag{3.1}$$

where A_0, $A_1(s)$, B_0 and $B_1(s)$ are n×n matrices satisfying the following condition.

CONDITION A.

i) $A_0 = \text{dia}\,(1, 1, \ldots, 1, 0)$,
ii) $B_1(\cdot)$ is continuous on $[-r,0]$,
iii) $A_{ij}(s) \equiv 0$ $i=1,2,\ldots,n$, $j=1,2,3,\ldots,n-1$
iv) $A_{in}(s)$ is continuous on $[-r,0]$, $i=1,2,\ldots,n-1$
v) $A_{nn}(s) = a|s|^{-\alpha}+B(s)$, $a>0$, $0<\alpha<1$ and $B(s)$ is continuous on $[-r,0]$ and $B(0)=0$.

Note that the aeroelastic system described in [4] is of the form (3.1) and satisfies the above conditions with $\alpha=1/2$, $n=8$ and $r=+\infty$. Although we are only considering the case where $r<+\infty$, the basic ideas below can be extended to cover the infinite delay problem described in [4].

Let, A_0, $A_1(\cdot)$, B_0, $B_1(\cdot)$ satisfy CONDITION A and define

$$D\phi(\cdot) = A_0\phi(0) + \int_{-r}^{0} A_1(s)\phi(s) \tag{3.2}$$

and

$$L\phi(\cdot) = B_0\phi(0) + \int_{-r}^{0} B_1(s)\phi(s). \tag{3.3}$$

Since A_0 is singular and $A_1(s)$ is integrable, the operator D is not atomic at s=0. For D and L defined by (3.2)-(3.3) we denote by A the corresponding operator (2.8)-(2.9).

Recall that A generates a C_0-semigroup on $\mathbb{R}^n \times L_p$ if and only if $\mathcal{D}(A)$ is dense, the resolvent set $\rho(A)$ is nonempty and the Cauchy problem

$$\frac{d}{dt}z(t) = Az(t) \quad z(0)=z_0 \in \mathcal{D}(A) \tag{3.4}$$

has a unique continuously differentiable solution on $[0, +\infty)$ (see [11] and page 102 in [13]). We shall use this equivalence to establish the following result.

THEOREM 2. If D and L are defined by (3.2)-(3.3) and CONDITION A is satisfied, then A generates a C_0-semigroup on $\mathbb{R}^n \times L_p$ for all p satisfying $1 \le p < 1/(1-\alpha)$.

Proof. The proof that the Cauchy problem (3.4) has continuously differentiable solution on $[0, +\infty)$ can be found in [14]. Since it is fairly technical and we are limited by space, we shall not repeat it here. It remains to show that A is densely defined and has non-empty resolvent.

Let $B_i : \mathbb{R}^n \times W^{1,p} \to R$ be the linear functional defined by

$$B_i(\eta, \phi(\cdot)) = \langle \eta - D\phi(\cdot), e_i \rangle_{R^n}$$

where e_i i=1,2....,n is the standard unit vector in $\mathbb{R}^n$. The domain of A is equal to $\cap_{i=1}^n \mathcal{V}(B_i)$, where $\mathcal{V}(B_i)$ denotes the null space of B_i. It follows from [3] that $\mathcal{D}(A)$ is dense if and only if each non-trivial linear combination $\Lambda=\sum_{i=1}^n w_iB_i$ is unbounded on $\mathbb{R}^n\times L_p$. If any of the w_i, i=1,2,...,n-1 are non zero then the proof of Lemma 2.2 in [6] can be used to show that Λ is unbounded. On the other hand if $w_i=0$, i=1,2,...,n-1, then $\Lambda=w_nB_n$. However, since $p<1/(1-\alpha)$, it follows that $w_nB_n(\eta,\phi(\cdot)) = w_n\eta-w_n\int_{-r}^0 A_{nn}(s)\phi_n(s)ds$ is not bounded on $\mathbb{R}^n\times L_p$ (note that $A_{nn}(s) \notin L_{p'}$, where $1/p+1/p'=1$). Therefore, $\mathcal{D}(A)$ is dense in $\mathbb{R}^n\times L_p$.

To establish that $\rho(A)$ is not empty, recall that $\lambda\epsilon\rho(A)$ if and only if $\Delta(\lambda)=\lambda D(e^{\lambda\cdot}I)-L(e^{\lambda\cdot}I)$ is non singular. Let $F(\lambda)$ denote the n×n diagonal matrix $F(\lambda) = dia(\lambda^{-1},\lambda^{-1},...,\lambda^{-1},\lambda^{-\alpha})$ and observe that (for $1\leq p<1/(1-\alpha)$) as $\lambda\to+\infty$

$$F(\lambda)\lambda D(e^{\lambda\cdot}I)\to dia(1,1,...,1,a_\infty)$$

where $a_\infty\neq 0$ and

$$F(\lambda)L(e^{\lambda\cdot}I)\to 0.$$

Therefore, as $\lambda\to+\infty$ the matrix $F(\lambda)\Delta(\lambda)$ converges to a non-singular matrix and hence for sufficiently large λ, $\Delta(\lambda)$ must be non singular. This completes the proof.

We conclude this paper by noting that the above ideas can be extended and applied to a larger class of equations than those considered here. A detailed treatment of the more general problem will appear elsewhere.

REFERENCES

[1] H.T. Banks and J.A. Burns, An abstract framework for approximate solutions to optimal control problems governed by hereditary systems, Proceedings: International Converence on Differential Equations, Academic Press, (1975), 10-25.

[2] Ju. Borisovic and A.S. Turbabin, On the Cauchy problem for linear nonhomogeneous differential equations with retarded arguments, Soviet Math. Dokl., 10 (1969), 401-405.

[3] R.C. Brown and A.M. Krall, Ordinary differential operators under Stieltjes boundary conditions, Trans. Amer. Math. Soc., 198 (1974), 73-92.

[4] J.A. Burns, E.M. Cliff and T.L. Herdman, A state-space model for an aeroelastic system, Proceedings: 22nd IEEE CDC, San Antonio, Texas (1983), 1074-1077.

[5] J.A. Burns and T.L. Herdman, Adjoint semigroup theory for a class of functional differential equations, SIAM J. Math. Anal. 5 (1976), 729-745.

[6] J.A. Burns, T.L. Herdman and H.W. Stech, Linear functional differential equations as semigroups on product spaces, SIAM J. Math. Anal. 14 (1983), 98-116.

[7] J.A. Burns, T.L. Herdman and H.W. Stech, The Cauchy problem for linear functional differential equations, Integral and Functional Differential Equations, Lecture Notes in Pure and Applied Mathematics Vol. 67, T.L. Herdman, S.M. Rankin and H.W. Stech, eds. Marcel Dekker, 1981, 139-149.

[8] M.C. Delfour, The largest class of hereditary systems defining a C_0-semigroup on the product space, Canad. J. Math., 32 (1980), 969-978.

[9] M.C. Delfour and S.K. Mitter, Controllability, observability and optimal feedback control of hereditary systems, SIAM J. Control, 10 (1972), 298-328.

[10] J.K. Hale and K.R. Meyer, A class of functional differential equations of neutral type, Mem. Amer. Math. Soc., 76 (1967).

[11] E. Hille, Une generalization du probleme de Cauchy, Ann. Inst. Fourier, 4 (1952), 31-48.

[12] F. Kappel and Kang pei Zhang, On neutral functional differential equations with nonatomic difference operator, Report No. 38, Institute for Mathematics, University of Graz (1984).

[13] A Pazy, Semigroups of Linear Operators and Applications to Partial Differential Equations, Springer-Verlag, Applied Mathematical Sciences 44, New York, 1983.

[14] J. Turi, Ph.D Thesis, Mathematics Department, Virginia Polytechnic Institute and State University, (In Preparation).

[15] R.B. Vinter, On a problem of Zabczk concerning semigroups generated by operators with non-local boundry conditions, Publication 77/8, Department of Computing and Control, Imperial College of Science and Technology, London (1977).

[16] D.V. Widder, The Laplace Transform, Princeton University Press, Princeton, 1946.

[17] Kang pei Zhang, On a neutral equation with nonatomic D-operator, Ph.D. Thesis, Institute for Mathematics, University of Graz (1983).

This paper is in final form and no version of it will be submitted for publication elsewhere.

Trends in the Theory and Practice of Non-Linear Analysis
V. Lakshmikantham (Editor)

PERIODICITY IN LINEAR VOLTERRA EQUATIONS

T. A. Burton

Department of Mathematics
Southern Illinois University
Carbondale, Illinois 62901

In this paper we present a unified theory of the existence of periodic solutions of linear ordinary differential equations and linear Volterra integro-differential equations.

1. INTRODUCTION

It is known [4] that solution spaces of the pair of n-dimensional systems

$$x' = Ax$$

and

$$x' = Ax + \int_0^t C(t-s)x(s)ds$$

or the pair

$$x' = B(t)x$$

and

$$x' = B(t)x + \int_0^t D(t,s)x(s)ds$$

are virtually indistinguishable:

(a) There are n linearly independent solutions on $[0,\infty)$.
(b) Every solution is a linear combination of those n.
(c) The variation of parameters formulae are similar.

Our goal here is to show similarity of limit sets. We look at a variety of differential equations generalizing

$$x' = Ax + f(t)$$

with A a constant $n \times n$ matrix and $f(t+T) = f(t)$ for some $T > 0$. It is assumed that the principal matrix solution of the homogeneous system is $L^1[0,\infty)$ and conclude four things:

(a) $x(t)$ is bounded on $[0,\infty)$.
(b) $x(t)$ approaches a periodic function.
(c) The periodic function has an integral formula.
(d) The periodic function satisfies another differential equation (called a limiting equation). Thus, the principal matrix solution being $L^1[0,\infty)$ yields the existence of a periodic solution of the limiting equation.

2. CONSTANT COEFFICIENTS

Let A be an $n \times n$ real constant matrix and let $f: (-\infty,\infty) \to R^n$ be continuous with $f(t+T) = f(t)$ for some $T > 0$ and all t. We are interested in the limiting behavior of solutions of

(1) $$x' = Ax + f(t), \quad x(0) = x_0.$$

The homogeneous equation $q' = Aq$ has principal matrix solution e^{At} and the variation of parameters formula for (1) is

$$x(t) = e^{At}x_0 + \int_0^t e^{A(t-s)}f(s)ds. \tag{2}$$

The following result is well-known in principle and gives a very compact and useful expression for the limit set of solutions of (1). It is the pattern for quite general equations and is verified by direct computation.

THEOREM 1. If the characteristic roots of A all have negative real parts then e^{At} is $L^1[0,\infty)$ and the following hold concerning the solution of (1).

(a) $x(t)$ is bounded on $[0,\infty)$.

(b) As $n \to \infty$, then

$$x(t+nT) \to \int_{-\infty}^t e^{A(t-s)}f(s)ds \overset{\text{def}}{=} y(t)$$

on $(-\infty,\infty)$ and $y(t)$ is a T-periodic function.

(c) $y(t)$ also satisfies (1). Thus, we say (1) is its own limiting equation.

3. PERIODIC COEFFICIENTS

Let B be an $n \times n$ matrix of continuous functions on $(-\infty,\infty)$, $f: (-\infty,\infty) \to R^n$ be continuous, let f and B be T-periodic, and consider the system

$$x' = B(t)x + f(t), \quad x(0) = x_0. \tag{3}$$

From Floquet theory there is an $n \times n$ nonsingular differentiable and T-periodic matrix P and a constant $n \times n$ matrix J with

$$P(t)e^{Jt}$$

being the principal matrix solution of $p' = B(t)p$ and the variation of parameters formula for (3) is

$$x(t) = P(t)e^{Jt}x_0 + \int_0^t P(t)e^{J(t-s)}P^{-1}(s)f(s)ds. \tag{4}$$

If

$$P(t)e^{Jt} \text{ tends to zero as } t \to \infty,$$

then

$$P(t)e^{Jt} \text{ is } L^1[0,\infty).$$

The following result is also well-known in principle and can be verified by direct computation.

THEOREM 2. If

$$P(t)e^{Jt} \text{ is in } L^1[0,\infty)$$

then the following statements are true concerning the solution of (3).

(a) $x(t)$ is bounded on $[0,\infty)$.

(b) As $n \to \infty$, then

$$x(t+nT) \to \int_{-\infty}^t P(t)e^{J(t-s)}P^{-1}(s)f(s)ds \overset{\text{def}}{=} y(t)$$

on $(-\infty,\infty)$ and $y(t)$ is a T-periodic function.

(c) $y(t)$ satisfies (3).

4. VOLTERRA CONVOLUTION SYSTEMS

Let A be an $n \times n$ matrix of real constants, $f: (-\infty,\infty) \to R^n$ be continuous and T-periodic, C be an $n \times n$ matrix of continuous functions which are $L^1[0,\infty)$, and consider the system

$$x' = Ax + \int_0^t C(t-s)x(s)ds + f(t), \quad x(0) = x_0. \tag{5}$$

In a fairly general setting, a necessary condition for a differential equation to have a T-periodic solution is that $x(t+T)$ be a solution whenever $x(t)$ is a solution; and that condition can not be verified for (5). Moreover, it is shown in [5] that (5) can not have a periodic solution unless it is reducible to an ordinary differential equation.

The solution of (5) is expressed by the variation of parameters formula

$$x(t) = Z(t)x_0 + \int_0^t Z(t-s)f(s)ds \tag{6}$$

where $Z(t)$ is the unique $n \times n$ matrix satisfying

$$Z' = AZ + \int_0^t C(t-s)Z(s)ds, \quad Z(0) = I.$$

Because of the form of (6), Theorems 1 and 2 extend trivially to (5) and we have the following result.

THEOREM 3. If Z and C are $L^1[0,\infty)$ then the following statements hold concerning solutions $x(t)$ of (5).
(a) $x(t)$ is bounded on $[0,\infty)$.
(b) As $n \to \infty$, then

$$x(t+nT) \to \int_{-\infty}^t Z(t-s)f(s)ds \overset{\text{def}}{=} y(t)$$

for $-\infty < t < \infty$ and y is a T-periodic function.
(c) $y(t)$ is a T-periodic solution of

$$y' = Ay + \int_{-\infty}^t C(t-s)y(s)ds + f(t)$$

which is called the limiting equation for (5).

Numerous conditions are known to ensure that $Z \in L^1[0,\infty)$. The Paley-Wiener result for integral equations was extended by Grossman and Miller [13] as follows.

THEOREM. Let $C \in L^1[0,\infty)$. Then $Z \in L^1[0,\infty)$ if and only if $\det [sI - A - L(C)] = 0$ has no solution for $\text{Re } s \geq 0$.

This is, of course, a transcendental equation. Some sufficient conditions and some necessary conditions for the determinant condition to hold have been given by Brauer [2] and by Jordan [15].

Frequently the property $Z \in L^1$ can be verified by Liapunov's direct method (cf. Chapters 2, 5, 6, and 7 of [3], [4], and [10]).

5. VOLTERRA NONCONVOLUTION SYSTEMS

We now consider the system

$$x' = B(t)x + \int_0^t D(t,s)x(s)ds + f(t), \quad x(0) = x_0 \tag{7}$$

in which B and D are $n \times n$ matrices of continuous functions, $f\colon (-\infty,\infty) \to R^n$ is continuous and T-periodic, $B(t+T) = B(t)$, $D(t+T, s+T) = D(t,s)$, and

$$\lim_{n\to\infty} \int_{-nT}^t |D(t,s)|ds = \int_{-\infty}^t |D(t,s)|ds$$

exists and is continuous on $(-\infty,\infty)$. Equation (7) is radically different from any of the previous ones because we are forced to go outside the system (7) to find the full solution in the variation of parameters formula

$$x(t) = R(t,0)x_0 + \int_0^t R(t,s)f(s)ds \tag{8}$$

where

$$\partial R(t,s)/\partial s = -R(t,s)B(s) - \int_s^t R(t,u)D(u,s)du, \quad R(t,t) = I. \tag{9}$$

For details see [3] or [16].

Now x(t) will be bounded if R(t,0) is bounded and if

$$\int_0^t |R(t,s)|ds$$

is bounded. In that case we are interested in the limit set of x(t). In each of the previous three equations which we have considered there was a variation of parameters formula of the form of (8) and the crucial property for proving the existence of a periodic limit set was that $R(t+T, s+T) = R(t,s)$. The theory can be unified because the same is true for (8). A proof is found in [7].

LEMMA. Let B and D be continuous with $B(t+T) = B(t)$ and $D(t+T, s+T) = D(t,s)$. The unique matrix R(t,s) satisfying (9) also satisfies $R(t+T, s+T) = R(t,s)$.

THEOREM 4. Let the conditions with (7) hold and suppose that $R(t,0) \to 0$ as $t \to \infty$ and

$$\int_0^t |R(t,s)|ds \le M$$

for some M and all $t \ge 0$. Then the following statements are true concerning the solution x(t) of (7).

(a) x(t) is bounded on $[0,\infty)$.

(b) As $n \to \infty$, then

$$x(t+nT) \to \int_{-\infty}^t R(t,s)f(s)ds \overset{\text{def}}{=} y(t)$$

and y(t) is T-periodic.

(c) y(t) is a solution of

$$y' = B(t)y + \int_{-\infty}^t D(t,s)y(s)ds + f(t),$$

called the limiting equation for (7).

To show

$$\int_0^t |R(t,s)|ds \le M$$

one may apply a theorem of Perron (cf. [14; p. 152]) as follows. Construct a Liapunov functional V for

$$x' = B(t)x + \int_0^t D(t,s)x(s)ds + g(t)$$

for arbitrary bounded continuous g and obtain $V' \le -\alpha|x| + K|g(t)|$ (cf. Chapters 2, 5, 6, and 7 of [3] and [10]) for α and K positive. Use a result in [9] to show x(t) bounded. From (8) we see

$$\int_0^t R(t,s)g(s)ds$$

is bounded for all such g. Perron's theorem now yields

$$\int_0^t |R(t,s)|ds \le M.$$

Nonlinear periodic results are found in [1], [3], [6], [8], [11], and [12].

REFERENCES

[1] Arino, O., Burton, T. A., and Haddock, J.R., Periodic solutions of functional differential equations, to appear.

[2] Brauer, F., Asymptotic stability of a class of integrodifferential equations, J. Differential Equations 28(1978) 180-188.

[3] Burton, T. A., Volterra Integral and Differential Equations (Academic Press, New York, 1983).

[4] ______, Structure of solutions of Volterra equations, SIAM Rev. 25(1983) 343-364.

[5] ______, Periodic solutions of linear Volterra equations, Fumkcial. Ekvacioj, to appear.

[6] ______, Periodic solutions of nonlinear Volterra equations, ibid, to appear.

[7] ______, Periodicity and limiting equations in Volterra systems, Bollettino UMI, to appear.

[8] ______, Periodic solutions of integrodifferential equations, to appear.
[9] Burton, T. A., Huang, Q., and Mahfoud, W. E., Liapunov functionals of convolution type, J. Math. Anal. Appl., to appear.
[10] Burton, T. A. and Mahfoud, W. E., Stability criteria for Volterra equations, Trans. Amer. Math. Soc. 279(1983) 143-174.
[11] Furumochi, T., Periodic solutions of periodic functional differential equations, Funkcial. Ekvacioj 24(1981) 247-258.
[12] ________, Periodic solutions of functional differential equations with large delays, ibid 25(1982) 33-42.
[13] Grossman, S. I. and Miller, R. K., Nonlinear Volterra integrodifferential equations with L^1-kernels, J. Differential Equations 13(1973) 551-566.
[14] Hale, J. K., Ordinary Differential Equations (Wiley, New York, 1969).
[15] Jordan, G. S., Asymptotic stability of a class of integrodifferential systems, J. Differential Equations 31(1979) 359-365.
[16] Miller, R. K., Nonlinear Volterra Integral Equations (Benjamin, Menlo Park, Calif., 1971).

The final (detailed) version of this paper has been submitted for publication elsewhere.

Trends in the Theory and Practice of Non-Linear Analysis
V. Lakshmikantham (Editor)

A NONLINEAR DIFFUSION EPIDEMIC SYSTEM WITH BOUNDARY FEEDBACK*

V. Capasso   L. Maddalena

Dipartimento di Matematica
Università di Bari
70100 BARI
Italy

An epidemic system is studied which is described mathematically by a non linear degenerate parabolic equation and a semilinear parabolic equation coupled via a positive feedback boundary condition of an integral type. Positivity properties are given and the asymptotic behaviour of the system is studied. Sufficient conditions for the existence and the asymptotic stability of a unique nontrivial endemic state are stated.

1. INTRODUCTION.

In the last decade a large amount of literature has been devoted to the analysis of reaction-diffusion systems which describe population dynamics, and in particular epidemic systems. More recently [5,6,8] special emphasis has been given to the nonlinear diffusion operator Δu^m, $m > 1$ to describe the spatial spread of a biological population. This operator, in contrast with the linear diffusion case Δu, includes, among others the interesting feature of a finite speed of propagation of any perturbation which arises in a bounded proper subset of the habitat Ω (usually an open bounded subset of $\mathbb{R}^n$). This is the main motivation of this paper in connection with the analysis of epidemic systems.

Another important feature we wish to stress here is the possibility that the coupling of the two equations may occur also via the boundary condition. In fact if we refer to the case of the spread of oral-fecal transmitted infectious diseases in habitats along the sea shores [2,4], the interaction of the infectious agent, and the human infective population, which may be assumed to be isolated, usually occurs through the sea shore, via an integral operator. This case has been studied with linear diffusion in [2]. Here we shall analyze the same system in presence of a nonlinear diffusion of the infectious agent.

In Section 2 the mathematical equations are presented; existence, uniqueness and comparison results are stated. In Section 3 the asymptotic behaviour is analyzed, and sufficient conditions are given for the existence and the asymptotic stability of a unique nontrivial steady state. For detailed proofs of the theorems we refer to [3].

(*) Work performed under the auspices of GNAFA-CNR (L.M.) and GNFM-CNR (V.C.) with the financial support of the Program "Control of Infectious Diseases" (CNR-Italy) and MPI-Italy.

2. THE MATHEMATICAL MODEL.

We consider the following system of a nonlinear diffusion equation coupled with a semilinear parabolic equation via the boundary:

$$(2.1)\quad \begin{cases} \dfrac{\partial u_1}{\partial t}(x;t) = \Delta u_1^m(x;t) - a_{11}u_1^p(x;t), & \text{in } \Omega \times (0,+\infty) \\ \dfrac{\partial u_2}{\partial t}(x;t) = d\Delta u_2(x;t) - a_{22}u_2(x;t) + g(u_1(x;t)), & \text{in } \Omega \times (0,+\infty) \end{cases}$$

$$(2.1b)\quad \frac{\partial}{\partial \nu} u_1^m(x;t) + \alpha(x)u_1^m(x;t) = \int_\Omega k(x,x')u_2(x';t)dx'; \quad \frac{\partial u_2}{\partial \nu} = 0, \quad \text{in } \partial\Omega \times (0,+\infty)$$

subject to the usual nonnegative initial conditions,

$$(2.1b)\quad u_i(x;0) = u_i^o(x) > 0, \qquad \text{in } \Omega, \ i = 1,2.$$

Here Ω is an open bounded domain in $\underline{\underline{R}}^n (n = 1,2,3,\ldots.)$, with a sufficiently smooth boundary $\partial\Omega$; Δ is the usual Laplace operator; $\frac{\partial}{\partial \nu}$ denotes the outward normal derivative on $\partial\Omega$; a_{11} and a_{22} are two positive constants. Moreover $\alpha : \partial\Omega \to \underline{\underline{R}}_+$ is assumed to be sufficiently smooth, and the kernel $: \partial\Omega \times \Omega \to \underline{\underline{R}}_+$ is a given nonnegative continuous function. $g : \underline{\underline{R}}_+ \to \underline{\underline{R}}_+$ is a twice continuosly differentiable function satisfing the following assumptions:

a) g is monotonically increasing;
b) $g(0) = 0$;
c) g is strictly concave in $(0,+\infty)$;
d) $0 < g'_+(0) < +\infty$
e) $\lim_{z\to+\infty} g(z)/z < a_{11}\, a_{22}\, /a_{12}$.

We shall assume that $m > 1$, and $p > m$, (the case $m = 1$ has been treated in [2]).

Remark 2.1. We wish to comment here that, if we set $k(x) := \int_\Omega k(x,x')dx'$, $x \in \partial\Omega$, we shall assume that the boundary $\partial\Omega$ is divided in two sections, $\partial_1\Omega$ and $\partial_2\Omega$ such that $\alpha(x) = k(x) = 0$, for $x \in \partial_1\Omega$; and $\alpha(x) > 0$, $k(x) > 0$, for $x \in \partial_2\Omega$. This is related to the physical meaning of $\partial_2\Omega$ as the sea-shore of the city, through which the positive feedback occurs; while on the rest of the boundary $\partial_1\Omega$ it is assumed complete isolation. We shall assume for simplicity that $0 < \bar{b} < \frac{\alpha(x)}{k(x)} < \bar{\bar{b}} < +\infty$.

We consider problem (2.1) in the Banach space $X = C(\bar{\Omega}) \times C(\bar{\Omega})$ of continuous functions, with the supremum norm $\|u\| = \|u_1\| + \|u_2\|$, where $\|u_i\| = \max_{x\in\bar{\Omega}} |u_i(x)|$, $i = 1,2$, for any $u = (u_1,u_2)' \in X$. We shall denote by X_+ the positive cone of X; i.e. $X_+ := \{u = (u_1,u_2)' \in X | \ u_i(x) \geq 0, \ i = 1,2\}$. In the sequel we shall make use of the following notations; if $u,v \in X$, we set $u \leq v$ iff $u(x) \leq v(x)$, for any $x \in \bar{\Omega}$; we set $u < v$ iff $u \leq v$ and $u \neq v$; we set $u \ll v$ iff $u_i(x) < v_i(x)$ for any $x \in \bar{\Omega}$. Thus X is an ordered Banach space with the partial order $\leq$. In the sequel we shall denote by $Q_T : \Omega \times (0,T)$ if $T > 0$.

It is well known that systems of the form (2.1) do not admit solutions in a classical sense but only in some weak sense. It can be shown that the following theorem holds, [3, 7].

Theorem 2.1. - If $u^o := (u_1^o,u_2^o)' \in X_+$ a unique $u(t) := (u_1(t), u_2(t))' \in X$, continuous in $t \in [0,+\infty)$ exists such that for any $T > 0$,

(i) the gradients $\nabla u_1^m(x;t)$ and $\nabla u_2(x;t)$ exist in Q_T such that

$$\int_{Q_T} [u_1^{2m} + (\nabla u_1^m)^2]\, dxdt < +\infty, \qquad \int_{Q_T} [u_2^2 + (\nabla u_2)^2]\, dxdt < +\infty$$

(ii) $$\int_\Omega u_1(x;T)\eta(x;T)dx = \int_\Omega u_1^o(x)\eta(x;0)dx + \int_{Q_T} [u_1\eta_t - \nabla u_1^m \nabla\eta - a_{11}u_1\eta]dxdt +$$
$$+\int_0^T dt \int_{\partial\Omega} dx\, \eta(x;t)\left(\int_\Omega k(x,x')u_2(x';t)dx' - \alpha(x)u_1^m(x;t)\right)$$

for any $\eta \in C^2(\bar{Q}_T)$.

(iii) $$\int_\Omega u_2(x;T)\eta(x;T)dx = \int_\Omega u_2^o(x)\eta(x;0)dx + \int_{Q_T} [u_2\eta_t - d\Delta u_2 \nabla_\eta - a_{22}u_2\eta + g(u_1)\eta]dxdt$$

for any $\eta \in C^2(\bar{Q}_T)$

Thanks to Theorem 2.1 it makes sense to define a solution of problem (2.1), (2.1b), (2.1o) corresponding to $u^o \in X$, as a function $u(t) \in X$, $t \in \underline{\underline{R}}_+$, such that (i), (ii), (iii) hold.

In the same way we may define a subsolution $\underline{u}(t) \in X$ (resp. a supersolution $\bar{u}(t) \in X$, continuous in $t \in \underline{\underline{R}}_+$ of problem (2.1), (2.1b), (2.1o) if (i), (ii), (iii) are satisfied with $\leq$ (resp. $\geq$) in place of $=$ in (ii), (iii); for any $\eta \in C^2(\bar{Q}_T)$, $\eta \geq 0$. With these definitions in mind one can prove [3] the following comparison theorem.

<u>Proposition 2.1.</u> - Let $\underline{u}(t)$, $\bar{u}(t) \in X$, continuous in $t \in \underline{\underline{R}}_+$, be a subsolution and a supersolution respectively of problem (2.1), (2.1b), (2.1o) such that $\underline{u}_i^o = \underline{u}_i(x;0) \leq \bar{u}_i(x;0) = \bar{u}_i^o$, in Ω, $i = 1,2$; then $\underline{u}(t) \leq \bar{u}(t)$, $t \in \underline{\underline{R}}_+$.

<u>Corollary</u> - If $u^o \in X_+$, then $u(t) \in X_+$, $t \in \underline{\underline{R}}_+$.

A more stringent result can be shown [3] ,

<u>Theorem 2.2.</u>- If $u^o \in X_+ - \{0\}$, then a $T^* > 0$ exists such that $u(t) \gg 0$ for $t > T^*$ (T^* depends in general upon u^o).

Due to Theorem 2.2 we may state that after some time $T^* > 0$ the solution $u(t)$, $t \in \underline{\underline{R}}_+$ of problem (2.1), (2.1b), (2.1o) will be positive in the whole $\bar{\Omega}$; hence system (2.1) is not any more degenerate after time T^*. It can then be shown [2] that the solution $u(t)$ is classical after T^*. In the sequel we will assume that this is the case; i.e.

$$u \in (C^{2,1}(\Omega \times (T^*, +\infty), \underline{\underline{R}}) \cap C^{1,0}(\bar{\Omega} \times (T^*, +\infty), \underline{\underline{R}}))^2$$

The steady states of system (2.1), (2.1b) are the solutions of the following nonlinear elliptic system

$$(2.2) \qquad \begin{cases} 0 = \Delta u_1^m - a_{11}u_1^p, & \text{in } \Omega \\ 0 = d\Delta u_2 + g(u_1) - a_{22}u_2, & \text{in } \Omega \end{cases}$$

with boundary condition

(2.2b) $$\frac{\partial}{\partial\nu} u_1^m + \alpha u_1^m = k * u_2, \qquad \frac{\partial u_2}{\partial\nu} = 0 \qquad \text{in } \partial\Omega$$

We shall say that $\underline{u} \in X$, $\underline{u} = (\underline{u}_1, \underline{u}_2)' \gg 0$ (resp. $\overline{u} \in X$, $\overline{u} = (\overline{u}_1, \overline{u}_2)' \gg 0$) is a subsolution (resp. supersolution) of system (2.2)(2.2b) iff (2.2), (2.2b) are satisfied in a classical sense with $\leq$(resp. $\geq$) instead of the equality (here we denote by $[k * u_2](x) = \int_\Omega k(x,x')u_2(x')dx'$).

If we denote by $u(t,u^o)$ the solution of problem (2.1), (2.1b), (2.1o) with initial condition $u^o \in X$, due to the comparison theorem (Prop. 2.1) we may state the following:

Theorem 2.3. -Let $\underline{u}$ (resp. $\overline{u}$) be a subsolution (resp. a supersolution) of the nonlinear elliptic system (2.2), (2.2b) with $0 \ll \underline{u} < \overline{u}$. Then $u(t,\underline{u})$ is nondecreasing in $t \in \underline{\underline{R}}_+$, (while $u(t,\overline{u})$ is nondecreasing in $t \in \underline{\underline{R}}_+$) and converges to a solution $0 \ll \phi \in X_+$, of system (2.2), (2.2b), in the sup-norm.

We wish now to point out that, if we refer to the strictly positive solutions of system (2.2), (2.2b), this problem is "equivalent" to the elliptic system

(2.3) $$\begin{cases} 0 = \Delta v_1 - a_{11} v_1^{p/m} , & \text{in } \Omega \\ 0 = d\Delta v_2 + g(v_1^{1/m}) - a_{22}v_2 , & \text{in } \Omega \end{cases}$$

with boundary condition:

(2.3b) $$\frac{\partial}{\partial\nu} v_1 + \alpha v_1 = k * v_2, \qquad \frac{\partial v_2}{\partial\nu} = 0 , \qquad \text{in } \partial\Omega$$

in the following sense: if $0 \ll u = (u_1, u_2)' \in X_+$ is a solution of system (2.2), (2.2b), then $0 \ll v = (v_1, v_2)' \in X_+$ is a solution of system (2.3), (2.3b) if $v_1 = u_1^m$, $v_2 = u_2$ and viceversa if $u_1 = v_1^{1/m}$, $u_2 = v_2$.

Observe that the same holds for sub- and supersolutions.

Referring now to the elliptic system (2.3), (2.3b) it can be shown by classical methods [9] that whenever a nontrivial subsolution $\underline{v}$ and a supersolution $\overline{v}$ exist such that $\underline{v} \leq \overline{v}$ then a strictly positive minimal solution and a maximal solution of such system exist. Moreover due to the concavity of g it can also be shown [2] that these two solutions must coincide; this unique solution will obviously be strictly positive. We may finally state the following.

Theorem 2.4. - If a subsolution $\underline{v}$ and a supersolution $\overline{v}$ of problem (2.3), (2.3b) exist such that $0 \ll \underline{v} \leq \overline{v}$, then a unique solution $\psi \gg 0$ exists for problem (2.2), (2.2b) .

3. ASYMPTOTIC BEHAVIOUR.

Theorems 2.3 and 2.4 will now be applied to show the asymptotic behaviour of the solutions of system (2.1), (2.1b). To this aim we need to construct a couple of strictly positive sub- and super-solutions of the system (2.2), (2.2b). Consider the following linear eigenvalue problem

$$(3.1)\quad \begin{cases} \Delta\phi_1 - \varepsilon^{p/m-1}\phi_1 = \lambda\phi_1 , & \text{in } \Omega \\ d\Delta\phi_2 - g(\varepsilon^{1/m})\varepsilon^{-1}\phi_1 - a_{22}\phi_2 = \lambda\phi_2 , & \text{in } \Omega \end{cases}$$

with the boundary condition

$$(3.1b)\quad \frac{\partial\phi_1}{\partial\nu} + \alpha\phi_1 = k * \phi_2, \quad \frac{\partial\phi_2}{\partial\nu} = 0 , \qquad \text{in } \partial\Omega$$

for some $\varepsilon > 0$.

In [2]it was shown that the eigenvalue problem (3.1), (3.1b) admits a dominant eigenvalue $\lambda_1 \in \mathbb{R}$; the associated eigenvector $\phi = (\phi_1, \phi_2)' \in X$ can be chosen to be strictly positive and with sup-norm $\|\phi\| = 1$. These facts allow us to state the following main theorem [3].

Theorem 3.1. - If, for some $\varepsilon > 0$, the eigenvalue problem (3.1), (3.1b) admits a positive eigenvalue, then a strictly positive equilibrium solution of system (2.1), (2.1b) exists which is globally asymptotically stable in $X_+ - \{0\}$.

Proof.- Let $\lambda_1 > 0$ be the positive eigenvalue of the problem (3.1), (3.1b) and $\phi_1 = (\phi_1, \phi_2)'$ the associated eigenvector. For any $\rho \in (0,\varepsilon)$ $\rho\phi$ is a subsolution of system (2.3), (2.3b). Hence $\psi = ((\rho\phi_1)^{1/m}, \phi_2)'$ is a subsolution of (2.2), (2.2b). On the other hand it is always possible to choose a $\xi = (\xi_1, \xi_2)'$, $\xi \gg 0$ such that

$$\frac{1}{a_{22}} g(\xi_1^{1/m}) < \xi_2 < \frac{\alpha(x)}{k(x)} \xi_1^m, \qquad x \in \partial\Omega$$

thus being a supersolution of system (2.2), (2.2b).
Now if the initial state $u^0 \in X_+$ is nontrivial, a $T^* > 0$ exists such that $u(t, u^0) \gg 0$ for any $t \gg T^*$.
We can then choose a subsolution $\psi \gg 0$ and a supersolution ξ of (2.2), (2.2b) such that

$$\psi \leqslant u(T^*, u^0) \leqslant \xi$$

Hence

$$u(t,\psi) \leqslant u(t+T^*, u^0) \leqslant u(t,\xi), \quad t > 0$$

from which the theorem follows.

References.

[1] D.G. Aronson, M.G. Crandall and L.A. Peletier: Stabilization of solutions of a degenerate nonlinear diffusion problem, Nonlinear Analysis T.M.A., 6 (1982), 1001-1022.

[2] V. Capasso, K. Kunish and W. Shappacher: A reaction-diffusion system with positive boundary feedback. Application to a class of epidemic systems, to appear (1984).

[3] V. Capasso and L.Maddalena: A degenerate nonlinear diffusion system with boundary feedback, to appear.

[4] V. Capasso and L. Maddalena: Convergence to equilibrium states for a reaction-diffusion system modelling the spatial spread of a class of bacterial and viral diseases, J. Math. Biology, 13 (1981), 173-184.

[5] W.S.C. Gurney and R.M. Nisbet: The regulation of inhomogeneous populations, J. Theor. Biol., 52 (1975), 441-457.

[6] E. Gurtin and R.C. MacCamy: On the diffusion of biological populations, Math. Biol., 33 (1977), 35-49.

[7] L. Maddalena: Existence, uniqueness and qualitative properties of the solution of a degenerate nonlinear parabolic system, to appear.

[8] A. Okubo:Diffusion and Ecology Problems: Mathematical Models, Springer-Verlag Heidelberg, 1980.

[9] P. Puel: Existence, comportment a l'infinì et stabilitè dans certain problems quasilinéaire elliptiques et paraboliques d'ordre 2, Ann. Sc. Norm. Sup. Pisa 3 (1976), 89-119.

The final (detailed) version of this paper will be submitted for publication.

Trends in the Theory and Practice of Non-Linear Analysis
V. Lakshmikantham (Editor)

APPLICATIONS OF ADMISSIBILITY FOR VOLTERRA INTEGRAL EQUATIONS*

Herminio Cassago, Jr.

Department of Mathematics
The University of Texas at Arlington
Arlington, Texas 76019
U.S.A.

and

Instituto de Ciências Matemáticas de São Carlos
U.S.P.
Brasil

1. INTRODUCTION

The question of admissibility for differential equation was introduced by Massera-Schäffer [6] who considered the systems

(1.1) $$\dot{x} = A(t)x$$

(1.2) $$\dot{x} = A(t)x + b(t)$$

Corduneanu in [1] extended the result of Massera-Schäffer by considering more general perturbations, namely, by considering

(1.3) $$\dot{x} = A(t)x + f(t,x(t)).$$

Later Corduneanu in [3] extended the above results to the case of abstract equations, namely, he considered the equations

(1.4) $$Lx = g$$

(1.5) $$Lx = g + f.$$

He needed an extension of the concept of admissibility, so he utilized the one defined by Cushing in [4], [5]. Moreover, he established sufficient conditions for existence of a family F of solutions in a given Banach space D.

In this paper we will use this results to Volterra integral equation given by

(1.6) $$x(t) = \int_0^t k(t,s)x(s)ds + x_0 + f(t,x(t)).$$

We use the family F of solutions for (1.6) to obtain that the set F is homeomorphic a some subspace of $\mathbb{R}^n$. Moreover, we study the relationship between two of these equations, where we utilize the dimension of the subspaces of solutions, which contain the asymptotic equivalence as particular cases. These results extend a previous result of the author in [7].

*This research was supported in part by "Conselho Nacional de Pesquisas" - Brasil- under proc. 20.2059/83-MA.

2. PRELIMINARIES

The symbol $\|\cdot\|$ will denote a norm in $X = \mathbb{R}^n$. $C_c(J,X)$ denotes the space of all continuous functions from $J = [0,\infty)$ into X, with the topology of uniform convergence on any compact subset of J. If B is a Banach space, $|\cdot|_B$ denotes its norm. We say that a Banach space B is stronger than $C_c(J,X)$ if B is algebraically contained in $C_c(J,X)$ and convergence in B implies convergence in $C_c(J,X)$.

In the sequel, all Banach spaces considered are implicitly assumed to be stronger than $C_c(J,X)$.

We consider the Volterra integral equations

$$(2.1) \qquad (Lz)(t) = z_0$$

$$(2.2) \qquad (Lw)(t) = b(t) + w_0$$

$$(2.3) \qquad (Lx)(t) = f(t,x(t)) + x_0$$

where $(Lu)(t) = u(t) - \int_0^t k(t,s)u(s)ds$; $z,w,x \in X$; $z_0,w_0,x_0 \in X$ are constant vectors; $k(t,s)$, $n\times n$, continuous matrix for $s,t \in J$; $b(\cdot) \in C_c(J,X)$ and $f(\cdot,x(\cdot)) \in C_c(J,X)$ whenever $x(\cdot) \in C_c(J,X)$.

A pair (B,D) of Banach spaces is *admissible* with respect to equation (2.2) if, for every $b \in B$, there exists at least one $w_0 \in \mathbb{R}^n$, such that (2.2) has its solution in D.

If D is a Banach space, we say that x is a D-solution of some equation when $x \in D$ and is a solution. X_{0D} denotes the linear subspace of initial values $z(0)$ of D-solutions of (2.1). X_{1D} represents any subspace of X complementary of X_{0D}.

Now we write the Corduneanu's results given in [3] for our case.

Lemma 2.1. Let (B,D) be admissible for (2.2) and assume that $X = X_{0D} \oplus X_{1D}$. Then (2.2) has a unique D-solution for each $b \in B$ such that $w_0 \in X_{1D}$.

This Lemma leads us to define two operators:

The first operator A from B onto X_{1D}, given by $Ab = \xi_1$, where ξ_1 is the unique element of X_{1D} such that $Lw = b + \xi_1$ has its solution in D. It is easy to see that A is a linear operator.

The second operator T from B into D, given by $Tb = x$, with $Lx = b + Ab$. The linearity of the operator T follows from the linearity of L and A.

The admissibility theory for equations of the form (2.2) is closely related to the continuity properties of both operators A and T.

Lemma 2.2. Let (B,D) be admissible for (2.2). Then both operators A and T are continuous.

The next lemma is analogous to Massera-Schäffer's result for differential equations.

Lemma 2.3. Let (B,D) be admissible for (2.2) then the D-solution of (2.2), corresponding to a pair (b,w_0), with $b \in B$ and $w_0 \in X_{0D}$, satisfies the inequality

$$|w|_D \leq C_0 \|w_0\| + K|b|_B$$

where C_0 and K are independent of (b,w_0).

In the sequel, if (B,D) is admissible for (2.2), C_0 and K always represent the constants of Lemma 2.3. Given a number $\rho > 0$, let $S_{D,\rho} = \{\phi \in D: |\phi|_D \leq \rho\}$. The next theorem is the generalization of Theorem 1 in [1].

Theorem 2.4. Suppose that the equation

$$(2.4) \qquad (Lx)(t) = f(t,x(t)) + \xi_0 + A(f(t,x(t))); \quad \xi_0 \in X_{0D}$$

has the following properties:

(i) (B,D) is admissible for (2.2)

(ii) There exists $\rho > 0$ so that $f(\cdot,x(\cdot)) \in B$ if $x(\cdot) \in S_{D,\rho}$

(iii) There exists λ, $0 < \lambda < K^{-1}$, such that

$$(2.5) \qquad |f(\cdot,x(\cdot))-f(\cdot,y(\cdot))|_B \leq \lambda|x(\cdot)-y(\cdot)|_D, \text{ for all } x(\cdot),y(\cdot) \in S_{D,\rho}.$$

If $\|\xi_0\|$ and $|f(\cdot,0)|_B$ are sufficiently small such that

$$(2.6) \qquad C_0\|\xi_0\| + K|f(\cdot,0)|_B \leq (1-\lambda K)\rho,$$

then, there exists a unique solution $x \in S_{D,\rho}$ of equation (2.5).

3. MAIN RESULTS

Under the hypothesis of Theorem 2.4, if $\sigma > 0$ is a number such that $C_0\sigma + K|f(\cdot,0)|_B \leq \rho(1-\lambda K)$, for each $\xi_0 \in V_\sigma = \{\xi \in X_{0D}: \|\xi\| < \sigma\}$, we represent by $x(\cdot,\xi_0)$ the unique D-solution of (2.4) in $S_{D,\rho}$. Let $F = \{x(\cdot,\xi) \in D: \xi \in V_\sigma\}$ provided with the induced topology of D. The following theorem connects V_σ with F and its section $F(0) = \{x(0,\xi) \in X: x(\cdot,\xi) \in F\}$.

Theorem 3.1. Suppose that the conditions of Theorem 2.4 are satisfied. Let $\sigma > 0$ be a number such that $C_0\sigma + K|f(\cdot,0)|_B \leq \rho(1-\lambda K)$. Then the mappings:

$$U : \xi \in V_\sigma \longrightarrow x(\cdot,\xi) \in F$$
$$U_0 : x(\cdot,\xi) \in F \longrightarrow x(0,\xi) \in F(0)$$
$$H : x(0,\xi) \in F(0) \longrightarrow \xi \in V_\sigma$$

are homeomorphisms.

Proof. U is continuous. In fact, for each $\xi_0 \in V_\sigma$ and each $u \in S_{D,\rho}$, Lemma 2.3 ensures that the unique solution x of $(Lx)(t) = f(t,u(t)) + \xi_0 + A(f(t,u(t)))$ satisfies: $|x(\cdot)|_D \leq C_0\|\xi_0\| + K|f(\cdot,u(\cdot))|_B$. Now we define an operator M_{ξ_0} on $S_{D,\rho}$ by $x = M_{\xi_0}u$. The condition (iii) of Theorem 2.4 implies that M_{ξ_0} is a uniform contraction with respect to ξ_0. It follows from Lemma 2.3 that $M_{\xi_0}u$ is continuous in ξ_0 for each fixed u in $S_{D,\rho}$. Relations (2.5) and (2.6) imply that $M_{\xi_0}S_{D,\rho} \subset S_{D,\rho}$. Thus by uniform contraction principle, the fixed point

$x(\cdot,\xi_0)$ of M_{ξ_0} depends continuously on ξ_0.

The mapping H is continuous as a restriction of the projection from X onto X_{OD}.

The continuity of U_0 follows from the fact: D is stronger than $C_c(J,X)$. A simple analysis on the diagram

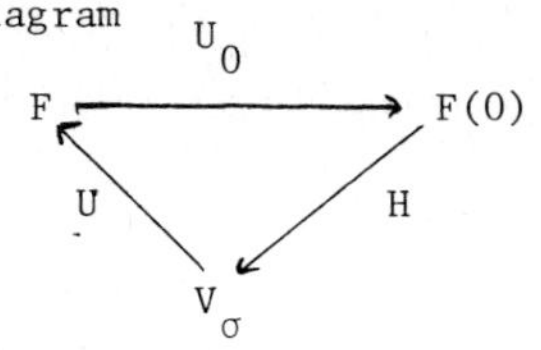

shows that each mapping is invertible, its inverse being the composition of the other two. This proves Theorem 3.1.

Consider now the equations

(3.1) $$(Lx)(t) = f_1(t,x(t)) + \xi_0 + A(f_1(t,x(t))), \quad \xi_0 \in X_{OD},$$

(3.2) $$(Ly)(t) = f_2(t,y(t)) + \eta_0 + A(f_2(t,y(t))), \quad \eta_0 \in X_{OD}.$$

Definition 3.2. Let p and q integers such that $0 \leq p \leq n$, $0 \leq q \leq n$. Equations (3.1) and (3.2) are (p,q)-related with respect to the ordered pair $[D_1,D_2]$ of Banach spaces if:

(i) There exists a family of F_p of D_1-solutions of (3.1) which depends upon p parameters;

(ii) For each solution $x = x(t)$ of (3.1) in F_p, there corresponds a family G_q of solutions $y = y(t)$ of (3.2) which depends upon q parameters;

such that the family of difference $x-y$ is bounded in D_2 and is homeomorphic to subspace q-dimensional.

We adopt the convention that a family which depends upon 0 parameters must consist of at least one member.

Remark. If $\dim X_{OD} = p$, Theorem 2.4 implies the existence of a family F_p of D-solutions of equation (3.1) which depends upon p parameters.

Theorem 3.3. Suppose that equations (3.1) and (3.2) satisfy the following conditions:

(a) (B,D_i) is admissible for (2.2), $i = 1,2$;

(b) There exist $\rho_i > 0$ such that $f_i(\cdot,u_i(\cdot)) \in B$ if $u_i(\cdot) \in S_{D_i,\rho_i}$, $i=1,2$;

(c) There exist λ_i, $0 < \lambda_i < K^{-1}$, such that

(3.3) $$|f_1(\cdot,x(\cdot))-f_1(\cdot,y(\cdot))|_B \leq \lambda_1|x(\cdot)-y(\cdot)|_{D_1}, \text{ for all } x(\cdot),y(\cdot) \in S_{D_1,\rho_1}$$

(3.4) $$|f_2(\cdot,u(\cdot)+x(\cdot))-f_2(\cdot,w(\cdot)+x(\cdot))|_B \leq \lambda_2|u(\cdot)-w(\cdot)|_{D_1\cap D_2},$$

for all $u(\cdot),w(\cdot) \in S_{D_1\cap D_2,\rho_2}$, $x(\cdot) \in S_{D_1,\rho_1}$;

(d) $\dim X_{OD_1} = p$ and $\dim X_{OD_1\cap D_2} = q$.

If $\alpha_i = |f_i(\cdot,0)|_B$ and σ are sufficiently small so that $C_0\sigma + K\alpha_i \leq (1-\lambda_i K)\rho_i$, then the equations (3.1) and (3.2) are (p,q)-related with respect to $[D_1, D_1 \cap D_2]$.

Proof. Theorem 2.4 implies the existence of a family F_p of D_1-solutions of (3.1) which depends upon p parameters. Let $\bar{x} \in F_p$, the change of variable $u = y-\bar{x}$ leads to the equation

$$(3.5) \qquad (Lu)(t) = F(t,u(t)) + u_0 + A(F(t,u(t)))$$

where $F(t,u) = f_2(t,u+\bar{x}) - f_1(t,\bar{x})$ and $u_0 = \eta_0 - \xi_0$. Theorem 2.4 now implies that there exists a family $\tilde{G}_q$ of $D_1 \cap D_2$-solutions of (3.5) which depends upon q parameters, and Theorem 3.1 implies that $\tilde{G}_q$ is homeomorphic to subspace q-dimensional. For each $u \in \tilde{G}_q$, $y = u + \bar{x}$ is a solution of equation (3.2) which depends on q parameters, such that $y - x$ is in $S_{D_1 \cap D_2, \rho_2}$. This completes the proof of Theorem 3.3.

ACKNOWLEDGEMENT

The author wishes to thank Professors C. Corduneanu and S. R. Bernfeld for their various suggestions.

REFERENCES

[1] Corduneanu, C., Sur certains systems differentiels non-lineaires, An. Sti. Univ. "Al. I. Cuza." Iasi, Sect, I, 6 (1960) 257-260.

[2] ________, Integral Equations and Stability of Feedback Systems (Academic Press, New York, 1973).

[3] ________, Bounded Solutions for certain systems of differential or functional equations - 1º Gruppo di Seminari e Instituti Matematici Italiani (1978).

[4] Cushing, J. M., Admissible operators and solutions of perturbed operator equations, Funk. Ekvacioj 19 (1976) 79-84.

[5] ________, Strongly admissible operators and Banach space solutions of non-linear equations, Funk. Ekvacioj 20 (1977) 237-245.

[6] Massera, J. L. and Schäffer, J. J., Linear differential equations and functional Analysis I, Ann. of Math. 67 (1958) 517-573.

[7] Onuchic, N. and Cassago Jr., H. Asymptotic behavior at infinity between the solutions of two systems of ordinary differential equations, J. Math. An. and Appl., to appear.

[8] Onuchic, N. and Táboas, P. Z., Qualitative properties of nonlinear ordinary differential equations, Proceesings of the Royal Edinburgh, 79A (1977) 79-85.

The final (detailed) version of this paper will be submitted for publication elsewhere.

Trends in the Theory and Practice of Non-Linear Analysis
V. Lakshmikantham (Editor)

MULTIPLE SOLUTIONS FOR A DIRICHLET PROBLEM WITH JUMPING NONLINEARITIES

Alfonso Castro and R. Shivaji

Department of Mathematics
Southwest Texas State University
San Marcos, Texas 78666
U.S.A.

Here we consider the Dirichlet problem

$$-u''(x) = g(u) - c, \quad x \in (0,1) \qquad (*)$$
$$u(0) = u(1) = 0$$

where $c > 0$ (constant) and $g(u)$ satisfy

$$\lim_{u\to-\infty}(g(u)/u) = m < \infty, \quad \lim_{u\to\infty}(g(u)/u) = \infty. \qquad (A)$$

Also for the sake of simplicity in the computations in section 3 we assume

$$g''(u) \geq 0 \text{ and } g'(0) \geq 0. \qquad (B)$$

We prove that for such a class of g's (*) has multiple solutions when c is large enough. In particular, our results hold when $g(u) = \exp(u)$ or when $g(u) = (u + \alpha)^\rho$ for $u \geq 0$, where $\rho > 1$ and $\alpha \geq 0$.

1. INTRODUCTION

In the spirit of extending the classical results of [1], in [7] A. Lazer and J. Mckenna studied problem (*) under the assumption that $m < \pi^2$ and $n^2\pi^2 < g(u)/u < (n + 1)^2\pi^2$ for u large and positive. In this paper we treat the case when $g(u)/u$ "jumps" over infinitely many eigenvalues. Namely we prove the following two results:

Theorem 1.1 <u>If</u> $m < 2^{\frac{1}{2}}\pi^2$, <u>then there exists</u> $\{c_n\}$ <u>such that</u> $\{c_n\}$ <u>is increasing and if</u> $c > c_n > 0$ <u>then</u> (*) <u>has</u> $2n + 1$ <u>solutions</u> v_j, $j = 1, 2, \ldots, n$; z_j, $j = 0, 1, \ldots, n$ <u>where</u> $v_j'(0) > 0$, $z_j'(0) < 0$ <u>and both</u> v_j <u>and</u> z_j <u>have</u> j <u>interior zeroes</u>.

Theorem 1.2 <u>If</u> $m \geq 2^{\frac{1}{2}}\pi^2$, <u>then given</u> k <u>there exists</u> $c_k > 0$ <u>such that if</u> $c > c_k$ <u>then</u> (*) <u>has at least</u> k <u>solutions</u>.

Our proofs are based on the so called quadrature method introduced for positive solutions in [5] and extended in [8] for other solutions. We describe relevant aspects of the method in section 2 and prove the theorems in section 3.

For the case when $m = \infty$ see [2] and references cited therein. For higher dimensional related results see [4], [6] and [3], among others.

2. QUADRATURE TECHNIQUE

Let us rename $g(u) - c = f(u)$. Multiplying (*) by u' and integrating we have

$$2F(u) + (u')^2 = \text{constant}, \tag{2.1}$$

where

$$F(u) = \int_0^u f(t)\,dt. \tag{2.2}$$

Since (*) is autonomous, it follows easily that if u(x) is a solution so is u(1 - x). Further we have

Lemma 2.1 _If_ u(x) _is_ _a_ _solution_ _of_ (*) _such_ _that_ u'(a) = 0 _then_ u(a - x) = u(a + x).
Proof: Let w(x) = u(a - x) and z(x) = u(a + x). Since both w and z satisfy -v" = f(v) with v(0) = u(a), v'(0) = 0, then the result follows.

Non-positive solutions. By Lemma 2.1 it follows that if u is a non-positive solution of (*) then u is symmetric with respect to $x = \frac{1}{2}$ (see fig. A). Hence from (2.1) we have

$$-u'(x) = (2(F(q) - F(u)))^{\frac{1}{2}}, \quad x \in [0,\tfrac{1}{2}], \quad u(\tfrac{1}{2}) = q. \tag{2.3}$$

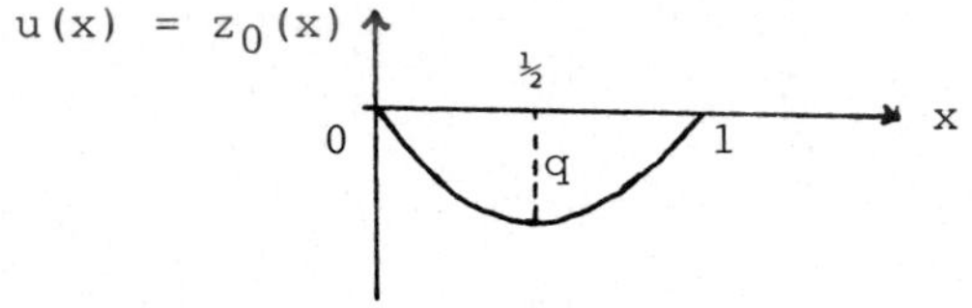

Figure A

Therefore from (2.3)

$$-\int_0^{u(x)} (F(q) - F(s))^{-\frac{1}{2}} ds = 2^{\frac{1}{2}}x, \quad x \in [0,\tfrac{1}{2}] \tag{2.4}$$

and so q has to satisfy

$$L(q) \equiv -\int_0^q (F(q) - F(s))^{-\frac{1}{2}} ds = 2^{-\frac{1}{2}}. \tag{2.5}$$

Thus, if there exists a q < 0 satisfying (2.5) then there exists a non-positive solution u(x) for (*) given by (2.4). We will prove the existence of such a q in section 3.

Solutions with one interior zero. We shall consider solutions u(x) of the form $v_1(x)$ (fig. B). Study of solutions of the form $z_1(x)$ (fig. B) follow immediately since $z_1(x) = v_1(1 - x)$.

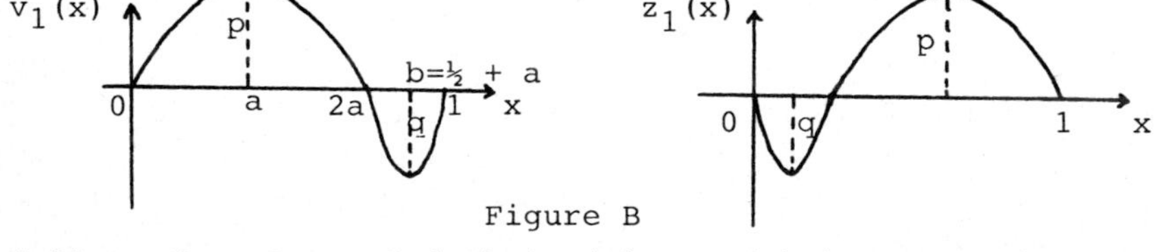

Figure B

It follows from Lemma 2.1 that $u(x) = v_1(x)$ is symmetric about a and $\frac{1}{2} + a$, where u(2a) = 0. Also from (2.1) we have

$$F(p) = F(q) \tag{2.6}$$

where u(a) = p, $u(\frac{1}{2} + a) = q$. Furthermore

$$2^{\frac{1}{2}}x = \int_0^{u(x)} (F(p) - F(s))^{-\frac{1}{2}}ds, \quad x \in [0,a] \tag{2.7}$$

$$2^{\frac{1}{2}}(1 - x) = \int_{u(x)}^{0} (F(q) - F(s))^{-\frac{1}{2}}ds, \quad x \in [\tfrac{1}{2} + a,1]. \tag{2.8}$$

Hence p and q must satisfy also

$$G_1(p,q) \equiv \int_0^p (F(p) - F(s))^{-\frac{1}{2}}ds - \int_0^q (F(q) - F(s))^{-\frac{1}{2}}ds$$
$$= 2^{-\frac{1}{2}}. \tag{2.9}$$

The above equation follows by adding (2.7) evaluated at $x = a$ and (2.8) evaluated at $x = \frac{1}{2} + a$. Therefore, if there exists $p > 0$ and $q < 0$ satisfying (2.6) and (2.9) then $u(x) = v_1(x)$ will be given by (2.7)-(2.8). We shall prove the existence of such a p, q when c is large enough in section 3.

<u>Solutions with j interior zeroes</u>. For solutions with j interior zeroes the analysis is quite similar. However when j is even we no longer have $z_j(x) = v_j(1 - x)$. In fact when j is even $v_j(1 - x) = v_j(x)$.

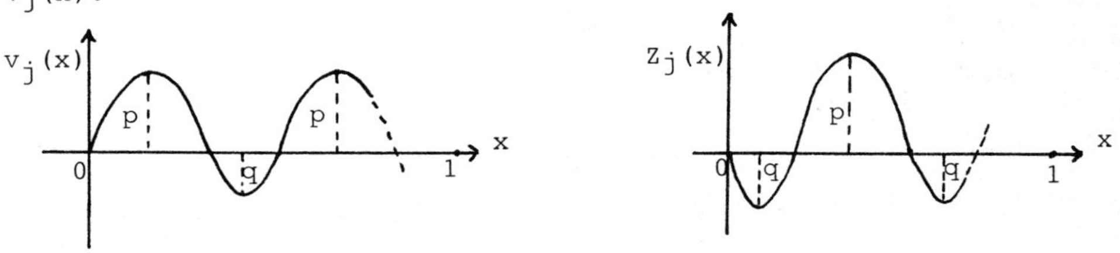

Figure C

It turns out that v_j's and z_j's will exists provided there exist $p > 0$ and $q < 0$ satisfying (2.6) and

$$G_j(p,q) \equiv P_j \int_0^p (F(p) - F(s))^{-\frac{1}{2}}ds - Q_j \int_0^q (F(q) - F(s))^{-\frac{1}{2}}ds$$
$$= 2^{-\frac{1}{2}}, \tag{2.10}$$

where $P_j = Q_j = \frac{1}{2}(j + 1)$ for j odd, and $P_j = \frac{1}{2}j + 1$, $Q_j = \frac{1}{2}j$ (for the v_j's), $P_j = \frac{1}{2}j$, $Q_j = \frac{1}{2}j + 1$ (for the z_j's) for j even. We shall prove the existence of such a p, q when c is large enough in section 3.

3. PROOF OF THEOREMS.

Since c is to be chosen large enough we may assume that $f(0) < 0$. Because of (A)-(B), F(u) has the form

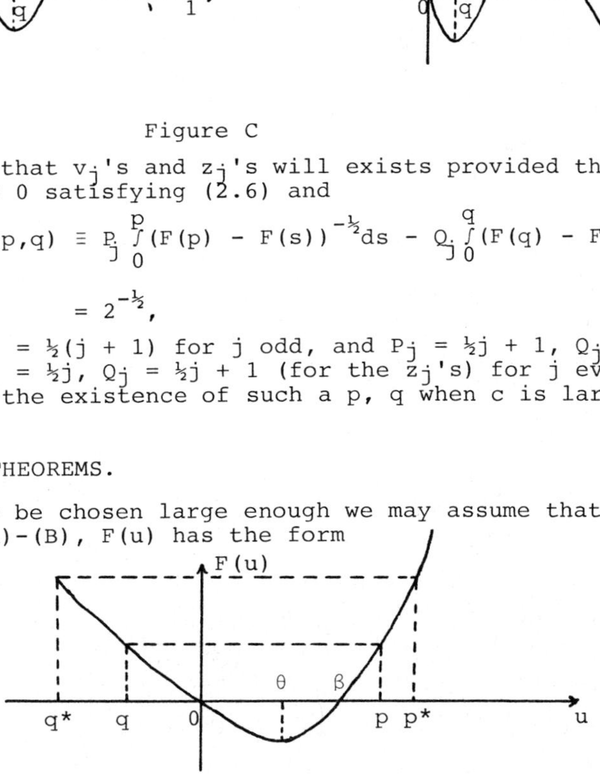

Figure D

where $f(\theta) = 0$, $F(\beta) = 0$. In the case $m \geq 0$ we set $q^* = -\infty$, while if $m < 0$ we set q^* to be the largest negative number such that $f(q^*) = 0$. Let

$$I(p) \equiv \int_0^p (F(p) - F(s))^{-\frac{1}{2}} ds, \quad p \in [\beta, p^*],$$

where $p^* = \infty$ if $m \geq 0$ while p^* is such that $F(p^*) = F(q^*)$ if $m < 0$. We also let

$$J(q) \equiv \int_0^q (F(q) - F(s))^{-\frac{1}{2}} ds, \quad q \in [q^*, 0].$$

Now we shall prove properties of $I(p)$ and $J(q)$ from which our theorems easily follow.

Imitating the proof of Theorem 2.9 and Theorem 3.1 of [5], it follows that

$$\lim_{q \to 0} J(q) = 0, \tag{3.1}$$

$$\lim_{q \to -\infty} -J(q) = \pi^2/m \qquad \text{if } m \geq 0, \tag{3.2}$$

$$\lim_{q \to q^*} -J(q) = \infty \qquad \text{if } m < 0. \tag{3.3}$$

Lemma 3.1. <u>For</u> I, β, <u>and</u> c <u>as above we have</u>

$$\frac{d[I(\beta)]}{dc} < 0.$$

Proof: From the definition of I we have

$$I'(\beta) = \int_0^1 (H(\beta) - H(\beta v))[-F(\beta v)]^{-3/2} dv,$$

where $H(t) = F(t) - \frac{1}{2} t f(t)$. Since $H'(t) = \frac{1}{2}[f(t) - tf'(t)]$, $H''(t) = -\frac{1}{2} t f''(t)$, and $f''(t) \geq 0$, we infer that $H(\beta) \leq H(\beta v)$ for $0 \leq v \leq 1$. Hence $I'(\beta) < 0$. Since

$$F(\beta) = \int_0^\beta g(s) ds - c\beta = 0,$$

differentiating with respect to c we have $(d\beta/dc) = \beta/f(\beta) > 0$ which proves the result.

Lemma 3.2. $\lim_{c \to \infty} I(\beta) = 0$.

Proof: Because of (B) we have $-F(s) \geq -F(\theta)s/\theta$ for $s \in [0, \theta]$, and $-F(s) \geq -F(\theta)(\beta - s)/(\beta - \theta)$ for $s \in [\theta, \beta]$. Thus

$$\begin{aligned} I(\beta) &= \int_0^\beta (-F(s))^{-\frac{1}{2}} ds \\ &\leq (\theta/-F(\theta))^{-\frac{1}{2}} [\int_0^\theta s^{-\frac{1}{2}} ds] + ((\beta - \theta)/-F(\theta))^{-\frac{1}{2}} [\int_\theta^\beta (\beta - s)^{-\frac{1}{2}} ds] \\ &= 2\beta(-F(\theta))^{-\frac{1}{2}}. \end{aligned} \tag{3.4}$$

By the convexity of g and the definition of θ we have $-F(\theta) \geq \frac{1}{2}[c - g(0)]\theta$. This combined with (3.4) gives

$$I(\beta) \leq 8^{\frac{1}{2}} \beta(\theta[c - g(0)])^{-\frac{1}{2}}. \tag{3.5}$$

An elementary calculation using the convexity of g shows that $\beta < 3\theta$. Also since $c = g(\theta)$, $I(\beta) \leq 9(\theta/(g(\theta) - g(0))^{\frac{1}{2}}$. Therefore by hypothesis (A) the claim of the lemma follows.

Now we prove Theorems 1.1 and 1.2. We choose p and q so that they satisfy (2.6), namely $F(p) = F(q)$. From (3.1) and Lemma 3.2 we have $G_j(\beta,0) < 1/\sqrt{2}$ when c is chosen larger than some c_j. By Lemma 3.1 the c_j's are non-decreasing. From (3.2)-(3.3) we have

$$\lim_{q\to q^*} G_j(p,q) > 1/\sqrt{2} \quad \text{if} \quad m < 2^{\frac{1}{2}}\pi^2,$$

and hence Theorem 1.1 is proved. If $m \geq 2^{\frac{1}{2}}\pi^2$ we have

$$\lim_{q\to q^*} G_j(p,q) \geq Q_j\pi^2/m > 1/\sqrt{2}$$

for j large enough, and hence Theorem 1.2 is proved.

REFERENCES:

[1] Ambrosetti A. and Prodi G., On the inversion of some differentiable mappings with singularities between Banach spaces, Ann. Mat. Pura Appl. 93 (1972), 213-246.

[2] Castro A. and Lazer A. C., On periodic solutions of weakly coupled systems of differential equations, Boll. Un. Mat. Ital., (5), 18-B (1981), 733-742.

[3] Kannan R. and Ortega R., Superlinear elliptic boundary value problems, to appear.

[4] Kazdan J. and Warner F., Remarks on some quasilinear elliptic equations, Comm. Pure Appl. Math. 28 (1975), 567-597.

[5] Laetsch T., The number of solutions of a nonlinear two point boundary value problem, Indiana Univ. Math. J. 20 (1970), 1-13.

[6] Lazer A. C. and Mckenna P. J., On the number of solutions of a nonlinear Dirichlet problem, J. Math. Anal. Appl. 84 (1981), 282-294.

[7] Lazer A. C. and Mckenna P. J., On a conjecture related to the number of solutions of a nonlinear Dirichlet problem, to appear.

[8] Shivaji R., Perturbed bifurcation theory for a class of autonomous ordinary differential equations, in: Lakshmikantham V. (ed.), Trends in theory and practice of nonlinear differential equations (Marcel Dekker, Inc., New York and Basel, 1984).

This paper is in final form and no version of it will be submitted for publication elsewhere.

Trends in the Theory and Practice of Non-Linear Analysis
V. Lakshmikantham (Editor)

SOME ESTIMATES FOR A SYSTEM OF MULTIPLE REACTIONS

Jagdish Chandra
US Army Research Office
Box 1221
Research Triangle Park, NC 27709

Paul Davis*
Mathematical Sciences Department
Worcester Polytechnic Institute
Worcester, MA 01609

*supported by the US Army Research Office under contract DAAG29-81-D-0018

Various studies of a certain system of autocatalytic reactions in a spatially uniform environment have shown the importance of the concentration of the pool chemical in determining the frequency of oscillation of the concentration of the autocatalytic intermediates. Here we give bounds on this important variable in the more general spatially distributed case.

1. Introduction

The system of reactions

(1) $A + X \rightarrow B + 2X$

(2) $X + Y \rightarrow B + 2Y$

(3) $A + Y \rightarrow B$

has been studied in a variety of contexts. With the pool chemical A held constant, Frank-Kamenetskii [2] proposed these reactions as an explanation for certain phenomena observed in hydrocarbon combustion. Lotka [5] cited the same set of reactions as an example of the possibility of self-sustained chemical oscillations.

Much later, Gray and Aarons [3] observed numerically that the neutrally stable oscillations studied by Frank-Kamenetskii and Lotka are apparently replaced by a single stable periodic oscillation in X and Y when the concentration of the pool chemical A is allowed to vary.

Motivated by the numerical study of Gray and Aarons, the authors used multi-scale asymptotic expansions to construct the transition from an initial pseudo-equilibruim toward the periodic solution guaranteed by a bifurcation analysis. That analysis clearly showed the pitfalls of the pseudo-steady state assumption of constant A . Both the amplitudes and the frequency of the oscillations in the autocatalytic intermediates depend upon the concentration of A . For example, the period of the oscillations in X and Y can double during the transition to the periodic steady state.

Having seen the strong influence of A in these explicit calculations for a spatially uniform model, we were led to seek more information about the behavior of A in a general spatially distributed system without the assumption of slow consumption of A . This note summarizes such general bounding results, identifies the features of the present kinetics that are key to the analysis, and suggests methods for more general kinetic schemes.

2. Spatially Distributed Model

A choice of appropriate time and concentration scales allows one to write dimensionless equations governing the reactions (1-3) in a region Ω ; see [1].

(4) $$\mathcal{L}_a a := \partial a/\partial t - D_a\nabla^2 a = -\varepsilon(x + y)a$$

(5) $$\mathcal{L}_x x := \partial x/\partial t - D_x\nabla^2 x = K(a - y)x$$

(6) $$\mathcal{L}_y y := \partial y/\partial t - D_y\nabla^2 y = K^{-1}(x - a)y$$

Here a, x, y are dimensionless concentrations, K and ε are dimensionless reaction rate parameters, and D_a, D_x, D_y are dimensionless diffusion coefficients. We do not assume ε is small.

Assuming that the walls of the reaction vessel are permeable to A, X, and Y leads to the boundary conditions

(7) $$a + \eta_a\partial_n a = 1, x + \eta_x\partial_n x = x_b, \; y + \eta_y\partial_n y = y_b \quad \text{on} \quad \partial\Omega, \; t > 0,$$

where η_a, η_x, η_y are dimensionless permeability coefficients and $1, x_b, y_b$ are the dimensionless concentrations of A, X, and y outside the reaction vessel; $x_b, y_b \geq 0$. For simplicity, we assume spatially uniform initial conditions

(8) $$(a, x, y) = (a_0, x_0, y_0) \quad \text{at} \quad t = 0, \; r \in \Omega,$$

where r is the spatial variable.

3. Bounds on the Concentrations

Our basic tool is construction of quasi-monotone nonlinearities (see [4]) which are upper or lower bounds for the kinetics terms on the right-hand side of (4-6). Solutions of these modified problems are then upper or lower bounds on the solution of (4-8).

In particular, if smooth solutions of (4-8) are bounded above by $\bar{a}, \bar{x}, \bar{y}$, then solutions of

(9) $$\mathcal{L}_a \underline{a} \leq -\varepsilon(\bar{x} + \bar{y})\underline{a}$$

(10) $$\mathcal{L}_x \underline{x} \leq K(\underline{a} - \bar{y})\underline{x}$$

(11) $$\mathcal{L}_y \underline{y} \leq K^{-1}(\underline{x} - \bar{a})\underline{y}$$

$$\underline{a} + \eta\,\partial_n \underline{a} \leq 1 \quad \text{on} \quad \partial\Omega, \quad \text{etc.}$$

$$(\underline{a}, \underline{x}, \underline{y}) \leq (a_0, x_0, y_0) \quad \text{at} \quad t = 0$$

bound solutions of (4-8) from below. Conversely, given á priori lower bounds on (a, x, y), we may construct a similar system of inequalities whose solutions bound those of (4-8) from above. Note that the particular form of these kinetics terms uncouples (9-11) into three linear problems to be solved in succession.

We summarize our bounding results without proof.

Theorem 1: Bounded, smooth solutions of (4-8) are non-negative.

This non-negativity result is a consequence of the homogeniety of each nonlinear kinetics term in its diagonal variable. The reaction term in (9) in homogeneous in a because it governs the consumption of A while those in (10-11) are homogeneous because X and Y are autocatalytic species.

Theorem 2: For $r \in \Omega$ and $t \geq 0$, $a(r,t) \leq \max(1, a_0)$.

Theorem 3: Solutions $\bar{x}, \bar{y}$ of

$$\mathcal{L}_x \bar{x} \geq K \max(1, a_0)\bar{x},$$
$$\mathcal{L}_y \bar{y} \geq K^{-1}\bar{x}\,\bar{y}$$
$$\bar{x} + \eta_x \partial_n \bar{x} \geq x_b, \quad \bar{y} + \eta_y \partial_n \bar{y} \geq y_b \quad \text{on} \quad \partial\Omega$$
$$\bar{x}(r, 0) \geq x_0, \quad \bar{y}(r, 0) \geq y_0, \quad r \in \Omega$$

bound x, y from above.

Theorem 4: Let $u(r, t)$ satisfy

$$\nabla^2 u = 0, \quad u + \eta_a \partial_n u = 1 \quad \text{on} \quad \partial\Omega .$$

Let μ be the smallest eigenvalue of

$$D_a \nabla^2 \phi + \mu\, \phi = 0, \quad \phi + \eta_a \partial_n \phi = 0 \quad \text{on} \quad \partial\Omega,$$

and let $\phi(r)$ be the corresponding eigenfunction scaled to satisfy

$$\phi(r) \geq \sup_{\Omega}(a_0 - u(r)), \quad r \in \Omega .$$

Then $a(r, t) \leq u(r) + \phi(r)e^{-\mu t}$.

Theorem 4 provides a bound on $a(r, t)$ in the general case that is reminiscent of the asymptotic approximation for the spatially uniform model in [1]. It provides a pointwise estimate of the equilibrium pool chemical concentration, which determines oscillation frequencies and amplitudes, as well as an estimate of the speed of approach to the equilibrium.

Given positive bounds on x and y , one may also construct alternating pincer bounds on $a(r, t)$ in the manner of Weyl [6] et al. Such a sequence, which alternately provides upper and lower bounds, is particularly useful if the reaction kinetics are not of first order; e.g., if (9) is replaced by

$$\mathcal{L}_a a = -\varepsilon(x^\ell a^m + y^n a^p)$$

for some positive constants ℓ, m, n, p . One may also treat

$$\mathcal{L}_a a = -g(x, y, a)$$

providing g is nonnegative, nondecreasing in each of its variables, and homogeneous in a . For simplicity, we shall state the result for the original system.

Theorem 5: Let $\underline{x}, \underline{y}, \bar{x}, \bar{y}$ be positive (not necessarily constant) lower and upper bounds on x, y . Let $\underline{f}(a) = \varepsilon(\underline{x} + \underline{y})a$, $\bar{f}(a) = -\varepsilon(\bar{x} + \bar{y})a$. Define the sequence $\{a_n\}$ by $a_0 = 0$,

$$\mathcal{L}_a a_n = \underline{f}\,(a_{n-1}) , \quad n = 1, 3, 5, \ldots,$$
$$\mathcal{L}_a a_n = \bar{f}\,(a_{n-1}) , \quad n = 2, 4, 6, \ldots .$$

Then we obtain

$$a_0 \leq a_2 \leq \ldots \leq a(r, t) \leq \ldots \leq a_2 \leq a_1 .$$

Of course, in the special case of the first-order kinetics occuring here, one may solve $\mathcal{L}_a a = \underline{f}(a)$ and $\mathcal{L}_a a = \bar{f}(a)$ directly to obtain bounds on a . The iteration scheme is attractive for more general kinetic schemes because each iterate is the solution of a linear problem while the analogs of $\mathcal{L}_a a = \underline{f}(a)$, etc. are non-linear.

References

[1] Chandra, J. and Davis, P. W., The effect of pool chemical variation on an autocatalytic reaction, submitted.

[2] Frank-Kamenetskii, D. A., Periodic processes in the kinetics of oxidation reactions, Akademia Nauk SSSR. Doklady. 25 (1939), 671-672.

[3] Gray, B. and Aarons, L., Small parasitic and chemical oscillations, Faraday Society Trans. (1974), 129-136.

[4] Lakshmikantham, V. and Leela, S., Differential and integral inequalities: theory and applications, Academic Press, New York, 1969.

[5] Lotka, A. J., Undamped oscillations derived from the law of mass action, J. Am. Chem. Soc. 42 (1920), 1595-1599.

[6] Weyl, H., Concerning the differential equations of some boundary layer problems, Proc. Nat. Acad. Sci. 27 (1941), 578-583.

The detailed version of this paper will be submitted for publication elsewhere.

Trends in the Theory and Practice of Non-Linear Analysis
V. Lakshmikantham (Editor)

QUALITATIVE PROBLEMS FOR SOME HYPERBOLIC EQUATIONS

C. Corduneanu and H. Poorkarimi

Department of Mathematics
University of Texas at Arlington
Arlington, TX 76019
U.S.A.

In [2], A. N. Tikhonov and A. A. Samarskii investigate a mathematical model for the dynamics of gas absorbtion. It is assumed that a mixture of air and gas passes through a tube filled with an absorbent. If the axis of the tube is taken as x-axis, and by t one denotes the time, then the following notations will be used:

1) $a(x,t)$ - the quantity of gas absorbed by a unit volume of the absorbent;

2) $u(x,t)$ - the concentration of the gas in the pores of the absorbent (in the layer x, at time t);

3) ν - the velocity of the mixture (it is assumed large enough, so that diffusion can be neglected);

4) y - the concentration of gas in "equilibrium" with the quantity a of gas absorbed;

5) β - the kinetic coefficient;

6) f - a characteristic of the absorbent (in general, a nonlinear function).

The basic equations describing the dynamics of the gas absorbtion are:

$$\text{(S)} \qquad \begin{cases} a_t + u_t + \nu u_x = 0, \\ a_t = \beta(u - y), \\ a = f(y), \end{cases}$$

and they must be considered in the semi-strip

$$(\Delta) \qquad 0 \le x \le \ell, \quad t \ge 0,$$

where ℓ denotes the length of the tube through which the mixture passes.

Some supplementary conditions have to be imposed on (S), if we want a unique solution of that system. For instance, the following conditions make sense (in both physical and mathematical interpretation) in regard to the system(S):

$$\text{(C)} \qquad \begin{cases} a(x,0) = 0, \quad 0 \le x \le \ell, \\ u(0,t) = u_0, \qquad t \ge 0. \end{cases}$$

In case the nonlinearity f is such that $f(y) = \gamma^{-1}y$, where γ^{-1} is the so-called Henry's coefficient, from (S) and (C) one obtains through elimination

(1) $$u_{xt} + \beta\nu^{-1}u_t + \beta\gamma u_x = 0,$$

with the following data on characteristics:

(2) $$\begin{cases} u(x,0) = u_0 e^{-\beta\nu^{-1}x}, & 0 \le x \le \ell \\ u(0,t) = u_0, & t \ge 0. \end{cases}$$

The problem of finding a solution to equation (1), under conditions (2), is known in the theory of partial differential equations as a Darboux-Goursat problem. Such problems have been considered for the first time by Riemann, in connection with a representation formula for the solution of Cauchy's problem for hyperbolic (linear) equations. To the best of our knowledge, such problems did not appear directly in a physical phenomenon until recently (see [2] for further references).

It is, therefore, of interest to see whether qualitative properties (such as boundedness, periodicity, almost periodicity, transiency, or existence of the limit as t tends to infinity) can be secured, under adequate conditions, for the solutions of hyperbolic equations, with data on characteristics. It should be noticed that in [2], an asymptotic investigation is conducted, after the reduction (practically) of the partial differential equation to an ordinary differential equation.

We shall deal in the remaining part of this paper with hyperbolic equations of the form

(E) $$u_{xt} + a(x,t)u_x + b(x,t)u_t = c(x,t,u)$$

where a, b, and c are defined for $(x,t) \in \Delta$, and c is defined for any real u. Of course, the solution is sought in Δ, and in order to assure uniqueness we shall impose conditions on the characteristics:

(3) $$\begin{cases} u(x,0) = \phi(x), & 0 \le x \le \ell, \\ u(0,t) = u_0(t), & t \ge 0. \end{cases}$$

Conditions on the functions involved in (E) and (3) will be stated further, in accordance with the type of solution we are looking for.

Let us point out that under all kind of conditions adopted in this paper on $c(x,t,u)$, the linear case

(4) $$c(x,t,u) = c(x,t)u + d(x,t)$$

will always be covered.

Going back now to the equation (E), in which a nonlinearity is involved in $c(x,t,u)$, we notice that without real loss of generality we can assume $b(x,t) \equiv 0$ in Δ. In other words, the equation (E) assumes a simplified form, namely

(E') $$u_{xt} + b(x,t)u_x = c(x,t,u).$$

Indeed, if in (E) one substitutes

(5) $$u = v\,\exp\{-\int_0^x b(\xi,t)d\xi\},$$

then the equation in v has exactly the form (E)- of course, with different coefficients - but there is no derivative v_t. We leave to the reader the task to carry out the calculations, and formulate the exact conditions on $b(x,t)$, such that (5) makes sense.

Nevertheless, it is appropriate to notice that the exponential factor in formula (5) is bounded in Δ when $b(x,t)$ is bounded there, it is periodic in t if $b(x,t)$ is periodic etc. In other words, the substitution (5) does not modify the asymptotic behavior of the solutions of the "simplified" equation.

In regard to the almost periodicity of solutions for (E') with respect to t, it is indicated also to consider that equation in the whole strip

$$(\Delta') \qquad 0 \le x \le \ell, \quad -x < t < \infty.$$

In this case, the only condition we need to determine a unique solution in (Δ') is

$$(3') \qquad u(0,t) = u_0(t), \quad -\infty < t < \infty.$$

Before discussing the existence problem for solutions to the equation (E'), it is appropriate to transform this equation, together with data on characteristics, into an integral equation of Volterra type (in two variables). Of course, one obtains different Volterra equations, for different problems envisaged with respect to the equation (E').

Let us consider first the equation (E') in Δ, under conditions (3). If we denote $u_x = v$, then equation (E') can be rewritten as

$$(6) \qquad v_t + b(x,t)v = c(x,t,u).$$

This looks like an ordinary differential equation in v, and one obtains

$$(7) \qquad \begin{aligned} v(x,t) &= v(x,0)\exp\{-\int_0^t b(x,\tau)d\tau\} + \\ &\int_0^t \exp\{-\int_\tau^t b(x,\theta)d\theta\}c(x,\tau,u(x,\tau))d\tau. \end{aligned}$$

Since $v(x,0) = u_x(x,0) = \phi'(x)$, $0 \le x \le \ell$, one obtains after integration of both members of (7) from 0 to x:

$$(8) \qquad \begin{aligned} u(x,t) &= u_0(t) + \int_0^x \phi'(\xi)\exp\{-\int_0^t b(\xi,\tau)d\tau\}d\xi \\ &+ \int_0^x\int_0^t \exp\{-\int_\tau^t b(\xi,\theta)d\theta\}c(\xi,\tau,u(\xi,\tau))d\xi d\tau. \end{aligned}$$

The presence of the exponential factor under the double integral in (8) helps consistently in regard to the existence of bounded solutions, provided we impose an adequate condition on $b(x,t)$, In order to handle equation (8) in view of obtaining bounded solutions in Δ, we will assume that

$$(9) \qquad b(x,t) \ge m > 0 \text{ in } \Delta.$$

Under assumption (9), the exponential factor has fast decay at infinity, a fact that enables us to prove boundedness of all solutions of the equation (E') in Δ, under very mild extra assumptions. It is worth noticing that, in equation (E'), the sign of the coefficient $b(x,t)$ is very important for the existence of bounded solutions.

An alternate condition to (9) is

(10) $$b(x,t) \leq -m < 0 \text{ in } \Delta.$$

Under assumption (10), the equation (6) cannot be "solved" by the formula (7) since the integral appearing in the right hand side may be unbounded (and it actually is unbounded in Δ, as very simple examples show). Instead, one can use the formula

(11) $$v(x,t) = -\int_t^\infty \exp\{\int_t^\tau b(x,\theta)d\theta\}c(x,\tau,u(x,\tau))d\tau,$$

which finally leads to the Volterra equation

(12) $$u(x,t) = u_0(t) - \int_0^x\int_t^\infty \exp\{\int_t^\tau b(\xi,\theta)d\theta\}c(\xi,\tau,u(\xi,\tau))d\xi d\tau$$

As one can see from (12), the condition $u(x,0) = \phi(x)$ does not have any role in this case, and the only condition to be considered in association with the equation (E') in Δ, under assumption (10), is the second condition (3), namely $u(0,t) = u_0(t)$, $t \geq 0$.

If we want to consider the equation (E') in the whole strip Δ', say under assumption (9), and with condition (3'), then the integral equation generated by these data will have the form

(13) $$u(x,t) = u_0(t) + \int_0^x\int_{-\infty}^t \exp\{-\int_\tau^t b(\xi,\theta)d\theta\}c(\xi,\tau,u(\xi,\tau))d\xi d\tau.$$

To summarize the above discussion on how to reduce our problems for (E') to integral equations, we must say that equations (8), (12), and (13), under adequate hypotheses, constitute the object of our investigation. Of course, more detailed discussion will be necessary to prove the equivalence of each of these equations with the partial differential (E'), under the respective conditions on characteristic lines. It is worth noticing that the exponential factor under the integral makes possible to obtain good estimates in the semi-strip Δ or in the whole strip Δ', in regard to the existence of bounded solutions.

We shall begin the investigation of existence of solutions with the case leading to the integral equation (8). The following result can be obtained by means of successive approximations.

Theorem 1. Assume the following conditions hold true in regard to the equation (8):

a) $u_0(t)$ is continuous and bounded on the positive half-axis $t \geq 0$;

b) $\phi(x)$ is continuously differentiable on $[0,\ell]$;

c) $b(x,t)$ is continuous on Δ, and verifies the inequality (9);

d) $c(x,t,u)$ is continuous on $\Delta \times R$, with $c(x,t,0)$ bounded on Δ, and verifies the Lipschitz condition.

(14) $$|c(x,t,u) - c(x,t,v)| \leq L|u-v|,$$

where L is a positive constant.

Then there exists a unique continuous solution $u(x,t)$ of equation (8), defined on Δ, and bounded there.

Proof (sketch). The method of successive approximations, starting with the function

$$u_1(x,t) = u_0(t) + \int_0^x \phi'(\xi)\exp\{-\int_0^t b(\xi,\tau)d\tau\}d\xi$$
$$+ \int_0^x\int_0^t \exp\{-\int_\tau^t b(\xi,\theta)d\theta\}c(\xi,\tau,0)d\xi d\tau,$$

leads to the sequence $\{u_n(x,t)\}$, $n \geq 1$, such that the following inequalities are satisfied in Δ:

(15)
$$|u_{n+1}(x,t) - u_n(x,t)| \leq \int_0^x\int_0^t \exp\{-m(t-\tau)\}|c(\xi,\tau,u_n(\xi,\tau)) - c(\xi,\tau,u_{n-1}(\xi,\tau))|d\xi d\tau$$

for any $n \geq 1$. Taking (14) into account, one obtains from (15)

(16)
$$|u_{n+1}(x,t) - u_n(x,t)| \leq L\int_0^x\int_0^t \exp\{-m(t-\tau)\}|u_n(\xi,\tau) - u_{n-1}(\xi,\tau)|d\xi d\tau$$

in Δ, for any $n \geq 1$. Proceeding by induction upon n, (16) leads to the following estimate in Δ:

(17)
$$|u_{n+1}(x,t) - u_n(x,t)| \leq A(\frac{Lx}{m})^n \frac{1}{n!},$$

where $A = \sup|u_1(x,t)|$ in Δ. Of course, (17) implies the uniform convergence in Δ of the sequence $\{u_n(x,t)\}$, which in turn implies the existence of a continuous solution (in Δ) $u(x,t)$ for the equation (8). From (17) one easily sees that this solution is bounded in Δ.

Uniqueness is proven in the standard manner, using the same successive approximations.

Let us formulate now, as a Corollary to Theorem 1, a result of existence and uniqueness for the solution of the problem (E'), (3). Of course, this result will be also based on the equivalence of the problem (E'), (3) to the integral equation (8), a matter that does not rise any difficulty under the assumptions of Theorem 1.

Corollary. Consider the problem (E'), (3), and assume that all conditions of Theorem 1 hold true. Moreover, let $u_0(t)$ be continuously differentiable on R_+. Then, there exists a unique solution of (E'), satisfying the conditions on the characteristics (3). This solution is bounded in Δ.

In order to obtain existence of bounded solutions for the equation (E'), under the basic assumption (10), we shall need only the second condition (3). As pointed out above, the integral equation (12) is the adequate tool to investigate the problem in this case.

The following result can also be obtained by the method of successive approximations.

Theorem 2. Consider equation (12), and assume the following conditions hold true:

a) $u_0(t)$ is continuous and bounded on the positive half-axis;

b) $b(x,t)$ is continuous on Δ, and verifies the inequality (10);

c) $c(x,t,u)$ is continuous on $\Delta \times R$, with $c(x,t,0)$ bounded in Δ, and verifies the Lipschitz condition (14).

Then there exists a unique solution of the integral equation (12), continuous on Δ, and bounded there.

The proof, by the method of successive approximations, can be carried out without difficulty. One can start with

$$u_1(x,t) = u_0(t) - \int_0^x\int_t^\infty \exp\{\int_t^\tau b(\xi,\theta)d\theta\}c(\xi,\tau,0)d\xi d\tau$$

which is obviously bounded in Δ. Let us point out that, on behalf of (10), the exponential factor under the double integral is dominated by $\exp\{-m(\tau-t)\}$, which easily leads to the convergence of that integral.

Corollary. Consider the equation (E'), under condition $u(0,t) = u_0(t)$, and assume that all conditions of Theorem 2 hold true. Then, there exists a unique solution of the problem, bounded in Δ.

Finally a result similar to those given in Theorems 1 and 2 can be obtained for equation (13), considered in the strip Δ'.

Theorem 3. Assume the following hypotheses in regard to equation (13):

a) $u_0(t)$ is continuous and bounded on the real axis;

b) $b(x,t)$ is continuous on Δ', and satisfies there the inequality (10);

c) $c(x,t,u)$ is continuous on $\Delta' \times R$, with $c(x,t,0)$ bounded in Δ', and such that (14) holds true.

Then, there exists a unique continuous and bounded solution (in Δ') of the equation (13).

We omit the proof of Theorem 3, which is basically the same as in the case of Theorems 1 and 2.

Corollary 1. Consider the equation (E'), under condition (3'), and assume that hypotheses of Theorem 3 are satisfied. Then there exists a unique bounded solution (in Δ') of the problem.

Corollary 2. If $u_0(t)$, $b(x,t)$, and $c(x,t,u)$ are periodic (almost periodic) in t, then the unique bounded solution of (13) enjoys the same property. Let us notice that, in case of periodicity, the period has to be the same for all the functions involved. Otherwise, the bounded solution might result only almost periodic in t, even though each function involved into the equation (13) is periodic.

In concluding this brief presentation of the results, we want to point out the fact that boundedness of solutions can be achieved even under less restrictive assumptions on the data. For instance, the boundedness of $u_0(t)$, [or $c(x,t,0)$] in R_+, could be easily replaced by a condition of the form

$$\sup_{t \in R_+} \int_t^{t+1} |u_0(s)|ds < +\infty.$$

See [1] for results of this nature.

REFERENCES

[1] Corduneanu, C., Integral Equations and Stability of Feedback Systems (Academic Press, New York, 1973).

[2] Tikhonov, A.N. and A.A. Samarskii, Equations of Mathematical Physics (Pergamon Press, The MacMillan Co., New York, 1963).

Trends in the Theory and Practice of Non-Linear Analysis
V. Lakshmikantham (Editor)
Elsevier Science Publishers B.V. (North-Holland), 1985

VISCOSITY SOLUTIONS OF HAMILTON-JACOBI EQUATIONS IN BANACH SPACES

Michael G. Crandall
Department of Mathematics
University of Wisconsin
Madison, Wisconsin
U.S.A.

Pierre-Louis Lions
Université de Paris-IX
place de Lattre-de-Tassigny
75775 Paris Cedex 16
France

INTRODUCTION

We consider Hamilton-Jacobi equations (or HJE's) of the form

$$\text{(HJ)} \qquad H(x,u,Du) = 0 \text{ in } \Omega$$

where Ω is an open subset of a Banach space V, V^* is the dual of V, $H \in C(V \times \mathbf{R} \times V^*)$ and Du denotes the Fréchet derivative of a function $u:\Omega \to \mathbf{R}$. A function u is a classical solution of (HJ) in Ω if u is continuously Fréchet differentiable on Ω and the equation is satisfied pointwise. As is well-known, even if $V = \mathbf{R}^n$, the notion of a classical solution is too restrictive to admit the "solutions" of HJE's which are important in the areas in which they arise - in particular, the "value" functions of control theory, the calculus of variations and differential games are usually nonclassical solutions of HJE's.

A mathematical theory of (HJ) in finite dimensions with the scope to accommodate applications has been perfected in recent years following the initial development of the theory of "viscosity solutions" of HJE's by Crandall and Lions [3], [4] (see also [2], [15]). The review article [9] outlines the basic finite dimensional theory and literature up to its time. Since then, the works Ishii [12], [13], [14] and Crandall and Lions [6], [7], [8] (among others) have advanced the basic theory still further. Indeed, the results of [6], which were the topic of the first author's lecture at the symposium to which this volume corresponds, partly stimulated the interesting work of Ishii [14] and this latter work, in its turn, was taken into account in the presentation of the first basic results in infinite dimensional spaces in Crandall and Lions [7], [8]. It is our goal here to sketch the basic definitions and some existence and uniqueness theorems in Banach spaces contained in Crandall and Lions [7], [8].

HJE's IN REGULAR SPACES - PRELIMINARIES

In this section we will assume that V possesses the Radon-Nikodym property (or "V is RNP"). The form of this property of relevance here is the following: If B is a closed ball in V and $\varphi:B \to \mathbf{R}$ is continuous and bounded, then for every $\varepsilon > 0$ there is an element x^* of V^* with norm less that ε such that $\varphi + x^*$ attains its maximum value over B at some point of B. The fact that RNP spaces have this property is due to Stegall [17] (and to Ekeland and Lebourg [10] in a particular case adequate for most applications). We recall that reflexive spaces and separable dual spaces are RNP. This result is what makes it reasonable to define viscosity solutions of (HJ) in an RNP space V just as one does in the finite dimensional case. To this end, if $u \in C(\Omega)$ and $y \in \Omega$, the sub- and superdifferentials $D^-u(y)$ and $D^+u(y)$ of u at y are defined (as in finite dimensions) by:

$$D^-u(y) = \{p \in V^* : \limsup_{x \to y} \frac{u(x) - u(y) - (p, x - y)}{|x - y|} \geq 0\},$$

and

$$D^+u(y) = \{p \in V^* : \liminf_{x \to y} \frac{u(x) - u(y) - (p, x - y)}{x - y} \leq 0 \}$$

where (p,x) denotes the value of $p \in V^*$ at $x \in V$.

Definition: Let Ω be an open subset of V and $H:\Omega \times \mathbf{R} \times V^* \to \mathbf{R}$. Then $u \in C(\Omega)$ is a viscosity subsolution (supersolution) of $H = 0$ in Ω if $H(x,u(x),p) \leq 0$ (respectively, $H(x,u(x),p) \geq 0$) for every $x \in \Omega$ and $p \in D^+u(x)$ (respectively, $p \in D^-u(x)$). Finally, a viscosity solution of $H = 0$ is a $u \in C(\Omega)$ which is both a viscosity subsolution and a viscosity supersolution.

In what follows we will use $|\ |$ to denote the norm of V, the norm of V^* and the absolute value on $\mathbf{R}$. We will also assume the existence of a Lipschitz continuous function $d:V \times V \to \mathbf{R}$ with the following properties: The mappings $x \to d(x,y)$ and $y \to d(x,y)$ are each Fréchet differentiable off the diagonal $x = y$ and there are numbers $k, K > 0$ such that

(1) $$k|x - y| \leq d(x,y) \leq K|x - y| \text{ for } x, y \in V.$$

For example, if the norm of V is differentiable on $V\backslash\{0\}$, then $d(x,y) = |x - y|$ has the desired properties. Moreover, if d has the desired properites, then so does $(x,y) \to d(x - y,0)$. Let us remark, however, that functions d which are not functions of $x - y$ must be considered to achieve full generality in what comes later. When such a function d exists we can express the notions of sub- and supersolutions in a form that is more convenient for some purposes. For example, in this event, u is a viscosity subsolution of $H = 0$ if and only if

If $\varphi \in C(\Omega)$ is differentiable at each point of Ω, $y \in \Omega$ is a point of continuity of $D\varphi$, $r_0 > 0$ and

(2) $$u(y) - \varphi(y) > \sup\{u(x) - \varphi(x): r < |x - y| \leq r_0\}$$

for $0 < r < r_0$, then $H(y,u(y),D\varphi(y)) \leq 0$.

The corresponding characterization for supersolutions, which arises upon flipping the inequalities, also holds.

We are using the term "regular space" to mean an RNP space which admits a function d with the properties above. The Radon-Nikodym property is not required to establish the equivalence of (2) and the definition of a supersolution, but without the Radon-Nikodym property the notion of viscosity solutions discussed here is not useful.

Working with the form (2) of the notion of a viscosity subsolution, one can establish the following stability property of the class of viscosity subsolutions in regular spaces:

Stability: Assume that $u_n \in C(\Omega)$ is a viscosity subsolution of an equation $H_n = 0$ for $n = 1,2,\ldots$. Assume, moreover, that and u_n and H_n converge to $u \in C(\Omega)$ and $H:V \times \mathbf{R} \times V^* \to \mathbf{R}$ in the following way: Every point $x \in \Omega$ has a neighborhood N on which $u_n \to u$ uniformly and whenever $x_n \to x \in \Omega$, $r_n \to r \in \mathbf{R}$ and $p_n \to p \in V^*$, then $\liminf_{n\to\infty} H_n(x_n,r_n,p_n) \geq H(x,r,p)$. Then u is a viscosity subsolution of $H = 0$.

The analogous property holds for supersolutions, and this stability is important for the proofs of the existence results presented following the uniqueness discussion.

We have uniqueness and existence results for (HJ) and for the corresponding evolution form

(E) $$u_t + H(x,t,u,Du) = 0 \text{ in } \Omega \times (0,T),$$

where $T > 0$ and $H \in C(V \times [0,T] \times \mathbf{R} \times V^*)$. Of course, the equation (HJ) is completely general and so contains the particular form (E) (using $V \times \mathbf{R}$ as the basic space) and so viscosity solutions, etc., are well-defined for E, but the results concerning (E) rely on its particular structure. Let us place our hypotheses on the Hamiltonian $H(x,t,r,p)$ as it appears in (E) and then understand that these assumptions when imposed on H in (HJ) are just what arises for a t-independent function. We suppose:

(H1) The mapping $p \to H(x,t,r,p)$ is uniformly continuous on bounded sets of V^* uniformly for bounded x,r and $t \in [0,T]$.

(H2) There is a constant c such that $r \to H(x,t,r,p) - cr$ is nondecreasing in r for all $(x,t,p) \in V \times [0,T] \times V^*$.

(H3) There is an everywhere differentiable Lipschitz continuous function $\nu: V \to [0,\infty)$ such that $\liminf_{|x| \to \infty} \nu(x)/|x| > 0$ and a function $\sigma: [0,\infty) \times [0,\infty) \to [0,\infty)$ which is nondecreasing in both arguments, satisfies $\sigma(0+,R) = 0$ for each $R > 0$ and

(3) $$H(x,t,r,p) - H(x,t,r,p + \lambda D\nu(x)) \leqslant \sigma(\lambda, \lambda + |p|)$$

for $(x,t,r,p) \in V \times [0,T] \times \mathbf{R} \times V^*$ and $\lambda \geqslant 0$.

Observe that (3) holds whenever ν is Lipschitz continuous, differentiable and H is uniformly continuous in p on bounded sets of V^* uniformly in the other arguments.

Finally, if d is the function whose existence is assumed satisfying (1), etc., then we will assume that

(H4) There is a function $m: [0,\infty) \to [0,\infty)$ satisfying $m(0+) = 0$ such that

(4) $$H(y,t,r,-\lambda D_y d(x,y)) - H(x,t,r,\lambda D_x d(x,y)) \leqslant m(\lambda d(x,y) + d(x,y))$$

for $x, y \in V$, $x \neq y$, $t \in [0,T]$, $r \in \mathbf{R}$, and $\lambda \geqslant 0$.

UNIQUENESS IN REGULAR SPACES

We formulate a typical uniqueness result in the guise of a comparison theorem. Of course, comparison theorems imply more than uniqueness for they may be used to establish continuity of solutions with respect to various data. In fact, the result stated below follows at once from what is proved in [7], where the results are formulated to exhibit a more subtle continuity of solutions with respect to variations of the Hamiltonian than can be deduced from the statement of Theorem 1. We refer the interested reader to [7].

Theorem 1: Let H satisfy (H1) - (H4). Let $u, v \in C(\overline{\Omega} \times [0,T])$ be uniformly continuous in $x \in \overline{\Omega}$ uniformly in $t \in [0,T]$ and be, respectively, a viscosity subsolution and a viscosity supersolution of (E) on $\Omega \times [0,T]$. Let $u(x,t) \leqslant v(x,t)$ for $x \in \Omega$ and $t = 0$ and for $x \in \partial\Omega$ and $t \in [0,T]$. Assume, moreover, that $u(x,t) - v(x,t) \to u(x,0) - v(x,0)$ as $t \to 0+$ uniformly on bounded subsets of Ω. Then $u \leqslant v$ on $\overline{\Omega} \times [0,T]$.

The corresponding result for (HJ) is:

Theorem 2: Let $H \in C(V \times \mathbf{R} \times V^*)$ satisfy (H1) - (H4) and $c > 0$ in (H2). Let $u, v \in C(\overline{\Omega})$ be uniformly continuous on $\overline{\Omega}$ and be, respectively, a viscosity subsolution and a viscosity supersolution of $H = 0$ in Ω. Let $u(x) \leqslant v(x)$ for $x \in \partial\Omega$. Then $u \leqslant v$ on $\overline{\Omega}$.

EXISTENCE IN REGULAR SPACES

The existence results require a bit more on the Hamiltonians and are for the case $\Omega = V$. We will use the assumption :

(H5) H is uniformly continuous on bounded subsets of $V \times [0,T] \times \mathbf{R} \times V^*$.

and in the case of (HJ) we suppose (for example) that there is a $\kappa \in (0,1]$, C_1, C_2, $C_3 \geq 0$ such that $C_1\kappa < 1$ and for every x, y $\in$ V with $x \neq y$ and $r \in \mathbf{R}$ we have

(H6) $H(y,r,-\lambda D_y d(x,y)) - H(x,r,\lambda D_x d(x,y)) \leq C_1\lambda d(x,y) + C_2 d(x,y)^\kappa + C_3$ for $0 \leq \lambda$.

Theorem 3. (i) Let $\varphi \in C(V)$ be uniformly continuous and $H \in C(V \times [0,T] \times \mathbf{R} \times V^*)$ satisfy (H2) - (H5). Then there is a unique $u \in C(V \times [0,T])$ which is uniformly continuous on V uniformly in $t \in [0,T]$ and uniformly continuous on bounded subsets of $V \times [0,T]$ and which is a viscosity solution of (E) on $V \times (0,T)$ satisfying $u(x,0) = \varphi(x)$ on V.
(ii) If $H \in C(V \times \mathbf{R} \times V^*)$ satisfies (H2) with $c > 0$ and (H3) - (H6), then there is a unique uniformly continuous viscosity solution of $H = 0$ on V.

Results less general than Theorems 1 - 3 have been proven in the finite dimensional case $V = \mathbf{R}^n$ in, e. g., [1], [6], [13], [14], [16] and [18]. Of course, there are many variants of these results. We mention that there is an example in [8] with $V = \mathbf{R}$ where (H2) holds with $c = 1$ and (H3) - (H5) are satisfied as well as (H6) except that $C_1 = \nu = 1$ (so $C_1\nu < 1$ does not hold), and (HJ) does not have a uniformly continuous viscosity solution.

The method used in [8] to prove the existence results is the following: The relationship between viscosity solutions and differential games is exploited to express solutions of truncated and regularized problems as value functions, thereby obtaining existence, and then the more precise error estimates of [7] are used to obtain solutions in general via limiting processes. Here is where the stability result plays a role. The relationship between viscosity solutions and control theory was pointed out in P.-L. Lions [15] using the dynamic programming principle and the analogous relationships for the case of differential games used in [7] were discussed in Evans and Souganidis [11] (where one also finds earlier references to this topic).

REMARKS ABOUT GENERAL SPACES

If V is not RNP, one does not expect to be able to use the notion of viscosity solutions defined above. However, using the notion of ε-approximate sub- and superdifferentials (see [10]) we introduced another notion in [7] which coincides with that given above in many cases. Results like those above can be established for this notion without assuming that V is RNP. Moreover, existence can be established via the connection with differential games in completely general spaces for sufficiently restricted Hamiltonians. However, there is an example of a simple Hamiltonian in [8] with $V = L^1$ (which does not admit a differentiable function d satisfying (1)) for which the comparison results do not not hold.

BIBLIOGRAPHY

[1] Barles, G., Ann. Inst. Henri Poincare Anal. non Lin., (to appear 1984).
[2] Crandall, M. G., L. C. Evans and P. L. Lions, Some properties of viscosity solutions of Hamilton-Jacobi equations, Trans. Amer. Math. Soc., 282 (1984), 487 - 502.
[3] Crandall, M. G., and P. L. Lions, Condition d'unicité pour les solutions généralisées des équations de Hamilton-Jacobi du premier ordre, C. R. Acad. Sci. Paris 292 (1981), 183 - 186.
[4] Crandall, M. G. and P. L. Lions, Viscosity solutions of Hamilton-Jacobi equations, Trans. Amer. Math. Soc. 277 (1983), 1 - 42.

[5] Crandall, M. G. and P. L. Lions, Solutions de viscosité non bornées des équations de Hamilton-Jacobi du premier ordre, C. R. Acad. Sci. Paris 298 (1984), 217 - 220.

[6] Crandall, M. G., and P. L. Lions, On existence and uniqueness of solutions of Hamilton-Jacobi equations, to appear in Non. Anal. Theor. Meth. Appl.

[7] Crandall, M. G., and P. L. Lions, Hamilton-Jacobi equations in Banach Spaces, Part I: Uniqueness of Viscosity Solutions Solutions, to appear in J. Func. Anal.

[8] Crandall, M. G., and P. L. Lions, Hamilton-Jacobi equations in Banach spaces, Part II: Existence of Viscosity Solutions, in preparation.

[9] Crandall, M. G. and P. E. Souganidis, Developments in the theory of nonlinear first order partial differential equations, in Differential Equations, I. W. Knowles and R. T. Lewis eds. (North Holland, Amsterdam, 1984).

[10] Ekeland, I. and G. Lebourg, Generic Fréchet differentiability and perturbed optimization in Banach spaces, Trans. Amer. Math. Soc. 224 (1976), 193 - 216.

[11] Evans, L. C. and P. E. Souganidis, Differential games and representation formulas for solutions of Hamilton-Jacobi-Isaacs equations, to appear in Indiana J. Math.

[12] Ishii, H., Uniqueness of unbounded solutions of Hamilton-Jacobi equations, Indiana Univ. Math. J., to appear.

[13] Ishii, H., Remarks on the Existence of Viscosity Solutions of Hamilton-Jacobi Equations, Bull. Facul. Sci. Eng., Chuo University, 26 (1983), 5-24.

[14] Ishii, H., Existence and uniqueness of solutions of Hamilton-Jacobi equations, preprint.

[15] Lions, P. L., Generalized Solutions of Hamilton-Jacobi Equations, (Pitman, London, 1982).

[16] Lions, P. L., Existence results for first-order Hamilton-Jacobi equations, Richerche Mat. Napoli, 32 (1983), 1 - 23.

[17] Stegall, C., Optimization of functions on certain subsets of Banach spaces, Math. Annal. 236 (1978), 171 -176.

[18] Souganidis, P. E., Existence of viscosity solutions of Hamilton-Jacobi equations, J. Diff. Eq., to appear.

Sponsored in part by the United States Army under Contract No. MCS-8002946 and in part by the National Science Foundation under Grant No. MCS-8002946.

The final version of this paper will be submitted for publication elsewhere.

Trends in the Theory and Practice of Non-Linear Analysis
V. Lakshmikantham (Editor)

MAXIMAL REGULARITY FOR ABSTRACT DIFFERENTIAL EQUATIONS AND APPLICATIONS TO THE EXISTENCE OF PERIODIC SOLUTIONS

G. Da Prato

Scuola Normale Superiore
56100 PISA
ITALY

We study periodic solutions of nonlinear equations by linearization.

1. INTRODUCTION

Consider the equation:

(1.1) $\quad Au + Bu = v$

where A and B are linear closed (generally unbounded) operators in a Banach space X. We say that we have maximal regularity for problem (1.1) if, for any $v \in X$, there exists a unique strict solution $u \in D(A) \cap D(B)$.
In general we do not have maximal regularity. Consider in fact the problem:

$$(1.2) \quad \begin{cases} -u'(t) + Bu(t) = v(t) \\ u(0) = 0 \end{cases}$$

where B generates a strongly continuous semi-group e^{tB} in a Banach space E. Set $Au = -u'$, $(Bu)(t) = Bu(t)$ and $X = C([0,T];E)$ (the set of all continuous mappings $[0,T] \to E$). Then, it is well known that problem (1.2) has a unique "mild" solution:

$$(1.3) \quad u(t) = \int_0^t e^{(t-s)B} v(s)ds$$

but we do not have, in general, $u', Bu \in C([0,T];E)$.
Maximal regularity for eq. (1.1) has been extensively studied in [1]. Here several spaces X in which maximal regularity holds, are constructed by using interpolation theory.
Assume now that we have maximal regularity for problem (1.1) and consider the nonlinear equation

(1.4) $\quad Au + f(u) = v$

where $f \in C^1(D(B);X)$, $f(0) = 0$, $f'(0) = B$.
Set

(1.5) $\quad F(u) = Au + f(u) \qquad \forall u \in D(A) \cap D(B)$

then $F'(0) = A + B$ and F is a local homeomorphism from $D(A) \cap D(B)$ into X. Thus, be the implicit functions theorem, there exists a solution of Eq. (1.4) for $|v|_X$ small.
In this paper we shall apply the previous argument to periodic problems. We remark that in several papers maximal regularity has been used for nonlinear Cauchy problems (see for instance [2],[6],[7]).

2. LINEAR PROBLEM

Let E be a Banach space, $B:D(B) \subset E \to E$ a linear closed operator. We shall assume that:

(2.1)
a) There exists $\xi \in \mathbb{R}$ and $\theta \in]\frac{\pi}{2}, \pi[$ such that the spectrum $\sigma(B)$ of B is included in the sector

$$S_{\xi,\theta} = \{\lambda \in \mathbb{C} ;\ |\arg(\lambda - \xi)| < \theta\}$$

b) There exists a constant M such that

$$|(\lambda - B)^{-1}| \leqslant \frac{M}{|\lambda - \xi|} \qquad \forall \lambda \in S_{\xi,\theta}$$

We do not assume that $D(B)$ is dense in E ; however a semi-group e^{tB} can be still defined by the Dunford integral

$$e^{tB} = \frac{1}{2\pi i} \int_\gamma e^{\lambda t}(\lambda - B)^{-1} dx$$

where γ is a suitable path in $S_{\xi,\theta}$ (see [8]).
We recall now a result on evolution equations:

PROPOSITION 2.1. Assume (2.1), let $g \in C^\alpha([0,2\pi];E)$ (the set of all mappings $[0,2\pi] \to E$, α-hölder continuous) and $g(0) = 0$. Let ϕ be the mild solution of the problem:

$$(2.2) \qquad \begin{cases} \phi'(t) = B\phi(t) + g(t) \\ \phi(0) = 0 \end{cases}$$

then we have

(2.3) $\quad \phi \in C^{1,\alpha}([0,2\pi];E) \cap C^\alpha([0,2\pi];D(B))$

(2.4) $\quad \phi' \in B([0,2\pi];D_B(\alpha,\infty))$

where $D_B(\alpha,\infty)$ is the Lions interpolation space $(D(B),E)_{1-\alpha,\infty}$ (see [5]) and $B([0,2\pi];D_B(\alpha,\infty))$ denotes the set of all mappings $[0,2\pi] \to D_B(\alpha,\infty)$ bounded.
We remark that (2.3) is proved in [1] and that (2.4) is proved in [8].
In the sequel we shall set $\phi = e^{tB} \star g$ and $C^\alpha_\#([0,2\pi];E) =$
$= \{ u \in C^\alpha([0,2\pi];E) ; u(0) = u(2\pi)\}$

We can prove now the main result of this section.

THEOREM 2.2. Assume (2.1) and that 1 belongs to the resolvent set $\rho(e^{2\pi B})$ of $e^{2\pi B}$.
Let $f \in C^{\alpha}_{\#}([0,2\pi];E)$, then there exists a unique strict solution u of the problem:

$$(2.5)\qquad \begin{cases} u'(t) = Bu(t) + f(t) \\ u(0) = u(2\pi) \end{cases}$$

Moreover $u \in C^{1,\alpha}_{\#}([0,2\pi];E) \cap C^{\alpha}_{\#}([0,2\pi];D(B))$.

Proof. It is easy to see that problem (2.5) has a unique "mild" solution u given by

$$(2.6)\qquad u = u_1 + u_2 + u_3$$

where

$$(2.7)\qquad u_1 = e^{tB} \star (f(\cdot) - f(0))$$

$$(2.8)\qquad u_2(t) = e^{tB}(1 - e^{2\pi B})\tilde{x}\ ,\quad \tilde{x} = \int_0^{2\pi} e^{(2\pi - s)B}(f(s) - f(0))ds$$

$$(2.9)\qquad u_3 = -B^{-1}f(0)\ .$$

Clearly $u_3 \in C^{1,\alpha}_{\#}([0,2\pi];E) \cap C^{\alpha}_{\#}([0,2\pi];D(B)) \overset{def}{=} Z$ and $u_1 \in Z$ by Proposition 2.1.
Finally we have

$$u_2(t) = e^{tB}(1 - e^{2\pi B})^{-1}\tilde{x}\ ,\quad \tilde{x} = \phi(2\pi)$$

where ϕ is the solution of (2.2) with $g(t) = f(t) - f(0)$.
By (2.4) we have

$$\phi'(2\pi) = B\phi(2\pi) + f(2\pi) - f(0) = B\phi(2\pi) \in D_B(\alpha,\infty)$$

that implies $u_2 \in Z$ #

REMARK 2.3. If $f \in C^{\alpha}([0,2\pi];E)$ but $f(0) \neq f(2\pi)$ the mild solution of (2.5) does not belong to Z in general as the following example shows. Let $f(t) = ty$, then the solution of (2.5) is given by

$$u(t) = e^{tB}(e^{2\pi B} - 1)^{-1}y + (1 + tB)B^{-2}y \quad \#$$

3. NONLINEAR PROBLEM

We are here concerned with the problem:

$$(3.1)\qquad \begin{cases} u' = Bu + F(u) + \psi(t) \\ u(0) = u(2\pi) \end{cases}$$

we assume

$$(3.2)\qquad \begin{cases} \text{a) } B \text{ verifies hypotheses (2.1)} \\ \text{b) } F \in C^3(D(B);E) \\ \text{c) } \psi \in C^{\alpha}_{\#}([0,2\pi];E) \end{cases}$$

(3.2) $\left\{ \text{d) } 1 \in \rho(e^{2\pi B}) \right.$

We are looking for solutions of problem (3.1) "near" 0.

PROPOSITION 3.1. There exist $\delta > 0$, $\eta > 0$ such that if $|\psi|_{C^\alpha([0,2\pi];D(B))} \leqslant \delta$ then there exists a unique solution u of (3.1) such that:

(3.3) $$u \in C^{1,\alpha}_{\#}([0,2\pi];E) \cap C^{\alpha}_{\#}([0,2\pi];D(B))$$

(3.4) $$|u|_{C^{1,\alpha}([0,2\pi];E)} + |u|_{C^{\alpha}([0,2\pi];D(B))} \leqslant \eta$$

Proof. Let $X = C^{1,\alpha}_{\#}([0,2\pi];E) \cap C^{\alpha}_{\#}([0,2\pi];D(B))$ and $Y = C^{\alpha}_{\#}([0,2\pi];E)$. Let γ be the mapping:

(3.5) $$\gamma : X \to Y \ , \quad u \to \gamma(u)$$

with

(3.6) $$\gamma(u) = u' - Bu - F(u) \ .$$

Problem (3.1) is equivalent to the equation

(3.7) $$\gamma(u) = \psi \ .$$

Moreover it is easy to check that $\gamma \in C^1(X;Y)$ and

$$\gamma'(0) \cdot v = v' - Bv \ .$$

By Theorem 2.2 it follows that $\gamma'(0)$ is an homeomorphism of X onto Y so that the implicit function theorem implies existence and uniqueness of a solution of equation (3.5) for $|\psi|_Y$ small #

EXAMPLE 3.2. Consider the problem:

(3.8) $$\begin{cases} u_t = u^2_{xx} + u_{xx} + \psi(t,x) \\ u(t,x) = u(t,\pi) = 0 \\ u(0,x) = u(2\pi,x) \end{cases}$$

where $\psi \in C^1([0,2\pi] \times [0,\pi])$ and ψ is periodic in t.
Set $E = C([0,\pi])$ and

(3.9) $$Bu = u_{xx} \ , \quad D(B) = \{u \in C^2([0,\pi]); u(0) = u(1) = 0\}$$

(3.10) $$F(u) = u^2_{xx}$$

Then B, F and ψ verify the hypotheses of Proposition 3.1. Thus, if

$$\sup_{t,s \in [0,2\pi]} \ \sup_{x \in [0,\pi]} \frac{|\psi(t,x) - \psi(s,x)|}{|t-s|^\alpha}$$

is sufficiently small, then problem (3.8) has a regular solution.

4. INTEGRAL EQUATIONS

Consider the equation

(4.1) $$u'(t) = Bu(t) + \int_0^t K(t-s)u(s)ds + f(t) .$$

We assume

(4.2)
$$\begin{cases} \text{a) } B \text{ verifies (2.1)} \\ \text{b) } K(t) \in \mathcal{L}(D(B);E) \quad \forall t \in [0,T] \\ \text{c) For any } x \in D(B),\ K(\cdot)x \text{ is absolutely Laplace transformable in } E \text{ and the Laplace transform } \hat{K}(\lambda)x \text{ is analytical in } S_{\xi,\theta} \\ \text{d) There exists an increasing function } N:[0,\theta_0[\to [0,+\infty[\text{ such that} \\ \quad \|\hat{K}(\lambda)x\| \leqslant \dfrac{N(\theta')}{|\lambda|}|Bx| \quad , \quad x \in D(B) \\ \quad \text{if } \lambda \in \overline{S}_\theta \text{ and } \theta' \in [0,\theta[\ . \end{cases}$$

Under hypotheses (4.2) there exists a resolvent operator $R(t)$ whose Laplace transform $F(\lambda)$ is given by $F(\lambda) = (\lambda - B - \hat{K}(\lambda))^{-1}$ ([4]). Moreover the following regularity result holds ([3]).

PROPOSITION 4.1. Assume (4.2), let $g \in C^\alpha([0,2\pi];E)$ and $g(0) = 0$. Let ϕ be the solution of the problem:

(4.3)
$$\begin{cases} \phi'(t) = B\phi(t) + \int_0^t K(t-s)\phi(s)ds + g(t) \\ \phi(0) = 0 \ . \end{cases}$$

Then (2.3) and (2.4) hold.

Now, proceeding as in Theorem 2.2 we can prove the result:

THEOREM 4.2. Assume (4.2) and that $1 \in \rho(R(2\pi))$. Let $f \in C^\alpha_\#([0,2\pi];E)$. Then there exists a unique periodic strict solution u of Eq. (4.1). Moreover $u \in C^{1,\alpha}_\#([0,2\pi];E) \cap C^\alpha_\#([0,2\pi];D(B))$.

EXAMPLE 4.3. Consider the problem:

(4.4)
$$\begin{cases} u_t = u_{xx} + e^{-t} \star u_{xx} + f(t,x) \\ u(0,x) = u(T,x) & x \in [0,\pi] \\ u(t,0) = u(t,\pi) = 0 \ . \end{cases}$$

Let $E = C([0,2\pi])$ and B be the same as in Example 3.2. We have ([3])

$$F(\lambda) = \begin{cases} \dfrac{1}{\lambda} & \text{if } \lambda = -2 \\ \dfrac{\lambda+1}{\lambda+2}\left(\dfrac{\lambda^2+\lambda}{\lambda+2} - B\right)^{-1} & \text{if } \lambda \neq -2 \end{cases}$$

and it is easy to check that

$$\operatorname{Sup} \sigma(R(t)) < 0 \qquad \forall t > 0$$

so that Theorem 4.2 applies #

We remark that, arguing as in Section 3, nonlinear integrodifferential equations can also be considered.

REFERENCES

[1] Da Prato, G. and Grisvard, P., Sommes d'opérateurs linéaires et équations differentielles opérationnelles, J. Math. Pures Appl. 54 (1975) 305-387

[2] Da Prato, G. and Grisvard, P., Equations d'évolution abstraites nonlinéaires de type parabolique, Ann. Mat. Pura Appl. 120 (1979) 329-396

[3] Da Prato, G. and Iannelli, M., Existence and regularity for a class of integrodifferential equations of parabolic type, to appear in: Jour. Math. Anal. Appl.

[4] Grimmer, R.C. and Kappel, F., Series expansions for resolvents of Volterra integrodifferential equations in Banach spaces, to appear

[5] Lions, J.L., Théorèmes de trace et d'interpolation (I), Ann. Sc. Norm. Pisa 13 (1959) 389-403

[6] Lunardi, A., Analyticity of the maximal solution of an abstract nonlinear parabolic equation, Nonlinear An. 6 (1982) 503-521

[7] Sinestrari, E., Continuous Interpolation spaces and spatial regularity in nonlinear Volterra integrodifferential equations, J. Integral Equations, 5 (1983) 283-308

[8] Sinestrari, E., On the abstract Cauchy problem of Parabolic type in spaces of continuous functions, to appear in: Journal of Math. An. and Appl.

This paper is in final form and no version of it will be submitted for publication elsewhere.

Trends in the Theory and Practice of Non-Linear Analysis
V. Lakshmikantham (Editor)

A FINITE ELEMENT ERROR ESTIMATE FOR REGULARIZED COMPRESSIBLE FLOW

Hung Dinh and Graham F. Carey

Aerospace Engineering/Engineering Mechanics Department
The University of Texas at Austin
Austin, Texas
U.S.A.

INTRODUCTION

A wide class of compressible flow problems can be described by the full potential equation. This governing equation is nonlinear and may be of mixed type -- elliptic in the subsonic flow region and hyperbolic in any supersonic regions (see, for example, Bers [1958], von Mises [1958]). Many numerical studies have been made using both finite difference and finite element methods for approximate solution of full potential problems. However, there appear to be no error analyses of the approximate methods to date. In the case of the mixed subsonic-supersonic flow the operator is non-monotone and standard techniques of finite element error analysis are not applicable. Even in the case where the flow is entirely subsonic the problem is difficult, since the analysis must incorporate the constraint that the flow remain subsonic. In the present analysis, we further restrict the problem formulation by considering a class of regularized flows: this may be associated with the choice of a fictitious gas that still presents a viable approximation of the real gas flow.

The idea of regularizing the flow by introducing a fictitious gas stems from the early analytical studies of Chaplygin [1902] and later by von Karman [1941]. It has recently been applied also in numerical schemes for shock-free airfoil design using both finite difference methods (Fung et al. [1980]) and finite element methods (Pan and Carey [1984]). Here we present finite element error estimates for this regularized problem and the most commonly used low-degree triangular elements.

FORMULATION

Let ρ be the density of the gas and $\underset{\sim}{q}$ the velocity with $\underset{\sim}{q} = \underset{\sim}{\nabla}\phi$ for potential ϕ. Conservation of mass implies $\underset{\sim}{\nabla}\cdot\rho\underset{\sim}{q} = \underset{\sim}{0}$ where $\rho = \rho(q)$ can be determined from the equation of state and momentum equation. Using the adiabatic equation of state for the gas, we obtain

$$\underset{\sim}{\nabla}\cdot[(1 - \frac{\gamma-1}{2}(\underset{\sim}{\nabla}\phi)^2)^{1/\gamma-1}\underset{\sim}{\nabla}\phi] = 0 \qquad (1)$$

where γ is the gas constant ($\gamma = 1.4$ for air). Note that the choice $\gamma = -1$ yields the well-known minimal surface equation and is also the Chaplygin gas or "tangent" gas of von Karman. As such it represents (asymptotically in Mach number) an accurate approximation to the real gas density relation.

A weak formulation corresponding to (1) is obtained from the stationary condition for the variational functional

$$J(v) = \int_\Omega [1 - (\frac{\gamma-1}{2})(\underset{\sim}{\nabla}v)^2]^{\gamma/\gamma-1}\,dx \qquad (2)$$

over admissible functions v. We have shown elsewhere (Dinh and Carey, 1984) that J is well defined over $W^{1,p}$ if $\gamma > 1$ or $\gamma \leq -1$ and on H^1 if $\gamma \leq -1$. Moreover, we can also verify strict convexity and establish existence and uniqueness of a solution minimizing J provided $\gamma \leq -1$.

In the approximate problem we discretize Ω and Ω_h and construct an appropriate piecewise-polynomial subspace $H^h \subset H^1$ where h denotes the mesh parameter. Then for $\gamma \leq -1$ we may demonstrate existence and uniqueness of a solution to the approximate problem.

ERROR ANALYSIS

Lemma. Let $\rho_h = \rho(\underset{\sim}{\nabla}\phi_h)$, for approximation ϕ_h and

$$E_h = \{\int_{\Omega_h} \rho_h |\underset{\sim}{\nabla}(\phi - \phi_h)|^2 dx\}^{1/2} \tag{3}$$

If $\phi \in W^{1,\infty}(\Omega) \cap H^r(\Omega)$ with $r \geq k+1$ where k is the element degree, then $E_h \leq Ch^k$, constant $C = C(\phi)$.

Proof: Let v_h be an arbitrary function in H^h with $v_h = \phi_h$ on $\partial\Omega_h$. Then, using the variational equation and its approximation, we can show

$$E_h^2 = \int_{\Omega_h} \rho_h \underset{\sim}{\nabla}(\phi-\phi_h)\cdot\underset{\sim}{\nabla}(\phi-v_h)dx + \int_{\Omega_h} (\rho_h-\rho)\underset{\sim}{\nabla}\phi\cdot\underset{\sim}{\nabla}(v_h-\phi_h)dx \tag{4}$$

where $v_h - \phi_h$ is extended by zero on $\Omega - \Omega_h$. The integrals in (4) can be bounded to obtain (with $\nu(\phi) < 1$, constant)

$$E_h^2 \leq E_h |\phi - v_h|_{1,\Omega_h} + \nu(\phi) E_h\{E_h + |\phi - v_h|_{1,\Omega_h}\}$$

whence

$$E_h \leq \frac{1+\nu(\phi)}{1-\nu(\phi)} \inf_{v_h \in H^h} |\phi - v_h|_{1,\Omega_h}$$

and from interpolation theory we obtain the desired result. ∎

Theorem. Under the regularity assumptions in the above Lemma and provided $||\underset{\sim}{\nabla}\phi_h||_{\infty,\Omega_h} \leq C$, for $C > 0$ constant, we obtain the optimal estimate

$$|\phi - \phi_h|_{1,\Omega_h} \leq Ch^k \tag{5}$$

Proof: Direct expansion of the H^1-seminorm and some elementary calculus yield

$$|\phi - \phi_h^2|_{1,\Omega_h} \leq CE_h^2$$

Using the above Lemma, we obtain the stated estimate. ∎

It remains for us to verify the condition $||\underset{\sim}{\nabla}\phi_h||_{\infty,\Omega_h} \leq C$ for the standard elements. For brevity we state here without proof a related inequality (obtained using the Lemma and stated regularity of ϕ) - see Dinh and Carey [1984]: For ϕ satisfying the above regularity condition, there exists a constant $C = C(\phi) > 0$ such that for element Ω_e and regular discretizations

$$\frac{1}{\text{meas}(\Omega_e)} \int_{\Omega_e} \rho_h |\underset{\sim}{\nabla}\phi_h|^2 dx \leq C \tag{6}$$

It is then straightforward to verify that the boundedness condition used in the theorem holds for linear elements and a similar result can be shown to hold for quadratic elements. These estimates have been corroborated in numerical studies (Dinh and Carey, [1984]).

ACKNOWLEDGMENTS:

This research has been supported in part by the Department of Energy.

REFERENCES

[1] Bers, L., Mathematical Aspects of Subsonic and Transonic Gas Dynamics, Interscience, New York, 1958.

[2] Chaplygin, S.A., On Gas Jets, Sci. Mem., Moscow Univ. Math. Phys. Sec. 21, pp. 1-121, 1902 (trans.: NACA Tech. Note 1063, 1944).

[3] Dinh, H. and G.F. Carey, Approximate Analysis of Regularized Compressible Flow Using a Fictitious Gas Approach, J. Nonlinear Analysis (submitted Jan., 1984).

[4] Fung, K.Y., Sobieczky, H. and Seebass, R., Shock-Free Wing Design, Vol. 18, 10, 1153-1158, 1980.

[5] Karman, Th. von, Compressibility Effects in Aerodynamics, J. for Aero. Sci., 8, 337, 356, 1941.

[6] Mises, R. von, Mathematical Theory of Compressible Fluid Flow, Academic Press, New York, 1958.

[7] Pan, T.T. and G.F. Carey, "Finite Element Calculation of Shock-Free Airfoil Design, Int. J. Numer. Meth. Fluids (in press), 1984.

The final (detailed) version of this paper has been submitted for publication elsewhere.

Trends in the Theory and Practice of Non-Linear Analysis
V. Lakshmikantham (Editor)

SOME RESULTS ON NON-RESONANT NON-LINEAR DELAY DIFFERENTIAL EQUATIONS

Lance D. Drager
Department of Mathematics
Texas Tech University
Lubbock, Texas

William Layton[1]
School of Mathematics
Georgia Institute of Technology
Atlanta, Georgia

We study the non-linear delay differential equation $x'(t) + g(x(t),x(t-\tau)) = f(t)$ under a non-resonance condition which assures the existence of a unique bounded solution. Using the algebra structure of the space of bounded continuous functions we investigate the properties of this solution. We discuss some generalizations and the initial value problem.

In this paper, we will give a brief outline of some recent results on delay differential equations and integro-differential equations which satisfy a non-resonance condition.

We will denote by BC^0 the space of bounded continuous functions $\mathbb{R} \to \mathbb{R}$, equipped with the supremum norm $||\cdot||$. BC^1 will denote the space of functions in BC^0 which have one derivative in BC^0. We define $x_\tau(t) = x(t-\tau)$.

Initially, we consider the (scalar) delay differential equation

(1) $$x'(t) + g(x(t), x(t-\tau)) = f(t).$$

where τ is a fixed real number (not necessarily positive), $g: \mathbb{R}^2 \to \mathbb{R}$, and $f \in BC^0$. We are looking for solutions of (1) which are defined and bounded on the whole t-axis.

To state our non-resonance assumption, let $R = \{(a,b) \in \mathbb{R}^2 \mid |a| \leq |b|\}$. We say that $g: \mathbb{R}^2 \to \mathbb{R}$ satisfies condition (NR) if g is C^1 and $\nabla g(\mathbb{R}^2)$, the image of the gradient of g, is a positive distance away from R (see Fig. 1). We will make some motivational remarks about this condition below. The basic theorem is

Theorem (1): Let τ be fixed but arbitrary and assume g satisfies condition (NR). Then for every $f \in BC^0$ there is a unique bounded solution x of $x' + g(x,x_\tau) = f$. x is in BC^1 and we write $x = S(f)$ to indicate the dependence on f. S will be called the solution operator of (1). S is a (non-linear) bijection $BC^0 \to BC^1$.

This theorem and some of the results discussed here were discussed in [3], but under the strong additional assumption that $\nabla g(\mathbb{R}^2)$ is not only bounded away from R but is also bounded (which implies g is Lipschitz). This additional assumption was also made in our talk at this conference, but we are now able to eliminate it.

With Theorem (1) in mind, the following remarks may help to motivate condition

(NR). A very special case of (1) is the equation

(2) $$x'(t) + ax(t) + bx(t-\tau) = 0 \quad a,b \in \mathbb{R}$$

where $\nabla g(\mathbb{R}^2) = \{(a,b)\}$. This equation can be studied by the classical techniques, from which it follows that (2) has a unique bounded solution (namely the trivial solution) for all τ if and only if $(a,b) \notin R' = \{(a,b) \mid |a| < |b| \text{ or } a = -b\}$. Theorem (1) says that the existence and uniqueness of the bounded solution persists if $\nabla g(\mathbb{R}^2)$ is allowed to spread out from a point, with the additional restriction that $\nabla g(\mathbb{R}^2)$ stays away from the boundary of R' (see [7] for results on bounded solutions of the forced version of (2)). We also remark that g can satisfy condition (NR) if g is independent of the second variable, so the results here apply to ordinary differential equations $x' + g(x) = f$, where condition (NR) reduces to the requirement that g' is bounded away from zero, which is closely related to the conditions of Corduneanu [2]. The techniques used to prove Theorem (1) are in the spirit of [9]. Condition (NR) can also be thought of as a monotonicity condition and could be somewhat weakened, but the present formulation seems geometrically appealing.

We will give a skeletal outline of the proof of Theorem (1) (see also [3]), for which we will refer to the notation of Fig. 1. We first consider the linear operator $L_a: BC^1 \to BC^0$ defined by $L_a x = x' + ax$ $(a \in \mathbb{R})$. From (very) elementary differential equations we see that L_a is invertible if $a \neq 0$ and we have the formulas

(3) $$(L_a^{-1}f)(t) = \begin{cases} \int_0^\infty e^{-as} f(t-s)\,ds, & a>0 \\ -\int_{-\infty}^0 e^{-as} f(t-s)\,ds, & a<0 \end{cases}$$

from which it follows that

(4) $$||L_a^{-1}f|| \leq \frac{1}{|a|}\,||f||.$$

For purposes of the proof we consider $f \in BC^0$ as fixed and for definiteness we consider the case where $\nabla g(\mathbb{R}^2)$ lies to the right of R.

For any $a \neq 0$, we can rewrite (1) as

$$L_a x = x' + ax = ax - g(x,x_\tau) + f$$

or

(5) $$x = L_a^{-1}[ax - g(x,x_\tau)] + L_a^{-1}f.$$

Conversely, if $x \in BC^0$ satisfies (5) for <u>some</u> $a \neq 0$ we can reverse the steps and show that x is a solution of (1). Define $N_a(x) = ax - g(x,x_\tau)$ and $T_a(x) = L_a^{-1}N_a(x) + L_a^{-1}f$. We have shown that the following conditions on $x \in BC^0$ are equivalent.

(6)
a) x is a solution of (1)
b) for all $a \neq 0$, x is a fixed point of T_a
c) for some $a \neq 0$, x is a fixed point of T_a.

Now since g satisfies condition (NR), we can find $r>0$ as in Fig. 1. For $\rho>0$, let $Q(\rho) = \{(\xi,\eta)\in \mathbb{R}^2 \mid |\xi|, |\eta| \leq \rho\}$ and $B(\rho) = \{x\in BC^0 \mid ||x|| \leq \rho\}$. Since $\nabla g(Q(\rho))$ is compact, we can choose $s(\rho)>r$ as in Fig. 1. An application of the mean value theorem and consideration of the geometry of Fig. 1 gives the estimate

(7) $$||N_a(x) - N_a(y)|| \leq K(a,\rho)||x-y|| \quad x,y\in B(\rho)$$

where $K(a,\rho) = \max\{|a-r|, |a-s(\rho)|\}$. This gives

(8) $$||N_a(x)|| \leq K(a,\rho)||x|| + C, \quad x\in B(\rho)$$

where $C = |g(0,0)|$.

We can now establish an a priori estimate for solutions of (1). If x is a solution of (1), $x = T_a(x)$ for all a, so we choose $\rho \geq ||x||$, $a = \frac{1}{2}(r+s(\rho))$ and apply (4) and (8).

After some manipulation, we get the a priori estimate

(9) $$||x|| \leq (1/r)[||f|| + C]$$

if x is a solution of (1).

Now to prove the existence and uniqueness of the bounded solution of (1), fix ρ with $\rho \geq (1/r)[||f|| + C]$ and let $\bar{a} = (\frac{1}{2})(r+s(\rho))$.

It can then be shown that $T_{\bar{a}}$ maps $B(\rho)$ to itself and (using (7)) that $T_{\bar{a}}$ is a contraction on $B(\rho)$. Thus $T_{\bar{a}}$ has a unique fixed pt in $B(\rho)$.

By (6) and the a priori estimate, any fixed point of $T_{\bar{a}}$ must be in $B(\rho)$, so $T_{\bar{a}}$ has a unique fixed point, which proves Theorem (1).

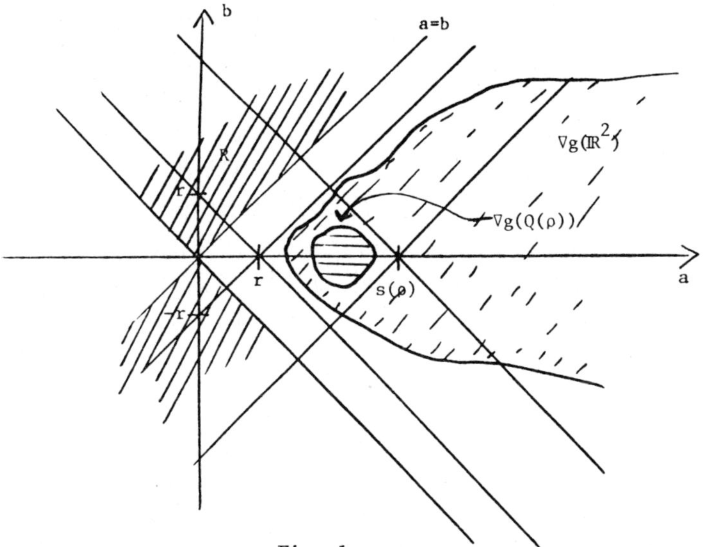

Fig. 1

If $\nabla g(\mathbb{R}^2)$ lies to the right [resp. left] of R we will say that g is forward [backward] non-resonant or FNR [BNR].

Now, we can make the following simple observation. If $A \subseteq BC^0$ is a closed subspace which satisfies the conditions (A1)-(A3) below, it is easy to see that if $f \in A$, $T_a(A) \subseteq A$. It then follows from the proof of Theorem (1) and the contraction mapping lemma that the bounded solution of (1) is in A. Thus, we have

Theorem (2): Let τ be fixed and assume that g satisfies condition (NR). Let $A \subseteq BC^0$ be a closed subspace satisfying the following conditions:

(A1) if $x \in A$, $x_\tau \in A$

(A2) $L_a^{-1} A \subseteq A$, for all $a>0$ [resp. $a<0$] if g is FNR [BNR]

(A3) if $x,y \in A$, $g(x,y) \in A$.

Then $S(A) \subseteq A$, ie. if $f \in A$, the bounded solution of (1) is in A.

From (3), condition (A2) is closely related to translation invariance of A. In fact, if A is contained in the uniformly continuous functions and is invariant under the appropriate translation semi-group, (A2) follows. In cases where A is not contained in the uniformly continuous functions it is often possible to give an *ad hoc* argument.

Since we are dealing with bounded functions an (easy) standard argument shows that condition (A3) is satisfied if A is a subalgebra of BC^0 and either A contains the constants or $g(0,0) = 0$. This is proved by approximating g by polynomials on the image of (x,y).

We will give some examples of the application of Theorem (2). If g satisfies condition (NR) and τ is arbitrary, it is easily checked that the following classes of functions are closed subalgebras of BC^0 satisfying (A1)-(A3), so if f is in the given class, the bounded solution of (1) is in the same class.

a) Constant functions

b) Periodic functions of a fixed period

(10) c) Uniformly almost periodic functions. It can be shown in this case that $\mathrm{mod}(S(f)) = \mathrm{mod}(f)$, where $\mathrm{mod}(f)$ denotes the frequency module of f. (see also [4],[5],[8] for results on Besicovitch almost periodic solutions of (1)).

d) Weakly almost periodic functions (see [6] for relevant material on this class of functions).

e) Functions which approach a steady state at ∞, ie. functions f for which $\lim_{t\to\infty} f(t)$ exists.

By a refinement of the techniques above, a similar treatment can be given of the case of multiple delays or distributed delays (including infinite delay). In fact, consider the integro-differential equation

(11) $$x'(t) + \int g(x(t), x(t-s))\,d\mu(s) = f(t)$$

where we assume that μ is a finite (positive) Borel measure on $\mathbb{R}$ (possibly supported on the whole line). It can be shown that if g satisfies condition (NR), (11) has a unique bounded solution and Theorem (2) applies. In particular, if f is in one of the classes of functions in (10), the bounded solution of (11) is in the same class. Another closely related equation to which these techniques can be applied is

(12) $$x'(t) + \int x(t-s)d\mu(s) + g(x(t)) = f(t)$$

where we assume that μ is a finite signed Borel measure. (12) is essentially the equation (scalar version) considered by Alexiades [1]. The non-resonance condition for (12) is, as in [1], $\int d|\mu| < \inf |g'|$.

The algebra structure of BC^0 can be used to obtain another class of results on the solutions of non-resonant equations. For simplicity, we restrict the discussion to equation (1).

As a motivating example, assume that

(13) a) g is FNR and $\tau \geq 0$

b) $g(0,0) = 0$

and let $K = \{f \in BC^0 | f(t) = 0 \text{ for } t \leq 0\}$.

It is easy to see that K is a closed subalgebra of BC^0 which is invariant under forward translations, and it is easy to see that under the assumptions in (13) K satisfies conditions (A1)-(A3). Thus, if $f \in K$ the bounded solution $x = S(f)$ of (1) is also in K. This can be interpreted as a causality result: no forcing is applied to the system until $t = 0$ and the system does not depart from the equilibrium solution $x = 0$ until $t = 0$. From (3) we see that the conditions (13) a) are natural for causality (essentially (13) b) just says that the equilibrium solution is $x \equiv 0$). This naturally leads us to ask if the operator S is causal under the conditions (13) a), ie. if $f_1(t) = f_2(t)$ for $t \leq 0$, is it true that $S(f_1)(t) = S(f_2)(t)$ for $t \leq 0$? (We have shown this if $f_1 \equiv 0$). This is indeed true, and the crucial observation in proving it is that K is actually an <u>ideal</u> in BC^0. If $f_1(t) = f_2(t)$ for $t \leq 0$, then $f_2 = f_1 + h$ where $h \in K$, so the result follows from the following theorem.

<u>Theorem (3)</u>: Let τ be fixed and assume g satisfies condition (NR). Let $I \subseteq BC^0$ be a closed ideal which satisfies conditions (A1) and (A2). Then if $f \in BC^0$ and $h \in I$ we have $S(f+h) = S(f) + \tilde{h}$, where $\tilde{h} \in I$.

The theorem is proved by applying arguments similar to the above to the linearization of (1) and then integrating (in BC^0).

Another interesting application of Theorem (3) is the following. Let τ be arbitrary, assume g satisfies condition (NR) and suppose that f_1 and f_2 are asymptotic to each other at $+\infty$, ie. $\lim_{t\to\infty} |f_1(t)-f_2(t)| = 0$. It then follows that the solutions $x_1 = S(f_1)$, $x_2 = S(f_2)$ are asymptotic at $+\infty$. For example, if f is asymptotically almost periodic, so is the bounded solution of (1).

The results above tell us something about the particular solution of (1) that is defined and bounded for all t. What can we say about the behavior of the rest of the solutions of (1), such as the solution of an initial value problem? The following result shows that the behavior at ∞ of the solution to an initial value problem for (1) is the same as that of S(f).

Theorem (4): Let $\tau \geq 0$ be fixed and assume that g is FNR. Assume that f is continuous and bounded on $[0,\infty)$ and that the initial function φ is continuous on $[-\tau,0]$. Then the unique maximally defined solution x of the initial value problem

$$(14)\qquad \begin{cases} x(t) = \varphi(t), & -\tau \leq t \leq 0 \\ x'(t) + g(x(t), x(t-\tau)) = f(t), & t \geq 0 \end{cases}$$

is defined and bounded on $[-\tau,\infty)$. Furthermore, if x_1, x_2 are two solutions of (14) for different initial functions φ_1, φ_2, there are constants C, $\lambda>0$ so that

$$|x_1(t) - x_2(t)| \leq Ce^{-\lambda t}.$$

As usual, x'(0) in (14) is interpreted as the right handed derivative. If $\tau \leq 0$ and g is BNR there is a corresponding result as $+ \to -\infty$.

The following remarks may help to make this theorem reasonable. It is plainly correct for the very special case of the equation $x' + ax = f$ (where FNR means $a>0$) and its not difficult to prove it for the non-linear ordinary differential equation. Further, the FNR region (to the right of R) is known to be the region for exponential stability of the trivial solution of $x' + ax + bx_\tau = 0$ (see [7]).

Theorem (4) is proved by a contraction mapping argument in the same spirit as the proof of Theorem (1).

As an example, if f is almost periodic (or even asymptotically almost periodic) the solution of (14) is asymptotically almost periodic.

[1]Work of the second author partially supported by NSF grant MCS 8202025.

References

[1] Alexiades, V., Almost periodic solutions of an integro-differential system with infinite delay, Nonlinear Analysis, Thy. Methods and Appls. 5 (1981) pp. 401-410.

[2] Corduneanu, C., Systémes differentiels admettant des solutions bornées, C.R. Acad. Sci. Paris, 245 (1957), pp. 21-24. Erratum, Ibid p. 1095.

[3] Drager, L and Layton, W., Non-Linear Delay Differential Equations and Function Algebras, Differential Equations, I. Knowles and R. Lewis (eds.), North-Holland, Amsterdam, 1984.

[4] Drager, L. and Layton, W., Non-resonance in functional equations with small time lag, to appear in proc. of Third International Conf. on Functional Differential Systems and Related Topics, Blazjewko, Poland, May 1983.

[5] Drager, L. and Layton, W., On non-linear difference approximations to non-linear functional equations, Libertas Mathematica, 3 (1983) (Arlington, TX).

[6] Eberlein, W.F., Abstract ergodic theorems and weak almost periodic functions, Trans. Amer. Math. Soc. 67 (1949), pp. 217-240.

[7] Hale, J.K., Theory of Functional Differential Equations, 2nd ed, Springer, NY, 1977.

[8] Layton, W., Existence of almost periodic solutions to delay differential equations with Lipschitz non-linearities, to appear.

[9] Mawhin, J., Contractive mappings and periodically perturbed conservative systems, Séminarie de Mathematiques Appliquées et Mécanige, Rapport 80, Louvain-la-Neuve, Presses Universitaires, 1974.

The final (detailed) version of this paper will be submitted for publication elsewhere.

Trends in the Theory and Practice of Non-Linear Analysis
V. Lakshmikantham (Editor)

ASYMPTOTICS OF NUMERICAL METHODS FOR NONLINEAR EVOLUTION EQUATIONS

Lance D. Drager
Mathematics Department
Texas Tech. University
Lubbock, Texas

William Layton*
School of Mathematics
Georgia Institute of Technology
Atlanta, Georgia

Robert M. M. Mattheij
Mathematics Institute
Catholic University
Nijmegen, The Netherlands

We consider the evolution equation

$$(E)\quad \frac{du}{dt} = Au + f(t,u), \quad u(0) = u_0$$

in a Hilbert space H, where A is assumed to generate a C^0 semigroup. The asymptotic-in-time behavior of solutions to (E) is studied together with the asymptotic-in-time behavior of methods for approximating (E).

1. INTRODUCTION

We consider in this report how well various discretizations model the time-asymptotic behavior of solutions to the following evolution equation: $u: [0,\infty) \to H$,

$$u'(t) = Au + f(t,u), \quad u(0) = u_0 \in H \tag{1.1}$$

Here, $\{H, <\cdot,\cdot>, ||\cdot||\}$ is a Hilbert space, $A: \mathcal{D}(A) \subset H \to H$ is a, possibly unbounded, operator to which the spectral theorem (in one of its many forms) applies. Further, A is assumed to be semibounded:

$$<Aw,w> \leq \alpha ||w||^2, \quad \forall w \in \mathcal{D}(A) \subset H. \tag{1.2}$$

$f: [0,\infty) \times H \to H$ is a gradient map satisfying the monotonicity condition:

$$f_0 \leq \frac{<f(t,u)-f(t,v),u-v>}{||u-v||^2} \leq f_1, \quad \forall u,v \in H,\ t \in [0,\infty).$$

In the last section, f will take a more special form: $f(t,u) = g(u) + \gamma(t)$ where g is a substitution ("Nemyetskii") operator.

We restrict ourselves to a common and representative class of discretization methods: projection type in the "spacial" variables coupled with the θ-method in time. Let $0 < \Delta t < 1$ be given, $t_n = n\Delta t$, $S \subset \mathcal{D}(A) \subset H$ be a subspace and $P: H \to S$ a projection operator.

*Work of the second author partially supported by N.S.F. grant M.C.S. 8202025.

Then, $S \ni U_n \cong u(t_n)$ is calculated via

$$\frac{U_{n+1}-U_n}{\Delta t} = \tilde{A}U_{n+\theta} + Pf_{n+\theta}(U_n), \quad n = 0,1,2,\ldots . \tag{1.3}$$

Here, $\tilde{A} = PAP$, $U_{n+\theta} = \theta U_n + (1-\theta)U_{n+1}$, and $f_{n+\theta}(U_n) = \theta f(t_n,U_n) + (1-\theta)f(t_{n+1},U_{n+1})$. We assume $\tilde{A}$ is semibounded on S

$$\langle\tilde{A}\phi,\phi\rangle \le \tilde{\alpha}\,||\phi||^2, \quad \forall\phi \in S, \tag{1.4}$$

e.g., for the Galerkin method P is the orthogonal projection into S so (1.4) follows from (1.2) with $\tilde{\alpha} = \alpha$. Since we are concerned with unconditionally stable schemes, we restrict θ: $0 \le \theta \le 1/2$.

In Section 2, we present error estimates for the method (1.3) that are uniform in time. Many other researchers have obtained such estimates for specific problems beginning in 1949 with Loud [13]. Later this was taken up by e.g., Dahlquist [5], Nevanlinna [15], Layton and Mattheij [12], and others for multistep methods for O.D.E.'s. An extensive literature exists for related estimates for parabolic P.D.E.'s, see e.g., Schultz [17, Thm. 9.12], Axelsson [1], Gekeler [8], with much less work along these lines on hyperbolic systems of P.D.E.'s, Gustafsson [9], Layton [10], [11]. Other references can be obtained by consulting these papers or the articles by Axelsson or Nevanlinna in this proceedings.

In Sections 3 and 4, we consider the more difficult and important question of how well the asymptotic behavior of (1.3) actually represents that of (1.1). To study this, it is convenient to work directly with the limiting equations associated with (1.1) and (1.3). The cases when the forcing term is asymptotically periodic or almost periodic or tends to a limiting value at infinity are considered.

Due to space limitations sometimes only the critical steps of proofs are given.

2. UNIFORMLY VALID ERROR ESTIMATES

Theorem 2.1. There is a $\phi(t)$ such that for Δt sufficiently small w.r.t. f_0, f_1,

$$\begin{aligned} ||u(t_n)-U_n|| \le C\phi(t)\,[\,&||u(0)-U_0|| \\ &+ \max_{0\le s\le t_n} \{(\Delta t)^k\,||D_t^{(k+1)}u(s)|| \\ &+ ||(A-\tilde{A})u(s)|| + ||(I-P)f(s,u(s))||\,\}], \end{aligned}$$

where $k = 2$ if $\theta = 1/2$ and $k = 1$ otherwise. Here, as $t \to \infty$,

$$\phi(t) = \begin{cases} O(e^{\beta t}), & \text{if } \beta \equiv \alpha + f_1 > 0 \\ O(t), & \text{if } \beta = 0 \\ O(1), & \text{if } \beta < 0. \end{cases}$$ □

The bulk of the previous theorem follows from standard arguments (e.g., the articles by Axelsson [2] and Nevanlinna [16]). Noteworthy, however, is the fact that <u>no smoothness assumptions on $f(u)$ are required: f need not even be C^0 in u!</u>

We will give a very brief sketch of the proof of the above to indicate how optimal convergence rates can be obtained even with, e.g., discontinuous nonlinearities.

<u>Proof</u>. Let $u_n \equiv u(t_n)$, $e_n = u_n - U_n$. e_n then satisfies the equation ($\tilde{\alpha} + \mu < 0$)

$$\begin{aligned} \frac{e_{n+1}-e_n}{\Delta t} - (\tilde{A}+\mu I)[e_{n+\theta}] &= P[\theta(f_n-\mu)(u_n)-\theta(f_n-\mu)(U_n) \\ &+ (1-\theta)(f_{n+1}-\mu)(u_{n+1}) - (1-\theta)(f_{n+1}-\mu)(U_{n+1})] \\ &+ r_{n+\theta} \equiv R_{n+\theta} \end{aligned} \tag{2.1}$$

where $r_{n+\theta} = \theta r(t_n) + (1-\theta)r(t_{n+1})$ is given by

$$\begin{aligned} r_{n+\theta} = \frac{u_{n+1}-u_n}{\Delta t} - u'_{n+\theta} + (A-\tilde{A})u_{n+\theta} \\ + (I-P)[\theta f(t_n,u_n) + (1-\theta)f(t_{n+1},u_{n+1})]. \end{aligned}$$

By regarding the R.H.S. of (2.1) as a forcing term (2.1) can be summed and estimated (via the spectral mapping theorem) to give:

$$\sup_{0\le n<\infty} \|e_n\| \le \|e_0\| + \kappa \sup_{0\le n<\infty} \|R_{n+\theta}\| \tag{2.2}$$

where (we only concern ourselves with one possibility to abbreviate the proof) $\kappa \le |\tilde{\alpha}+\mu|^{-1}$, provided $|\Delta t(\mu+\tilde{\alpha})\theta| \le 1$.

Now, the nonlinear terms in $R_{n+\theta}$ can be estimated by using Mawhin [14; Lemma 1], as

$$\|(f_n-\mu)(u_n)-(f_n-\mu)(U_n)\| \le \max\{|f_0-\mu|,|f_1-\mu|\}\,\|u_n-U_n\|$$

so that (2.2) becomes

$$\begin{aligned} \sup_{0\le n<\infty}\|e_n\| \le \|e_0\| + \frac{\max\{|f_0-\mu|,|f_1-\mu|\}}{|\tilde{\alpha}+\mu|} \sup_{0\le n<\infty}\|e_n\| \\ + \kappa \sup_{0\le n<\infty}\|r_{n+\theta}\|. \end{aligned}$$

The theorem follows by noticing that there is a choice of μ compatible with the above operations that makes

$$\max\{|f_0-\mu|,|f_1-\mu|\} < |\alpha + \mu|,$$

see Layton and Mattheij [12] for the details of the previous argument. □

We stress that such uniformly valid error estimates are only the first step to understanding the time asymptotics of numerical methods. A far more important question is: How does the difference equations model the qualitative asymptotic behavior of the differential equation? We will attempt to give some partial results in the next two sections.

3. LIMITING EQUATIONS: CONTINUOUS AND DISCRETE

To study the qualitative properties of the asymptotic behavior of an evolution problem, independent of the initial transients, it is useful to pass to the associated limiting equation. Next, we describe briefly how these arise for (1.1), (1.3). For more details on limiting equations, see the article by Burton [3]. We will suppress the (subordinate) role of the spatial discretization in the following.

Let μ be such that $\mu + \alpha < 0$, the solution $u(t)$ to (1.1) satisfies

$$u(t) = e^{(A+\mu I)t} u_0 + \int_0^t e^{(A+\mu I)(t-s)} [f(s,u(s))-\mu u(s)]ds.$$

Letting $u_m(t) = u(t+t_m)$, $t_m \to +\infty$ we find that (under reasonable conditions on f), by passing to a subsequence if necessary, $u_m(t) \to u^*(t)$ in $L^\infty(-\infty,\infty;H)$ satisfying

$$u^*(t) = \int_{-\infty}^t e^{(A+\mu I)(t-s)} [f^*(s,u(s))-\mu u^*(s)]ds \tag{3.1}$$

with $f^*(s,z) = \lim_{m\to\infty} f(s+t_m,z)$. Likewise, with $\tilde{\alpha}+\mu < 0$, $\{U_n\}$ satisfies

$$U_n = A^n U_0 + \sum_{j=1}^n A^{j-1} \zeta_j,$$

where $A = r(A) = p(A)/q(A)$, $p(z) = 1 + \Delta t\theta(z+\mu)$, $q(z) = 1 - \Delta t(1-\theta)(z+\mu)$ and $\zeta_n = \Delta t q(A)^{-1}[f(U_n)-\mu U_n]_{n+\theta}$. Letting

$$\{U_n^{(m)}\} = \{U_{n+m}\}, \qquad \zeta_n^{(m)} = \zeta_{n+m}$$

and $m \to \infty$ gives that $\{U_n^{(m)}\} \to \{U_n^*\}$ the solution of

$$U_n^* = \sum_{j=1}^\infty A^{j-1} \zeta_{n-j}^* \tag{3.2}$$

where

$$\zeta_n^* = \lim_{m\to\infty} \zeta_n^{(m)}$$

is simply ζ_n with f replaced by f*. The significance of the limiting equations (3.1), (3.2) is explained by the following

Theorem 3.1. As $t \to \infty$, $u(t) \to u^*(t)$, in H. As $n \to \infty$, $U_n \to U_n^*$ in H. □

By modifiying the proof of Theorem 2.1 as in Drager and Layton [7], it follows that U_n^* is an optimal approximation to u^* on the entire real line:

Theorem 3.2.

$$\sup_{-\infty<t<\infty} \|u^*(t_n)-U_n^*\| \leq C(\Delta t)^k \sup_{t\in\mathbb{R}} \|D_t^{(k+1)}u^*\|$$

where $k = 2$ if $\theta = 1/2$, $k = 1$ otherwise. □

Thus, the behavior of the discrete limiting equation is within $O(\Delta t)^k$ of that of the continuous limiting equation on all of $\mathbb{R}$.

4. QUALITATIVE BEHAVIOR

We begin this section with a simple example. Consider the linear equation $u^{*\prime} + u^* = f^*e^{i\lambda t}$, and the difference approximation ($\theta = 0$)

$$\frac{U_{n+1}^*-U_n^*}{\Delta t} + U_{n+1}^* = f^*e^{i\lambda t_n}, \quad t_n = n\Delta t, \; n \in \mathbb{Z}. \tag{4.1}$$

$u^*(t)$ is periodic with period $2\pi\lambda$. The substitution $U_n^* = U^*e^{i\lambda t_n}$ gives

$$U_n^* = f^*\left[\frac{e^{i\lambda\Delta t}-1}{\Delta t} + e^{i\lambda\Delta t}\right]^{-1}e^{i\lambda t_n}.$$

Thus, the solution to (4.1) is the restriction to the mesh points of a periodic function with the same period as $u^*(t)$. However, $\{U_n\}$ will be a periodic function of n only when $k2\pi\lambda(\Delta t)^{-1}$ is an integer for $k \in \mathbb{Z}$. If this is not true, then $\{U_n\}$ will be quasi-periodic as a discrete process (Corduneanu [4]) even though it is really periodic as a function of t.

The most mathematically convenient way around this is to regard U as a function of a continuous variable $U = U(t)$ and read, e.g., (4.1) as

$$\frac{U^*(t+\Delta t)-U^*(t)}{\Delta t} + U^*(t+\Delta t) = f^*e^{i\lambda t}, \quad t \in \mathbb{R}.$$

We adopt this convention without further comment.

To elucidate the main ideas we concentrate on the case of a forced, autonomous equation:

$$f(t,u) = g(u) + \gamma(t)$$

where $\gamma(t)$ is, e.g., asymptotically almost periodic, asymptotically periodic in t or asymptotically constant as a map $\gamma: [0,\infty) \to H$. The convergence result in Theorem 3.2 immediately gives a result asymptotically in Δt.

Theorem 4.1. Suppose $\gamma(t)$ is asymptotically almost periodic in t. Then, for $\lambda \in \mathbb{R}$,

$$\lim_{\Delta t \to 0} \lim_{T \to \infty} \frac{1}{2T} \int_{-T}^{T} e^{-i\lambda t}(u_n^*(t) - U_n^*(t))dt = 0.$$

I.e., $\{U_n\}$ is asymptotically (in Δt), asymptotically (in t_n) almost periodic. □

Let $\Lambda = \text{mod}\ \{\lambda_j\}$ be the module of real numbers generated by the Fourier exponents of the almost periodic part of $\gamma(t)$, i.e., of $\gamma^*(t)$.

Theorem 4.2.

(a) U^* exists and is unique in $L^\infty(\mathbb{R};H)$.

(b) If γ is asymptotically almost periodic, then $U^*(t)$ is almost periodic.

(c) The Fourier exponents of $U^*(t)$ lie in $\Lambda = \text{mod}\ \{\lambda_j\}$.

Corollary 4.1. If $\gamma(t)$ is asymptotically almost periodic, then $\{U(t)\}$ is asymptotically (in t_n only!) almost periodic.

The Fourier exponents of the almost periodic part of $\{U(t)\}$ lie in the module generated by those of the almost periodic part of γ. □

Corollary 4.2. If $\gamma(t)$ is asymptotically periodic, then $\{U_n\}$ is the restriction to $\{t_n, n \in \mathbb{Z}^+\}$ of an asymptotically periodic function with the same asymptotic period. □

Proof of Theorem 4.2 (sketch). $U^*(t)$ satisfes $U^*(t) = F(U^*) + \Gamma^*(t)$ where

$$\Gamma^*(t) = \Delta t \sum_{j=1}^{\infty} A^{j-1} q(A)^{-1} \gamma^*(t_n)_{n+\theta} \in L^\infty(\mathbb{R};H),$$

μ is a free parameter such that $\tilde{\alpha}+\mu < 0$ and

$$F(U) = \sum_{j=1}^{\infty} A^{j-1} \Delta t q(A)^{-1} [g(U_n) - \mu U_n]_{n+\theta}.$$

A calculation similar to the proof of Theorem 2.1 reveals that for μ sufficiently negative F is a contraction on $L^\infty(\mathbb{R};H)$. Thus (a) is proven.

For (b) note that $\gamma^*(t)$ is the almost periodic part of γ. Thus, Γ^* is almost periodic with the same Fourier exponents as γ^*. Let V denote the closed subspace $V = \text{closure}\ [\text{span}\{e^{i\lambda t} | \lambda \in \Lambda\}]$, with the closure taken in the uniform norm. Then, a rather involved calculation, similar in spirit to Drager and Layton [6] or [7] reveals that $F: V \to V$.

Since $\Gamma^* \in V$, this implies that $U^* \in V$ since U^* is the limit of the contraction iterates. This simultaneously proves (b) and (c). □

Consider the case when u approches a limiting value at $t \to \infty$. Specifically, assume

$$\gamma(t) \to \gamma_\infty \in H \text{ as } t \to \infty. \tag{4.2}$$

Then, the following proposition is well-known to the differential equations/semigroup folklore.

Proposition 4.1. Under (4.2), $u(t) \to u_\infty \in H$ as $t \to \infty$. u is the unique solution in H of

$$Au_\infty + g(u_\infty) + \gamma_\infty = 0. \tag{4.3}$$

□

The following theorem shows that the approximation models this behavior accurately.

Theorem 4.3. (a) The approximate solution approaches a limit $U(t_n) \to U_\infty$ as $t_n \to \infty$.

(b) Assuming $D_t^{(k+1)}u \in L^\infty(0,\infty;H)$ $\|u_\infty - U_\infty\| = O(\Delta t^k)$.

(c) $U_\infty \in H$ is determined uniquely by the equations:

$$AU_\infty + g(U_\infty) + \gamma_\infty = 0. \tag{4.4}$$

(d) Thus, $U_\infty \equiv u_\infty$.

Proof. (a) follows from the almost periodic result in Theorem 4.2 on the limiting equation by taking $\Lambda = \text{mod}\,\{0\} = \{0\}$. (b) is a direct consequence of the basic error estimate in Theorem 3.2. (c) is a consequence of (a) since if $U(t_n) \to U_\infty$, $U(t_{n+1}) - U(t_n) \to 0$. (d) follows since (4.4), (4.3) are exactly the same equation and (by an energy argument) its solution is unique. □

Thus, the discrete-time, continuous space approximation produces the exact limit at infinity. If the fully discrete approximation (1.3) is used the equation for U_∞ must be modified to read: $U_\infty \in S \subset H$ $AU_\infty + P(g(U_\infty) + \gamma_\infty) = 0$, which gives an quasioptimal approximation in S to u_∞.

REFERENCES

[1] O. Axelsson, Error Estimates for Galerkin Methods for Quasilinear Parabolic and Elliptic Differential Equations in Divergence Form, Numer. Math., V. 28 (1977), 1-14.

[2] O. Axelsson, article in this proceedings.

[3] T. Burton, Periodic Solutions of Volterra Equations, article in this proceedings.

[4] C. Corduneanu, Almost Periodic Discrete Processes, Libertas Math., V. 2 (1982), 159-169.

[5] G. Dahlquist, Error Analysis for a Class of Methods for Stiff, Nonlinear, Initial Value Problems, pp. 60-74 in L.N.M. Vol. 506, Springer-Verlag, New York, 1976.

[6] L. Drager and W. Layton, Qualitative Properties of Bounded Solutions to Nonresonant Delay Equations and Algebras and Ideals of

Bounded Continuous Functions.

[7] L. Drager and W. Layton, On Nonlinear Difference Approximations to Nonlinear Functionals Differential Equations, Libertas Math. V. 3 (1983), 45-65.

[8] E. Gekeler, A-Convergence of Finite Difference Approximations of Parabolic Initial-Boundary Value Problems, SIAM J.N.A., V. 12 (1975), 1-12.

[9] B. Gustafsson, On Difference Approximations to Hyperbolic Differential Equations Over Long Time Intervals, SIAM J.N.A., V. 6 (1969), 508-522.

[10] W. Layton, Local Error Estimates for Finite Difference Approximations to Hyperbolic Equations for Large Time, Proc. A.M.S., to appear.

[11] W. Layton, On the Behavior over Long Time Intervals of Finite Difference and Finite Element Approximations to First Order, Hyperbolic Systems, Comp. and Math. w. Appls., to appear.

[12] W. Layton and R. Mattheij, Estimates over Infinite Intervals of Approximations to Initial Value Problems, Report 8338, Math. Dept., Cath. Univ. Nijmegen, The Netherlands, 1983.

[13] W. S. Loud, On the Long-Run Error in the Numerical Solution of Certain Differential Equations, J.M.P., V. 28 (1949), 45-49.

[14] J. Mawhin, Semilinear Equations of Gradient Type in Hilbert Space and Applications to Differential Equations, pp. 269-282, in: Nonlinear Diff. Eqns.: Invariance Stability and Bifurcation, Academic Press, New York, 1971.

[15] O. Nevanlinna, On the Behavior of Global Errors at Infinity in the Numerical Integration of Stable Initial Value Problems, Numer. Math., V. 28 (1977), 445-454.

[16] O. Nevanlinna, article in this proceedings.

[17] M. H. Schultz, Spline Analysis, Prentice Hall, Englewood Cliffs, N.J., 1973.

The final (detailed) version of this paper will be submitted for publication elsewhere.

Trends in the Theory and Practice of Non-Linear Analysis
V. Lakshmikantham (Editor)

EXPONENTIAL DICHOTOMY OF NONLINEAR SYSTEMS OF ORDINARY DIFFERENTIAL EQUATIONS

S. Elaydi*

University of Colorado
Colorado Springs, Colorado 80933 U.S.A.

O. Hajek

Case Western Reserve University
Cleveland, Ohio 44106 U.S.A.

INTRODUCTION

A dichotomy, exponential or ordinary, is a type of conditional stability. Roughly speaking, a linear differential equation possesses a dichotomy if there exists an invariant splitting or a continuous decomposition of the Euclidean space into stable and unstable subspaces. The concept of dichotomy was first formulated by Massera and Schaffer [16, 17, 18] who demonstrated its effectiveness in dealing with the problems of asymptoticity, boundedness and admissability in linear differential equations. Later Daleckii and Krein [8] published a book which includes a chapter on dichotomy. This was followed by the book of Fink [10] in which the author investigated exponential dichotomies of almost periodic systems. An excellent exposition on the subject was given in 1978 by Coppel [6]. Some material on dichotomy may be also found in an earlier book by Coppel [5]. A notable contribution was made by Sacker and Sell 24, 25, 26 , Palmer [20, 21, 22, 23], Lazer [11], Berkey [1] and Zikov [28]. In [15], [19] Martin and Muldowney, respectively, introduced a generalized dichotomy which incorporates both exponential and ordinary dichotomies. Another generalization of ordinary dichotomy was given recently by Cecchi et al in [2]. The later notion is called L^P- dichotomy, $1 \leqslant P \leqslant \infty$.

For nonlinear systems, however, the concept of dichotomy, to the best of our knowledge, has not appeared previously in the literature. The main purpose of this paper is to give an account on this subject. It should be pointed out here that some relevant work on nonlinear systems was given by Chang [3], Corduneanu [7], Fink [10], Palmer [21], and Martin [15].

The paper is divided into two parts. Part I is a survey of known results on exponential dichotomies of linear systems. Part II consists of new results on exponential dichotomies of nonlinear systems.

PART I: DICHOTOMIES OF LINEAR SYSTEMS

GENERAL THEORY

Consider the system

$$x' = A(t)\,x \tag{I.1}$$

where $A(t)$ is a continuous matrix on $t \in J = (\omega_-, \omega_+) \subset R$ (usually $J = [0,\infty]$, $[-\infty, 0]$ or R). Let $X(t)$ be a fundamental matrix of (I.1) with $X(0) = I$. Then equation (I.1) is said to have an exponential dichotomy [5, 6, 8, 10, 17] if there exists a projection P and positive constants $\sigma_1, \sigma_2, K_1, K_2$ so that

*Part of this work was done while the first author was visiting Case Western Reserve University.

$$|X(t)\,PX^{-1}(s)| \leqslant K_1 e^{-\sigma_1 (t-s)}, \quad t \geqslant s$$
$$|X(t)\,(I-P)\,X^{-1}(s)| \leqslant K_2 e^{-\sigma_2 (s-t)}, \quad t \leq s \tag{I.2}$$

The equation (I.1) is said to have an ordinary dichotomy if the inequalities (I.2) hold with $\sigma_1 = \sigma_2 = 0$.

Condition (I.2) is equivalent to the following:

$$|X(t)\,P\,\theta| \leqslant L_1 e^{-\sigma_1 (t-s)} |X(s)\,P\theta|, \; t \geqslant s$$
$$|X(t)\,(I-P)\,\theta| \leqslant L_2 e^{-\sigma_2 (s-t)} |X(s)\,(I-P)\,\theta|, \; t \leqslant s \tag{I.3}$$

$$|X(t)\,PX^{-1}(t)| \leqslant M \quad \text{for all } t \in J. \tag{I.4}$$

If the projection P on R^n has rank r, then condition (I.3) states that the solution space of (I.1) has two supplementary subspaces F_1 and F_2 (of dimension r and n-r) of solutions x(t) that

$$|x(t)| \leqslant L_1 e^{-\sigma_1 (t-s)} |x(s)|, \; t \geqslant s, \; x(t) \in F_1$$

$$|x(t)| \leqslant L_2 e^{-\sigma_2 (s-t)} |x(s)|, \; t \leqslant s, \; x(t) \in F_2$$

The supplementary projections onto these subspaces are the functions $X(t)\,PX^{-1}(t)$ and $X(t)\,(I-P)\,X^{-1}(t)$. Condition (I.4) states, simply, that these projections are bounded on J.

We now state the generalized dichotomy of Martin and Muldowney [15], [18]. Let β_1, β_2 be continuous real valued functions on J. Then the equation (I.1) is said to have $(\beta_1-\beta_2)$ dichotomy if there exists a projection P on R^n such that

$$|X(t)\,PX^{-1}(s)| \leqslant K_1 \exp\left(\int_s^t \beta_1\right) \quad \text{if} \quad t \geqslant s$$
$$|X(t)\,(I-P)\,X^{-1}(s)| \leqslant K_2 \exp\left(\int_s^t \beta_2\right) \text{ if } t \leqslant s \tag{I.5}$$

Notice that if β_1 and β_2 are constants, then condition (I.5) defines an exponential dichotomy. On the other hand if $\beta_1 = \beta_2 = 0$, then we have an ordinary dichotomy.

Now we turn to the L^P- dichotomy of Cecchi et al [2] which generalizes the ordinary dichotomy. We say that equation (I.1) has an L^P- dichotomy if there exist a constant K and a projection Q such that

$$\left[\int_0^t |X(t)\,QX^{-1}(s)|^P ds\right]^{\frac{1}{P}} \leqslant K \quad \text{if} \quad s \leqslant t$$
$$\left[\int_0^t |X(t)\,(I-Q)\,X^{-1}(s)|^P ds\right]^{\frac{1}{P}} \leqslant K \quad \text{if} \quad t \leqslant s \tag{I.6}$$

when $1 \leqslant P < \infty$, or

$$|X(t)\,QX^{-1}(s)| \leqslant K \quad \text{if } s \leqslant t$$
$$|X(t)\,(I-Q)\,X^{-1}(s)| \leqslant K \quad \text{if } t \leqslant s \tag{I.7}$$

when P = ∞. Condition (I.7) or the L^{∞} dichotomy is the ordinary dichotomy. In this paper we will not discuss further these two generalizations of dichotomy.

Let us go back to discuss the exponential dichotomy of (I.1). Now if A(t) = A is a constant matrix, then the system (I.1) has an exponential dichotomy if all the eigenvalues of A are off the imaginary axis. It has an ordinary dichotomy if all the eigenvalues of A with zero real part are semisimple.

If A(t) is periodic, then by Floquet theory we have $X(t) = Q(t)\, e^{Bt}$, where Q(t) is periodic and B is constant. In this case $x' = A(t)x$ possesses an exponential dichotomy if $y' = By$ has an exponential dichotomy. Thus the case of periodic systems may be reduced to the autonomous linear systems.

Note that Q(t) must satisfy the equation

$$Q'(t) = A(t)\,Q(t) - Q(t)\,B(t) \tag{I.8}$$

In [13] Markus generalized the above idea to the notion of kinematic similarity between any two nonautonomous systems. The system (I.1) is said to be kinematically similar to

$$y' = B(t)y \tag{I.9}$$

if there exists an invertible, continuously differentiable matrix Q(t) such that Q(t) and $Q^{-1}(t)$ are bounded and satisfy (I.8). It is clear that the change of variables $x(t) = Q(t)\,y(t)$ changes (I.1) into (I.9). If Y(t) is a fundamental matrix of (I.9), then $Y(t) = Q^{-1}(t)\,X(t)$. Hence if (I.1) and (I.9) are kinematically similar, then (I.1) has an exponential dichotomy iff (I.9) has an exponential dichotomy. Similar conclusions hold for ordinary dichotomy.

In the general case when A(t) is neither periodic nor constant, eigenvalues fail as a general theoretical tool for dichotomy. The following example due to Markus and Yamake [14] demonstrates the above mentioned fact. Consider $x' = A(t)x$, where

$$A(t) = \begin{pmatrix} -1 + \frac{3}{2}\cos^2 t & 1 - \frac{3}{2}\cos t \sin t \\ -1 - \frac{3}{2}\sin t \cos t & -1 + \frac{3}{2}\sin^2 t \end{pmatrix}$$

Then for all t, the eigenvalues of A(t) are the constants $\frac{1}{4}(-1 \pm \sqrt{7}\, i)$. However, the norm of the solution $x(t) = (-\cos t, \sin t)\, e^{t/2}$ increases to ∞ as t → ∞.

Hence the condition on the eigenvalues of constant matrices are not strong enough to insure dichotomy in the case of nonconstant matrices. Nevertheless, it is possible to salvage some thing. In [4] Chang and Coppel gave the following theorem:

Theorem I.1 Let A(t) be continuously differentiable and $|A(t)| \leqslant M$. If A(t) has ℓ eigenvalues whose real part $\leqslant -\alpha < 0$ and $n-\ell$ eigenvalues whose real part $\geqslant \beta > 0$, then for $0 < \epsilon < \min(\alpha, \beta)$, there is a constant $\delta = \delta(M, \alpha + \beta, \epsilon)$ such that if

$|A'(t)| \leqslant \delta$, then $x' = A(t)x$ satisfies an exponential dichotomy (I.2) where the projection $P = \begin{pmatrix} I_\ell & 0 \\ 0 & 0 \end{pmatrix}$ and K_i depend only on M, $\alpha + \beta$ and ϵ.

In order to obtain criteria for exponential dichotomy that does not require the smallness of A'(t), Lazer [11] introduced the notion of row dominance.

Definition I.2 [11]. Let $A(t) = (a_{ij}(t))$ be a matrix so that

$$|\operatorname{Re} a_{ii}(t)| \geq \sum_{i=j} |a_{ij}(t)| + \delta \tag{I.10}$$

for some $\delta > 0$, all i, and $t \in J$, then A(t) is said to be row dominant.

Theorem I.3 [11]. Let A(t) be a bounded continuous n x n matrix which satisfies the row dominance condition (I.10). If $\operatorname{Re} a_{ii}(t) < 0$ for exactly ℓ subscript i, then the equation (I.1) has an exponential dichotomy (I.2) with $\sigma_1 = \sigma_2 = \delta$ and $P = \begin{pmatrix} I_\ell & 0 \\ 0 & 0 \end{pmatrix}$.

We remark here that Palmer [20] has shown that for real systems one still obtains exponential dichotomy if (I.10) holds with $\delta = 0$ and $\inf |\det A(t)| > 0$. It is clear that Palmer's criterion is more general than (I.10). One may observe using Geschorgin's theorem [10], that condition (I.10) implies that all of the eigenvalues of A(t) are off the imaginary axis.

For diagonal dominance and its connection with dichotomy, the reader may consult Berkey [1].

Exponential dichotomy and admissability

The question of admissability in differential equations has been considered first by Massera and Schaffer [16]. The following theorem relates dichotomy with admissability in linear systems and may be found in Coppel [6].

Theorem I.4 [6]. The differential equation (I.1) has an exponential dichotomy iff the inhomogeneous equation $y' = A(t)y + f(t)$ has at least one bounded solution for every $f \in F$, where F is the Banach space of all locally integrable vector functions f with the norm

$$\|f\|_F = \sup_{t \geq 0} \int_t^{t+1} |f(s)|\, ds$$

In order to extend the preceding theorem to more general function spaces, one has to impose further condition on equation (I.1). One such condition which is popular in the literature is the condition of bounded growth. We say that equation (I.1) has bounded growth on an interval J if for some fixed $h > o$, there exists a constant $C \geq 1$ such that every solution of (I.1) satisfies

$$|x(t)| \leq C|x(s)| \tag{I.11}$$

for $s, t \in J$, $s \leq t \leq s + h$.

It is easy to show [6] that condition (I.11) is equivalent to the following condition:

$$|X(t)X^{-1}(s)| \leq Ke^{\sigma(t-s)} \qquad t \geq s, \tag{I.12}$$

where K and σ are real constants. Condition (I.11) or (I.12) holds if $\int_t^{t+h} |A(s)|\, ds$ is bounded or more generally if $\int_t^{t+h} \tau[A(s)]\, ds$ is bounded, where $\tau(A) = \lim_{h \to 0^+} \frac{|I+hA| - 1}{h}$ is the Losinskii logarithmic norm [5]. We remark here that condition (I.11) is equivalent to the condition of reducibility of (I.1) [6] ((I.1) is reducible if it is kinematically similar to $y = B(t)y$, where $B(t) = \begin{pmatrix} B_1(t) & 0 \\ 0 & B_2(t) \end{pmatrix}$.

Under the bounded growth condition, we have the following theorem which is due to Maizel [12] and Massera and Schaffer [18].

Theorem I.5. Suppose that (I.1) has a bounded growth. Then (I.1) has an exponential dichotomy iff the inhomogeneous equation $y' = A(t)y + f(t)$ has at least one bounded solution for every $f \in G$, where G is the Banach space of all bounded continuous vector functions f with the norm

$$\| f \|_G = \sup_{t \geqslant 0} \| f(t) \|.$$

Chang [3] and Palmer [20] investigated the admissability problem and its connection to dichotomy for the more general systems

$$y' = A(t)y + f(t, y) \qquad \text{(I.13)}$$

A special case of (I.13) was studied in [10] where both A(t) and f(t, y) are assumed to be almost periodic.

Theorem (I.6) [10]. Suppose that (I.1) has an exponential dichotomy such that A(t) is almost periodic. If f(t, y) is almost periodic in t, uniformly in x and it satisfies a Lipschitz condition, then (I.13) has an almost periodic solution.

When $f(t,y) = f(t)$ in (I.13), Traple [27] extended Theorem (I.6) to weak almost periodic functions and solutions.

Roughness of exponential dichotomy

Massera and Schaffer [16, 17, 18] and Daleckii and Krein [8] have shown that the property of exponential dichotomy is rough in the sense that it is not destroyed by small perturbations of the coefficient matrix.

Theorem (I.7). If (I.1) has an exponential dichotomy on R^+ and $\sup |B(t)| \leqslant L$, then $y' = [A(t) + B(t)] y$ has an exponential dichotomy on $t \in R^+$.

For a more precise statement of the above theorem, the reader may consult [6].

Part II. Exponential dichotomy of nonlinear systems

Consider the following nonlinear differential equation

$$x' = f(t, x) \qquad \text{(II.1)}$$

where $f \in C [R\times R^n, R^n]$, $f_x \equiv \frac{\partial f}{\partial x}$ exists and continuous on $R\times R^n$, $f(t, o) \equiv 0$.

Equation (II.1) may be written in the form

$$x' = f_x(t, o)\, x + g(t, x) \qquad \text{(II.2)}$$

Throughout this part it is always assumed that $| g(t, x)| \leqslant \epsilon$, $| g(t, x) - g(t, y)| \leqslant \epsilon |x-y|$.

The linear part of (II.2) is

$$y' = f_x(t, o)\, y \qquad \text{(II.3)}$$

and its variational equation relative to the solution $x(t, x_0) \equiv x(t, o, x_0)$ is

$$Z' = [f_x \;\; t, x(t, x_0)]\, Z \tag{II.4}$$

The fundamental matrix $X(t, x_0)$ of (II.4) is given by $X(t, x_0) = \frac{\partial}{\partial x_0} [\, x(t, x_0)\,]$.

We are ready to give the definition of dichotomy in nonlinear systems.

Definition II.1. The system (II.1) possesses a global exponential dichotomy in variation on R if there exist projections $\hat{P}_i$ (i=1, 2) with $D\hat{P}_1(0) = I - D\hat{P}_2(0)$ and such that

$$| X(t, \hat{P}_i x_0)\, P_i X(s, \hat{P}_i x_0) | \quad K_i e^{-\sigma_i (t-s)} \quad \text{if} \quad (-1)^i (s-t) \geqslant 0 \tag{II.5}$$

where $P_i = D\hat{P}_i(o)$, $\sigma_1 > 0$, $\sigma_2 < 0$.

If $\sigma_i = 0$ (i=1, 2), then (II.5) gives ordinary dichotomy in variation. In this paper we will discuss only exponential dichotomy of (II.1) or (II.2). To do this we need some preliminary results. The reader may compare our techniques with those in [9] (this proceedings).

Lemma II.2. If the system (II.3) has an exponential dichotomy on R (I.2), then for any $a \in R^n$ there is a unique solution to the integral equation.

$$x(t) = X(t)\, Pa + \int_{-\infty}^{\infty} K(t, s)\, g(s, x(s))\, ds \qquad (t \in R),$$

where

$$K(t, s) = \begin{cases} X(t)\, PX^{-1}(s) & \text{if } -\infty < s \leqslant t \\ -X(t)\,(I-P)\, X^{-1}(s) & \text{if } t \leqslant s < \infty \end{cases}$$

Furthermore, x(t) is a solution of (II.1) which is bounded on R^+.

Similarly, the integral equation

$$x(t) = X(t)\,(I-P)\, a + \int_{-\infty}^{\infty} K(t, s)\, g(s, x(s))\, ds$$

has a unique solution, which is also a solution of (II.2) bounded on R^-.

Lemma II.3. Let $x_1(t)$ and $x_2(t)$ be two solutions of (II.1) whose difference is bounded on R^+, then $| x_1(t) - x_2(t) | \leqslant L e^{-\beta(t-s)} | x_1(s) - x_2(s) |$, $s \leqslant t$, provided that the system (II.3) has an exponential dichotomy on R.

Corollary II.4. If the system (II.3) has an exponential dichotomy on R, $x_1(t)$ and $x_2(t)$ are solutions of (II.1) whose difference is bounded on R, then $x_1(t) \equiv x_2(t)$. In particular, the only solution of (II.1) which is bounded on R is the trivial one.

We now give a crucial result.

Theorem II.5. Suppose that the system (II.3) has an exponential dichotomy on R. Then for any solution x(t) of (II.1), there exist solutions $y_1(t)$ and $y_2(t)$ of (II.1) such that

(i) $y_1(t)$ is bounded on R^+

(ii) $y_2(t)$ is bounded on R^-

(iii) $x(t) - y_1(t) - y_2(t)$ is bounded on R.

Furthermore, $y_1(t)$ and $y_2(t)$ are uniquely determined by x(t) and the above three requirements.

To prove this theorem one defines $y_1(t)$ and $y_2(t)$ in the following way

$$y_1(t) = X(t)\left[Px(o) - \int_{-\infty}^{0} PX^{-1}(u)\, g(u, x(u))\, du\right] + \int_{-\infty}^{\infty} K(t, u)\, g(u, y_1(u))\, du, \tag{II.6}$$

$$y_2(t) = X(t)\left[(I-P)\, x(o) + \int_{0}^{\infty} (I-P)\, X^{-1}(u)\, g(u, x(u))\, du\right] + \int_{-\infty}^{\infty} K(t, u)\, g(u, y_2(u))\, du \tag{II.7}$$

We are now ready to give the main theorem on this part.

Theorem II.6. If the sytem (II.3) has an exponential dichotomy on R, then the system (II.1) has an exponential dichotomy in variation, provided that $| f_x(u, z) - f_x(u, o) | \leqslant \epsilon$.

The projections in the exponential dichotomy in variations (II.5) may be defined by

$\hat{P}_i(x_o) = y_i(o) \quad (i = 1, 2)$, where $y_i(t)$ are defined in (II.6) and (II.7).

Corollary II.7. Under the hypothesis of the above theorem, the following hold:

(i) The zero solution of $z' = f_x(t, y_1(t))\, z$ is positively exponentially stable.

(ii) The zero solution of $z' = f_x(t, y_2(t))\, z$ is negatively exponentially stable.

Proofs of the second part of this paper and further results will appear somewhere else.

REFERENCES:

[1] Berkey, D.D., Comparative exponential dichotomies and column diagonal dominance, J. Math. Anal. and Appl. 55 (1976) 140-149.

[2] Cecchi, M., Marini, M. and Zezza, P.L., Asymptotic properties of the solutions of nonlinear equations with dichotomies and applications, Bolletino U.M.I., Analisi, Funzionale e Appl. 1 (1982) 209-234.

[3] Chang, K.W., Perturbations of nonlinear differential equations, J. Math. Anal. and Appl. 34 (1971) 418-428.

[4] Chang, K.W. and Coppel, W.A., Singular perturbations of initial value problems over a finite interval, Arch. Rat. Mech. Anal. 32 (1969) 268-28.

[5] Coppel, W.A., Stability and asymptotic behaviour of differential equations (Heath Math. Monographs, Boston, 1965).

[6] Dichotomies in Stability Theory (Springer-Verlag Lecture Notes in Math. 629, New York, 1978).

[7] Corduneanu, C., Sur Certains systems differentiels non-lineaires, An. Sti. Univ. πAl. I. Cuzaπ Iasi, Sect. I, 6 (1960) 257-60.

[8] Daleckii, Ju. L. and Krein, M.G., Stability of solutions of differential equations in Banach spaces. (American Math. Transl., Providence, Rhode Island, 1974).

[9] Drager, L.D., and Layton, W., Some results on non resonant nonlinear delay differential equations, This Proceedings.

[10] Fink, A.M., Almost Periodic Differential Equations (Lecture Notes in Math. 334, Springer-Verlag, New York 1974).

[11] Lazer, A.C., Characteristic exponents and diagonally dominant linear differential systems, J. Math. Anal. and Appl. 35 (1971) 215-229.

[12] Maizel, A.D., On the stability of solutions of systems of differential equations, Ural. Politehn., Inst. Trudy 51 (1954) 20-50 (Russian).

[13] Markus, L., Continuous Matrices and the stability of differential systems, Math. Z. 62 (1955) 310-319.

[14] Markus, L. and Yamake, Global stability criteria for differential systems, Osaka Math. J. 12 (1960) 305-317.

[15] Martin, R.H., Conditional stability and separation of solutions to differential equations, J. Diff. Equations 13 (1973) 81-105.

[16] Massera, J.L. and Schaffer, J.J., Linear differential equations and functional analysis, I, Ann. of Math. 67 (1958) 517-573.

[17] ____________________, Linear differential equations and functional analysis, III, Ann. of Math. 69 (1959) 535-574.

[18] Linear Differential Equations and Function Spaces (Academic Press, New York, 1966).

[19] Muldowney, J.S., Dichotomies and asymptotic behaviour for linear differential systems, Trans. Amer. Math. Soc. (to appear).

[20] Palmer, K.J., A diagonal dominance criterion for exponential dichotomy, Bull. Austral. Math. Soc. 17 (1977) 263-274.

[21] __________, Exponential dichotomy, integral separation and diagonalizability of linear systems of ordinary differential equations, J. Differential Equations 43 (1982) 184-203.

[22] __________, A generalization of Hartman's linearization theorem, J. Math. Anal. and Appl. 41 (1973) 753-758.

[23] __________, Exponential separation, exponential dichotomy and spectral theory for linear systems of ordinary differential equations, J. Differential Equations 46 (1982) 324-345.

[24] Sacker, R.J., and Sell, G.R., Existence of dichotomies and invariant splittings for linear differential systems, I, J. Differential Equations 15 (1974) 429-458.

[25] ________________________________, Existence of dichotomies and invariant splittings for linear differential systems, II, J. Differential Equations 22 (1976) 478-496.

[26] ________________________________, Existence of dichotomies and invariant splittings for linear differential systems, III, J. Differential Equations 22 (1976) 497-522.

[27] Traple, J., Weak almost periodic solutions of differential equations, J. Differential Equations. 45 (1982) 199-206.

[28] Zikov, V.V., Some questions of admissability and dichotomy, the method of averaging, Izv. Akad. Nauk. SSSR Ser. Math. 40 (1976) 1380-1408.

This paper is in final form and no version of it will be submitted for publication elsewhere.

Trends in the Theory and Practice of Non-Linear Analysis
V. Lakshmikantham (Editor)

A FLOW ASSOCIATED WITH A SEMIFLOW

Saber Elaydi
Department of Mathematics, University of
Colorado at Colorado Springs, Colorado
80933, U.S.A.

and

Saroop K. Kaul
Department of Mathematics and Statistics
University of Regina, Regina, Saskatchewan
S4S 0A2, Canada

Abstract. We consider semiflows (X,P,π) whose phase semigroup P is a replete semigroup of a generative group T. There is associated with a flow, in a natural way, an inverse system, with indexing set P, whose inverse limit X_∞ is non-empty if each π^t is onto. There is a naturally defined action ρ of T on X_∞ giving a flow (X_∞,T,ρ). The purpose of this paper is to study the structure of this flow as well as its relationship to the semiflow giving rise to it. We identify some dynamical properties such as disjointedness, minimality, almost periodicity, Liapunov stability etc. that may be lifted from a semiflow to its inverse limit flow.

INTRODUCTION

A dynamical system is a triple $X = (X,T,\pi)$ where X is a topological space, T is a topological group and $\pi: X \times T \to X$ is a continuous function satisfying (i) $\pi(x,e) = x$, for any $x \in X$ and the identity e in T, and (ii) $\pi(\pi(x,t),s) = \pi(x,ts)$ for any t, $s \in T$ and $x \in X$. For convenience we shall write xt for $\pi(x,t)$. We call a dynamical system continuous if $T = R$, the group of reals, and discrete if $T = Z$, the group of integers. A semi-dynamical system in a triple (X,P,π) where X is a topological space, P is a replete semigroup [4] in a topological group T, and π is a continuous function as above satisfying (i) and (ii).

In this paper we study semi-dynamical systems for which T is a generative group, that is T is isomorphic to $R^m \times Z^n \times K$, where K is a compact abelian group, and m and n are non-negative integers, and P is a cone in T.

There is associated, in a natural way, an inverse system [3] with any semi-dynamical system (X,P,π) in such a way that if its inverse limit X_∞ is not empty it admits an action ρ of the whole group T containing P, such that (X_∞,T,ρ) is a dynamical system. The purpose of this paper is to study properties of this associated dynamical system and its relationship to the given semidynamical system. Part of this program for continuous semiflows has already been carried out in [1] and [2]. For lack of space we give here only main definitions and results. The details will appear elsewhere.

1. Let $X = (X,P,\pi)$ be a semidynamical system, where P is a cone in a generative group T, that is, $P = Q \times K$, and $Q = Q' \cap R^n \times Z^m$, where Q' is a cone in R^{n+m}. Given any $t,s \varepsilon P$ we say $t \geq s$ if $t-s \varepsilon P$. It is easy to see that $\geq$ is a direction on P. For any $t \varepsilon P$ let $X_t = X$ and for $t \geq s$, $t,s \varepsilon P$, let $\pi_s^t: X_t \to X_s$ be given by $\pi_s^t = \pi^{t-s}$, where $\pi^t(x) = \pi(x,t)$. Then $\{X_t, \pi_s^t, t \geq s\}$ is the naturally associated inverse system, let $X_\infty = \text{Inv Lim}\{X_t, \pi_s^t, t \geq s\} \subset \Pi_{t \varepsilon P} X_t$ be its inverse limit [3]. Then $x = (x_t)_{t \varepsilon P} \varepsilon X_\infty$ if and only if for any $t \geq s$, $\pi_s^t(x_t) = x_s$. If $X_\infty \neq \phi$ we define a function $\rho: X_\infty \times T \to X_\infty$ as follows:

Given $x = (x_t) \varepsilon X_\infty$, define $\rho(x,r) = y = (y_t)$, where (i) if $r \varepsilon P$, $y_t = x_{t-r}$ if $r \leq t$ and $y_t = x_0(r-t)$ if $t < r$; (ii) if $r \varepsilon -P$, $y_t = x_{t+r}$ and (iii) if $r \varepsilon T$ then since $r = r_1 - r_2$ for some r_1, $r_2 \varepsilon P$ [4] $y = \rho(\rho(x,r_1),-r_2)$. One can show that ρ is well defined even though the decomposition $r = r_1 - r_2$ in (iii) is not unique.

DEFINITION

$X = (X,P,\pi)$ has no <u>start points</u> if given any $x \varepsilon X$ there exists a $y \varepsilon X$ and $t \varepsilon P$ so that $yt = x$, t not the identity.

Suppose X is a locally compact T_2 (Hausdorff) space, and $X^* = X \cup \{\infty\}$ be the one point compactification of X. Define $\pi^*: X^* \times P \to X^*$ by setting $\pi^*(\infty,t) = \infty$ for all $t \varepsilon P$ and $\pi^* = \pi^*|_{X \times P}$, the restriction of π^* to $X \times P$, elsewhere. If π^* is continuous we say that X is extendable to the semiflow X^*.

THEOREM (1.1)

If π^t is onto for each $t \varepsilon P$ for a semiflow (X,P,π), then the following statements are true.

(1) $X_\infty \neq \phi$ and (X_∞,T,ρ) is a flow.

(2) If X is a metric space, then X_∞ is metrizable.

(3) Through every point $x \varepsilon X$ there passes a principal negative solution [7].

THEOREM (1.2)

If a semiflow X extends to X^* and X is locally compact T_2 then X_∞ is locally compact T_2.

2. In view of Theorem (1.1) we analyse the relationship of π^t's being onto with other properties.

DEFINITION

Let $F: X \times P \to X$ be the set valued function defined by $F(x,t) = \{y \varepsilon X: yt = x\}$. For $A \subset X$ and $S \subset P$ set $F(A,S) = \cup \{F(x,t): x \varepsilon A, t \varepsilon S\}$

THEOREM (2.1)
A semidynamical system (X,Z^+,π) has no start points if and only if π^n is onto for each $n\varepsilon Z^+ = \{n\varepsilon Z: n \geq 0\}$.

This result does not extend to higher dimensions as the following easy example shows.

EXAMPLE (2.2)
Let $X = \{(x,y)\varepsilon \mathbb{R}^2: x > 0\}$, let $P = \{(m,n)\varepsilon Z^2: m \geq 0, n \geq 0\}$ and define $\pi((x,y),(m,n)) = (x+m,y)$. Then (X,P,π) has no start points; however, note that $\pi^{(m,n)}$ is not onto if $m > 0$ and $n > 0$, i.e. $(m,n)\varepsilon \text{Int } P$.

THEOREM (2.2)
Let (X,P,π) be a semiflow, where P is a cone in Z^n. Then π^t is onto for each $t\varepsilon P$ if and only if X has no start points and π^s is onto for some $s\varepsilon \text{Int } P(\text{Int } P = Z^n \cap \text{Int } Q$, where $P = Z^n \cap Q$, Q a cone in $\mathbb{R}^n)$.

THEOREM (2.3)
Let $X = (X,P,\pi)$ be a semiflow, where X is locally compact T_2 and P is a cone in $Z^n \times \mathbb{R}^m$. Then the following are equivalent:

(1) X is extendable to X*.

(2) The attainability sets F(A,S) are compact if A and S are compact.

THEOREM (2.4)
Let $X = (X,P,\pi)$ be a semiflow, where X is locally compact T_2 and P is a cone in $\mathbb{R}^n$. If X extends to X* and π^s is onto for some $s\varepsilon \text{Int } P$, then π^t is onto for each $t\varepsilon P$.

The following example shows that the converse of Theorem (2.4) is not true by Theorem (2.3) because clearly the attainability sets are not always compact.

EXAMPLE (2.5)
Let $X = \{(a,b)\varepsilon \mathbb{R}^2: a \leq 0, 0 \leq b \leq 1\} \cup \{(a,0): a > 0\}$ with the relative topology from $\mathbb{R}^2$. In the semiflow (X,P,π), π^t is a translation for each $t\varepsilon P = R^+$ as shown in the picture below:

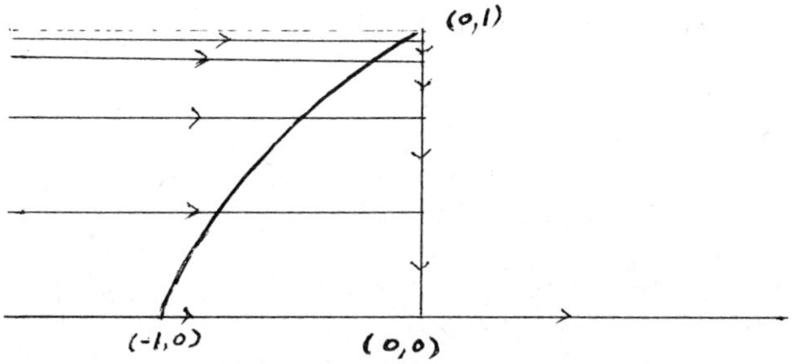

Figure 1

The attainability set F((0,0),1) is shown by the "arc".

THEOREM (2.6)

This example illustrates Theorem (4.2) of McCann [6,p.46] because N((0,0)) is 1.

THEOREM (2.7)

Let $X = (X,P,\pi)$ be a semiflow where X is locally compact T_2 and P is a cone in $R^n \times Z^m$. Then the following are pairwise equivalent.

(1) The attainability sets F(A,S) are non-empty and compact for non-empty and compact sets $A \subset X$ and $s \subset P$.

(2) F(x,t) is compact for each $(x,t) \varepsilon X \times P$ and F is upper semi-continuous.

(3) X extends to X*, X contains no start points and π^s is onto for some $s \varepsilon$ Int P.

3. Let (X,P,π) be a semiflow with X locally compact T_2. Assume also that condition (1) of Theorem (2.7) holds.

THEOREM (3.1)

If X is P-minimal them (X_∞,T,ρ) is minimal.

THEOREM (3.2)

Let X be a metric space. Then (X,P,ρ) is P-equicontinuous if and only if (X_∞,T,ρ) is P-equicontinuous.

REMARK (3.3)

Theorem (3.2) hold also if in its statement the word equicontinuity is replaced by any of the following: (a) almost periodic, (b) recursive, (c) Liapunov stable, (d) strong characteristic 0, (e) regular and (f) weakly equicontinuous.

THEOREM (3.4)

If two semiflows (X,P,π) and (X',P,π') are disjoint then so are (X_∞,T,ρ) and (X'_∞,T,ρ').

REFERENCES

[1] Elaydi, S., "Semidynamical systems with non-unique global backward extensions", Funkc. Ekvac. 26(1983), 173-187.

[2] Elaydi, S., "Semidynamical systems with non-unique global backward extensions II; the negative aspects", Funkc. Ekvac (to appear).

[3] Dugundji, J., Topology, Allyn and Bacon, Inc. Boston (1966).

[4] Gottschalk, W.M., and Hedlund, G.A., Topological Dynamics, Amer. Math. Soc. Colloq. Pub., 36 Providence, R.I. (1955).

[5] Kaul, S.K.,"Recurrence and almost periodicity in transformation groups", To appear in Applicable Analysis.

[6] McCann, R., "Negative escape time in semi-dynamical systems", Funkc. Ekvac; 20(1977), 39-47.

[7] Saperstone, S.H., Semidynamical systems in infinite dimensional spaces", Applied Math. Sc. Vol.(23), Springer Verlag (1981).

The final (detailed) version of this paper will be submitted for publication elsewhere.

Trends in the Theory and Practice of Non-Linear Analysis
V. Lakshmikantham (Editor)

A METHOD OF FINDING CRITICAL POINTS OF NONLINEAR FUNCTIONALS

Alexander Eydeland

Mathematics Research Center
University of Wisconsin-Madison
Madison, Wisconsin 53705
U.S.A.

In this paper the method of transformation of the objective functional is presented as a general approach to a problem of finding critical points of nonlinear functionals. The method is applied to a large variety of problems, among them convex and non-convex problems, differentiable and non-differentiable problems, problems with and without constraints. The method can be used for finding extremum points, as well as saddle points of functionals.

1. INTRODUCTION

Throughout the paper we consider the following problem:

Find a stationary point of the functional

$$\Phi(u) = \sum_{i=1}^{m} \int_{D_i} f_i(x,u,\nabla u)dx, \qquad u \in B . \tag{1}$$

Here B is a subset of a Hilbert space H; the choice of B and H depend on the nature of the problem and on the constraints. The sets D_i are convex subsets of $\mathbf{R}^n$. Note that by no means the functional (1) is the most general functional to which our method can be applied. However we prefer to consider this form of the functional in order to simplify the presentation and because all our examples are of this form.

A standard way of solving the problem (1) is to replace it by a sequence of simpler problems, the solutions of which converge to a solution of (1). In most methods, like gradient or Newton's procedures, these simpler problems are obtained by taking one or two terms of Taylor's expansion of the functional (1). In this paper we investigate a method of simplification different from Taylor's expansion, a method of transformation of the objective functional. In order to introduce the idea of the method we start with its simplest version.

2. SIMPLE FORM OF THE METHOD

Let all the functions f_i, $i = 1,\dots,m$, from (1) be differentiable, convex in u and ∇u, and non-negative. Our goal is to find a point of minimum of the functional (1) over the set B, provided that this point of minimum exists. The main assumption is that there exist functions $g_i(\alpha)$, $i = 1,\dots,m$, $\alpha \in \mathbf{R}^+$ which we call transforming functions, such that the composition functions $g_i(f_i(x,u,\nabla u))$, $i = 1,\dots,m$, are simpler than the original functions $f_i(x,u,\nabla u)$ (at this moment we may assume that they are quadratic functions in u and ∇u, in Section 3 we shall give a more detailed definition). The following example will illustrate the above-said.

<u>Example 1</u>. (The minimal surface problem). We consider the problem of finding a point of minimum of the functional

$$\Phi(u) = \int_D \sqrt{1 + |\nabla u|^2}\, dx, \qquad u \in B\,,$$

$$B = \{u \in W_1^1(D),\ u = \phi \quad \text{on} \quad \partial D\}\,.$$

The functional (2) is of the form (1) with $f_1(\nabla u) = \sqrt{1 + |\nabla u|^2}$. If we take a transforming function $g_1(\alpha) = \alpha^2$, then the composition function will be $g_1(f_1(\nabla u)) = 1 + |\nabla u|^2$, a quadratic function in ∇u.

The question now arises, how to use the fact of existence of these simplifying functions $g_i(\alpha)$. More precisely, we want to construct a procedure where the $k + 1$-st approximation u^{k+1} to a solution u^* of (1) is determined as a point of minimum of the functional

(3) $$\Gamma^k(u) = \sum_{i=1}^m \int_{D_i} \{\beta_i^k(x) g_i(f_i(x,u,\nabla u))\} dx, \quad u \in B_1 \subset B\,.$$

Here $\beta_i^k(x)$, $i = 1,\ldots,m$, are some coefficients which depend on the k-th approximation $u^k(x)$. The subset B_1 is dense in B and the functional (3) is differentiable on B_1 for every choice of coefficients β_i. By the assumptions made above the problem of finding a point of minimum of the functional (3) is simpler than the original problem (1). Now we define the coefficients $\beta_i^k(x)$, $i = 1,\ldots,m$, $k = 0,1,\ldots$. If $g_i(\alpha) \in C^1(\mathbf{R}^+)$ and

(4) $$g_i(f_i(x,u,\nabla u)) > \delta > 0, \quad i = 1,\ldots,m, \quad x \in D_i, \quad u \in B\,,$$

then

$$\beta_i^k(x) = 1/g_i'(f_i(x,u^k,\nabla u^k)), \quad i = 1,\ldots,m, \quad k = 0,1,\ldots\,.$$

Thus we obtain the following iterative procedure:

u^0 <u>is an arbitrary point from</u> B_1, <u>for</u> $k = 0,1,\ldots$
u^{k+1} <u>is a point of minimum over</u> B_1 <u>of the functional</u>

$$\Gamma^k(u) = \sum_{i=1}^m \int_{D_i} \{g_i(f_i(x,u,\nabla u))/g_i'(f_i(x,u^k,\nabla u^k))\} dx\,.$$

The procedures of this type were investigated extensively in [2], where it was proved that for a large class of functions $g_i(\alpha)$

(6) $$\Phi(u^k) \downarrow \min_{u \in B} \Phi(u)\,.$$

Under additional conditions on $\Phi(u)$, such as strict convexity, by a standard argument we can obtain from(5) the convergence of u^k to the point of minimum u^*. In particular, for the minimal surface problem from Example 1 we obtin that the iterative procedure

$u^0 \in W_1^2(D)$; <u>for</u> $k = 0,1,\ldots$

u^{k+1} <u>is a point of minimum of</u>

$$\Gamma^k(u) = \int_D \{|\nabla u|^2/2\sqrt{1 + |\nabla u^k|^2}\}dx, \quad u \in W_1^2(D) ,$$

generates a minimizing sequence for the functional (2), i.e. $u^k \to u^*$ in the W_1^1-norm, provided that a point of minimum u^* of (2) exists and belongs to $C^1(\bar{D})$, see [2].

3. GENERAL PROCEDURE

In this section we consider a wider class of transforming functions g_i. Let the functions f_i, $i = 1,\ldots,m$, from (1) satisfy the same conditions as in Section 2 and let our goal be to find a minimum point of the function (1), provided that this point exists. We now assume that there eixst functions $g_i(x,z,\bar{p},\alpha)$, $i = 1,\ldots,m$, $z \in \mathbf{R}$, $\bar{p} \in \mathbf{R}^n$, $\alpha \in \mathbf{R}^+$, such that the composition functions $g_i(x,u,\nabla u,f_i(x,u,\nabla u))$ are simpler than the functions $f_i(x,u,\nabla u)$, $i = 1,\ldots,m$. The term simpler throughout this paper means that a stationary point of the functional

$$\sum_{i=1}^m \int_{D_i} \{\beta_i(x)g_i(x,u,\nabla u,f_i(x,u,\nabla u)) \tag{7}$$
$$+ \text{(quadratic and linear terms in } u, \nabla u)\}dx$$

can be found faster and by more efficient means than a stationary point of the functional (1) for any choice of coefficients $\beta_i(x)$ from a certain set of functions (usually it is the set of non-negative continuous functions). As in Section 1 we use the above property of the functional (7) to construct an iterative scheme which generates a minimizing sequence for the functional (1):

$u^0 \in B_1 \subset B$; <u>for</u> $k = 0,1,\ldots$

u^{k+1} <u>is a point of minimum over</u> B_1 <u>of the functional</u>

$$\Gamma^k(u) = \sum_i \int_{D_i} \{[g_i(x,u,\nabla u,f_i(x,u,\nabla u)) - u\frac{\partial g_i}{\partial z}(x,u^k,\nabla u^k,f_i(x,u^k,\nabla u^k)) \tag{8}$$
$$- \nabla u\cdot \frac{\partial g_i}{\partial \bar{p}}(x,u^k,\nabla u^k,f_i(x,u^k,\nabla u^k))] \Big/ \frac{\partial g_i}{\partial \alpha}(x,u^k,\nabla u^k,f_i(x,u^k,\nabla u^k))\}dx$$

Note that for each k the functional in (8) is of the form (7) and therefore by our assumptions can be easily minimized. We refer the reader to [2] for convergence results for (8) and some applications.

<u>Example 2</u>. Consider the problem (1) with the functional

$$\Phi(u) = \int_D \{1/2|\nabla u|^2 + \Psi(u)\}dx , \tag{9}$$

where $\Psi(u) = \int_0^u \psi(\zeta)d\zeta$ is a convex non-negative functional. We want to minimize the functional (9) over a hyperspace $B = \{u \in W_1^2(D),\ u = \phi \text{ on } \partial D\}$. This problem corresponds to the Dirichlet problem for the equation

$$-\Delta u + \psi(u) = 0 \text{ in } E, \quad u = \phi \text{ on } \partial D . \tag{10}$$

The functional (9) is of the type (1) with $f_1(\nabla u) = 1/2\ |\nabla u|^2$, $f_2(u) = \Psi(u)$. If we take $g_1(\alpha) = \alpha$, $g_2(x,\alpha) = \alpha - \Psi(u) + 1/2\ \Psi''(u^k(x))u^2$, then, as can be easily shown, each step of the procedure (7) is an iteration of Newton's method for the equation (9).

Example 3. (Minimal surface problem with constraints). In this example we minimize the functional

$$\phi(u) = \int_D \sqrt{1 + |\nabla u|^2}dx, \quad u \geqslant \Psi(x) \geqslant 0 \ ,$$

$$u \in W_1^2(D), \quad u = \phi \quad \text{on} \quad \partial D \ .$$

The above problem can be replaced by a problem of minimization of the penalty functional

(11) $$\Phi_\rho(u) = \int_D \sqrt{1 + |\nabla u|^2}\ dx + \int_D A_-(u - \Psi(x))dx \ ,$$

where $A_-(\xi) = \rho\xi^2$ for $\xi \leqslant 0$ and $A_-(\xi) \equiv 0$ for $\xi > 0$, $\rho >> 0$. The functional (11) is of the form (1) with $f_1(\nabla u) = \sqrt{1 + |\nabla u|^2}$, $f_2(x,u) = A_-(u - \Psi(x))$. If we take $g_1(\alpha) = \alpha^2$, $g_2(x,u,\alpha) = \alpha + A_+(u - \Psi(x))$, where $A_+(\xi) = A_-(-\xi)$, then by (8) we obtain a minimizing sequence $\{u^k\}$ for the problem (11), where for $k = 0,1,\ldots$ u^{k+1} is a minimum point of the quadratic functional

$$\int_D \{|\nabla u|^2 \ / \ 2\sqrt{1 + |\nabla u^k|^2} + \rho(u - \Psi(x))^2$$

$$- uA_+'(u^k(x) - \Psi(x))\}dx, \quad u \in W_1^2(D), \quad u = \phi \quad \text{on} \quad \partial D \ .$$

More details concerning this problem can be found in [4].

4. NON-DIFFERENTIABLE $\Phi(u)$

In this section we consider a problem (1) with

(12) $$\Phi(u) = \int_D \gamma(x)|\nabla u| + \text{quadratic and linear terms.}$$

It is clear that if we take $g_1(\alpha) = \alpha^2$ then the corresponding procedure (5) is a sequence of quadratic problems. The only obstacle is that now the condition (4) does not hold, so that denominators in (5) can vanish. To overcome this problem we modify the procedure (5) in the following way:

$$u^{k+1} \to \min_{u \in B_1} \sum_i \int_{D_i} g_i(f_i(x,u,\nabla u))/g_i'(\alpha_i^k(x)) \ ,$$

$$\alpha_i^k(x) = \max\{\delta^k, g_i'(f_i(x,u^k,\nabla u^k))\} \ ,$$

where $k = 0,1,\ldots$. In [3] we proved that if $\delta^k \downarrow 0$, $\sum \delta^k = \infty$ and $\sum (\delta^k)^2 < \infty$ then $\Phi(u^k) \downarrow \min \phi(u)$, $u \in B$.

Up to here we have considered only convex problems. Now we are going to study non-convex problems.

5. A VARIATIONAL PROBLEM WITH NON-CONVEX FUNCTIONAL BOUNDED FROM BELOW

To illustrate the ideas of our method in the non-convex case we consider a simple problem of finding a positive solution of a semilinear elliptic equation

$$\Delta u + g(u) = 0 \quad \text{in} \quad D, \quad u = 0 \quad \text{on} \quad \partial D \, , \tag{13}$$

where $g(u)$ is a smooth and positive function for $u > 0$, $g(u) = o(u)$ as $u \to 0$ and $g(u) \sim u^{\sigma}$, $0 < \sigma < 1$, as $u \to \infty$. The problem (13) is equivalent to the problem of finding a stationary point of the functional

$$\Phi(u) = \int_D \{1/2 \, |\nabla u|^2 - G(u)\} dx \, , \tag{14}$$

$$u \in W_1^2(D), \quad u = 0 \quad \text{on} \quad \partial D, \quad G(u) = \int_0^{\xi} g(\xi) d\xi \, .$$

The functional (14) is of the type (1) with $f_1(\nabla u) = 1/2 \, |\nabla u|^2$ and $f_2(u) = -G(u)$. Consider $g_1(\alpha) = \alpha$, $g_2(u,\alpha) = \alpha + G(u) + Cu^2$, where a constant C is chosen in such a way that $g_2(u,\alpha)$ is convex for $u \geqslant 0$, $\alpha > 0$ (this can be done since G is subquadratic for $u \to \infty$). In this case the procedure (8) is as follows:

<u>for</u> $k = 0,1,\ldots$ u^{k+1} <u>is a point of minimum of the functional</u>

$$\int_D \{1/2 \, |\nabla u|^2 + Cu^2 - u(G'(u^k) + 2Cu^k)\} dx \, ,$$

$u \in W_2^1(D)$, $u = 0$ on ∂D. In order not to converge to a trivial solution we have to start with u^0 such that $\Phi(u^0) < 0$, where the gunctional Φ is defined in (14). In order to prove the convergence of u^k to the solution u^* we use the Palais-Smale condition (similar to the way it was used in [7]). The complete proof and numerical results can be found in [5].

6. PERIODIC SOLUTIONS OF SECOND ORDER DYNAMICAL SYSTEMS

We consider a separable Hamiltonian system, i.e. the system

$$\ddot{x} + \nabla U(x) = 0, \quad x \in R^n \, . \tag{16}$$

Let for simplicity $U(x)$ be a convex and even function. Our goal is to find a periodic solution of (16). In order to find this solution we use the approach of M. Berger, see [1], and replace the problem (16) with a problem of finding a minimum point $x^*(t)$ of the functional

$$\Phi(x) = \int_0^{\pi} - U(x) dt \tag{17}$$

subject to the constraints

$$\int_0^{\pi} |\dot{x}|^2 dt = R, \quad x(0) = x(\pi) = 0 \, , \tag{18}$$

where R is a positive constant. It is clear that the vector function $\bar{x}(t) = x^*(t/\sqrt{\lambda})$ for $t \in [0,\pi]$, $\bar{x}(t) = -x^*(-t/\sqrt{\lambda})$ for $t \in [-\pi,\pi]$, where λ is the Lagrange multiplier for the problem (17-18), is a periodic solution of (16) with the period $T = 2\pi\sqrt{\lambda}$. The functional (17) is of the type (1) with $f_1(x) = -U(x)$. The difficulty of this case and its difference from the previous cases is that the function f_1 is concave. Fortunately we can still construct a procedure similar to (3) if we use concave transforming functions. Thus if $U(x)$ is a superquadratic function and if there exists a concave function $g(\alpha)$ such that $g(U(x))$ is quadratic then a sequence of linear eigenvalue problems

$$\underline{\text{for}}\quad k = 0,1,\ldots$$

$$x^{k+1}(t)\ \ \underline{\text{minimizes}}\ \ \int_0^{\pi} \{-g(U(x))/g'(U(x^k))\}dt,$$

$$x^{k+1}\ \ \underline{\text{satisfies the constraints (18)}},$$

generates a sequence which converges to the solution of the problem (17-18), see [6].

7. SADDLE POINTS

The results of the previous section indicate that the method of transformation of the objective functional can be used not only for finding extremum points but also for finding saddle points. Our hope is that it can be applied for more general saddle point problems, e.g. of the mountain path lemma type. The investigation of this type of problems is a subject of the forthcoming paper.

REFERENCES

[1] Berger, M., On periodic solutions of second order Hamiltonian systems, Journal of Math. Analysis and Applications 29 (1970) 512-522.

[2] Eydeland, A., A method of solving nonlinear variational problems by nonlinear transformation of the objective functional. Part 1, Numer. Math. 43 (1984) 59-82.

[3] Eydeland, A., A method of solving nonlinear variational problems by nonlinear transformation of the objective functional. Part 2, to appear.

[4] Eydeland, A., A procedure for solving the constrained minimal surface problem, to appear.

[5] Eydeland, A., Computational procedures for nonlinear non-convex variational problems, to appear.

[6] Eydeland, A., A method of finding periodic solutions of second order dynamical systems, to appear.

[7] Nirenberg, L., Variational and topological methods, Bull. of Am. Math. Soc. 4 (1981) 267-374.

This paper is in final form and no version of it will be submitted for publication elsewhere.

Trends in the Theory and Practice of Non-Linear Analysis
V. Lakshmikantham (Editor)

ASYPMTOTIC BEHAVIOR OF NONLINEAR FUNCTIONAL EVOLUTION EQUATIONS IN FADING MEMORY SPACES

W. E. Fitzgibbon
Department of Mathematics
University of Houston
Houston, Texas 77004

1. Introduction - Preliminaries

Let X denote a Banach space. If $r > 0$ and $\rho(\ )$ is a positive nondecreasing function on $(-\infty,0]$, let E denote the set of strongly measurable functions ϕ from $(-\infty,0]$ to X such that ϕ is continuous on $[-r,0]$ and $\phi\rho$ is integrable on $(-\infty,-r]$. The set E becomes a Banach space under the norm,

$$||\phi||_E = \sup_{\theta \in [-r,0]} ||\phi(\theta)|| + \int_{-\infty}^{-r} ||\phi(\theta)||\rho(\theta)d\theta .$$

We further require that:

(E1) There exists a $\gamma > 0$ such that $e^{-\gamma\theta}\rho(\theta)$ is nondecreasing on $(-\infty,-r]$.
(E2) $\rho(\theta) < 1$ for $\theta \in (-\infty,0]$.

If $u(\ ): (-\infty,T] \to X$ is continuous on $[0,T]$ and has the property that $u(\theta) = \phi(\theta)$ for $\theta \in (-\infty,0]$ where $\phi \in E$, then u_t is that element of E having pointwise a·e definition,

$$u_t(\theta) = \begin{cases} u(t+\theta) & \text{for } t+\theta \ge 0 , \\ \phi(t+\theta) & \text{for } t+\theta < 0 , \end{cases}$$

We shall be concerned with abstract nonlinear functional differential equations involving infinite delay and having the form,

(1.1a) $$x(\phi)(t) + A(t)x(\phi)(t) = F(t,x_t(\phi))$$
(1.1b) $$x_0(\phi) = \phi \in E.$$

where $\{A(t)|t \ge 0\}$ is a family of nonlinear operators mapping X to X and $\{F(t,\cdot)|t \ge 0\}$ is a family of mappings of E to X.

Specifically we require $\{A(t)|t \ge 0\}$ satisfy:

(A.1) $\{A(t)|t \ge 0\}$ be a family of m-accretive operators on X
(A.2) There exists a $\omega > 0$ so that $(A(t)-\omega I)$ is accretive for all $t > 0$
(A.3) $D = D(A(t))$ is independent of t.
(A.4) There exists an $L > 0$ so that $||A(t)x-A(\tau)x|| \le |t-\tau|L(||x||+||A(t)x||)$ for $t,\tau > 0$

The mappings $F: R^+ \times E \to X$ are required to satisfy:

(F.1) There exists $K > 0$ so that
$||F(t,\phi)-F(t,\phi)|| \le K||\phi-\psi||_E$ for $t > 0$, $\phi, \psi \in E$.

(F.2) $K(1+1/\gamma) < \omega$ where ω is the constant of (A.2) and γ is the exponential weight of (E.1).

Our purpose is to obtain an asymptotic convergence result for solutions to (1.1). In doing so we adapt a integral comparision technique of Redlinger [7] from the case of semilinear finite delay equations to nonlinear infinite delay. It in our opinion the Redlinger technique will prove to be a useful tool for nonlinear evolution equations.

2. Results

In the case of finite delay, Kartsatos and Parrott [6] introduce a notion of mild solutions to (1.1a-b). They use a fixed point argument to guarantee the existence of a function $x(\phi)(\ )$ which is continuous for $t \ge 0$ and has the property that $x(\phi)(\theta) = \phi(\theta)$ for $\theta \le 0$. If a strong solution to (1.1a-b) exists it can be shown to be $x(\phi)(\ )$. Moreover, the restriction of $x(\phi)(\ )$ to $[0,T]$ can be shown to be the limit for backward difference scheme

$$\text{(2.1)} \qquad \frac{x_i - x_{i-1}}{t_i^n - t_{i-1}^n} + A(t_i^n)x_i = F(t_i^n, x_{t_i^n}(\phi))$$

where $\{t_i^n\}$ is a sequence of partitions whose mesh converges to zero. Finally, they show that if $x(\phi)$ and $x(\psi)$ are mild solutions having initial data $\phi(\ )$ and $\psi(\ )$ then

$$\text{(2.2)} \qquad ||x(\phi)(t)-x(\psi)(t)|| \le ||\phi(0)-\psi(0)||e^{-\omega t} + \int_0^t e^{-\omega(t-s)}||F(s,x_s(\phi))-F(s,x_s(\psi))||ds .$$

With a suitable strengthening of hypotheses (A.1-4) analogous results can be obtained for the case at hand. In the case at hand we shall not be interested in existence theory but shall simply assume mild solutions in the sense of Kartsatos and Parrott [6] which satisfy (2.1) and (2.2).

Following Redlinger [7] we introduce the scalar integral operator S

(2.3) Let $g(\ )$, $p(\ )$ be functions mapping $[0,T)(0<T\le\infty)$ to R such that: (i) $g(s) \ge 0$ for $0 \le s < T$; (ii) $g(\ )$ is continuous and nonnegative on $(-\infty,T]$; and (iii) $w(\cdot) = y(s+\cdot)\rho(\cdot) \in L^1(-\infty,-r]$ for $s \in [0,T)$ then for $\tau > 0$

$$(Sy)(t) = p(t) + \int_\tau^t g(t-s)|y_s|ds \quad \text{for} \quad \tau < t < T$$

where

$$|y_s| = \sup_{\theta \in [-r,0]} y(s+\theta) + \int_{-\infty}^{-r} y(s+\theta)\rho(\theta)d\theta$$

The following comparision principle is a minor modification of a lemma of Redlinger.

LEMMA. Let $y(\ )$ and $z(\ )$ be continuous nonnegative functions on $(-\infty,T)$ $(0 \le T < \infty)$ if

$$y(t)-(Sy)(t) < z(t) - (Sz)(t) \qquad \text{for} \quad \tau < t < T$$

and

$$y(t) < z(t) \qquad \text{for} \quad -\infty < t < \tau$$

then

$$y(t) < z(t) \qquad \text{for} \quad -\infty < t < T$$

Proof. If we assume that $t_0 = \inf\{t: y(t) = z(t)\}$ we may observe that $z(t_0) = y(t_0) < z(t_0) + (Sy)(t_0) - (Sz)(t_0) < z(t_0)$ and reach a contradiction.

We now can obtain our convergence result.

Theorem. Assume that (A.1-4) and (F.1)-(F.2) are satisfied. If $\phi,\ \psi \in E$ are continuous and uniformly bounded then there exists a $D = D(\phi,\psi)$ and $\delta > 0$ so that

$$||x(\phi)(t) - x(\psi)(t)|| \le e^{-\delta t}D .$$

Proof. By virtue of 2.2, we have

$$(2.4) \qquad ||x(\phi)(t)-x(\psi)(t)|| \le e^{-\omega t}||\phi(0)-\psi(0)|| + \int_0^t e^{-\omega(t-s)}||F(s,x_s(\phi))-F(s,x_s(\psi))||ds$$

$$\le e^{-\omega t}||\phi(0)-\psi(0)|| + K\int_0^t e^{-\omega(t-s)}||x_s(\phi) - x_s(\psi)||_E ds$$

we set $z(t) = e^{-\delta t}D$ and require that $\delta < \min\{\gamma,\omega-K(1+1/\gamma)\}$ and observe that

$$(2.5) \qquad ||\psi(0)-\psi(0)||e^{-\delta t} + K\int_0^t e^{-\omega(t-s)}|z_s|ds$$

$$< ||\phi(0)-\psi(0)||e^{-\delta t} + e^{\delta r}K(1+1/\gamma)(\omega-\delta)^{-1}e^{-\delta t}D$$

We further require that $\delta > 0$ be chosen so that $e^{\delta r}K(1+1/(\gamma-\delta))(\omega-\delta)^{-1} < 1$. If $A \ge \sup_{\theta \in (-\infty,0]} ||\phi(\theta)-\psi(\theta)||$ and $D > A + e^{\delta r}K(1+1/\gamma-\delta)(\omega-\delta)^{-1}D$ then

$$||\phi(0)-\psi(0)||e^{-\delta t} + K\int_0^t e^{-\omega(t-s)}|z_s|ds = De^{-\delta t} < z(t)$$

Let $y(t) = ||x(\phi)(t) - x(\psi)(t)||$. From (2.4) and (2.5) we observe that $y(t)-(Sy)(t) \quad z(t)-(Sz)(t)$ and apply our lemma to reach the desired conclusion.

3. An Example

We consider the following partial integrodifferential equation:

(3.1a) $$\partial u/\partial t - a(t)b(\partial u/\partial x)\partial^2 u/\partial x^2 + c(t)u = \int_{\infty}^{t} g(t-s)f(u(x,s))ds$$

(3.1b) $$u(x,\theta) = \phi(x,\theta) \quad \text{for} \quad \theta \in (-\infty,0]$$

(3.1c) $$u(0,t) = u(1,t) = 0$$

We require that the coefficients of the right hand side satisfy the following: (i) $a(\ )$ is uniformly Lipschitz continuous and there exists an $\varepsilon_0 > 0$ so that $a(t) > \varepsilon_0$ for all $t \geq 0$; (ii) $b(\ ) \in c(R)$ and $b(x) > \varepsilon_0$ for all $x \in R$; (iii) $c(\ )$ is uniformly Lipschitz continuous and $c(t) > \omega > 0$.

The function $f(\ ): R \to R$ is required to be Lipschitz continuous with constant k_1. The function $g: [0,\infty) \to R$ is required to be continuous. We further assume there exists a function $\rho(\ ): (-\infty,0] \to R^+$ with $\rho(\theta) < 1$ and there are constants γ and k_2 such that $|g(-\theta)| \leq k_2\rho(\theta)$ for $\theta \in (-\infty,0]$ and that $j(\theta) = e^{-\gamma\theta}\rho(\theta)$ is increasing on $(-\infty,0]$.

We let $X = C(0,1)$ and define a family of nonlinear operators $\{A(t)|t > 0\}$ pointwise by $(A(t)u)(x) = a(t)b(u'(x))u''(x) + c(t)u(x)$ with $D(A(t)) = \{u \in C^2(0,1)\ u(0) = u(1) = 0\}$. Using the results of Burch and Goldstein [3] it is immediate that $\{A(t)\ t \geq 0\}$ satisfies conditions (A.1-4). The space of initial histories is chosen to be the space E of functions mapping $(-\infty,0]$ to X with norm $||\phi||_E = \sup_{\theta \in [-r,0]} ||\phi(\theta)|| + \int_{\infty}^{-r} ||\phi(\theta)||\rho(\theta)d\theta$ where $r > 0$. We also assume for convenience $r < 1$.

A mapping $F: E \to X$ is defined

$$F(\phi)(x) = \int_{-\infty}^{0} g(-\theta)f(\phi(x,\theta))d\theta$$

If k_1 and k_2 are such that $k_1k_2(1+1/\gamma) < \omega$ then it is not difficult to see that $F: E \to X$ satisfies (F1.2).

Thus we have provided convergence criteria for mild solutions of

$$\dot{x}(\phi)(t) + A(t)x(\phi)(t) = F(t,x_t(\phi))$$

which are abstract versions of (3.1a-c).

REFERENCES

[1] D. W. Brewer, A nonlinear semigroup for a functional differential equation, Trans. Amer. Math. Soc. 236 (1978) 173-191.

[2] ____________, The asymptotic stability of a nonlinear functional differential equation of infinite delay, Houston J. Math. 6 (1980) 321-330.

[3] B. C. Burch and J. A. Goldstein, Nonlinear semigroups and a problem in heat condution, Houston J. Math. 4 (1978) 311-328.

[4] L. C. Evans, Nonlinear evolution equations in an arbitrary Banach space, Israel J. Math. 36 (1976) 1-42.
[5] W. E. Fitzgibbon, Convergence theorems for semilinear Volterra equations with infinite delay, (preprint).
[6] A. G. Kartsotos and M. E. Parrott, A simplified approach to the existence and stability of a functional evolution equation in a general Banach space, Proc. Conf. on Operator Semigroups, Retzhof Austria 1983 (Pittman Press) (to appear).
[7] R. Redlinger, On the asymptotic behavior of a semilinear functional equation in a Banach space, J. Math. Anal. Appl. (to appear).

This paper is in final form and no version of it will be submitted for publication elsewhere.

Trends in the Theory and Practice of Non-Linear Analysis
V. Lakshmikantham (Editor)

FRIENDLY SPACES FOR FUNCTIONAL DIFFERENTIAL EQUATIONS WITH INFINITE DELAY

John R. Haddock*

Department of Mathematical Sciences
Memphis State University
Memphis, Tennessee 38152
U.S.A.

*Supported in part by NSF under Grant MCS-8301304

1. INTRODUCTION

In a paper published in 1980, Corduneanu and Lakshmikantham [7] surveyed results for functional differential equations (FDEs) with infinite (unbounded) delay. Included in the paper is a discussion of numerous topics and a list of 289 references.

Several interesting theorems and techniques for FDEs with infinite delay have been developed since [7] was submitted for publication, and one of the purposes of this paper is to provide an update of some of the results that have been obtained. It should be emphasized, though, that the presentation here will not be as extensive as in [7]. In fact, attention will be restricted to a discussion of certain phase spaces (i.e., spaces of initial functions) which are "friendly" with respect to current research involving existence, comparison theorems, convergence of solutions, and periodic solutions.

The choice of phase space for FDEs with finite delay (or for ordinary differential equations-ODEs) is standard. However the situation for infinite delay equations is quite different. For this case, there are several possibilities, and an underlying space usually is chosen in connection with the particular equation at hand.

2. ADMISSIBLE SPACES

During the past 5 or 6 years, several papers have been devoted to the study of general properties (sometimes referred to as axioms) of "admissible" spaces for infinite delay equations. Along these lines, we refer to Kaminogo [15], Sawano [24], Hale and Kato [12], Kappel and Schappacher [16], Schumacher [25], Kato [17,18], Naito [20,21] and Hino [13], to name a few sources. These references - many of which were motivated by earlier works of Coleman and Mizel [5], Driver [8] and Hale [11] - have examined fundamental properties of spaces in various forms, yet they essentially deal with the same underlying ideas. In the definition of admissible space given below, we adopt conditions which seem to be the easiest to understand and manage.

Let $|\cdot|$ denote any norm in R^n, and let B be a real vector space of functions mapping $(-\infty,0]$ into R^n with semi-norm $|\cdot|_B$. For ϕ and ψ in B, we say $\phi = \psi$ if $\phi(s) = \psi(s)$ for all $s \leq 0$, while ϕ is equavalent to ψ (written $\phi \sim \psi$) if $|\phi-\psi|_B = 0$. Let $\hat{B}$ denote the collection of equivalence classes with equivalence class of ϕ

denoted by $\hat{\phi}$. For $\hat{\phi}$ in $\hat{B}$, define $|\hat{\phi}|_{\hat{B}} = |\phi|_B$. Then $\hat{B}$ equipped with norm $|\cdot|_{\hat{B}}$ is a normed linear space. We assume throughout that $\hat{B}$ is a Banach space. For many of our examples, B itself is a Banach space, so $B = \hat{B}$.

If $x:(-\infty,A)\to R^n$, $-\infty < A \le \infty$, then, for any t in $(-\infty,A)$, define $x_t:(-\infty,0]\to R^n$ by $x_t(s) = x(t+s)$, $s \le 0$. Then x_t is the translate of x on $(-\infty,t]$ to $(-\infty,0]$. If $A \ge 0$, then x_0 is merely x restricted to $(-\infty,0]$.

We consider FDEs of the form

$$x' = f(t,x_t), \tag{2.1}$$

where ' denotes the right-hand derivative with respect to t and $f:D\to R^n, D \subset R\times B$.

DEFINITION 2.1. A space B (as defined above) is said to be <u>admissible</u> (with respect to (2.1)) whenever there exist continuous functions $K,M:[0,\infty)\to[0,\infty)$ and a constant $J > 0$ such that the following conditions hold:

if $x:(-\infty,A)\to R^n$ is continous on $[a,A)$ with x_a in B for some $a < A$, then, for all t in $[a,A)$,

(B1) x_t is an element of B;

(B2) x_t is continuous in t with respect to $|\cdot|_B$;

(B3) $|x_t|_B \le K(t-a) \max_{a\le s\le t} |x(s)| + M(t-a)|x_a|_B$; and

(B4) $|\phi(0)| \le J|\phi|_B$ for all ϕ in B.

The inequalities in (B3) and (B4) often are referred to as the fundamental inequalities for admissible spaces. (B4) implies that $\hat{\phi}(0)$ is well-defined and $\hat{\phi}(0) = \psi(0)$ for all ψ in the equivalence class $\hat{\phi}$.

REMARK 2.1. (i) Conditions (B1)-(B3) assure a Peano type existence result; that is, if D (defined above) is open and $f:D\to R^n$ is continuous, then any Cauchy problem

$$x' = f(t,x_t);\; x_{t_0} = \phi;\; (t_0,\phi) \text{ in } D, \tag{2.2}$$

possesses a continuously differentiable solution that satisfies (2.1) for all t in some interval $[t_0,A)$, $t_0 < A \le \infty$. (ii) Standard uniqueness, continual dependence and continuation results also can be obtained by employing (B1) - (B3).

The result stated in (i) above is attributed to Kaminogo [15]; whereas, Sawano [24] generally is credited with results related to topics listed in (ii). Along these lines, existence, uniqueness, continual dependence and continuation results are proven in Hale and Kato [12] and Kappel and Schappacher [16], while a nice summary of fundamental results is given by Naito [21, pp. 76-77]. Naito also provides a comparison of (B1)-(B4) with similar properties that have been used in the development of axioms for infinite delay spaces.

Before giving examples of admissible spaces, we briefly discuss an important "nonexample." Let BC be the space of bounded continuous functions that map $(-\infty,0]$ into R^n with sup norm

$$|\phi|_{BC} = \sup_{s\le 0}|\phi(s)|.$$

Although (B1), (B3) and (B4) hold with $K(t) = M(t) = J = 1$, (B2) does not hold. (This can be seen by examining the function $x(t) = \sin t^2$.) Hence BC is <u>not</u> an admissible space. Equally as important, this space is sometimes "unfriendly" with respect to (2.1). For instance, Seifert [27] has proven the existence of a continuous linear mapping $F: BC \to R^2$ and an initial function ϕ in BC such that the Cauchy problem

$$x' = F(x_t);\ x_0 = \phi,$$

does not prossess a solution - not even of Caratheodory type. On the other hand, BC often is useful for certain forms of (2.1), as is evidenced in many articles involving Volterra integrodifferential equations. An account of cases for which BC suffices is given in a recent paper by Sawano [25].

The remainder of this section consists of an introduction to an admissible space that we consider to be "friendly" with regards to (2.1), and it is this space around which we base the results stated in the final section. For the sake of completeness, we give an example of another important admissible space following the discussion of our main example below.

EXAMPLE 2.1. Let

$$\text{(2.3)}\qquad \begin{cases} g:(-\infty,0]\to[1,\infty) \text{ be a continuous nonincreasing} \\ \text{function on } (-\infty,0] \text{ such that } g(0) = 1 \end{cases}$$

and let C denote the space of continuous functions that map $(-\infty,0]$ into R^n such that

$$\sup_{s\le 0} |\phi(s)|/g(s) < \infty.$$

Then C_g equipped with norm

$$\text{(2.4)}\qquad |\phi|_g = \sup_{s\le 0} |\phi(s)|/g(s)$$

is a Banach space (cf., e.g., Corduneanu [6, Section 1.1]). Furthermore, properties (B1), (B3) and (B4) are satisfied, and we can choose $K(t) = M(t) = J = 1$. This follows since

$$|x_t|_g = \sup_{s\le 0} |x(t + s)|/g(s)$$

$$= \sup_{s \le t} |x(s)|/g(s - t)$$

$$\le \sup_{a \le s \le t} |x(s)|/g(s - t) + \sup_{s \le a} |x(s)|/g(s - t)$$

$$< \sup_{a \le s \le t} |x(s)| + |x_a|_g .$$

Since BC (g=1) is a special case, we see that spaces C_g are not necessarily admissible without additional restrictions being imposed. If we assume:

either

$$(2.5) \quad \begin{cases} [g(u + s)/g(s) - 1] \to 0 \text{ as } u \to 0 \text{ uniformly for } s \le 0, \text{ and} \\ \phi/g \text{ is uniformly continuous for all } \phi \text{ in } C_g \end{cases}$$

or

$$(2.6) \quad \phi \text{ is uniformly continuous for all } \phi \text{ in } C_g,$$

then it is straighforward to show that (B1)-(B4) hold. Hence, C_g with norm given by (2.4) is admissible.

From this point on, when we make a general reference to a space C_g, we shall assume that it is admissible with (2.3) holding.

Two important special cases are the space ($g = 1$) of bounded uniformly continuous functions with sup norm and the space ($g(s) = e^{-\gamma s}$, for some $\gamma > 0$) of all functions ϕ such that

$$e^{\gamma s}\phi(s) \text{ is uniformly continuous on } (-\infty, 0]$$

with norm

$$|\phi|_\gamma = \sup_{s \le 0} e^{\gamma s}|\phi(s)| .$$

One of the main thrusts of the present paper is to emphasize that these are not the only two C_g-type spaces of import. This will be discussed in more detail in the next section. In closing this section, we present an admissible space that has motivated much of the research in the last decade or so on infinite delay equations.

EXAMPLE 2.2. Let $g:(-\infty,0] \to (0,\infty)$ satisfy

$$(2.7) \quad \begin{cases} dg(s)/ds \ge 0 \text{ for } s \le 0 \text{ and} \\ \int_{-\infty}^{0} g(s)\,ds < \infty \end{cases}$$

and Let M denote the vector space of measurable functions that map $(-\infty,0]$ into R^n. Set

$$M_g = \{\phi \text{ in } M: \int_{-\infty}^{0} g(s)\, ds < \infty\} .$$

Then M_g equipped with semi norm

$$|\phi|_{M_g} = |\phi(0)| + \int_{-\infty}^{0} g(s)|\phi(s)|ds$$

and certain L^p-type counterparts are admissible, and have received a great deal of attention in the literature. See, for example, Saperstone [23], Hale and Kato [12] or Coleman and Mizel [5], and references therein. We do not give additional attention to M_g in this paper.

3. RESULTS FOR C_g SPACES

In this section, we announce three new results for FDEs as related to C_g spaces. The first of these (Theorem 3.1 below) holds for C_g spaces in general; whereas, the other two results require the restriction

(3.1) $$g(s) \to \infty \text{ as } s \to -\infty$$

REMARK 3.1. There are several points of interesting regarding condition (3.1).

(i) If a sequence $\{\phi_n\}$ in C_g converges to a function ϕ uniformly on compact sets, then ϕ is in C_g and $|\phi_n - \phi|_g \to 0$ as $n \to \infty$, when $g(s) = e^{-\gamma s}$ or $|\phi(s)|/g(s) \to 0$ as $s \to -\infty$.

(ii) For each $K,L > 0$, the set

$\{\phi \text{ in } C_g : |\phi(u)| \leq K \text{ and } |\phi(u) - \phi(v)| \leq L|u-v|, u,v \leq 0\}$

is compact.

(iii) C_g possesses a "fading memory" property.

(iv) C_g spaces with (3.1) satisfied often arise in a natural way for FDEs with infinite delay.

Several comments are in order. (a) (i) can be proven in a straightforward manner, and then be used to prove (ii). Along these lines, we refer to a recent paper of Arino, Burton and Haddock [1]. (b) Regarding (iii), Kato [17] defines an admissible space to be a fading memory space if the function M in property (B3) satisfies; $M(t) \to 0$ as $t \to \infty$. This holds for C_g if, for example, $g(s) = \exp(-\gamma s)$. However, we feel the condition $M(t) \to 0$ perhaps is too restrictive. In particular, we prefer to refer to any C_g as possessing a fading memory property whenever (3.1) holds. For this case, we have the compactness properties of (i) and (ii) above as well as the condition:

(3.2) $$\begin{cases} \text{if } x_t \text{ is in } C_g \text{ for } t \geq 0, \text{ then } x(t) \to 0 \text{ as } t \to \infty \\ \text{if and only if } |x_t|_g \to 0 \text{ as } t \to \infty. \end{cases}$$

Theorem 3.2 below provides a generalization of (3.2) with respect to solutions of FDEs. (c) To clarify (iv), consider the scalar nonlinear equation

(3.3) $$x' = -x^3 + \int_{-\infty}^{t} C(t-s)x^3(s)ds + F(t)$$

where $C:[0,\infty)\to R$ and $F:R\to R$ are continuous and $\int_0^\infty |C(t)|dt < k < \infty$

Now (3.3) can be written in the form (2.1) with

$$f(t,\phi) = -\phi^3(0) + \int_{-\infty}^{0} C(-s)\phi^3(s)ds.$$

From Burton and Grimmer [4], there exists a strictly decreasing function g which satisfies (2.3) and (3.1) such that

$$\int_{-\infty}^{0} |C(-s)|g^3(s)ds < k.$$

For this g, it is not overly difficult to establish fundamental existence, uniqueness, continual dependence and continuation result for (3.3) with respect to Cg. (Again, see [1].)

A COMPARISON RESULT

Theorem 3.1 below is reminiscent of comparison results given in Lakshmikantham and Leela [19]. We emphasize that this result does not require that g satisfy (3.1), and it is not stated in its most general form (see Remark 3.2).

Let $w:R^+\to R^+$ be continuous with t_0,u_0 in R^+, and u(t) be a maximal solution of

$$u' = w(t,u); \quad u(t_0) = u_0, t_0 \le t \le \infty.$$

Finally, let D^+ denote the upper-right derivative with respect to t, so that

$$D^+|x(t)| = \limsup_{h\to 0+} [|x(t+h)-x(t)|]/h.$$

THEOREM 3.1 (Haddock, Krisztin and Terjéki [9]). Suppose g satisfies (2.3) and x is continuous on $[t_0,A)$ with

(3.4) $$\begin{cases} D^+|x(t)| \le w(t,|x(t)|) & \text{for all } t \text{ in } [t_0,A) \\ \text{such that } |x(t)| = |x_t|_g. \end{cases}$$

Then $|x_{t_0}|_g \le u_0$ implies $|x_t|_g \le u(t)$ for all t in $[t_0,A)$.

A simple but important consequence of the above result concerns the case w = 0. In particular,

(3.5) $$\begin{cases} \text{if } |x(t)| = |x_t|_g \text{ implies } D^+|x(t)| \le 0, \\ \text{then } |x_t|_g \text{ is a nonincreasing function of } t. \end{cases}$$

In Example 3.1, we illustrate how Theorems 3.1 and 3.2 can be combined to produce information regarding behavior of solutions of FDEs.

A CONVERGENCE RESULT

THEOREM 3.2 (Haddock, Krisztin and Terjéki [9]). Suppose g is strictly decreasing on $(-\infty,0]$ such that (2.3) and (3.1) hold. Further, suppose $F:C\to R^n$ is continuous and maps bounded subsets of C_g into bounded subsets of R^n. Let x be a solution of the autonomous system

(3.6) $$x' = f(x_t)$$

such that x(t) is defined on an interval $[0,\infty)$. Then, for any constant c, $0 \leq c < \infty$,

(3.7) $$|x(t)| \to c \text{ as } t \to \infty \text{ if and only if}$$
$$|x_t|_g \to c \text{ as } t \to \infty.$$

REMARK 3.2. (i) Theorems 3.1 and 3.2 are not stated here in a final form. General results together with proofs will be given in the aforementioned reference [9]. We have used the present forms due to their simplicity and, more importantly, their ability to illustrate interesting relationships between the R^n and C_g norms of solutions of FDEs. (ii) Theorem 3.1 is in essence a Razumikhin-type comparison result. One (but not the most important) generalization of the theorem will be to extend it to include general (Liapunov) functions V (and not just $V = |x|$). For other recent Razumikhin results for infinite delay equations, we refer to Kato [18], Parrott [22] and Zhicheng [29]. (iii) Perhaps Theorem 3.2 is surprising since (3.7) holds for any constant c and not just c = 0. To obtain a similar result for finite delay equations requires additonal conditions and often a considerable effort (cf., e.g., Haddock and Terjéki [10] and references therein).

EXAMPLE 3.1. Consider the scalar nonlinear equation (3.3) with F(t) identically zero and assume

(3.8) $$\int_0^\infty |C(s)|ds < 1.$$

As was mentioned earlier in this section, we can choose a function g which satisfies the hypothesis of Theorem 3.2 such that

$$\int_0^\infty |C(-s)|g^3(s)ds < 1.$$

Choose C_g so that it is admissible; i.e., assume either (2.5) or (2.6) is satisfied. Then for any solution x of (3.3) defined on an interval $[t_0,A)$, we have

$$D^+|x(t))| \leq -|x^3(t)| + \int_{-\infty}^t |C(t-s)||x(s)|^3 ds$$

$$= -|x^3(t)| + \int_{-\infty}^0 |C(-s)|g^3(s)[|x_t(s)|^3/g^3(s)]ds$$

$$< -|x^3(t)| + |x_t|_g \int_{-\infty}^0 |C(-s)|g^3(s)ds$$

Clearly, $D^+|x(t)| \leq 0$ whenever $|x(t)| = |x_t|_g$. Since C_g is admissible, we can combine the statement following Theorem 3.1 with standard arguments (as in [24]) to see that $|x_t|_g$ is defined and nonincreasing on $[t_0,\infty)$. Thus, from Theorem 3.2, $|x(t)|$ tends to a constant as $t\to\infty$. Using condition (3.8) on Equation (3.3), it is elementary to prove that $x(t)\to 0$ as $t\to\infty$. We deduce that all solutions of (3.3) with (3.8) holding and F identically zero) tend to 0 as $t\to\infty$.

EXISTENCE OF A PERIODIC SOLUTION

Let $T > 0$ be given and consider the periodic ODE

$$x' = h(t,x), \tag{3.9}$$

where $h(t+T,x) = h(t,x)$ for all t in R and x in R^n. Under certain smoothness conditions, Yoshizawa [28, Theorem 29.3] employed uniform boundedness and uniform ultimate boundedness properties of solutions to prove the existence of a T-periodic solution of (3.1). Yoshizawa also pointed out that the result could be extended to finite delay equations whenever the delay r is not larger than the period T of the system (i.e., $T \geq r$). Recently, Burton [2] and [3, Section 8.6] defined the concepts of g-uniform boundedness and g-uniform ultimate boundedness and developed a counterpart of these results for infinite delay Volterra integrodifferential equations. However, he established (cf. [3, Theorem 8.6.6]) the existence of an mT-periodic (rather that T-periodic) solution for some integer $m > 0$.

Theorem 3.3 provides several improvements of results mentioned in the above paragraph. In particular, it places Burton's ideas in [2,3] in a general setting and, more significantly, it allows us to use $m = 1$ for the infinite delay case and to remove the restriction $T > r$ for finite delay equations. Before stating the result, we need to define three fundamental concepts.

We denote a solution of (2.1) "through" $(0,\phi)$ by $x(0,\phi)(t)$. Thus, $x_0(0,\phi) = \phi$.

DEFINITION 3.1. Solutions of (2.1) are C_g <u>uniform bounded</u> (at t=0) if for each $Q >$ there exists $H > 0$ such that

$$[\phi \text{ in } C_g,\ |\phi|_g < Q,\ t \geq 0] \text{ imply } |x(0,\phi)(t)| \leq H.$$

DEFINITION 3.2. Solutions of (2.1) are C_g <u>uniform ultimate bounded with bound</u> B (at t=0) if for each $A > 0$ there exists $K > 0$ such that

$$[\phi \text{ in } C_g,\ |\phi|_g < A,\ t \geq K] \text{ imply } |x(0,\phi)(t)| \leq B.$$

DEFINITION 3.3. Solutions of (2.1) <u>depend continuously on initial conditions in a set</u> $S \subset C_g$ if, for each $\varepsilon > 0$ and $J > 0$ there exists $\delta > 0$ such that

$$[\phi,\psi \text{ in } S \text{ and } |\phi-\psi|_g < \delta] \text{ imply } |x_J(0,\phi) - x_J(0,\psi)|_g < \varepsilon.$$

THEOREM 3.3 (Arino, Burton and Haddock [1]). Let g satisfy (3.1) and f satisfy

$$f(t+T,\phi) = f(t,\phi) \text{ for all } \phi \text{ in } C_g,\ t \text{ in } R \text{ and some } T > 0,$$

and suppose the following conditions hold.

(i) If ϕ is in C_g, then there is a unique solution $x(0,\phi)(t)$ of (2.1) on $[0,\infty)$.

(ii) Solutions of (2.1) are C_g-uniform bounded and C_g-uniform ultimate bounded with bound B (at t=0).

(iii) For each $\gamma > 0$, there is an $L > 0$ such that ϕ in C_g and $|\phi(u)| \leq \gamma$ on $(-\infty, 0]$ imply $|x'(0,\phi)(t)| \leq L$ on $[0,\infty)$.

(iv) For each $\gamma > 0$, solutions of (2.1) depend continuously on initial conditions in the set

$$\{\phi \text{ in } C_g : |\phi(u)| \leq \gamma \text{ for all } u \leq 0\}.$$

Then (2.1) possesses a T-periodic solution.

A proof of Theorem 3.3 and several examples and remarks concerning this theorem are provided in the reference cited above. The proof employs compactness properties such as in Remark 3.1- (ii), a fixed point theorem of Horn [14] and some fairly intricate constructions.

In view of the theorems, remarks and comments given in this paper, we feel justified in concluding that admissible spaces C_g with satisfying condition (3.1) are friendly spaces (referred to in the title) for functional differential equations with infinite delay.

REFERENCES

[1] O. Arino, T. Burton, and J. Haddock, Periodic solutions for functional differential equations, submitted for publication.

[2] T. Burton, Periodic solutions of nonlinear Volterra equations, Funkcialaj Ekvacioj, to appear.

[3] T. Burton, Volterra Integral and Differential Equations, Academic Press, New York, 1983.

[4] T. Burton and R. Grimmer, Oscillation, continuation, and uniqueness of solutions of retarded differential equations, Trans. Amer. Math. Soc., 169 (1973), 193-209.

[5] B. Coleman and V. Mizel, On the stability of solutions of functional differential equations, Arch. Rat. Mech. Anal., 30 (1968), 173-196.

[6] C. Corduneanu, Integral Equations and Stability of Feedback Systems, Academic Press, New York, 1973.

[7] C. Corduneanu and V. Lakshmikantham, Equations with unbounded delay: a survey, Nonl. Anal. TMA, 4 (1980), 831-877.

[8] R. Driver, Existence and stability of a delay-differential system, Arch. Rat. Mech. Anal., 10 (1962), 401-426.

[9] J. Haddock, T. Krisztin and J. Terjéki, manuscript in preparation.

[10] J. Haddock and J. Terjéki, Liapunov-Razumikhin functions and an invariance principle for functional differential equations, J. Diff. Eqs., 48 (1983), 95-122.

[11] J. Hale, Dynamical systems and stability, J. Math. Anal. Appl., 25 (1969), 39-59.

[12] J. Hale and J. Kato, Phase space for retarded equations with infinite delay, Funkcialaj Ekvacioj, 21 (1978), 11-41.

[13] Y. Hino, Stability properties for functional differential equations with infinite delay, Tohoku Math. J., 35 (1983), 597-605.

[14] W. Horn, Some fixed point theorems for compact maps and flows in Banach spaces, Trans. Amer. Math. Soc., 149 (1970), 391-404.

[15] T. Kaminogo, Knesner's property and boundary value problems for some retarded functional differential equations, Tohoku Math J., 30 (1978), 471-486.

[16] F. Kappel and W. Schappacher, Some considerations to the fundamental theory of infinite delay equations, J. Diff. Eqs., 37 (1980), 141-183.
[17] J. Kato, Liapunov's second method in functional differential equations, Tohoku Math J., 32 (1980), 487-497.
[18] J. Kato, Stability problem in functional differential equations with infinite delay, Funkcialaj Ekvacioj, 21 (1978), 63-80.
[19] V. Lakshmikantham and S. Leela, Differential and Integral Inequalities, Vol. II, Academic Press, New York, 1969.
[20] T. Naito, Fundamental matrices of linear autonomous retarded equations with infinite delay, Tohoku Math. J., 32 (1980), 539-556.
[21] T. Naito, On linear autonomous retarded equations with an abstract phase space for infinite delay, J. Diff. Eqs., 33 (1979), 74-91.
[22] M. Parrott, Convergence of solutions of infinite delay differential equations with an underlying phase space of continuous functions, in Ordinary and Partial Differential Equations, Springer Lecture Notes in Mathematics, 846 (1980), 280-289.
[23] S. Saperstone, Semidynamical Systems in Infinite Dimensional Spaces, Appl. Math. Sci. Series, Vol. 37, Springer-Verlag, New York, 1981.
[24] K. Sawano, Exponential asymptotic stability for functional differential equations with infinite retardations, Tohoku Math. J., 31 (1979), 363-382.
[25] K. Sawano, Some considerations on the fundamental theorems for functional differential equations with infinite delay, Funkcialaj Ekvacioj, 25 (1982), 97-104.
[26] K. Schumacher, Existence and continuous dependence for functional -differential equations with unbounded delay, Arch. Rat. Mech. Anal., 67 (1978), 315-335.
[27] G. Seifert, On Caratheodory conditions for functional differential equations with infinite delays, Rocky Mtn. J. Math., 12 (1982) 615-619.
[28] T. Yoshizawa, Stability Theory by Liapunov's Second Method, The Mathematical Society of Japan, Tokyo, 1966.
[29] W. Zhicheng, Comparison method and stability problem in functional differential equations, Tohoku Math. J., 35 (1983), 349-356.

This paper is in final form and no version of it will be submitted for publication elsewhere.

Trends in the Theory and Practice of Non-Linear Analysis
V. Lakshmikantham (Editor)

THE FIRST BOUNDARY VALUE PROBLEM FOR NONLINEAR DIFFUSION

Charles J. Holland
Code 411
Office of Naval Research
Arlington, VA 22217
and
James G. Berryman
Lawrence Livermore National Laboratory
P. O. Box 808, L-200
Livermore, CA 94550

Recent progress on understanding the asymptotic behavior of solutions to nonlinear diffusion equations in bounded domains is surveyed. Both positive and zero lateral boundary data have been considered.

INTRODUCTION:

This paper surveys recent progress in determining the asymptotic behavior of the solution $u = u(x,t)$ to the nonlinear diffusion equation

$$u_t = \Delta(u^m),\ m>0,\ \text{in } B\times(0,\infty), \tag{1}$$

B a bounded domain with smooth boundary, and smooth lateral boundary data

$$u(x,t) = F(x) > 0 \quad \text{on} \quad \partial B\times(0,\infty) \tag{2}$$

and smooth initial condition

$$u(x,0) = G(x) \geq 0 \quad \text{for} \quad x\in\bar{B}. \tag{3}$$

Equation (1) with boundary condition (2) and initial condition (3) is referred to as the first boundary value problem.

Nonlinear diffusion problems occur in the modelling of many physical processes. The case $m = .5$ corresponds to Okuda-Dawson diffusion in plasma physics while the case $m = 2.5$ occurs in electron heat conduction [1].

If $m = 1$, then problem (1) - (3) is linear diffusion and it is well-known that the solution $u(x,t)$ converges exponentially to the solution $v = v(x)$ of the stationary equation

$$\Delta(v^m) = 0, \tag{4}$$

with $m = 1$, and boundary data (2). In fact, there exists a function $H(\cdot)$ positive in B and a positive constant k such that

$$(\exp kt)(u(\cdot,t) - v(\cdot)) \to cH(\cdot)$$

in $H_0^{'}(B)$ for some constant c. Here $(\Delta+k)H=0$ in B, $H = 0$ on ∂B, and H is an eigenfunction corresponding to the principal eigenvalue k.

In [2] we have established that there is an analogous exponential convergence result for the nonlinear problem (1)-(3) with $m>0$ provided that the lateral boundary F is positive on ∂B. This result will be stated below in Theorem 1. However, if the lateral boundary data F is everywhere zero on ∂B, then exponential convergence does not hold. The zero boundary data case has been investigated by Aronson and Peletier [3] in the slow diffusion case ($m>0$) and by the present authors [4] in the fast diffusion case $0<m<1$. These results are reviewed in Theorems 2 and 3 below.

POSITIVE LATERAL BOUNDARY DATA ($F>0$ on $\bar{B}$):

Theorem 1. Assume $B\subset R^N$ with $N<6$. Let k, S be the principal eigenvalue and corresponding normalized eigenfunction to the eigenvalue problem

$$\Delta(mv^{m-1}S^*) = -k^*S^*,\ S^* = 0 \text{ on } \partial B. \tag{5}$$

Then there exists a constant c, which depends upon the initial data and may be zero, such that

$$(\exp kt)(u(\cdot,t) - v(\cdot)) \to cS(\cdot) \tag{6}$$

in $H_0'(B)$ as $t\to\infty$.

Remarks. The restriction that $N < 6$ may not be needed for the theorem, but is used in the proof in [2]. The estimate (6) might be guessed from formal linearization.

Outline of proof: The theorem is established by deriving a corresponding result for the difference u^m-v^m. Note that $\bar{u} = u^m$ satisfies

$$\Delta\bar{u} = (q+1)\ \bar{u}^q\bar{u}_t \quad \text{on} \quad Bx(0,\infty) \tag{7}$$

with $q = (1-m)/m$ and $\bar{v} = v^m$ satisfies $\Delta\bar{v} = 0$. Define $p = (\bar{u}-\bar{v})\exp kt$. It suffices to show that $p(\cdot,t) \to cS'(\cdot)$ in $H'(B)$ where S' is positive and satisfies

$$\Delta S' + k(q+1)v(x)^qS' = 0 \quad \text{on } B,\ S'=0 \text{ on } \partial B. \tag{8}$$

This is established by introducing an appropriate Liapunov functional (which is time dependent) and follows with some modifications to the approach in [4]. See the paper [2] for details.

ZERO LATERAL BOUNDARY DATA:

We now turn to the case of zero lateral boundary data. (Note that $v=0$ is then the solution to the stationary equation.) Here exponential convergence to the zero solution does not occur and the behavior depends dramatically on whether $m>1$ (slow diffusion) or $0<m<1$ (fast diffusion).

The slow diffusion case is considered first. Define $a=1/(m-1)$. Note that $(A+t)^{-a} f(x)$ is a separable solution to the problem (1), (2) for any constant A provided f satisfies the equation

$$\Delta(f^m) + af = 0 \text{ in } B,\quad f = 0 \text{ on } \partial B. \tag{9}$$

As shown in [3], there is a unique nontrivial solution to (9). The result in [3] is that the solution to (1)-(3) decays to zero looking like a separable solution to (1)-(2). This result is established using a comparison principle for weak solutions of (1)-(3) and thus an estimate on the rate of convergence to the separable solution is also obtained.

Theorem 2. There exists a constant c depending only on the initial data (3) such that

$$|(1+t)^a u(x,t) - f(x)| < cf(x)(1+t)^{-1} \text{ in } \bar{B}x(0,\infty). \qquad (10)$$

While difficult to prove, Theorem 2 provides a concise and precise analysis of the asymptotic behavior of solutions to the first boundary value problem in the slow diffusion case. Such a simple description is not possible in the fast diffusion case which we now consider.

Now let $a = 1/(1-m)$. Then separable solutions, if they exist, are of the form $f(x)(T^*-t)^a$ where f satisfies (9). Note that separable solutions have a finite extinction time T^* in that $f(x)(T^*-t)^a$ is positive in $Bx(0,T^*)$ and vanishes identically for $t = T^*$. Under certain conditions stated in Theorem 3 below, one is able to show that solutions u to (1)-(3) become extinguished "looking like" an appropriate separable solution with the same extinction time.

Theorem 3. Assume that there exists an extinction time T^* such that

(A) $u(x,t)$ is a classical solution of (1)-(3) on $B^*(0,T^*)$ which is positive on $Bx(0,T^*)$, and for any $T'<T^*$, u_t, u_{x_i}, $u_{x_i t}$, $u_{x_i x_i}$ are of class $C(\bar{B}x(0,T'))$.

(B) $u(x,T^*) = 0$ for all $x \in \bar{B}$.

(C) There exists a unique solution to (9) positive on B.

Let $0 < m < 1$ if $N < 2$ or $\frac{N-2}{N+2} < m < 1$ if $N > 2$.

Then

$$(T^*-t)^{-a} [u(\cdot,t) - f(\cdot)(T^*-t)^a] \to 0 \qquad (11)$$

in $H_0^1(B)$ as $t \to T^{*-}$.

Remarks. Let us now discuss the a priori assumptions. Earlier work of Sabinina [5] shows an extinction time T^* exists such that (B) holds for arbitrary smooth data if $N = 1$. If $N = 1$, existence and uniqueness is known for positive solutions to (9). If $N \geq 2$, existence is known providing $\frac{N-2}{N+2} < m < 1$. If $0 < m < \frac{N-2}{N+2}$, nonexistence of a classical solution is known to hold in star shaped regions while existence holds in annular shaped regions. Only recently have uniqueness/nonuniqueness results been obtained for the case $N > 2$. Uniqueness for the sphere was established in Gidas et al. [6]. If B is an annular region, nonuniqueness of positive solutions was established in case $N > 3$ and m "sufficiently close" to $\frac{N-2}{N+2}$ by Brezis and Nirenberg [7].

The lateral boundary regularity assumptions of (A) have not been rigorously established. However, in contrast to the case of slow diffusion, fast diffusion appears to have a smoothing effect.

ACKNOWLEDGMENTS:

Work of J.G.B. performed under the auspices of the U.S. Department of Energy by the Lawrence Livermore National Laboratory under Contract No. W-7405-ENG-48.

REFERENCES:

[1] Berryman, J.G., Evolution of a stable profile for a class of nonlinear diffusion equations. III. Slow diffusion on the line, J. Math. Phys. 21 (1980) 1326-1331.

[2] Holland, C.J. and Berryman, J.G., Exponential convergence for nonlinear diffusion problems with positive lateral boundary conditions, J. Math. Phys., in press.

[3] Aronson, D. and Peletier, L., Large time behavior of solutions of the porous medium equation in bounded domains, J. Diff. Equations 39 (1981) 378-412.

[4] Berryman, J.G. and Holland, C.J., Stability of the separable solution for fast diffusion, Arch. Rational Mech. Anal. 74 (1980) 379-388.

[5] Sabinina, E.S., A class of nonlinear degenerating parabolic equations, Sov. Math. Doklady 143 (1962) 495-498.

[6] Gidas, B., Ni, W.M., and Nirenberg, L., Symmetry and related properties via the maximum principle, Commun. Math. Phys. 68 (1979) 209-243.

[7] Brezis, H. and Nirenberg, L., Positive solutions of nonlinear elliptic equations involving critical Sobolev exponents, Commun. Pure Appl. Math. 36 (1983) 437-477.

The final (detailed) version of this paper has been submitted for publication elsewhere.

Trends in the Theory and Practice of Non-Linear Analysis
V. Lakshmikantham (Editor)

AN ASYMPTOTIC ANALYSIS OF A REACTION-DIFFUSION SYSTEM

F. A. Howes

Department of Mathematics
University of California, Davis
Davis, California 95616
U.S.A.

1. INTRODUCTION

In this note we outline an approach to the study of solutions of initial-boundary value problems for the singularly perturbed system $u_t + h(x,t,u) = \varepsilon^2 u_{xx}$ in a rectangle $\Pi \subset \mathbb{R}^2$, for small values of the perturbation parameter ε^2. The perturbation parameter is written as ε^2 since we anticipate the appearance of its (positive) square root in some of our estimates. Under certain assumptions on h solutions are shown to follow solutions of the initial value problem for the unperturbed system $U_t + h(x,t,U) = 0$ in $\bar{\Pi}$ as $\varepsilon \to 0^+$. The principal assumption is the existence of a solution U in $\bar{\Pi}$ which is stable and diagonally dominant in a sense to be made precise in the next section.

Such problems arise in many applied areas (cf. for example [2], [5], [10; Pt. II]), where they are usually referred to as "reaction-diffusion" problems. The assumed smallness of ε thus implies that diffusive or dissipative effects (as measured by the term $\varepsilon^2 u_{xx}$) are formally smaller than reactive or interactive effects (as measured by the term $u_t + h$). This is the physical reason why solutions of the reactive system $U_t + h(x,t,U) = 0$ approximate solutions of the reaction-diffusion system as $\varepsilon \to 0^+$. We illustrate some of these ideas in Section 3 with a two-component system from ecology (cf. [6]).

2. STATEMENT OF THE PROBLEM

Consider then the initial-boundary value problem

$$\text{(IBVP)} \quad \begin{aligned} &u_t + h(x,t,u) = \varepsilon^2 u_{xx}, \ (x,t) \text{ in } \Pi =: (a,b) \times (0,T), \\ &u(x,0) = \varphi(x), \quad x \text{ in } [a,b], \\ &u(a,t) - Pu_x(a,t) = A(t), \ u(b,t) + Qu_x(b,t) = B(t), \ t \text{ in } [0,T], \end{aligned}$$

where u, h, φ, A and B are n-vector-valued functions and P, Q are diagonal $(n \times n)$-matrices with constant entries $p_i > 0$, $q_i > 0$, respectively, for $i=1,\dots,n$. Formally neglecting ε^2 leads us to consider the initial value (reduced) problem

$$\text{(IVP)} \qquad U_t + h(x,t,U) = 0, \quad U(x,0) = \varphi(x),$$

which we assume has a smooth solution $U = U(x,t)$ for (x,t) in $\bar{\Pi}$. In general, it is not possible to ask that U satisfy either of the boundary conditions prescribed in (IBVP), since the boundaries $x = a$ and $x = b$ are characteristic curves of the first-order operator; however, the problem (IVP) is always well-posed (at least for small $t > 0$) since the base, $t = 0$, of Π is noncharacteristic (cf. [4; Chap. II]). Our assumption on (IVP) thus amounts to the assumption that this local solution exists for all t in $[0,T]$, for each x in $[a,b]$. In order that the function U can serve as a good approximation to a solution of (IBVP) in $\overline{\Pi}$ as $\varepsilon \to 0^+$, we must further require that U is asymptotically stable, in the sense that there exist a small positive constant δ and positive constants m_i such that for $i=1,\dots,n$

$$(\partial h_i/\partial u_i)(x,t,u) > m_i^2 > 0$$

for all (x,t,u) in $\mathcal{D} =: \overline{\Pi} \times \{u: |u_i - U_i(x,t)| \le \delta\}$. Since δ is small it is easy to see that this assumption obtains if $(\partial h_i/\partial u_i)(x,t,U_i(x,t)) > 0$ in $\overline{\Pi}$, for $i=1,\dots,n$. It turns out that this "local" stability is enough to guarantee the existence of a solution of (IBVP) which is uniformly close to $U(x,t)$ because the boundary conditions in (IBVP) impose a uniform bound on derivatives of any solution. Finally we assume that the function U is "diagonally dominant", in the sense that for $i,j=1,\dots,n$ there exist positive constants μ_{ij} such that

$$m_i^2 - \sum_{\substack{j=1 \\ j\neq i}}^{n} \mu_{ij} =: d_i > 0,$$

where $|(\partial h_i/\partial u_j)(x,t,u)| < \mu_{ij}$, for $j \neq i$ and all (x,t,u) in $\mathcal{D}$. Again since δ is small, it is enough to define the constants μ_{ij} $(j \neq i)$ as upper bounds on $|(\partial h_i/\partial u_j)(x,t,U(x,t))|$ in $\overline{\Pi}$. The precise result is contained in the following theorem.

<u>Theorem</u>. <u>Assume that the reduced problem</u> (IVP) <u>has a smooth asymptotically stable, diagonally dominant solution</u> $U = U(x,t)$ <u>in</u> $\overline{\Pi}$. <u>Then there exists an</u> $\varepsilon_0 > 0$ <u>such that the problem</u> (IBVP) <u>has a smooth solution</u> $u = u(x,t,\varepsilon)$ <u>whenever</u> $0 < \varepsilon \le \varepsilon_0$. <u>Moreover, for</u> (x,t) <u>in</u> $\overline{\Pi}$ <u>and</u> $i=1,\dots,n$ <u>we have the estimates</u>

$$\begin{aligned}|u_i(x,t,\varepsilon) - U_i(x,t)| &\le K(\varepsilon/\overline{m}) \exp[-(\overline{m}/\varepsilon)(x-a)] \\ &+ L(\varepsilon/\overline{m}) \exp[-(\overline{m}/\varepsilon)(b-x)] \\ &+ M\varepsilon^2,\end{aligned}$$

<u>where</u> $K = \max\{K_1,\dots,K_n\}$ <u>for</u> $K_i =: \max_{[0,T]} |A_i(t) - U_i(a,t) + p_i U_{i,x}(a,t)|$, $L = \max\{L_1, \dots, L_n\}$ <u>for</u> $L_i =: \max_{[0,T]} |B_i(t) - U_i(b,t) - q_i U_{i,x}(b,t)|$, $0 < \overline{m}^2 < \min\{d_1,\dots,d_n\}$, <u>and</u> M <u>is a known positive constant depending on</u> h <u>and</u> U.

<u>Proof</u>. The theorem follows by constructing lower and upper solutions of (IBVP) which contain the desired asymptotic behavior as $\varepsilon \to 0^+$; cf. [8]. To this end, we define for (x,t) in $\overline{\Pi}$ and $\varepsilon > 0$ the functions

$$\omega_i(x,t,\varepsilon) = U_i(x,t) - \lambda(x,\varepsilon) - M\varepsilon^2$$

and

$$\Omega_i(x,t,\varepsilon) = U_i(x,t) + \lambda(x,\varepsilon) + M\varepsilon^2,$$

for $i=1,\ldots,n$, where $\lambda =: K(\varepsilon/\bar{m})\exp[-(\bar{m}/\varepsilon)(x-a)] + L(\varepsilon/\bar{m})\exp[-(\bar{m}/\varepsilon)(b-x)]$. Then if we can show that as $\varepsilon \to 0^+$, $\omega_i(x,0,\varepsilon) \le \varphi_i(x) \le \Omega_i(x,0,\varepsilon)$ for x in $[a,b]$, $(\omega_i - p_i\omega_{i,x})(a,t,\varepsilon) \le A_i(t) \le (\Omega_i - p_i\Omega_{i,x})(a,t,\varepsilon)$ and $(\omega_i + q_i\omega_{i,x})(b,t,\varepsilon) \le B_i(t) \le (\Omega_i + q_i\Omega_{i,x})(b,t,\varepsilon)$ for t in $[0,T]$, and $\omega_{i,t} + h_i(x,t,u_1,\ldots,u_{i-1},\omega_i,u_{i+1},\ldots,u_n) \le \varepsilon^2\omega_{i,xx}$, $\Omega_{i,t} + h_i(x,t,u_1,\ldots,u_{i-1},\Omega_i,u_{i+1},\ldots,u_n) \ge \varepsilon^2\Omega_{i,xx}$ for (x,t) in Π and for all u_j, $j \ne i$, in $[\omega_j,\Omega_j]$, the theory tells us that (IBVP) has a solution $u = u(x,t,\varepsilon)$ as $\varepsilon \to 0^+$ satisfying $\omega_i(x,t,\varepsilon) \le u_i(x,t,\varepsilon) \le \Omega_i(x,t,\varepsilon)$ in $\bar{\Pi}$; cf. [1]. Now the algebraic inequalities follow directly from the form of ω_i and Ω_i; recall that $U_i(x,0) = \varphi_i(x)$. It is enough to verify that ω_i satisfies the required differential inequality since the verification for Ω_i follows by symmetry. Differentiating and substituting, we see that for u_j in $[\omega_j,\Omega_j]$, $j \ne i$,

$$\begin{aligned}\varepsilon^2\omega_{i,xx} - \omega_{i,t} - h_i = \varepsilon^2 U_{i,xx} - \varepsilon^2\lambda_{xx} - U_{i,t} \\ - h_i(x,t,u_1,\ldots,u_{i-1},U_i,u_{i+1},\ldots,u_n) \\ - (\partial h_i/\partial u_i)(x,t,u_1,\ldots,u_{i-1},U_i + \\ \theta(\omega_i - U_i),u_{i+1},\ldots,u_n)(\omega_i - U_i),\end{aligned}$$

where $0 < \theta < 1$. Now for ε sufficiently small, say $0 < \varepsilon \le \varepsilon_0$, the points $[x,t] =: (x,t,u_1,\ldots,u_{i-1},U_i+\theta(\omega_i-U_i),u_{i+1},\ldots,u_n)$ lie in $\mathcal{D}$, and therefore $(\partial h_i/\partial u_i)[x,t] \ge m_i^2 > 0$; in addition, owing to the diagonal dominance of U, we have that $|U_{i,t} - h_i(x,t,u_1,\ldots,u_{i-1},U_i,u_{i+1},\ldots,u_n)| \le (\lambda + M\varepsilon^2)\sum_{\substack{j=1\\ j\ne i}}^{n}\mu_{ij}$, since $-U_{i,t} = h_i(x,t,U_1,\ldots,U_n)$. Consequently we can continue with the inequality

$$\varepsilon^2\omega_{i,xx} - \omega_{i,t} - h_i \ge [d_i - \bar{m}^2]\lambda + [d_i M - \ell_i]\varepsilon^2 \ge 0$$

provided $M =: \max\{\ell_1/d_1,\ldots,\ell_n/d_n\}$ for $\ell_i =: \max_{\bar{\Pi}}|U_{i,xx}|$ since by definition $0 < \bar{m}^2 < \min\{d_1,\ldots,d_n\}$. Thus the functions ω_i satisfy the required differential inequalities for all (x,t) in Π and $0 < \varepsilon \le \varepsilon_0$. A similar argument shows that the Ω_i also satisfy the required differential inequalities, and so the theorem is proved.

Remark 1. The proof shows that it is possible to extend the theorem to an initial-boundary value problem for the more general equation $u_t + h(x,t,u) = \text{diag}\{\varepsilon^{\alpha_1},\ldots,\varepsilon^{\alpha_n}\}u_{xx}$, where $\alpha_i > 0$ for $i=1,\ldots,n$.

Remark 2. If one is interested in the existence of a solution of (IBVP) for only small values of $t > 0$, then the theorem can be proved without the assumptions of diagonal dominance and asymptotic stability. This follows by introducing the variables $w_i =: u_i e^{-mt}$, for $i=1,\ldots,n$, and noting that the solution of the corresponding initial value problem is diagonally dominant and asymptotically stable for a sufficiently large value of $m > 0$; cf. [11; Chap. 2]. Actually this substitution allows us to prove the theorem under the sole assumption that U exists in

$\overline{\Pi}$; however, the estimate is not very sharp since $u_i(x,t,\varepsilon) = \mathcal{O}(w_i(x,t,\varepsilon)e^{mt}) = \mathcal{O}(e^{mt})$ for large t.

Remark 3. We conclude this section by noting that the theorem obtains if the reduced solution U is only asymptotically stable, provided the system $u_t + h(x,t,u,\varepsilon) = \varepsilon^2 u_{xx}$ is weakly nonlinear in the sense that for $i=1,\ldots,n$, $U_{i,t} + h_i(x,t,u_1,\ldots,u_{i-1},U_i,u_{i+1},\ldots,u_n,\varepsilon) = \mathcal{O}(\varepsilon^2)$ for (x,t) in $\overline{\Pi}$ and u_j in $[\omega_j,\Omega_j]$, $j \neq i$. See [7].

3. AN EXAMPLE AND AN EXTENSION

Let us first illustrate the theory by looking for positive solutions of the two-component system from ecology (cf. [6])

$$
\text{(E)}\quad
\begin{aligned}
&u_t + u(a-bu-cv) = \varepsilon^2 u_{xx}, \quad 0 < x < 1, \quad t > 0,\\
&u(x,0) = \varphi(x),\ (u-p_1u_x)(0,t) = A_1,\ (u+q_1u_x)(1,t) = B_1,\\
&v_t + v(d-eu-fv) = \varepsilon^2 v_{xx}, \quad 0 < x < 1, \quad t > 0,\\
&v(x,0) = \psi(x),\ (v-p_2v_x)(0,t) = A_2,\ (v+q_2v_x)(1,t) = B_2,
\end{aligned}
$$

where $a,\ldots,f$ are positive constants and $\varphi(x) > 0$, $\psi(x) > 0$ for x in $[0,1]$. We begin by noting that $(0,0)$ is an asymptotically stable equilibrium point of the reduced system $U_t + U(a-bU-cV) = 0$, $V_t + V(d-eU-fV) = 0$, since $(\partial h_1/\partial u)(0,0) = a > 0$, $(\partial h_2/\partial v)(0,0) = d > 0$ and $(\partial h_1/\partial v)(0,0) = (\partial h_2/\partial u)(0,0) = 0$, for $h_1 =: u(a-bu-cv)$ and $h_2 =: v(d-eu-fv)$. If the initial data φ, ψ are such that

$$
\varphi(x) < (2af-cd)/\Delta, \quad \psi(x) < (2bd-ae)/\Delta, \tag{3.1}
$$

for $\Delta =: 4bf - ce \neq 0$, in $[0,1]$, then it follows that the solution (U,V) of the reduced system satisfying $(U(x,0),V(x,0)) = (\varphi(x),\psi(x))$ exists for all $t > 0$ and $\lim_{t\to\infty}(U,V) = (0,0)$. Moreover, the solution (U,V) is asymptotically stable and diagonally dominant. The reason for this is seen easily if we rewrite the inequalities in (3.1) in matrix form as

$$
M^{-1}\begin{pmatrix} a \\ d \end{pmatrix} > \begin{pmatrix} \varphi(x) \\ \psi(x) \end{pmatrix}, \tag{3.2}
$$

for $M^{-1} =: \Delta^{-1}\begin{pmatrix} 2f & -c \\ -e & 2b \end{pmatrix}$, and note that therefore

$$
M\left(M^{-1}\begin{pmatrix} a \\ d \end{pmatrix}\right) = \begin{pmatrix} a \\ d \end{pmatrix} > M\begin{pmatrix} \varphi(x) \\ \psi(x) \end{pmatrix} = \begin{pmatrix} 2b & c \\ e & 2f \end{pmatrix}\begin{pmatrix} \varphi(x) \\ \psi(x) \end{pmatrix}. \tag{3.3}
$$

The inequality (3.3) follows from (3.2) since M^{-1} is of monotonic type. (Recall that a square matrix K is said to be of monotonic type if by definition $K\binom{x}{y} > \binom{0}{0} \Rightarrow \binom{x}{y} > \binom{0}{0}$, which is equivalent to the implication $\binom{x}{y} > \binom{0}{0} \Rightarrow K^{-1}\binom{x}{y} > \binom{0}{0}$. A necessary and sufficient condition for K to be of monotonic type is that the elements of K^{-1} are all positive (cf. [3; pp. 42-47]); whence, M^{-1} is of monotonic type since $(M^{-1})^{-1} = M =: \begin{pmatrix} 2b & c \\ e & 2f \end{pmatrix}$.) Since $(0,0)$ is asymptotically stable and since (3.3) states that (U,V) is initially asymptotically stable and diagonally dominant, we conclude that (U,V) is asymptotically stable and diagonally dominant for <u>all</u> $t \geq 0$. Consequently the Theorem asserts that for initial data φ, ψ satisfying the inequalities in (3.1) the problem (E) has a solution (u,v) as $\epsilon \to 0^+$ such that

$$\lim_{\epsilon \to 0^+} (u,v)(x,t,\epsilon) = (U,V)(x,t) \quad \text{in } \overline{\Pi}$$

We note however that it is possible to relax the restrictions in (3.1) by requiring only that initially $a - bu - cv > 0$ and $d - eu - fv > 0$, or

$$a - b\varphi(x) - c\psi(x) > 0, \quad d - e\varphi(x) - f\psi(x) > 0,$$

for x in [0,1], that is,

(3.4) $$\varphi(x) < (af-cd)/\Delta', \quad \psi(x) < (bd-ae)/\Delta',$$

for $\Delta' =: bf - ce \neq 0$. This follows because under the inequalities in (3.4) $h_1/u > 0$ for $0 < u < \varphi(x)$ and $h_2/v > 0$ for $0 < v < \psi(x)$ in [0,1]. Essentially, the sharper inequalities in (3.4) take into account higher order terms in the functions h_1 and h_2, whereas the inequalities in (3.1) are based on linearization, as was the Theorem. Thus we can likewise improve the Theorem by including higher order terms in our definitions of asymptotic stability and diagonal dominance. For instance, suppose that $h(x,t,0) \equiv 0$ and that $(\partial h_i/\partial u_i)(x,t,0) > m_i^2 > 0$, $i=1,\ldots,n$, for (x,t) in $\overline{\Pi}$, that is, 0 is an asymptotically stable equilibrium of the unperturbed equation $U_t + h(x,t,U) = 0$. If the initial data φ is such that for each x in [a,b]

(3.5) $$\lambda_i h_i(x,0,\lambda_1,\ldots,\lambda_n) < 0$$

for all nonzero values of λ_j between 0 and $\varphi_j(x)$, $i,j=1,\ldots,n$, then the conclusion of the Theorem obtains. This result is well known for the pure initial value problem (IVP). The validity of its extension to (IBVP) is a direct consequence of the equivalence of results on the existence of invariant regions for (IVP) and (IBVP).

Remark. It is also possible to prove a result analogous to the Theorem when $(\partial^j h_i/\partial u_i^j)(x,t,0) \equiv 0$ for $j=0,\ldots,2q$ and $(\partial^{2q+1} h_i/\partial u_i^{2q+1})(x,t,0) > 0$ in $\overline{\Pi}$. The exponential terms in the estimates are then replaced by algebraic terms (cf. [7]) and so the convergence of the solutions of (IBVP) to the stable solution of (IVP) is slower.

4. CONCLUDING REMARKS

It is not difficult to extend the results of Section 2 to problems in which x is an N-dimentional vector and the term $\epsilon^2 u_{xx}$ is replaced by $\epsilon^2 \Sigma_{j=1}^N \partial^2 u/\partial x_j^2$; cf. [8]. However, difficulties do arise if we replace the Robin boundary conditions in (IBVP) with Dirichlet conditions or if we allow solutions of (IBVP) to have nonuniformities of the shock layer type at points in $(a,b) \times (0,T)$. The reason is that in a neighborhood of a (Dirichlet) boundary point or a "shock" point the solution differs from a stable reduced solution by an amount of order one. Thus we need to assume additional stability of the reduced solution in order to derive a result akin to the Theorem for such problems. These stability conditions are derived by O'Donnell [9] for the corresponding steady-state Dirichlet problem $\varepsilon^2 u_{xx} = h(x,u)$, $u(a,\varepsilon)$, $u(b,\varepsilon)$ prescribed, and we refer the interested reader to this paper for details.

ACKNOWLEDGMENTS

The author wishes to thank the typist, Ida Mae Zalac, for her fine secretarial work, as well as NSF for its support under grant no. DMS-8319783.

REFERENCES

[1] Amann, H., Periodic Solutions of Semilinear Parabolic Equations, in Nonlinear Analysis, ed. by L. Cesari et al., Academic Press, New York, 1978, pp. 1-29.

[2] Aris, R., The Mathematical Theory of Diffusion and Reaction in Permeable Catalysts, vol. II, Clarendon Press, Oxford, 1975.

[3] Collatz, L., The Numerical Treatment of Differential Equations, Springer-Verlag, Berlin, 1960.

[4] Courant, R. and Hillbert, D., Methods of Mathematical Physics, vol. II, Interscience, New York, 1962.

[5] Fife, P. C., Mathematical Aspects of Reacting and Diffusing Systems, Lecture Notes in Biomathematics, vol. 28, Springer-Verlag, New York, 1979.

[6] Hastings, A., Global Stability in Lotka-Volterra Systems with Diffusion, J. Math. Biology 6(1978), 163-168.

[7] Howes, F. A., Singularly Perturbed Semilinear Systems, Studies in Appl. Math. 61(1979), 185-209.

[8] Howes, F. A., Multi-Dimensional Initial-Boundary Value Problems with Strong Nonlinearities, Arch. Rational Mech. Anal.,

[9] O'Donnell, M. A., Boundary and Corner Layer Behavior in Singularly Perturbed Semilinear Systems of Boundary Value Problems, SIAM J. Math. Anal. 15(1984), 317-332.

[10] Smoller, J., Shock Waves and Reaction-Diffusion Equations, Springer-Verlag, New York, 1983.

[11] Sperb, R., Maximum Principles and Their Applications, Academic Press, New York, 1981.

This paper is in final form and no version of it will be submitted for publication elsewhere.

Trends in the Theory and Practice of Non-Linear Analysis
V. Lakshmikantham (Editor)

ON A NONLINEAR HYPERBOLIC INTEGRODIFFERENTIAL EQUATION WITH A SINGULAR KERNEL

W. J. Hrusa
Department of Mathematics
Carnegie-Mellon University
Pittsburgh, Pennsylvania 15213
U.S.A.

M. Renardy
Mathematics Research Center
University of Wisconsin
Madison, Wisconsin 53705
U.S.A.

We discuss local and global existence of classical solutions to a nonlinear hyperbolic Volterra equation which models the motion of a one-dimensional nonlinear viscoelastic solid. In contrast to most previous studies, we allow the kernel to be singular.

1. INTRODUCTION

Many constitutive models for viscoelastic materials lead to equations of motion which have the form of a quasilinear hyperbolic PDE perturbed by a dissipative integral term of Volterra type. In the recent literature, a number of existence theorems have been proved for such equations [1,2,7,10,11,12,15,18,19,23]. These papers establish the existence of classical solutions locally in time and (in some cases) globally in time if the given data are suitably small. For large data, global existence does not hold in general and shocks are expected to develop [4,6,16,17,22].

Common to all of the works referred to above is the assumption that the kernel in the integral term is sufficiently smooth on $[0,\infty)$. We are here interested in the possibility that this kernel is singular at 0. Kinetic theories for chain molecules [3,21,24] and some experimental data [13] suggest that this is a realistic possibility, at least for some viscoelastic materials.

Hannsgen and Wheeler [5] have shown (for the constant coefficient linear problem) that the solution operator has a certain compactness property if and only if the kernel is singular. Renardy [20] and Hrusa and Renardy [9] have studied linear wave propagation. They showed that certain singular kernels have a smoothing effect; the precise degree of smoothing depends crucially on the strength of the singularity in the kernel. These results suggest that, if anything, models with singular kernels should have "nicer" existence properties than those with regular kernels. However, this also indicates that one cannot expect the methods of previous existence proofs to extend to singular kernels. These proofs rely on an iteration scheme that treats the hyperbolic part as the principal term and the integral as a perturbation. This, of course, works irrespective of the sign of the integral. If, however, singular kernels lead to smoothing, then the opposite sign in the integral will lead to instantaneous blow up, and a local existence theorem cannot hold.

In this paper, we discuss the history-value problem

$$(1.1)\quad u_{tt}(x,t) = \varphi(u_x(x,t))_x + \int_{-\infty}^{t} a'(t-\tau)\psi(u_x(x,\tau))_x d\tau + f(x,t), \quad x \in [0,1],\ t \geq 0,$$

$$(1.2)\quad u(0,t) = u(1,t) = 0, \quad t \geq 0,$$

$$(1.3)\quad u(x,t) = v(x,t), \quad x \in [0,1], \quad t \leq 0,$$

where $\varphi, \psi : \mathbb{R} \to \mathbb{R}$ are assigned smooth constitutive functions, $a : (0,\infty) \to \mathbb{R}$ is a given kernel, f is a known forcing function, and v is a prescribed "history". Subscripts x and t denote partial differentiation and $'$ indicates the derivative of a function of a single variable. Throughout this paper, all derivatives should be interpreted in the sense of distributions.

Throughout our discussion, we assume that f is smooth on $[0,1] \times (-\infty,\infty)$ and that the history v satisfies equation (1.1) and the boundary conditions (1.2) for $t \leq 0$. This ensures that the data (f and v) are compatible with the boundary conditions and that derivatives of v as $t \uparrow 0$ are compatible with derivatives of u and $t \downarrow 0$. It is possible to remove the assumption that v satisfies the equation (provided f and v are compatible with the boundary conditions), with the result that certain derivatives of u may be discontinuous across $t = 0$.

The history-value problem (1.1),(1.2),(1.3) was studied by Dafermos and Nohel [2] in the case where a' is absolutely continuous on $[0,\infty)$. (Closely related problems with smooth kernels have also been studied by MacCamy [15], Dafermos and Nohel [1], Staffans [23], and Hrusa and Nohel [11]. See [10] for a summary of these works as well as a discussion of the physical interpretation of (1.1).) Like Dafermos and Nohel, we normalize a so that $a(\infty) = 0$ and assume $\varphi' > 0$, $\psi' > 0$, $\varphi' - a(0)\psi' > 0$. They require that the kernel a is strongly positive definite; for technical reasons we make the stronger assumption that a is positive, monotone decreasing, and convex. While they assume that $a, a', a'' \in L^1(0,\infty)$, we allow a' to have a singularity at 0, e.g. $a'(t) \sim t^{-\alpha}$ as $t \downarrow 0$ with $0 < \alpha < 1$.

We note that our local existence theorem also holds for Neumann or mixed boundary conditions, or for the all-space problem (i.e., x varies from $-\infty$ to ∞). The global theorem can also be generalized to other boundary conditions. For a Neumann problem, we need a trivial modification in the conclusion, due to the possibility of rigid motions which need not decay as $t \to \infty$. We do not known how to extend the global result to the all-space problem. Recent work on this problem by Hrusa and Nohel [11] makes essential use of the assumption that the kernel is regular.

The results which we present here are proved in [8]. The only other existence theorem for nonlinear models with singular kernels that we are aware of is a result of Londen [14] concerning the existence of weak solutions to an abstract integrodifferential equation. Londen's existence theorem can be applied to (1.1),(1.2),(1.3) in the special case where $\psi \equiv \varphi$; his assumptions require a' to have a singularity which is stronger than logarithmic.

2. LOCAL EXISTENCE

If the kernel is smooth on $[0,\infty)$ (e.g., if $a' \in AC_{loc}[0,\infty)$) then the iteration scheme which consists of solving

$$(2.1)\quad u_{tt}^{(n+1)}(x,t) = \varphi'(u^{(n)})u_{xx}^{(n+1)}(x,t) + \int_{-\infty}^{t} a'(t-\tau)\psi(u_x^{(n)}(x,\tau))_x d\tau + f(x,t), \qquad x \in [0,1], \quad t \geq 0,$$

$$(2.2)\quad u^{(n+1)}(0,t) = u^{(n+1)}(1,t) = 0, \quad t \geq 0,$$

$$(2.3)\quad u^{(n+1)}(x,t) = v(x,t), \quad x \in [0,1], \quad t \leq 0,$$

can be used to establish local existence of solutions provided that $\varphi' > 0$ and the data are sufficiently smooth. Observe that once $u^{(n)}$ has been determined, (2.1) is a (variable coefficient) linear wave equation for $u^{(n+1)}$ with forcing term $f(x,t) + \int_{-\infty}^{t} a'(t-\tau)\psi(u_x^{(n)}(x,\tau))_x d\tau$. Standard theory for linear hyperbolic equations, together with energy estimates and a contraction argument, can be used to prove convergence of the iterates on $[0,1] \times (-\infty,T]$ for sufficiently small $T > 0$. See Theorem 2.1 of [2]. (An abstract local existence result of Hughes, Kato, and Marsden can also be applied to problems of this type with smooth kernels. See [18] and [19].)

As noted in the Introduction, the above iteration scheme cannot be expected to work for singular kernels. Therefore, we use

$$(2.4) \qquad u_{tt}^{(n+1)}(x,t) = \varphi'(u_x^{(n)})u_{xx}^{(n+1)}(x,t) + \int_{-\infty}^{t} a'(t-\tau)\psi'(u_x^{(n)})u_{xx}^{(n+1)}(x,\tau)d\tau + f(x,t), \quad x \in [0,1], \quad t \geq 0.$$

in place of (2.1). Observe that the memory term in (2.4) involves $u^{(n+1)}$.

To compute the iterates generated by (2.4),(2.2),(2.3), we must solve a sequence of linear integrodifferential equations of the form

$$(2.5) \qquad u_{tt}(x,t) = \alpha(x,t)u_{xx}(x,t) + \int_{-\infty}^{t} a'(t-\tau)\beta(x,\tau)u_{xx}(x,\tau)d\tau + f(x,t),$$

together with the appropriate auxiliary conditions. An existence theorem for (2.5) is established in [8] under hypotheses which permit a' to have an integrable singularity at 0. The proof is accomplished by approximating (2.5) by a sequence of equations with regular kernels (for which existence of solutions is known) and using energy estimates, in conjunction with properties of strongly positive definite kernels, to show that the corresponding sequence of solutions converges to a solution of (2.5). See Theorem 3.1 of [8].

Using the existence result for (2.5) and the contraction mapping principle, it is shown in [8] that the iterates generated by (2.4), (2.2),(2.3) converge to a solution of (1.1),(1.2),(1.3) on $[0,1] \times (-\infty,T]$ for sufficiently small $T > 0$. The local existence result from [8] is stated below.

Theorem 1: Assume that

(i) $a, a' \in L^1(0,\infty)$, $a \geq 0$, $a' \leq 0$, $a'' \geq 0$ (in the sense of measures), and a'' is not a purely singular measure;

(ii) $\varphi, \psi \in C^3(\mathbb{R})$, $\varphi' > 0$, $\psi' > 0$;

(iii) $f, f_x, f_t \in L^\infty((-\infty,\infty); L^2(0,1)) \cap L^2((-\infty,\infty); L^2(0,1))$, $f_{tt} \in L^2((-\infty,\infty); L^2(0,1))$;

(iv) The given history v satisfies equation (1.1) and the boundary conditions for $t \leq 0$, v and its derivatives through third order belong to $L^\infty((-\infty,0]; L^2(0,1)) \cap L^2((-\infty,0]; L^2(0,1))$, $v_{xxtt} \in L^2((-\infty,0]; L^2(0,1))$.

Then, the history-value problem (1.1),(1.2),(1.3) has a unique solution u defined on a maximal time interval $(-\infty, T_0)$, $T_0 > 0$, such that u and its derivatives through third order belong to $L^\infty((-\infty,T]; L^2(0,1))$ for every $T < T_0$. Moreover, if

$$\text{ess-sup}_{t\in[0,T_0)} \int_0^1 \{u_{xxx}^2 + u_{xxt}^2 + u_{xtt}^2 + u_{ttt}^2\}(x,t)\,dx < \infty \tag{2.6}$$

then $T_0 = \infty$.

Remark: It is also shown in [8] that u satisfies an additional continuity property if $\hat{a}(i\omega)$ behaves suitably as $\omega \to \pm\infty$, where $\hat{a}$ denotes the Laplace transform of a.

3. GLOBAL EXISTENCE

Dafermos and Nohel [2] established a (small data) global existence theorem for (1.1),(1.2),(1.3) under assumptions which require $a'' \in L^1(0,\infty)$. Using Theorem 1 (in place of Theorem 2.1 of [2]) and Lemma 2.5 of [8] (in place of (3.2) of [2]) the global argument of Dafermos and Nohel can be carried out without assuming $a'' \in L^1(0,\infty)$.

Theorem 2: Assume that

(i) $a, a' \in L^1(0,\infty)$, $a \geq 0$, $a' \leq 0$, $a'' \geq 0$ (in the sense of measures), and a'' is not a purely singular measure;

(ii) $\varphi, \psi \in C^3(\mathbb{R})$, $\varphi'(0) > 0$, $\psi'(0) > 0$, $\varphi'(0) - a(0)\psi'(0) > 0$;

(iii) $f, f_x, f_t \in L^\infty((-\infty,\infty); L^2(0,1)) \cap L^2((-\infty,\infty); L^2(0,1))$, $f_{tt} \in L^2((-\infty,\infty); L^2(0,1))$, and the norms of f, f_x, f_t, f_{tt} in the indicated spaces are sufficiently small;

(iv) The given history v satisfies equation (1.1) and the boundary conditions for $t \leq 0$, v and its derivatives through third order belong to $L^{\infty}((-\infty, 0]; L^2(0,1)) \cap L^2((-\infty, 0]; L^2(0,1))$, $v_{xxtt} \in L^2((-\infty, 0]; L^2(0,1))$.

Then, the history-value problem (1.1),(1.2),(1.3) has a unique solution $u : [0,1] \times (-\infty,\infty) \to \mathbb{R}$ such that u and its derivatives through third order belong to $L^{\infty}((-\infty,\infty); L^2(0,1)) \cap L^2((-\infty,\infty); L^2(0,1))$. Moreover, u and its derivatives through second order converge to zero uniformly on $[0,1]$ as $t \to \infty$.

Remarks:

1. In assumption (iv) we did not require smallness of the norms. However, assumption (iii) and the fact that v satisfies the equation and boundary conditions for $t < 0$ imply that v is "small".

2. Theorem 2 applies without essential changes if Dirichlet conditions are replaced by Neumann or mixed conditions. In the case of Neumann conditions, the boundedness and decay statements apply to u minus its spatial mean value $\langle u \rangle$ which evolves according to the trivial equation

$$\frac{d^2}{dt^2} \langle u \rangle (t) = \langle f \rangle (t).$$

3. The question of global existence for the all-space problem is more difficult. Hrusa and Nohel [11] gave a proof for regular kernels. This proof, however, makes essential use of the assumption $a'' \in L^1(0,\infty)$ and does not appear generalizable to singular kernels.

4. It is conceivable that, for an appropriate class of singular kernels, global smooth solutions exist even for large data. However, we have not been able to verify this.

ACKNOWLEDGMENT: This research was sponsored by the United States Army under Contract No. DAAG29-80-C-0041. The work of the first author was carried out at the Mathematics Research Center of the University of Wisconsin-Madison and was partially supported by the National Science Foundation under Grant No. MCS-8210950. The work of the second author was partially supported by the National Science Foundation under grant Nos. MCS-8210950 and MCS-8215064.

REFERENCES

[1] Dafermos, C. M. and Nohel, J. A., Energy methods for nonlinear hyperbolic Volterra integrodifferential equations, Comm. PDE 4 (1979), 219-278.

[2] Dafermos, C. M. and Nohel, J. A., A nonlinear hyperbolic Volterra equation in viscoelasticity, Amer. J. Math., Supplement (1981), 87-116.

[3] Doi, M. and Edwards, S. F., Dynamics of concentrated polymer systems, J. Chem. Soc. Faraday 74 (1978), 1789-1832 and 75 (1979), 38-54.

[4] Gripenberg, G., Nonexistence of smooth solutions for shearing flows in a nonlinear viscoelastic fluid, SIAM J. Math. Anal. 13 (1982), 954-961.

[5] Hannsgen, K. B. and Wheeler, R. L., Behavior of the solutions of a Volterra equations as a parameter tends to infinity, J. Integral Equations (to appear).

[6] Hattori, H., Breakdown of smooth solutions in dissipative nonlinear hyperbolic equations, Q. Appl. Math. 40 (1982/83), 113-127.

[7] Hrusa, W. J., A nonlinear functional differential equation in Banach space with applications to materials with fading memory, Arch. Rational Mech. Anal. 84 (1983), 99-137.

[8] Hrusa, W. J. and Renardy, M., On a class of quasilinear partial integrodifferential equations with singular kernels, J. Differential Equations (submitted).

[9] Hrusa, W. J. and Renardy, M., On wave propagation in linear viscoelasticity, Q. Appl. Math. (submitted).

[10] Hrusa, W. J. and Nohel, J. A., Global existence and asymptotics in one-dimensional nonlinear viscoelasticity, Proc. 5th Symp. on Trends in Appl. of Pure Math. to Mech., Springer Lecture Notes in Physics #195, (1984), 165-187.

[11] Hrusa, W. J. and Nohel, J. A., The Cauchy problem in one-dimensional nonlinear viscoelasticity, J. Differential Equations (to appear).

[12] Kim, J. U., Global smooth solutions for the equations of motion of a nonlinear fluid with fading memory, Arch. Rational Mech. Anal. 79 (1982), 97-130.

[13] Laun, H. M., Description of the non-linear shear behavior of a low density polyethylene melt by means of an experimentally determined strain dependent memory function, Rheol. Acta 17 (1978), 1-15.

[14] Londen, S.-O., An existence result on a Volterra equation in a Banach space, Trans. Amer. Math. Soc. 235 (1978), 285-304.

[15] MacCamy, R. C., A model for one-dimensional nonlinear viscoelasticity, Q. Appl. Math. 35 (1977), 21-33.

[16] Malek-Madani, R. and Nohel, J. A., Formation of singularities for a conservation law with memory, SIAM J. Math. Anal. (to appear).

[17] Markowich, P. A. and Renardy, M., Lax-Wendroff methods for hyperbolic history value problems, SIAM J. Num. Anal. 21 (1984), 24-51.

[18] Renardy, M., Singularly perturbed hyperbolic evolutions problems with infinite delay and an application to polymer rheology, SIAM J. Math. Anal. 15 (1984), 333-349.

[19] Renardy, M., A local existence and uniqueness theorem for a K-BKZ fluid, Arch. Rational Mech. Anal. (to appear).

[20] Renardy, M., Some remarks on the propagation and non-propagation of discontinuities in linearly viscoelastic liquids, Rheol. Acta 21 (1982), 251-254.

[21] Rouse, P. E., A theory of the linear viscoelastic properties of dilute solutions of coiling polymers, J. Chem. Phys. 21 (1953), 1271-1280.

[22] Slemrod, M., Instability of steady shearing flows in a nonlinear viscoelastic fluid, Arch. Rational Mech. Anal. 68 (1978), 211-225.

[23] Staffans, O., On a nonlinear hyperbolic Volterra equation, SIAM J. Math. Anal. 11 (1980), 793-812.

[24] Zimm, B. H., Dynamics of polymer molecules in dilute solutions: viscoelasticity, flow birefringence and dielectric loss, J. Chem. Phys. 24 (1956), 269-278.

The final (detailed) version of this paper will be submitted for publication elsewhere.

Trends in the Theory and Practice of Non-Linear Analysis
V. Lakshmikantham (Editor)

BOUNDARY TRAJECTORIES OF GENERALIZED CONTROL SYSTEMS

Barbara Kaśkosz
Department of Mathematics
University of Rhode Island
Kingston, Rhode Island, 02881
U.S.A.

A theorem is presented which gives a necessary condition for a trajectory of a so called generalized control system to be a boundary trajectory. The theorem implies a maximum principle for differential inclusions with no assumptions at all on the right-hand side of the inclusion.

1. GENERALIZED CONTROL SYSTEMS

First we introduce a concept of a generalized control system in order to be able to treat together differential inclusions and ordinary control systems. In the next section we give a necessary condition for a trajectory of a generalized control system to be a boundary trajectory which we then apply to differential inclusions and ordinary control systems. The results presented below come mostly from the joint work of the author with S. Łojasiewicz [5], (also see [5] for proofs).

It is convenient to define a generalized control system as follows. Let S be a collection of functions $f(t,x)$ defined for $t\in[0,1]$, $x\in R^n$. We call S a generalized control system if the following conditions are satisfied:

(a) Every function $f(t,x)$ from S is measurable in t for each fixed x.

(b) For each bounded subset B of R^n there exists a function $r_B(\cdot)\in L^1[0,1]$ such that for every $f\in S$ the following inequality holds:

$$|f(t,x)-f(t,\bar{x})| \leqq r_B(t)\,|x-\bar{x}| \quad \text{for } x,\ \bar{x}\in B,\ \text{a.e. } t\in[0,1]$$

(c) For every two sequences $f_i\in S$, $A_i\subset[0,1]$, $i=1,2\ldots$, such that A_i are Lebesque measurable for $i=1,2\ldots.$, $A_i\cap A_j=\emptyset$ if $i\neq j$, $\bigcup_{i=1}^{\infty}A_i=[0,1]$ the following function:

$$\sum_{i=1}^{\infty} \chi_{A_i}(t) f_i(t,x)$$

belongs again to S, where $\chi_{A_i}(t)$ denotes the characteristic function of the set A_i.

Take an element $f\in S$ and $x_0\in R^n$ and consider the following initial problem:

$$\dot{x}(t)=f(t,x(t))$$
$$x(0)=x_0$$

If its solution exists over the whole interval $[0,1]$ we denote it $x_{f,x_0}(t)$ and call the trajectory of the system S corresponding to the element f and the initial condition x_0.

Let C, here and throughout the paper, be a given closed subset of R^n. The set of all points which can be reached at the moment 1 by trajectories of S starting from C; that is:

$$\mathcal{R}(C)= \{\ x_{f,x_0}(1)\ |\ f\in S,\ x_0\in C\ \}$$

is called the reachable set from C.

2. MAXIMUM PRINCIPLE FOR GENERALIZED CONTROL SYSTEMS

Let $g:R^n \to R^m$ be, here and throughout the paper, a given locally Lipschitz mapping. A trajectory $x_*(\cdot)$ is called boundary with respect to C and g if $x_*(0) \in C$ and $g(x_*(1))$ belongs to the boundary of the set $g(\mathcal{R}(C))$. The following theorem gives a necessary condition of the maximum principle type for a trajectory of S to be boundary.

Theorem 1. Let $x_*(\cdot)$ be a trajectory of S boundary with respect to C and g. Then for every element $f \in S$ such that:

$$\dot{x}_*(t)=f(t,x_*(t)) \quad \text{for a.e. } t\in[0,1] \tag{1}$$

there exist an absolutly continuousfunction $p(\cdot) : [0,1] \to R^n$ and a vector $v \in R^m$, $|v|=1$, such that:

$$-\dot{p}(t) \in p(t)\ \partial f(t,x_*(t)) \quad \text{for a.e. } t\in[0,1] \tag{2}$$

$$p(0) \in N_C(x_*(0)),\ p(1) \in v\partial g(x_*(1)) \tag{3}$$

and for every element $h \in S$ the following inequality holds:

$$<p(t),\ \dot{x}_*(t)> \geq <p(t),\ h(t,x_*(t))> \quad \text{for a.e. } t\in[0,1] \tag{4}$$

The symbol ∂ above denotes Clarke's generalized derivative with respect to x, ([1],[4]); that is, in (2) it means generalized Jacobian of f(t,x) with respect to x, in (3) generalized gradient of g. By $N_C(x)$ we denote Clarke's normal cone to the set C at the point x.

The theorem, as it will become clear in the last section, is a generalization of the Pontriagin's maximum principle: choice of an element f satisfying (1) corresponds to the choice of a control function generating the trajectory $x_*(\cdot)$, the relations (2), (3) to the adjoint equation with transversality conditions,finally, (5) corresponds to the maximum condition.

3. DIFFERENTIAL INCLUSIONS

Let us have a differential inclusion:

$$\begin{aligned}&\dot{x}(t) \in F(t,x(t))\\ &x(0) \in C \qquad , \quad x \in R^n,\ t\in[0,1]\end{aligned} \tag{5}$$

where F(t,x) is a given subset of R^n for every $x \in R^n$, $t\in[0,1]$. An absolutly continuous function $x(\cdot)$ is called a trajectory of (5) if it satisfies the initial condition and its derivative satisfies the inclusion almost everywhere in [0,1]. As before $\mathcal{R}(C)$ denotes the reachable set from C at the moment 1 ; that is:

$$\mathcal{R}(C) = \{x(1) \mid x(\cdot) \text{ a trajectory of (5) }\}$$

Choose for every bounded subset B of R^n a function $r_B(\cdot) \in L^1[0,1]$. Denote by S_F the set of all singled-valued selections f(t,x) of the multifunction F(t,x); that is, $f(t,x) \in F(t,x)$ for $x \in R^n$, $t\in[0,1]$, which are measurable in t, locally Lipschitz in x, with Lipschitz constants in every bounded set B not exceeding $r_B(\cdot)$. Clearly, S_F is a generalized control system, (perhaps empty), whose reachable set is contained in the reachable set of the inclusion (5). Theorem 1 implies easily the following necessary condition for a trajectory of (5) to be boundary.

Theorem 2. Let $x_*(\cdot)$ be a trajectory of (5) such that $g(x_*(1))$ belongs to the boundary of $g(\mathcal{R}(C))$. Then for every selection $f \in S_F$ such that:

$$\dot{x}_*(t) = f(t,x_*(t)) \quad \text{for a.e. } t\in[0,1]$$

there exist an absolutly continuous function $p(\cdot) : [0,1] \to R^n$ and a vector $v \in R^m$, $|v| = 1$, such that:

$$-\dot{p}(t) \in p(t)\ \partial f(t,x_*(t)),\ p(0) \in N_C(x_*(0)),\ p(1) \in v\ \partial g(x_*(1)),$$

and for any other selection $h \in S_F$ the following inequality holds:

$$<p(t),\ \dot{x}_*(t)> \geqq <p(t),\ h(t,x_*(t))> \quad \text{for a.e. } t \in [0,1] .$$

Theorem 2 does not require any assumptions on the right-hand side $F(t,x)$ of the inclusion. Of course, without some regularity assumptions on $F(t,x)$ it may happen that no trajectories of (5) exist, so the problem is empty or that no selections $f(t,x)$ of the required regularity exist, so the condition of Theorem 2 may be empty. The following hypotheses ensure that the condition of Theorem 2 is substantial.

(A_1) For every $t \in [0,1]$, $x \in R^n$ the set $F(t,x)$ is non-empty, compact, convex.

(A_2) $F(\cdot,x)$ is a measurable multifunction of t for each fixed $x \in R^n$; that is, for each closed subset M of R^n the set $\{\ t \in [0,1]\ |\ F(t,x) \cap M \neq \emptyset\}$ is Lebesque measurable.

(A_3) For each bounded subset B of R^n there exists a function $k_B(\cdot) \in L^1[0,1]$ such that:

$$\rho_H(F(t,x),\ F(t,\bar{x})) \leqq k_B(t)\ |x-\bar{x}| \quad \text{for } x,\ \bar{x} \in B\ ,\ t \in [0,1],$$

where ρ_H denotes the Hausdorff metric.

If the inclusion (5) satisfies (A_1)-(A_3) then it is actually equivalent to a generalized control system. Namely, let S_F be the set of all single-valued selections $f(t,x)$ of $F(t,x)$ which are measurable in t, locally Lipschitz in x, with Lipschitz constants on every bounded set B in R^n not exceeding $r_B(t)=4nk_B(t)$, where $k_B(t)$ is that from (A_3).Then for every trajectory $x(\cdot)$ of (5) there exists an element $f \in S_F$ such that:

$$\dot{x}(t)=f(t,x(t)) \qquad \text{for a.e. } t \in [0,1],$$

(see [5],[6] for detailes). Therefore, the set of trajectories of the inclusion (5) and of the generalized control system S_F coincide. In this case the following theorem can be derived from Theorem 2 :

Theorem 3. Assume that the right-hand side of the inclusion (5) satisfies (A_1)-(A_3). Let $x_*(\cdot)$ be a trajectory of (5) such that $g(x_*(1))$ belongs to the boundary of $g(\mathcal{R}(C))$.Then, for every selection $f \in S_F$ such that:

$$\dot{x}_*(t) = f(t,x_*(t)) \qquad \text{for a.e. } t \in [\,0,1] \tag{6}$$

there exist an absolutly continuous function $p(\cdot) : [0,1] \to R^n$ and $v \in R^m$, $|v|=1$, such that:

$$-\dot{p}(t) \in p(t)\ \partial f(t,x_*(t)),\ p(0) \in N_C(x_*(0)),\ p(1) \in v\ \partial g(x_*(1))$$

and

$$<p(t),\ \dot{x}_*(t)> = \max_{w \in F(t,x_*(t))} <p(t),\ w> \qquad \text{for a.e. } t \in [0,1] \tag{7}$$

The necessary condition of Theorem 3 is not empty since an element $f \in S_F$ satisfying (6) exists. Also, rather cumbersome inequality condition from Theorem 2 can be replaced by the maximum condition (7). The maximum principle of Theorem 3 is essentially different than Clarke's maximum principle for differential inclusions [3]. See [5] for a discussion.

4. ORDINARY CONTROL SYSTEMS

Consider a control system of the form:

$$\begin{aligned} &\dot{x}(t)=f(t,x(t),u(t)) \\ &x(0) \in C \qquad ,\ x \in R^n\ ,\ t \in [0,1] \end{aligned} \tag{8}$$

A control function $u(\cdot)$ is called admissible if it is measurable and $u(t) \in U(t)$ for a.e. $t \in [0,1]$, where $U(t) \subset R^k$ are given control sets. Again, $\mathcal{R}(C)$ denotes the reachable set at the moment 1 by trajectories corresponding to admissible control functions. Assume the following about the system:

(H_1) For each $x \in R^n$ the function $f(\cdot,x,\cdot)$ is $L \times B^k$ measurable, where $L \times B^k$ denotes the σ - algebra of subsets of $[0,1] \times R^k$ generated by products of Lebesque measurable subsets of $[0,1]$ by Borel subsets of R^k.

(H_2) For each bounded set B in R^n there exists a function $k_B(\cdot) \in L^1[0,1]$ such that for $t \in [0,1]$, x, $\bar{x} \in B$, $u \in U(t)$ the following holds:

$$|f(t,x,u)-f(t,\bar{x},u)| \leqq k_B(t)\,|x-\bar{x}|$$

(H_3) The sets $U(t)$ are bounded and the graph of the multifunction $U(\cdot)$ is $L \times B^k$ measurable.

Under the hypotheses for each admissible control function the right-hand side of the system $f(t,x,u(t))$ is measurable in t and locally Lipschitz in x, therefore the set:

$$S_U = \{\ f(t,x,u(t)) \mid u(\cdot) \text{ admissible }\}$$

is a generalized cotrol system. Theorem 1 implies easily the following

Theorem 4. Let $x_*(\cdot)$ be a trajectory of (8) such that $g(x_*(1))$ belongs to the boundary of the set $g(\mathcal{R}(C))$. Then for each control function $u_*(\cdot)$ such that:

$$\dot{x}_*(t)=f(t,x_*(t),u_*(t)) \quad \text{for a.e. } t \in [0,1]$$

there exist an absolutly continuous function $p(\cdot) : [0,1] \to R^n$ and $v \in R^m$, $|v|=1$, such that:

$$-\dot{p}(t) \in p(t)\ \partial f(t,x_*(t),u_*(t)),\ p(0) \in N_C(x_*(0)),\ p(1) \in v\ \partial g(x_*(1)),$$

$$<p(t),\ \dot{x}_*(t)> = \max_{u \in U(t)} <p(t),\ f(t,x_*(t),u)> \quad \text{for a.e. } t \in [0,1].$$

Theorem 4 coicides with Clarke's maximum principle for non-smooth control systems [2].Notice, that the control system (8) can be written as a differential inclusion in the following way:

$$\dot{x}(t) \in f(t,x(t),U(t))$$

$$x(0) \in C \tag{9}$$

where $f(t,x,U(t)) = \{\ f(t,x,u) \mid u \in U(t)\ \}$. With some additional regularity assumptions on the system, under which the Fillipov lemma can be applied, the sets of trajectories, hence also reachable sets of (8) and (9) coincide. Suppose we have a boundary trajectory $x_*(\cdot)$ of (8) so also of (9). Do Theorems 2 (or 3) and 4 applied to (9) and (8) respectively give the same? It is not clear. Theorem 2 gives that for every single-valued selection $h(t,x) \in f(t,x,U(t))$ which generates $x_*(\cdot)$ there exists an adjoint function $p(\cdot)$ which satisfies the maximum condition, while Theorem 4 says only that for every selection given by an open-loop control function $h(t,x)=f(t,x,u_*(t)) \in f(t,x,U(t))$ which generates $x_*(\cdot)$ there exists an appropriate $p(\cdot)$. So, perhaps, Theorem 2 applied to (9) gives a stronger necessary condition than Theorem 4 applied to (8). The author does not know if it can happen for a locally Lipschitz control system that by considering all selections of the right-hand side one gets a stronger necessary condition than when considering only those generated by open loop control functions. It can happen for a non-Lipschitz control system as in the example below.

Example. Consider the following control system on the plane:

$$\dot{x}_1 = u_1$$

$$\dot{x}_2 = \sqrt{|u_1|}\,\sqrt{|x_1|} + u_2$$

$$x_1(0)=x_2(0)=0 \quad , \quad u_1 \in [-1,\ 1],\ u_2 \in [-1,\ 0].$$

That is: $x=(x_1,x_2)$, $u=(u_1,\ u_2)$, $f(x,u)=(u_1,\ \sqrt{|u_1|}\sqrt{|x_1|} + u_2\ )$, $U=[-1,1] \times [-1,0]$, $C=\{0\}$. Let g be the identity mapping from R^2 to R^2. The differential inclusion corresponding to the system is:

$$F(x)=F(x_1,x_2)=\{ (v_1,v_2) \mid -1 \leqq v_1 \leqq 1, \sqrt{|x_1|}\sqrt{|v_1|} - 1 \leqq v_2 \leqq \sqrt{|x_1|}\sqrt{|v_1|} \}$$

Consider the trajectory $x_*(t)\equiv 0$. The only control function which generates this trajectory is $u_*(t)\equiv 0$, and the corresponding selection $f(x, u_*(t))\equiv 0$. The maximum principle holds for this selection with $p(t)\equiv(0,1)$. So if we use just selections corresponding to open-loop control functions, we cannot eliminate the trajectory as a candidate for being boundary. We can do it, however, applying Theorem 2 and using other selections.

Take the selection $h(x)=h(x_1,x_2)=(x_1,x_1)$ for x in a neighborhood of 0 and extend in any way to a Lipschitz selection of $F(x)$; for example:

$$h(x) = \begin{cases} (x_1,x_1) & \text{if } |x_1| \leqq 1/4 \\ (1/2 -x_1,\ 1/2 -x_1) & \text{if } 1/4 \leqq x_1 \leqq 1/2 \\ (1/2 +x_1,\ 1/2 +x_1) & \text{if } -1/2 \leqq x_1 \leqq -1/4 \\ (0,0) & \text{if } |x_1| \geqq 1/2 \end{cases}$$

Of course, $\dot{x}_*(t)=h(x_*(t))$, but no $p(\cdot)=(p_1(\cdot),p_2(\cdot))$ which has properties required in Theorem 2 exists. We have:

$$\partial h(0)=\begin{bmatrix} 1 & 0 \\ 1 & 0 \end{bmatrix} \text{ so the adjoint equation becomes: } \begin{cases} -\dot{p}_1(t)=p_1(t)+p_2(t) \\ -\dot{p}_2(t)=0 \end{cases}$$

But if $p(t)$ is to satisfy the inequality condition of Theorem 2 it must be $p_1(t)\equiv 0$. Indeed, take the following selections of $F(x)$:

$$h_1(x)=\begin{cases} (1/2 - |x_1|,\ 0) & \text{if } |x_1| \leqq 1/2, \\ (0,0) & \text{if } |x_1| \geqq 1/2 \end{cases} \qquad h_2(x)=\begin{cases} (-1/2+x_1,\ 0) & \text{if } |x_1| \leqq 1/2 \\ (0,0) & \text{if } |x_1| \geqq 1/2. \end{cases}$$

The inequality condition implies that $-1/2p_1(t)\leqq 0$ and $1/2p_1(t)\geqq 0$, hence $p_1(t)\equiv 0$.

But then the adjoint equation gives that $p_2(t)\equiv 0$ so $p(t)$ is trivial and it cannot satisfy the transversality conditions. Therefore, Theorem 2 does eliminate the trajectory $x_*(t)$ as a candidate for being boundary. Notice, that neither the control system (8) nor the corresponding inclusion (9) are Lipschitz, so of all theorems of the paper only Theorem 2 is applicable.

REFERENCES

[1] F.H.Clarke, Generalized gradients and applications, Trans. Amer. Math. Soc., 205(1975), 247-262.

[2] F.H.Clarke, The maximum principle under minimal hypotheses, SIAM J.Control Opt. 14(1976), 1078-1091.

[3] F.H.Clarke, Necessary conditions for a general control problem, Proc. Int. Sym. on the Calculus of Variations and Control Theory, D.L. Russel, ed., Academic Press, New York, 1976.

[4] F.H.Clarke, Optimization and nonsmooth analysis, Wiley-Intersc., New York 1983.

[5] B.Kaśkosz, S.Łojasiewicz, A maximum principle for generalized control systems, to appear in Nonlinear Analysis.

[6] S.Łojasiewicz, A.Pliś, R. Suarez, Necessary conditions for nonlinear control systems, to appear in J.Diff. Eqns.

This paper is in final form and no version of it will be submitted for publication elsewhere.

Trends in the Theory and Practice of Non-Linear Analysis
V. Lakshmikantham (Editor)
Elsevier Science Publishers B.V. (North-Holland), 1985

ISOLATION OF THE ZEROS OF A COMPLEX POLYNOMIAL BY EXPLORING FUNCTION STRUCTURE UNIQUENESS OF THE SOLUTION SET ESTABLISHED

Dorothea A. Klip

Department of Physiology and Biophysics
Department of Computer and Information Sciences
University of Alabama in Birmingham
Birmingham, Alabama
U.S.A.

The presented algorithm for simultaneous polynomial root isolation is based on the discrete tracing of the zero curves of the real (and imaginary) part of the function. Being guided by the structure of the function as presented by these curves is essential in attaining the desired characteristics of robustness, reliability and global optimal computational complexity. A mathematical condition for uniqueness of the tracing is optionally verified and - if not fulfilled in exceptional cases - uniqueness is established. The method immediately applies to transcendental functions in one complex variable in a region of analyticity.

INTRODUCTION

Algorithms for polynomial root isolation which do not depend on the deflation process, have the advantage that loss of accuracy inherent with the successive dividing out of linear factors, cannot occur. They have been studied for several decennia and we refer to standard texts for applied techniques and results [3,4]. They have not become popular due to the unduly long computing times. It seems appropriate to extend the characterization of those methods by the term 'simultaneous' to algorithms where the order in which the roots are found does not affect their precision. In this sense, the algorithm which was presented by us on several occasions [6,7,8] and is discussed in the sequel belongs to this class. Simultaneity in the literal sense can be achieved in an environment which allows parallel processing.
Although the text is concerned with the complex polynomial, the method directly carries over to analytic functions in some region of definition, with the exception of the determined condition at the boundary. The method is convergent, robust and reliable and one may contend that among search algorithms it is optimal in the time complexity dimension.
On mathematical grounds it appeared desirable to implement an absolute criterion for correctness. Verification respectively establishing correctness in case the tracing at first pass would not meet this criterion, is provided by a recently implemented subsystem. It makes the algorithm comparable with algebraic approaches [3] while maintaining its efficiency relative to computing time. We will refer to this part of the system as the 'uniqueness option'. We have kept the original approach, to which we will refer as the 'simple system', because in this setting there is greater flexibility, as will be discussed.
The general properties of the zero curves are expressed in a number of lemmas in 1. With the description of the topology of these curves established, an outline of the algorithm for the 'simple system' is given in 2. and results obtained with this system are given in the form of a listing of CPU times. In 3. the flexibility of the system by virtue of various available options is demonstrated. In 4. a few essential aspects of the package for multiple precision floating-point

arithmetic are mentioned. It accounts for the availability of virtually unlimited precision. In 5. the concept of uniqueness of the tracing is defined and the algorithm which establishes uniqueness is briefly discussed.

1. PROPERTIES OF THE LEVEL CURVES

We shall first consider those properties which hold for general analytic functions and then state the special additional properties of the polynomial function. Suppose $f(z)$ is analytic in a simply connected region of the complex plane. We separate f in its real and imaginary parts, as follows,
$w = f(z) = f(x+iy) = u(x,y) + iv(x,y) = u(z) + iv(z)$.
It is well-known [1] that u and v are harmonic functions of the real variables x and y and that the Cauchy-Riemann equations hold for their first partial derivatives. Since the mapping f is topological and conformal at each point where $f'(z)\neq 0$, the level curves $u=c_1$, $v=c_2$ are piecewise Jordan arcs and they form an orthogonal system in the region of definition, except for isolated singularities. The curves $u=0$, $v=0$ are of particular interest, since they contain the roots of f.
At an exceptional point the behavior of these curves can be described in precise terms. An exceptional point z_0 of $u=0$ is a multiple root of the analytic function $g(z)=f(z)-iv(z_0)$. Since roots of analytic functions have integer order, one may write $g(z)=(z-z_0)^m h(z)$, where $h(z)$ is analytic and $h(z_0)\neq 0$. In a δ-neighborhood of z_0, such that $h(z)\neq 0$ a single-valued analytic branch of $h(z)^{1/m}$ can be defined and one may write $g(z)=\zeta(z)^m$, where $\zeta(z)=(z-z_0)h(z)^{1/m}$.
Since $Re(f(z))=Re(g(z))$, the zeropoints of u along the circle $C: |z-z_0|=\delta'$ $(0<\delta'<\delta)$ are those values of z for which $\arg(\zeta(z))=(2k+1)\pi/2m$, from which it follows $\arg(z-z_0)=(2k+1)\pi/2m-\arg(h(z))/m$ $(k=0,\ldots,2m-1;\ 0<|z-z_0|<\delta)$ in those points. Along the same circle $v(z)$ assumes the value $v(z_0)$ 2m times and the set of straight lines tangent to these 4m arcs $u=0$, $v=v(z_0)$ at z_0 is described by the equations

$$z=z_0+te^{i\phi(k)} \quad (k=0,\ldots,4m-1;\ t\geq 0;\ \phi(k)=-\alpha+k\pi/2m;\ \alpha=\arg(h(z_0)/m)).$$

These findings are in agreement with the Principle of the Argument applied to $g(z)$ at z_0. They are summarized in

LEMMA 1. The zero curves $u=0$ and $v=0$ of an analytic function $f(z)$ are piecewise Jordan arcs. They form an orthogonal system, except in those points which are roots of f'. If k-1 is the multiplicity of z_0 as a root of f', then 2k arcs of $u=0$ (or of $v=0$) intersect at z_0 under equal angles. If z_0 is also a root of f, then both kinds of curves have a k-fold intersection point at z_0 such that there is rotational symmetry of the set of tangent lines at z_0.

DEFINITION 1. The set of points $\{z_0\}$ of the system of curves $u=0$ $(v=0)$ for which $f'(z_0)=0$, $f(z_0)\neq 0$ are called the branch points of the system.

The occurrence of branch points and 'near-branchpoints' is the major cause of non-uniqueness of the tracing, as will be explained in 5.
While the local structure is defined by lemma 1, the global properties follow most easily from the following powerful theorem.

THEOREM 1. The zero curves $u=0$, $v=0$ are piecewise arcs of steepest descent.
PROOF. Direction of steepest descent is defined by $-\mathrm{grad}|f(z)|$.

$\mathrm{grad}|f(z)|=((uu_x+vv_x)/(u^2+v^2)^{\frac{1}{2}},(uu_y+vv_y)/(u^2+v^2)^{\frac{1}{2}})$. In a point $u=0$, $v=v_0$ this vector reduces to $(v_x,v_y)=(v_x,u_x)$ (using a Cauchy-Riemann equation). If we let the x-axis of the coordinate system coincide with the direction of the tangent to $u=0$ at $(0,v_0)$ $(f'(0,v_0)\neq 0)$, then $u_x=0$ so that $\mathrm{grad}|f(z)|=(v_x,0)$, the increase of v in the direction of the u=0 arc. The gradient is defined except in the branch points and the multiple roots.

The openness of the system of zero curves is basic to the success of the tracing algorithm. Although almost trivial we state

LEMMA 2. The pointsets u=0, v=0 form an open system.
PROOF. A closed pathway of arcs of u=0 must contain a "smallest" pathway, one which bounds a region C in which u does not change sign. Suppose u>0 in C. Then u has a maximum m on the closure C^-. According to the Principle of the Maximum for harmonic functions m is taken on the boundary ∂C. However, on ∂C u=0, which is a contradiction.

By studying the mapping of the circle C: $|z-z_0|=\delta$ at a branch point z_0 it is seen

COROLLARY 1. The arcs of u=0 (v=0) which converge at a branch point z_0 are successively arcs of descent and of ascent relative to z_0.

The additional properties of the zero curves for a polynomial are based on the fact that a polynomial is a meromorphic function in the extended plane with ∞ its n-th order pole. The curves thus stretch to infinity and according to lemma 1 it is seen

LEMMA 3. Each system of curves u=0, v=0 of a polynomial of n-th degree has a set of n simple asymptotes. The 2n asymptotes intersect in the center M of the root configuration, $M:=-a_1/n$ (using the notation $f=z^n+\Sigma_{j=1}^{n} a_j z^{n-j}$ as a monic polynomial) in a rotational symmetric way.

On the basis of lemma 1 one may define continuation of a pathway at an exceptional point by that arc which enters in opposite direction. This enables us to introduce the concept 'branch'.

DEFINITION 2. A branch of u=0 is a connected set of arcs in the (non-extended) complex plane, such that the discontinuity of $\overrightarrow{du/ds}$ at an exceptional point z_0 of u=0 is removed by defining $(\overrightarrow{du/ds})_{z_0}=\lim_{u=z_0}\overrightarrow{du/ds}=-\lim_{u'=z_0}\overrightarrow{du'/ds}$, where u'=0 is that arc at z_0 which satisfies the last equation. We shall speak of r-branches and i-branches when referring to the set u=0 respectively v=0.

On the basis of lemmas 2 and 3 it is seen

COROLLARY 2. Each system u=0 (v=0) consists of n distinct branches.

Since zero arcs are paths of steepest descent, each branch must contain either a root or a branch point. Furthermore, since analytic functions do not have a local minimum, one may assert

LEMMA 4. Each pathway of arcs of descent terminates at a root.

Also, since branch points are roots of f',

LEMMA 5. The branch points are contained in any root enclosing convex region.

It may be elucidative relative to the first step of the algorithm to combine lemmas 3 and 5 and to explicitly mention

COROLLARY 3. The boundary of a convex region containing the roots is intersected 4n times by the system of branches such that the Principle of the Argument is fulfilled. Along the boundary of a large disk C: $|z-a_1/n|=R$ the distances between successive intersection points are approximately equal.

Necessarily, when following the boundary in a positive sense, each time one passes an intersection point, the function quadrant increases by 1.
The following lemma, which was already formulated by Liouville, is a consequence of the descent property.

LEMMA 6. If a branch contains more than one root, each pair of successive roots is separated by an odd number of branch points.

In case of real coefficients Im(f) contains the factor y. Since Re(f) and Im(f)/y contain only even powers of y, it follows

LEMMA 7. In case of real coefficients the x-axis is axis of symmetry relative to the r- and i-branches. It is also an i-branch.

From lemmas 6 and 7 Rolle's theorem follows as a special case.

2. THE 'SIMPLE ALGORITHM'

2.1 MAIN STEPS

Procedure: ROOT-ISOLATION - (simple algorithm)

Step 1: Initialization:
isolate 4n intersection points of branches with ∂C by calculating f at division points W_j
	calculate rootbound R; circle C:=(0,R);
	for j=1 to 4n do; find
	$W_j \in \partial C$ such that $\{q_j \mid q_j:=quadr(f(W_j))\}$
	satisfy $q_{j+1}-q_j=1$ (mod 4); end;

Step 2: next branch b_j (or next r-branch if only r-branches are traced) is defined by start points W_j, W_{j+1}	next-branch: for j=1 to 4n by 1 (or by 2) do; $V_1:=W_j$; $V_2=:W_{j+1}$; $q_1:=q_j$; $q_2:=q_{j+1}$;
Step 3: iterative tracing process: construct equilateral triangle V_1,V_2,V_3 After first triangle construction, continuation is simply achieved by reflection	next-tri: $V_3:=e^{i\pi/3}V_1+e^{-i\pi/3}V_2$; goto calc; reflect: if $q_3=q_2$ then $V_3:=V_3+V_1-V_2$; if $q_3=q_1$ then $V_3:=V_3+V_2-V_1$; calc: q_3:=quadr $(f(V_3))$; if $q_3=q_1$ or $q_3=q_2$ then go to reflect;
Step 4: since a different quadrant is found, triangle base V_1V_2 is bisected. Check on given break-off criterion	bisect: $V_m:=(V_1+V_2)/2$; q_m:=quadr$(f(V_m))$; if $q_m=q_1$ then $V_1:=V_m$;if $q_m=q_2$ then $V_2:=V_m$; if $\lvert V_1V_2\rvert>\varepsilon$ then goto next-tri;
Step 5: terminate tracing; check for smallest function value among V_1,V_2,V_3, $Z:=(V_1+V_2+V_3)/3$	Approx:= $\{V \mid S:=\{V_1,V_2,V_3,Z\};\ \lvert f(V)\rvert:=\min_{W\in S}\lvert f(W)\rvert\}$ end next-branch;
Step 6: order 4n (or 2n if only r-branches are traced) approximations into m sets of approximations to same root on the basis of δ	Analyse-results: construct $\{S_j \mid j=1$ to $m\}$ $S_j:=\{$Approx(root(j))$\}$; root(j):=mean$\{$Approx(root(j))$\}$;
Step 7: (optional) refine root(j) to desired accuracy	for j=1 to m do; call LAGUERRE(root(j)); end;
Step 8: (optional) under graph-option trace points are plotted	call DISPLAY; end ROOT-ISOLATION;

Outline of 'simple algorithm'

A rootbound is calculated from the well-known expression $R=\{2\max(|a_m|^{1/m})|1\leq m<n\}$. On the basis of Corollary 3 along the boundary ∂C of $C:=(0,R)$ f^m is calculated (and function quadrant retained) at regular intervals, while adjusting distances between division points if necessary, in order to separate the 4n intersection points of the branches with ∂C (step 1). These branches are subsequently traced by means of triangular enclosure (step 2), with the initial line segment V_1V_2 (with q_1 and q_2 the corresponding quadrants) as the base of the first triangle. Construction of the 3rd vertex occurs in the direction of descent, i.e. by choosing it to the left of the directed base V_1V_2 (step 3). Continuation occurs by choosing as new base the side whose endpoints have different quadrants, so that it is intersected by the branch. Since the pathways of descent terminate at a root, a branch of the other kind will after a number of steps be intersected. Then at V_3 a different quadrant is found. This induces bisection of the base (step 4). Eventually the triangle will fulfill a certain break-off criterion. This criterion is at the same time a measure for the distance to the root. The point corresponding with smallest function value is taken as the approximation to the root (step 5). In step 6 the 2k (or 4k if also i-branches are traced) approximations to a k-fold root are separated in sets of approximations to the same root on the basis of 'closeness', $|r_1-r_2|/|r_1|\leq\frac{1}{4}\delta$ for elements of the same set, and 'distinctness', $|r_1-r_2|/|r_1|>3/4\delta$ for elements of different sets. δ is a 'user' supplied estimate of the relative root separation. The average value of the elements of a set is taken as the approximation to the root. A warning is issued if separation on this basis failed.
The program is coded in Fortran IV for the IBM series.

2.2 THE 'SELF-CORRECTING' PROPERTY OF THE TRACING

An important aspect contributing to the efficiency of the algorithm is its self-correcting property. This phenomenon shows up in two ways. If for a tracepoint at a branch a wrong sign is found due to round-off (or if it is not determined because Re(f)=0), then it is easily seen that the arc is picked up again in the next two steps. The other self-correcting property concerns the selected precision. The criterion for a switch from lower to higher precision is done in the program on a uniform basis, i.e. on the basis of a certain vicinity to the root, as indicated by the length of the triangle side. Such a uniform criterion can be used without penalty because too low precision manifests itself by inconsistency. If in step 3 a different quadrant is found ($q_3\neq q_1$ or q_2) and if this is due to round-off, then in the next step bisection of the base is bound to yield a quadrant $q_m\neq q_1$ or q_2. The appearance of inconsistency is in fact a 'self-correcting' phenomenon. An obvious switch to higher precision is undertaken on a retrospective basis, until agreement between the results in both kinds of precision is attained. However, one must be prepared for the rare event that in the vicinity of a cluster an incorrect sign might yield a loop situation. Bisecting on an iterative basis will correct this, because this phenomenon can only occur if the distance to both arcs is of the same order. The ultimate available precision is the responsibility of the 'user' (see section 4).

2.3 PROOF OF CONVERGENCE

Convergence of the discrete tracing, as described, is proven if one can demonstrate that the next step either implies descent or induces bisection. If z_i and z_j are the intersection points with the traced arc at the base of the current resp. the next triangle, then descent implies $|f(z_i)|>|f(z_j)|$. When tracing a single arc, descent is inherent with the choice of the 3rd vertex (step 3), as a consequence of the topological and conformal properties of the mapping at all points of the arc. Since descent terminates at a root, eventually an arc of the other kind is encountered, which fact induces bisection (step 4). - It remains to investigate the unusual situations 1) at a cluster of roots the triangle may be intersected by a number of arcs, which fact will go unnoticed if the 3rd vertex yields a quadrant q_1 or q_2, 2) when tracing an r-branch, if, in the vicinity of a root of f', intersection with a number of r-arcs takes place. - For both events holds that continuation is always defined, because 2 of the 3 sides have an odd number of intersection points. A detailed analysis of the continuation at the hand of the structure of the branches will show that convergence in the above sense takes place in both cases.

2.4 COMPUTATIONAL COMPLEXITY EXPRESSED IN BASIC TIME UNITS

An estimate of the average time complexity of the 'simple algorithm' is made on the basis of the following simplifying assumptions 1) the branches may be replaced by their asymptotes over that part of the traced pathway which is outside the circle $C_2:=(0,M)$, where M is sharp as a rootbound for the polynomial f of degree n ($R/\bar{M}<2n$), 2) inside C_2 the roots are single and uniformly distributed 3) refinement of the approximation in the vicinity of a root is done on the basis of δ, the relative root separation parameter. ad 1) Since the initial triangle side $s_0=\pi R/2n$, one finds $(4n/\pi\sqrt{3})\log_2 2n$ as upperbound for the number of steps when tracing one branch. For 2n branches, with each polynomial evaluation requiring n multiplications in single precision, with $L(d_1)$ the time for one addition, this part requires $T_1\le 2n^3\log_2(2n)(L(d_1))^2$. ad 2) With $s=\pi M/2n$ at the boundary of C_2 no further bisections are assumed to occur up to a vicinity of the root. An average pathway of length M is traced in $8n/\pi\sqrt{3}$ steps. Presuming double precision arithmetic (or multiple precision arithmetic in case of extreme ill-condition) in this part of the tracing, with $L(d_2)$ required for one addition, one finds $T_2\le 4n^3(L(d_2))^2$. ad 3) Additional iteration, combined with bisecting will yield a triangle in an ε- neighborhood of the root z_0, where $\varepsilon=1/8\delta|z_0|$. If as an average $|z_0|=0.1M$, this step will require $T_3\le 6n\log_2(20\pi/(\delta n))(L(d_2))^2$.

2.5 ROBUSTNESS OF THE ALGORITHM

From the data listed in Table I it is apparent that the algorithm is very efficient. The most important contributing factors are the self-adjustment of precision, as discussed in 2.3, and the low precision over most of the traced pathway.

The data were obtained with the IBM 4381 and are compared with data from [3]. The roots were found (with a relative accuracy $\le$ .0001) by tracing r-branches only. Due to the symmetry mentioned in lemma 7, tracing in the upper halfplane suffices in the real case. This accounts for the time difference between the two kinds of coefficients. The test polynomials contained several multiple roots, e.g. for degree 20 (real coefficients) was selected $(z^2+1)^4(z^2+2)^2(z^2+5)^2(z-5)^2(z-500)^2$.

TABLE I. CPU TIMES FOR PROCEDURE ROOT-ISOLATION

degree	real coef. seconds	complex coef. seconds	Collins square free seconds
5	0.22	0.32	18
8	0.45	0.62	--
10	0.60	0.86	67
15	1.27	1.73	174
17	1.55	2.36	--
20	2.14	3.30	--

3. THE OPTIONS OF THE 'SIMPLE ALGORITHM'

3.1 GRAPH OPTION

The option to display certain trace points provides a heuristic means for checking correctness of obtained results. Although for the isolation of the roots tracing of one kind of branches is sufficient, one may prefer to trace both r- and i-branches under this option. The structure of a ninth degree polynomial is shown in fig. 1. Tracing can also be performed in any selected rectangular

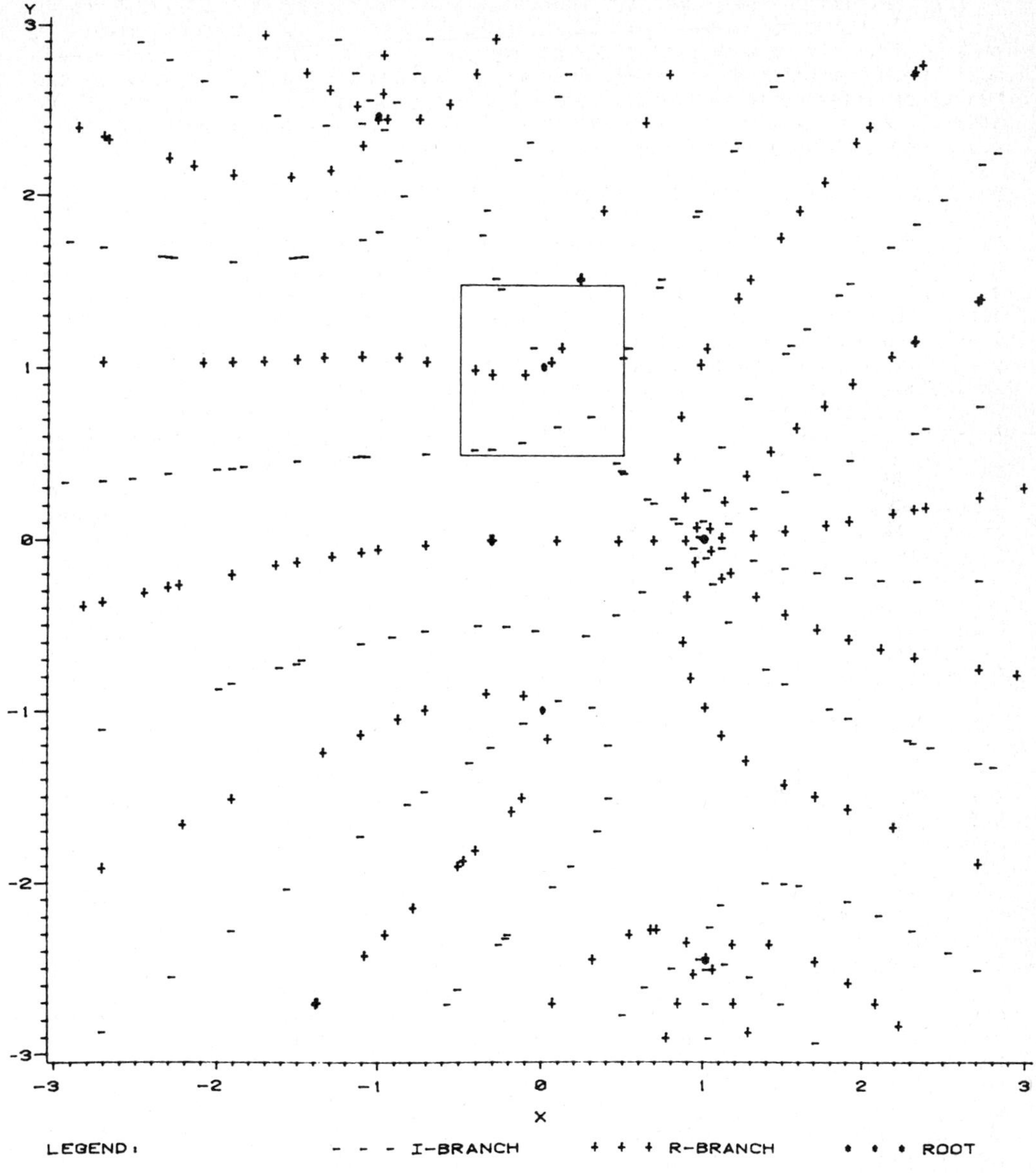

Figure 1. Tracepoints of r- and i-branches of $f:=(z^2+1)(z-1)^3(z^2+5+5i)^2$.
CPU time 1.85". Non-uniqueness (='step-over') occurs in framed area.

area. This amounts to looking at such an area on a larger scale. The marked section in fig. 1 is retraced and the result displayed in fig. 2. It is demonstrated how tracing of a small area may correct a 'step-over' to a different arc in the vicinity of a root of f', which step-over led to 'non-uniqueness' of the tracing.

3.2 OPTION FOR REFINEMENT OF THE ROOTS

One may choose to refine the roots to any preset precision by means of LAGUERRE iteration, which has 3rd order convergence. In order to ensure convergence, one must have a 'good' start point, i.e. it may turn out that the parameter δ must be more stringent. We shall see how under the uniqueness option this dilemma is resolved. Multiplicity of the roots will be determined by applying LAGUERRE iteration to the successive derivatives. This must yield agreement with the number of approximations (2k, alternatively 4k, for a k-fold root).

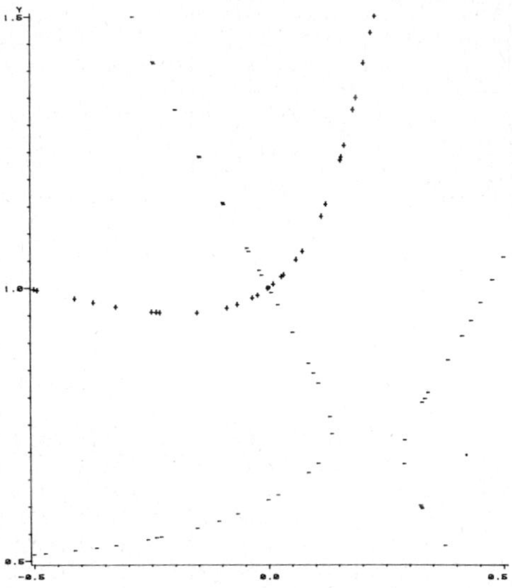

Figure 2. Retracing of the square in fig. 1 eliminated the 'step-over' due to the smaller stepsize.

4. BRENT'S SYSTEM FOR MULTIPLE PRECISION FLOATING-POINT ARITHMETIC

Through the implementation of Brent's package [2] we have been able to provide unlimited precision. The system has the convenient property - besides its ingenious design - that one may select for each run the desired ultimate precision. In addition we have provided the possibility to increase the precision dynamically on a local basis in the Laguerre iteration process. If it would fail to converge within a reasonable number of steps for certain roots, the attempt is repeated, using higher precision. The polynomial coefficients are processed by the MP system and thus can have any length or format. Also a product of input polynomials will be processed and expanded in MP mode.

5. THE UNIQUENESS OPTION

5.1 INTRODUCTION

In 2.3 the various conditions were mentioned which may cause a switch from the original branch to a different one as a consequence of the discrete tracing. At a branch point continuation is not uniquely defined and the arc selected by the algorithm depends on the relative location of triangle and branch point. This may cause certain arcs to be traced more than once, which implies that an equal number of arcs is not traced. This phenomenon is called 'non-uniqueness' of the tracing. An example was given in fig. 1. In a program which allows each single root to be found along 4 different pathways, it is extremely unlikely that for this reason not all roots would be found, especially since the occurrence of a branch point or 'near-branchpoint' is a rarety. From a mathematical point of view it is important to have a criterion for 'uniqueness', i.e. distinctness of the traced pathways. Fortunately for analytic functions the structure of the function itself, as expressed in lemma 1, provides such a criterion. The tracing is unique if at each isolated root the converging pathways enter under distinct angles. The system which checks uniqueness in this sense and establishes it when violated, is referred to as the uniqueness option.

5.2 ESTABLISHING UNIQUENESS

A more detailed account of the uniqueness procedure must be given elsewhere. Only a few important aspects will be mentioned. If uniqueness does not hold, then pairs of branches which entered under equal angle will be retraced simultaneously, while ensuring disjointness. In an area where the pathways start overlapping, it is essential for the separation procedure to have a criterion for the tracing of a single arc. This criterion could be derived by means of a descent function presented by Pomentale [9]. It is associated with his class of iteration functions for analytic functions $z_{i+1}=z_i+\phi_k(z_i)$ $(k=2,3,\ldots;$ $\phi_k=(k-1)(f'/f)^{(k-2)}/(f'/f)^{(k-1)})$. We were attracted to his approach, since the order of convergence k is independent of the multiplicity of the root. A parameter λ is introduced in the iteration function $\omega_\lambda=z-\lambda F(z)/(f'(z)-\lambda F(z))$ $(F:=f'+f/\phi_k)$, in order to provide for a descent property in the large $|f(\omega_\lambda)|<|f(z)|$, which holds for values of λ satisfying

$$0<\lambda\leq\min\{1,2|f'|^2/(M|f|+6|F||f'|)\}\quad (M=\max\{|f''(z+\theta(\omega_\lambda-z))|;\ 0\leq\theta\leq1,\ 0<\lambda\leq1\}) \qquad (1)$$

We were able to prove that replacing 1 by $\frac{1}{2}$ in the right-hand member of (1) yields a sufficient condition for tracing a single arc.
Elimination of the 'step-over' involves modification of at least one of the traced pathways. The pertinent points must be registered in order to retrieve them in case of subsequent retracing. This could efficiently be achieved by means of our listprocessing system VARLIST [5].

5.3 LOCAL TERMINATION CRITERION FOR THE LINEAR TRACING

The dilemma of finding a good start point for Laguerre iteration metioned in 3.2, could be resolved under the uniqueness option by means of the descent function. If λ is allowed to take the value 1 in (1), then the descent function is identical with the high-order iteration function and this will most likely provide an optimal switch point, also if Laguerre would remain the iteration function of choice. Therefore, when calculating the expression in (1) as soon as the length of the triangle side indicates a reasonable closeness to the root, a local termination criterion is obtained.

REFERENCES

[1] Ahlfors, L.V., Complex Analysis (McGraw-Hill, 1966).

[2] Brent, R.P., A Fortran Multiple Precision Arithmetic Package, ACM TOMS 4 (1978) 57-70.

[3] Collins, G.E., Infallible calculations of polynomial zeros, in: Rice, J.R. (ed), Math. Software III (Acad. Press, 1977).

[4] Henrici, P., Applied and Computational Complex Analysis I (Wiley & Sons, 1974).

[5] Klip, D.A., The variable cell-length listprocessor VARLIST, Proc. ACM Annual Conf. (ACM New York, 1974).

[6] Klip, D.A., Polynomial zero isolation by directed search, Abstracts SIAM Fall Meeting (1979).

[7] Klip, D.A., Solution of (complex) univariate polynomial equations. Abstracts SIAM National Meeting (1981).

[8] Klip, D.A., Solution of polynomial equations by directed search. Uniqueness of the tracing algorithm achieved. Abstracts SIAM National Meeting (1983).

[9] Pomentale, T., A class of iteration methods of holomorphic functions, Numer. Math. 18 (1971) 193-203.

This paper is in final form and no version of it will be submitted for publication elsewhere.

Trends in the Theory and Practice of Non-Linear Analysis
V. Lakshmikantham (Editor)

DISTAL, EQUICONTINUOUS, ZERO CHARACTERISTIC, AND RECURRENT DYNAMICAL SYSTEMS

Ronald A. Knight

Mathematics Division
Northeast Missouri State University
Kirksville, Missouri
U.S.A.

Distal, equicontinuous, and zero characteristic continuous and discrete dynamical systems are characterized using the fundamental relations. Recurrent dynamical systems are classified similary. Interrelationships among these flows are given.

INTRODUCTION

During the last three decades two major classes of transformation groups of interest to researchers have been the distal and equicontinuous groups. Midway in that period the author introduced the class of continuous flows of characteristic 0. Until recently no connections were obtained between the characteristic 0 class and either of the other two classes. Our purposes for both continuous and discrete flows is to give characterizations of these classes in terms of the fundamental relations, to classify recurrent motions, and to present several relationships among the classes.

The references contain the notations and definitions we shall use. A few are recapitulated below for convenience. Throughout the paper we assume (X,π) is a given continuous or discrete flow on a Hausdorff phase space X. The real or integral phase group will be designated by G. We denote the fundamental relations *orbit, orbit closure, limit, prolongation, prolongational limit, weak attraction, attraction, and strong attraction* by C, K, L, D, J, A_w, A, and A_s, respectively. A set M in X is said to be F-*stable* (*stable*) if for any neighborhood U of M there is a neighborhood V of M satisfying $F(V) = V \subset U$ $(C(V) \subset U)$. We call a set M in X (point p in X) a *saddle set* (*point*) provided there is a neighborhood U of M (of p) such that each neighborhood V of M (of p) contains a point x for which we have $C^+(x) \not\subset U$ and $C^-(x) \not\subset U$. A set M in X is *almost invariant* if there is a $T > 0$ such that $C(M) = M[0,T]$. If each neighborhood of a set M in X contains a subneighborhood of M of a property P, then M is said to be *approximated by sets of property* P.

For X noncompact locally compact we denote the extended flow on the one point compactification $X^* = X \cup \{\infty\}$ by (X^*,π^*). The product flow induced on $X^2 = X \times X$ by $(x,y)t = (xt,yt)$ for x, $y \in X$ and $t \in G$ is denoted by (X^2,π^2). The usual Δ designates the diagonal.

Each fundamental relation identified above is invariant in (X^2,π^2), D and J are closed in X^2, $K = C \cup L$, $D = C \cup J$, $L \subset J$, $D = D^{-1}$, $J = J^{-1}$, $\Delta \subset C \cap K \cap D$, and $A_s \subset A \subset A_w$. Whenever confusion might occur between the fundamental relations on (X,π) and (X^2,π^2), we shall use the notation F_2 to distinguish a fundamental relation F on X^2 from F on X.

DISTAL FLOWS

A flow (X,π) is said to be *distal* whenever $xF \to z$ and $yF \to z$ imply $x = y$ for any x, y, and z in X and any filter F in G. We shall say that a flow (X,π) on a uniform space is *strongly distal* provided, given any distinct x and y in X, there exists an index α with $C(x,y) \subset X^2 \setminus \alpha$. These notions are equivalent whenever X is compact.

The following theorem characterizes distal flows in terms of the fundamental relations. Note that each relation explicitly mentioned in the theorem is defined on the product flow. See Theorem 1 of [7] for a proof.

Theorem I. Consider the following statements:
(1) (X,π) is distal;
(2) $C(x,y) \subset X^2 \setminus V$ for some neighborhood V of Δ provided $x \neq y$;
(3) $xt_i \to z$ and $yt_i \to z$ imply $x = y$ for $x,y,z \in X$ and net (t_i) in G;
(4) $K(x,y) \cap \Delta \neq \emptyset$ implies $x = y$ for $x,y \in X$;
(5) $K(X^2 \setminus \Delta) = X^2 \setminus \Delta$;
(6) $L(X^2 \setminus \Delta) \quad X^2 \setminus \Delta$;
(7) $A_w(\Delta) = \Delta$;
(8) $L(X^2 \setminus \Delta) = X^2 \setminus \Delta$;
(9) $C(x,y)$ and Δ are separated by disjoint neighborhoods for $x \neq y$; and
(10) (X,π) is strongly distal.
Then, each of the following holds:
(a) 1 through 7 are pairwise equivalent and 8 through 10 imply 1.
(b) If each orbit closure of (X,π) is compact and X is locally compact (compact), 1 through 9 (1 through 10) are pairwise equivalent.

EQUICONTINUOUS FLOWS

Our next theorem characterizes equicontinuous flows in terms of the fundamental relations. In statements 7 and 8 the notation Y^* denotes the one point compactification of Y when Y is noncompact locally compact and denotes Y otherwise. Each relation explicitly mentioned in the theorem is defined on the product flow of (X,π).

Theorem II. Consider the following statements:
(1) (X,π) is pointwise equicontinuous;
(2) Δ is stable;
(3) $A_s(\Delta) = \Delta$;
(4) Δ is stable relative to A_w, A, or A_s;
(5) Δ consists of stable orbit closures;
(6) X^2 consists of stable orbit closures;
(7) (X^{2*},π^{2*}) is of characteristic 0;
(8) (X^{*2},π^{*2}) is of characteristic 0;
(9) $x_i \to x$, $y_i \to y$, $x_it_i \to z$, and $y_it_i \to z$ imply $x = y$ for $x, y, z \in X$, nets (x_i) and (y_i) in X, and net (t_i) in G;
(10) $D(x,y) \cap \Delta \neq \emptyset$ implies $x = y$ for $x, y \in X$;
(11) $D(\Delta) = \Delta$;
(12) $J(\Delta) \subset \Delta(J(\Delta) = \Delta$ provided each orbit closure of (X,π) is compact);
(13) $C(x,y)$ and Δ are separated by disjoint invariant neighborhoods for $x, y \in X$ and $x \neq y$; and
(14) (X,π) is uniformly equicontinuous.
Then, each of the following holds:

(a) 2 through 4 are pairwise equivalent, 9 through 13 are pairwise equivalent, 5 implies 2, and 2 implies 11.
(b) If each orbit closure of (X,π) is compact and X is locally compact, 1 through 8 are pairwise equivalent.
(c) If X is compact, the statements are pairwise equivalent.

See Theorem 2 of [7] for a proof of Theorem II.

CONNECTIONS OF CERTAIN RELATIONS TO THE FUNDAMENTAL RELATIONS

Let X be compact, and hence, be a uniform space in this section. The points x and y are *proximal* if given an index α of X, there exists a t in G such that (xt,yt) is in α. The set of proximal pairs is denoted by P and called the *proximal relation*. The succeeding theorem relates the proximal and orbit closure relations.

Theorem III. $P = K^{-1}(\Delta)$.

The *regionally proximal relation* Q is $\cap \{\overline{C(\alpha)} \mid \alpha \text{ is an index of } X\}$. The next theorem connects Q to the prolongational relation.

Theorem IV. $Q = D(\Delta)$.

Let $\mathcal{B}$ be the collection of closed invariant equivalence relations on X such that the flow induced on X/B is distal (equicontinuous) for $B \in \mathcal{B}$. Then, the *distal (equicontinuous) structure relation* $\mathcal{D}$ ($\mathcal{E}$) is defined as $\cap \mathcal{B}$. The relations $\mathcal{D}$ and $\mathcal{E}$ are the least such equivalence relations and are linked to the orbit closure and prolongational relations as follows.

Theorem V. $\mathcal{D}$ (or $\mathcal{E}$) is the least closed invariant equivalence relation on X such that $\mathcal{D} = K^{-1}(\mathcal{D})$ ($\mathcal{E} = D(\mathcal{E})$).

Similarly, let $\mathcal{Z}$ denote the collection of closed invariant equivalence relations on X such that the flow induced on X/Z is of characteristic 0 for $Z \in \mathcal{Z}$. We define the *characteristic 0 structure relation* $\mathcal{F}$ to be $\cap \mathcal{Z}$. Then, $\mathcal{F}$ is the least element of $\mathcal{Z}$ and is related to the orbit closure relation K on X and prolongation D_2 on X^2 as follows.

Theorem VI. $\mathcal{F}$ is the least closed invariant equivalence relation on X such that $D_2(\mathcal{F}) \subset K$.

FLOWS OF CHARACTERISTIC 0

Showing that $C \subset K \subset \overline{C} = \overline{K} = D$ relative to X^2 is not difficult. A flow is of *characteristic 0* whenever $D = K$, and so, in that case we have $\overline{C} = K = \overline{K} = D$. From these statements and [7] we obtain the following characterization of flows of characteristic 0 on arbitrary Hausdorff phase spaces in terms of the product flow.

Theorem VII. Consider the following statements:
(1) (X,π) is of characteristic 0;
(2) $K = D_2(D)$;
(3) $K = \overline{C}$;
(4) $K = \overline{K}$;
(5) $K = D_2(K)$;
(6) $D_2(\Delta) \subset K$; and
(7) $Q \subset K$.

Then, each of the following holds:
(a) 1 through 4 are pairwise equivalent.
(b) If each orbit closure is compact, 1 through 6 are pairwise equivalent.

(c) If X is compact, the statements are pairwise equivalent.

By using Theorem VII to characterize the product flow (X^{*2},π^{*2}) of Theorem II (8), we obtain additional characterizations of the equicontinuity of (X,π).

RECURRENT FLOWS

Throughout this section we assume X is locally compact. Also, we denote the *sets of critical and recurrent points* of the flow (X,π) (or (X^*,π^*)) by S and Y (S^* and Y^*), respectively. A point x is called *recurrent* (*uniformly recurrent*) if, given any neighborhood U of x, there is a $T \geq 0$ (and a neighborhood V of x) such that $U \cap y[0,T] \neq \emptyset$ ($V \cap y[0,T] \neq \emptyset$) for every $y \in C(x)$ ($y \in C(z)$ where $z \in V$).

The following theorem classifies the orbit closures of a recurrent flow on a locally compact space. See [6] for a proof of the continuous flow case.

Theorem VIII. Each orbit closure of a recurrent flow on a locally compact space is either stable or else a saddle set.

A flow of characteristic 0 on a locally compact space with compact orbit closures is recurrent (see [1] and [2]). A flow (X,π) is of characteristic 0 with compact orbit closures if and only if it is a recurrent flow with no saddle orbit closures. For every neighborhood V of any orbit closure in a recurrent flow (X,π) we have $A_w^+(V) = C(V)$ and, according to the next theorem, those which are stable are the ones approximated by compact neighborhoods V such that $A^+(V) \cup A_w^+(V^o) = C(V)$. For a proof of the continuous case see [6].

Theorem IX. The critical set and each noncritical orbit closure of a recurrent flow (X,π) with either X or S compact are stable if and only if they are approximated by compact neighborhoods V all satisfying one of the following equivalent conditions:

(1) $C(V) = C(V^o) \cup A^+(\partial V)$,

(2) $A_w^+(X \setminus V) \cap A_w^+(\partial V) \subset A_w^+(V^o)$, or

(3) V is almost invariant.

Corollary. For a recurrent flow (X,π) a noncritical orbit closure or a compact component S_o of S isolated from $S \setminus S_o$ is

(1) stable if and only if it is approximated by compact almost invariant neighborhoods,

(2) a saddle set if and only if it has a neighborhood U such that each compact subneighborhood V contains at least one point x satisfying both $C(x) \subset X \setminus V^o$ and $C(x) \not\subset U$, and

(3) uniformly recurrent if and only if if is stable.

We conclude this section with a generalization of the Cycle Stability Theorems of [3] and [4] to recurrent motions. For S compact (noncompact) we let $\tilde{X}$ be the quotient space of X (of X^*) determined by the relation $\Delta \cup \{(x,y) \in B \times B : B$ is a component of $S \cup \partial Y$ ($S^* \cup \partial Y^*$)$\}$. The flow induced on $\tilde{X}$ is denoted by $(\tilde{X},\tilde{\pi})$ and the recurrent (critical) set is denoted by $\tilde{Y}$ (by $\tilde{S}$). A relation induced on $\tilde{X}$ by a relation F on X will be denoted by $\tilde{F}$. We shall use C rather than $\tilde{C}$ on $\tilde{X} \setminus (\tilde{S} \cup \partial\tilde{Y})$ since $C = \tilde{C}$ for such points. The set $\tilde{Y}$ is closed in $\tilde{X}$ with totally disconnected boundary $\tilde{S} \cup \partial\tilde{Y}$, and each point of the set $\tilde{S} \cup \partial\tilde{Y}$ represents a component of $S \cup \partial Y$ (of $S^* \cup \partial Y^*$). Thus, the following theorem classifies a flow (X,π).

Theorem X. The following hold:

(1) each orbit $C(x)$ contained in $\tilde{Y}^o \setminus \tilde{S}$ is either
 (a) a stable closed orbit,
 (b) a saddle set with a saddle set orbit closure, or
 (c) a saddle set with stable orbit closure;

(2) each component of $\tilde{Y}^o \cap \tilde{S}$ is either
 (a) stable or
 (b) a saddle set; and

(3) each point x of $\partial\tilde{Y}$ satisfies one of the following relative to $\tilde{X} \setminus \tilde{Y}^o$:
 (a) x is a positively or negatively asymptotically stable isolated point of $\partial\tilde{Y}$;
 (b) x is a positively or negatively stable point of $\partial\tilde{Y}$ not isolated in $\partial\tilde{Y}$;
 (c) x is a saddle point;
 (d) x is an isolated point of $\partial\tilde{Y}$ and x has a neighborhood V such that $\mathcal{J}^+(y) = \{x\}$ or $\mathcal{J}^-(y) = \{x\}$ for each point y of $V \setminus \{x\}$ but $\mathcal{J}^+(x) \neq \{x\} \neq \mathcal{J}^-(x)$; or
 (e) x is the limit of a net of distinct critical points in $\partial\tilde{Y}$.

INTERRELATIONSHIPS AMONG THE PRECEDING CLASSES OF FLOWS

Our first diagram specifies the relationships for X compact.

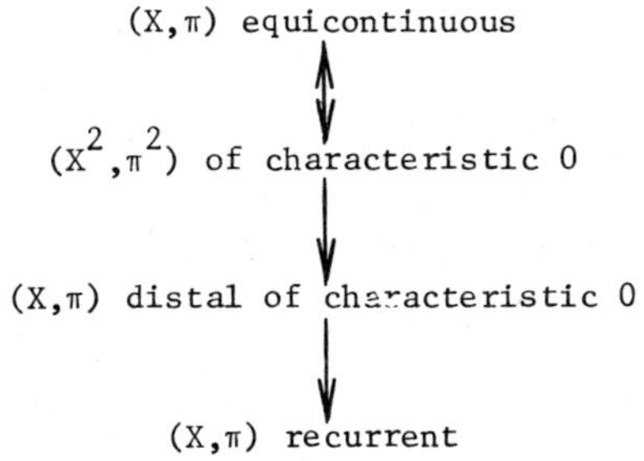

The next diagram conveys relationships for X noncompact locally compact and each orbit closure of (X,π) compact.

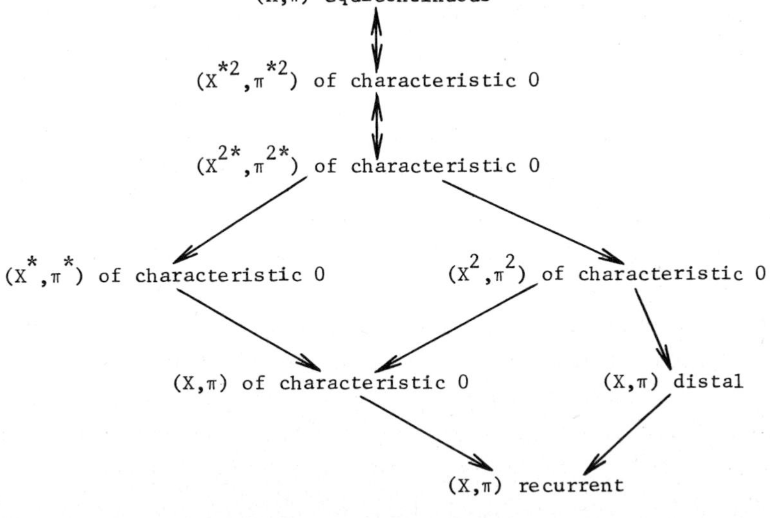

The following diagram states the relationships for X noncompact locally compact.

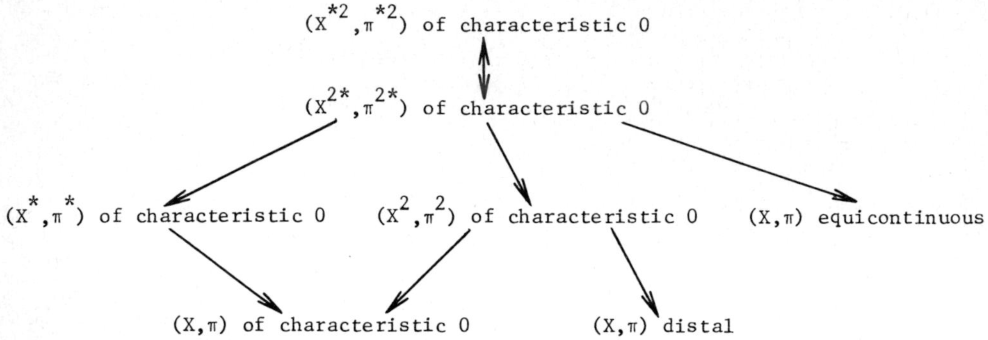

Finally, whenever X is an arbitrary Hausdorff phase space, the following diagram presents the only relationship.

$$(X^2,\pi^2) \text{ of characteristic } 0 \longrightarrow (X,\pi) \text{ distal of characteristic } 0$$

Examples showing that the converses of the implications are false in the diagrams above can be constructed. See [6] and [7] for some of them.

REFERENCES

[1] Knight, R., Structure and characterization of certain continuous flows, Funkcialaj Ekvacioj 17 (1974) 223-230.

[2] Knight, R., Prolongationally stable discrete flows, Fundamenta Mathematicae 108 (1980) 137-144.

[3] Knight, R., Compact dynamical systems, Proc. Amer. Math. Soc. 72 (1978) 501-504.

[4] Knight, R., Compact discrete flows, Fundamenta Mathematicae 118 (1983) 183-190.

[5] Knight, R., Dynamical systems satisfying certain stability and recursive criteria, Proc. Amer. Math. Soc. 85 (1982) 373-380.

[6] Knight, R., Classification of recurrent flows, unpublished manuscript (1984).

[7] Knight, R., Certain almost periodic transformation groups and the fundamental relations, unpublished manuscript (1984).

[8] Knight, R., A characterization of recurrent motions, Bull. Australian Math. Soc. 28 (1983) 1-4.

This paper is in final form and no version of it will be submitted for publication elsewhere.

Trends in the Theory and Practice of Non-Linear Analysis
V. Lakshmikantham (Editor)

ON THE ASYMPTOTIC BEHAVIOR OF SOLUTIONS TO NONLINEAR VOLTERRA EQUATIONS

Kazuo Kobayasi

Sagami Institute of Technology
1-1-25 Tsujido-Nishikaigan, Fujisawa 251, Japan

The asymptotic behavior of solutions as $t \to \infty$ of the nonlinear Volterra equation in a real Banach space is investigated. Our theorems will work in a uniformly convex Banach space which partly contains L^p spaces.

1. INTRODUCTION

We study the asymptotic behavior of solutions as $t \to \infty$ of the nonlinear Volterra equation

(V) $$u(t) + b*Au(t) \ni g(t), \qquad t \in R^+ = [0,\infty)$$

where $b:R^+ \to R$ is a given kernel, A is an m-accretive (possibly multivalued) operator in a real Banach space X, $*$ denotes the convolution $b*z(t) = \int_0^t b(t-s)z(s)ds$, and $g:R^+ \to X$ is a given function. The existence and uniqueness of solutions of (V) has been studied by many authors. Among other things, Crandall and Nohel [8] considered (V) under the following assumption on the kernel b and the function g:

(H1) $$b \in AC_{loc}(R^+; R),\ b(0) > 0,\ b' \in BV_{loc}(R^+; R)$$
$$g \in W^{1,1}_{loc}(R^+; X),\ g(0) \in \overline{D(A)}.$$

In what follows we assume without loss of generality that $b(0) = 1$. (Otherwise replace $b(t)$ and A by $b(t)/b(0)$ and $b(0)A$, respectively.) Under this assumption they showed that equation (V) and the initial value problem

(E) $$du(t)/dt + Au(t) + G(u)(t) \ni f(t), \quad t \in R^+,$$
$$u(0) = u_o$$

are equivalent so long as *strong solutions* of respective equations are concerned, where

(1.1) $$G(u)(t) = k(0)u(t) + \int_0^t u(t-s)dk(s)$$

(1.2) $$f(t) = g'(t) + k*g'(t) + k(t)g(0)$$

(1.3) $\quad u_o = g(0)$

(1.4) $\quad b(t) + k*b(t) = 1, \quad t \in R^+.$

Concerning the equivalence of (V) and (E) also see [12]. (See [8, p.315] concerning the definition of strong solutions of (V) and (E).)

We say that $u \in C(R^+; \overline{D(A)})$ is an integral solution of (E) if the restriction u^T of u to [0,T] is an integral solution in the sense of Benilan [2] of the initial value problem: $du^T/dt + Au^T \ni h$, $0 \le t \le T$, $u^T(0) = u_o$, for all $T > 0$, where $h(t) = f(t) - G(u^T)(t)$.

Under condition (H1) it is proved in [8] that (E) has a unique integral solution and that if, moreover, $g' \in BV_{loc}(R^+; X)$, $g(0) \in D(A)$ and X is reflexive then the integral solution of (E) is a strong solution of (E) and hence a strong solution of (V) by the above equivalence of (V) and (E). Thus we will adopt the integral solution of (E) as a *generalized solution* of (V).

Our aim is to study the asymptotic behavior of the generalized or strong solution of (V). This problem has been treated in a Hilbert space setting by several authors, e.g. [1], [6], [7], [9], [10], while we consider it in a real Banach space. To this end we shall impose the following assumption (H2) on the kernel b which guarantees order preserving properties of solutions of (V). This requirement is useful and natural in the application of the physical model of heat flow in a material with memory (see [6]).

(H2) b is completely positive on [0,T] for all $T > 0$ and $b_\infty = \lim_{t\to\infty} b(t) > 0$.

It is known [1],[6] that (H2) is equivalent to the fact that k defined by (1.4) is nonnegative and nonincreasing and $k \in L^1(R^+; R)$.

2. STATEMENT OF RESULTS

We begin by stating two results (Theorems 2.1 and 2.2 below) of [11].

THEOREM 2.1. Let X be reflexive and satisfy Opial's condition (see [11]). Let u be the generalized solution of (V) and let (H1) and (H2) be satisfied. If $g' \in L^1(R^+; X)$, $A^{-1}0 = \{x \in D(A); Ax \ni 0\} \neq \emptyset$ and

(2.1) $\quad \lim_{t\to\infty}|u(t+h) - u(t)| = 0 \quad$ for each $h > 0$

holds, then u(t) converges weakly as $t\to\infty$ to a point of $A^{-1}0$.

THEOREM 2.2. Let X be an arbitrary Banach space. Let u be the generalized solution of (V) and let (H1) and (H2) be satisfied. If the resolvent $(I + rA)^{-1}$

of A is compact for some $r > 0$, $g' \in L^1(R^+; X)$, $A^{-1}0 \neq \emptyset$ and (2.1) holds, then $u(t)$ converges strongly as $t \to \infty$ to a point of $A^{-1}0$.

REMARK. When X has a weakly sequentially continuous duality mapping, Theorem 2.1 remains valid even if (2.1) is replaced by the condition that $\text{w-lim}_{t\to\infty}(u(t+h) - u(t)) = 0$ for each $h > 0$.

Opial's condition plays a decisive role in the proof of Theorem 2.1 in [11], while L^p spaces do not satisfy Opial's condition if $p \neq 2$, so that we can not use Theorem 2.1 in an application of L^p space settings. Hence it is worth while to obtain theorems which are applicable to these cases.

THEOREM 2.3. Let X be uniformly convex. Let u be the generalized solution of (V). Let $A^{-1}0 \neq \emptyset$ and let $P:X \to A^{-1}0$ be the nearest point mapping. Assume that (H1), (H2) and one of the following conditions hold.

(H3) $\text{w-lim}_{n\to\infty} x_n = x \in A^{-1}0$ and $\lim_{n\to\infty} Px_n = z$ imply that $x = z$.

(H4) $X = L^{2k}$, $k = 1,2,\dots$.

If $g' \in L^1(R^+; X)$ and (2.1) holds, then $u(t)$ converges weakly as $t \to \infty$ to a point of $A^{-1}0$.

REMARK. If $A^{-1}0$ is a singleton, then (H3) holds automatically. Condition (H3) was first considered by Hirano and condition (H4) was considered by Bruck and Reich in the study of nonlinear mean ergodic theorems.

Finally, under different assumptions we state another asymptotic theorem which extends some results of Hirano [9] to uniformly convex Banach spaces. The assumptions are the followings:

(C1) A is single-valued and strictly accretive.

(C2) $b \in AC_{loc}(R^+; R)$, $b(0) > 0$ and $b_\infty = \lim_{t\to\infty} b(t)$ exists.

(C3) $b' \in L^r(R^+; R)$ for some $1 \leq r < \infty$.

(C4) $g \in W^{1,1}_{loc}(R^+; X)$

(C5) There exists $z_0 \in X$ such that $\lim_{t\to\infty} \int_t^{t+h} |g'(s) - z_0| ds = 0$ for each $h > 0$.

THEOREM 2.4. Let X be uniformly convex and let (C1)-(C5) be satisfied. Let u be a bounded strong solution of (V) satisfying: (i) there exist $w_0 \in X$ and $p \in [1, r/(r-1))$ such that $Au - w_0 \in L^p(R^+; X)$; and (ii) $y_0 = b(0)^{-1}(z_0 - (b_\infty -$

$b(0))w_o) \in R(A)$. Then $(1/t)\int_h^{t+h} u(s)ds$ is weakly convergent to a point of $A^{-1}y_o$ uniformly in $h \geq 0$ as $t \to \infty$.

COROLLARY 2.5. If the assumptions of Theorem 2.4 hold with $z_o = w_o = 0$, then $u(t)$ itself converges weakly as $t \to \infty$ to a point of $A^{-1}0$.

REMARK. The assumption that A is single-valued is not essential. In Theorem 2.4 one assumes that (V) has a strong solution. If, moreover, one assumes in Theorem 2.4 that $b' \in BV_{loc}(R^+; R)$, $g' \in BV_{loc}(R^+; X)$ and $g(0) \in D(A)$, then the existence of a strong solution of (V) is ensured (see Section 1).

3. PROOF OF THEOREMS 2.3 AND 2.4

In order to prove Theorem 2.3 we start from

LEMMA 3.1. Under the assumptions of Theorem 2.3 we have that $\lim_{t\to\infty} |u(t) - T(h)u(t)| = 0$ for each $h \geq 0$, where T is the contraction semigroup generated by $-A$.

PROOF. By definition u is an integral solution of (E). The fundamental property of integral solutions leads to

$$|u(t+h) - T(h)u(t)| \leq \int_0^h |f(t+\xi) - G(u)(t+\xi)|d\xi.$$

Since $G(u)(t) = k(t)u(t) + \int_0^t (u(t-s) - u(t))dk(s)$, it follows that

$$|u(t+h) - T(h)u(t)| \leq \int_t^{t+h} (|f(\xi)| + C\,k(\xi))d\xi + \int_0^h \phi(t+\xi)d\xi \tag{3.1}$$

where $C = \sup_{t\geq 0}|u(t)|$ and $\phi(t) = -\int_0^t |u(t-s) - u(t)|dk(s)$. (Note that $u(t)$ is bounded on $[0,\infty)$ by [11, Lemma 3.2].) Since $k \in L^1(R^+; R)$, (2.1) and the Lebesgue convergence theorem imply that $\lim_{t\to\infty} \phi(t) = 0$ and hence $\lim_{t\to\infty} \int_0^h \phi(t+\xi)d\xi$ for each $h \geq 0$. We also note that $f \in L^1(R^+; X)$ by the assumption that $g' \in L^1(R^+; X)$. Hence, from (2.1) and (3.1) we have the result.

LEMMA 3.2. Under the assumptions of Theorem 2.3 we have that $\lim_{t\to\infty} Pu(t)$ exists.

PROOF. For $y \in A^{-1}0$ set $r(y) = \lim_{t\to\infty}|u(t) - y|$, which exists (e.g. see [11, Lemma 3.2]). Put $d = \inf\{r(y);\ y \in A^{-1}0\}$. Since X is uniformly convex, there exists a unique $z \in A^{-1}0$ such that $r(z) = d$. We show that $\lim_{t\to\infty} Pu(t) = z$. If this were false, then $|Pu(t_n) - z| \geq \varepsilon$ for some $\varepsilon > 0$ and $\{t_n\}$ with $t_n \to \infty$. Let δ denote the modulus of convexity of X. We can choose $M > d$ such that

$M(1 - \delta(\varepsilon/M)) < d$. Let an integer N be such that $|u(t_n) - Pu(t_n)| \le |u(t_n) - z| \le M$ for $n \ge N$. It follows that

$$(3.2) \qquad |u(t_n) - (Pu(t_n) + z)/2| \le M(1 - \delta(\varepsilon/M)), \qquad n \ge N.$$

Now, put $w = (Pu(s) + z)/2$. Since $A^{-1}0$ is convex, w is a point of $A^{-1}0$. By [11, Lemma 3.1] we obtain

$$|u(t) - w| + k*|u - w|(t) - |u(s) - w| - k*|u - w|(s)$$
$$\le \int_s^t |f(\xi) - G(w)(\xi)|d\xi, \qquad 0 \le s \le t.$$

Note that $|G(w)(\xi)| = |k(\xi)w| \le k(\xi)|u(s) - z| \le \text{Const.}k(\xi)$. Hence, passing $t \to \infty$ and then $s \to \infty$ in the above inequality gives

$$(1 + \int_0^\infty k(\xi)d\xi)d \le (1 + \int_0^\infty k(\xi)d\xi)r(w)$$
$$\le \liminf_{s \to \infty} |u(s) - w| + \int_0^\infty k(\xi) \limsup_{s \to \infty} |u(s-\xi) - w|d\xi.$$

But, since $|u(s-\xi) - w| \le |u(s-\xi) - u(s)| + |u(s) - z|$, (2.1) and the above inequality give that $d \le \liminf_{s\to\infty} |u(s) - (Pu(s) + z)/2|$. This contradicts (3.2). The proof is complete.

PROOF OF THEOREM 2.3. Let $\omega_w = \{y \in X;\ y = \text{w-}\lim_{n\to\infty} u(t_n)$ for some $\{t_n\}$ with $t_n \to \infty\}$. By Lemma 3.1 and [3], ω_w is a subset of the set of fixed points of $T(h)$ for each $h > 0$. Hence, $\omega_w \subset A^{-1}0$. It suffices to show that ω_w is a singleton. If (H3) holds, ω_w is a singleton by virtue of Lemma 3.2. Next, assume that (H4) holds. It is easy to see that L^{2k} spaces have the property that if $\lim_{n\to\infty} |u_n - rv|$ exists for $0 \le r \le 1$ then this limit also exists for all $r \ge 0$. Therefore, $\lim_{t\to\infty} |su(t) + (1 - s)y_1 - y_2|$ exists for all $s > 0$ and $y_1, y_2 \in A^{-1}0$ because $\lim_{t\to\infty} |u(t) - ((s - 1)y_1 + y_2)/s|$ exists for $s \ge 1$. That ω_w is a singleton follows from the proof of [4, Lemma 2.3]. The proof is complete.

Next, we prove Theorem 2.4. Let u denote a strong solution of (V). We may again assume that $b(0) = 1$. Then, by differentiating (V) in t we have

$$u'(t) + Bu(t) = g'(t) - z_o - b'*(Au - w_o)(t) + (b_\infty - b(t))w_o \qquad \text{a.e.} t$$

where $Bv = Av + (b_\infty - 1)w_o - z_o$ for $v \in D(B) = D(A)$. B is m-accretive and $B^{-1}0 = \{y_o\}$ by (ii) of Theorem 2.4.

LEMMA 3.3. Under the assumptions of Theorem 2.4 we have that $\lim_{t\to\infty} |u(t+h) - S(h)u(t)| = 0$ for each $h \ge 0$, where S is the contraction semigroup generated by $-B$.

PROOF. It follows that

$$|u(t+h) - S(h)u(t)|$$

$$(3.3) \quad \le \int_0^h |g'(t+\xi) - z_o - b'*(Au - w_o)(t+\xi) + (b_\infty - b(t+\xi))w_o| d\xi$$

$$\le \int_t^{t+h} (|g'(\xi) - z_o| + |b_\infty - b(\xi)||w_o|) d\xi + \int_t^{t+h} |b'*(Au - w_o)(\xi)| d\xi$$

The first term of the right hand of (3.3) goes to 0 as $t \to \infty$ by (C2) and (C5). To estimate the second term of the right hand of (3.3), let $1/q = (1/p) + (1/r) - 1$. Then we have $1 \le q < \infty$. By Young's inequality

$$|b'*(Au - w_o)|_{L^q(R^+;\, X)} \le |b'|_{L^r(R^+;\, R)} |Au - w_o|_{L^p(R^+;\, X)}.$$

Hence (C3) and (i) of Theorem 2.4 imply that $b'*(Au - w_o) \in L^q(R^+; X)$. This gives $\int_t^{t+h} |b'*(Au - w_o)(\xi)| d\xi$ goes to 0 as $t \to \infty$. Thus we have the result.

LEMMA 3.4. Under the assumptions of Theorem 2.4, u is asymptotically uniformly continuous, i.e. given $\varepsilon > 0$, there exist $\delta = \delta(\varepsilon) > 0$ and $\rho(\varepsilon) > 0$ such that $|u(t) - u(s)| < \varepsilon$ whenever $|s - t| < \delta$ and $s, t > \rho$.

PROOF. We have for $0 \le t$ and $0 \le s \le 1$

$$u(t+s) - u(t)$$

$$= g(t+s) - g(t) - \int_0^t (b(t+s-\xi) - b(t-\xi))(Au(\xi) - w_o) d\xi$$

$$+ \int_t^{t+s} b(t+s-\xi)(Au(\xi) - w_o) d\xi - \int_t^{t+s} b(\xi) w_o d\xi$$

and hence

$$(3.4) \qquad |u(t+s) - u(t)| \le I_1 + I_2 + I_3 + I_4$$

where

$$I_1 = |g(t+s) - g(t)|, \qquad I_2 = \int_0^t |b(t+s-\xi) - b(t-\xi)||Au(\xi) - w_o| d\xi,$$

$$I_3 = \int_t^{t+s} |b(t+s-\xi)||Au(\xi) - w_o| d\xi, \qquad I_4 = |w_o| \int_t^{t+s} |b(\xi)| d\xi.$$

We first estimate I_1, I_3 and I_4: (a) $I_1 = |\int_t^{t+s} g'(\xi) d\xi| \le \int_t^{t+s} |g'(\xi) - z_o| d\xi + s|z_o|$; (b) $I_3 \le |b|_{L^\infty} \int_t^{t+s} |Au(\xi) - w_o| d\xi$; (c) $|I_4 - s|b_\infty||w_o|| \le |w_o| \int_t^{t+s} |b(\xi) - b_\infty| d\xi$. Therefore, by assumptions (C1), (C2) and (i), for each $\varepsilon > 0$ there exist $\delta_1 = \delta_1(\varepsilon) > 0$ and $\rho = \rho(\varepsilon) > 0$ such that $I_1 + I_3 + I_4 < \varepsilon/2$ whenever $s < \delta_1$ and $t > \rho$.

Next, we estimate I_2. We first consider the case that $p > 1$. If $1/p' = 1 - 1/p$, then by Hölder's inequality

$$\int_0^\infty |b(\xi+s) - b(\xi)|^{p'} d\xi \le (2|b|_{L^\infty})^{p'-r} \int_0^\infty |b(\xi+s) - b(\xi)|^r d\xi$$

$$\le \text{Const.} \int_0^\infty \left|\int_0^s b'(\xi+\tau)d\tau\right|^r d\xi \le \text{Const.} \int_0^\infty s^{r/r'} \int_0^s |b'(\xi+\tau)|^r d\tau d\xi$$

$$= \text{Const.}\ s^{r/r'} \int_0^s\int_0^\infty |b'(\xi+\tau)|^r d\xi d\tau \le \text{Const.}\ s^r \int_0^\infty |b'(\xi)|^r d\xi$$

where $1/r' + 1/r = 1$. Hence

$$I_2 \le \left(\int_0^t |b(t+s-\xi) - b(t-\xi)|^{p'} d\xi\right)^{1/p'} \left(\int_0^t |Au(\xi) - w_0|^p d\xi\right)^{1/p}$$

$$\le \left(\int_0^\infty |b(\xi+s) - b(\xi)|^{p'} d\xi\right)^{1/p'} |Au - w_0|_{L^p(R^+;\ X)} \le \text{Const.}\ s^{r/p'}.$$

Secondly, when $p = 1$, the uniform continuity of b on R^+ by virtue of (C2) implies

$$I_2 \le m(s)|Au - w_0|_{L^1(R^+;\ X)}$$

where m is the modulus of uniform continuity of b. Therefore, we have that each $\varepsilon > 0$, there exists $\delta_2 = \delta_2(\varepsilon) > 0$ such that $I_2 < \varepsilon/2$ for $0 \le s < \delta_2$ and $t \ge 0$. Consequently, by (3.4) we obtain that $|u(t+s) - u(t)| < \varepsilon$ whenever $0 \le s < \delta = \min\{\delta_1, \delta_2\}$ and $t > \rho$.

LEMMA 3.5. (Bruck and Passty [5]) Let X be uniformly convex. Let $u \in L^\infty(R^+;\ X) \cap L^1_{loc}(R^+;\ X)$. Suppose that u is asymptotically uniformly continuous and $\lim_{t\to\infty}|u(t+h) - S(h)u(t)| = 0$ for each $h \ge 0$, where S is the contraction semigroup generated by $-B$. If $(1/T_n)\int_{a_n}^{a_n+T_n} u(t)dt$ converges weakly to x as $n \to \infty$ (where $T_n \to \infty$ as $n \to \infty$), then $x \in B^{-1}0$.

PROOF OF THEOREM 2.4. This theorem can be easily proved by Lemmas 3.3, 3.4 and 3.5 and the fact that $B^{-1}0$ is a singleton.

PROOF OF COROLLARY 2.5. We shall show that $|u(t+s) - u(t)| \to 0$ as $t \to \infty$ for each $s \ge 0$, for this is a Tauberian condition for almost-convergence of $\{u(t)\}$ (e.g. see [4, p.115]). To prove this we recall the proof of Lemma 3.4. By (3.4) it suffices to show that $I_i \to 0$ as $t \to \infty$ for each $s \ge 0$, $i = 1, 2, 3, 4$. From the estimates (a) and (b) in the proof of the lemma we have that $I_3 \to 0$ as $t \to \infty$ and that if $z_0 = w_0 = 0$ then $I_4 = 0$ and $I_1 \to 0$ as $t \to \infty$.

Next, $I_2 \to 0$ as $t \to \infty$ will follow from the fact that $I_2 \in L^q(R^+;\ R)$ where $1/q = 1/p + 1/r - 1$ as I_2 is a continuous function in t. But, this fact follows from Young's inequality

$$|I_2|_{L^q(R^+;\ R)} \le |b(\cdot+s) - b|_{L^r(R^+;\ R)} |Au|_{L^p(R^+;\ X)}$$

$$\le s|b'|_{L^r}|Au|_{L^p}.$$

Thus the proof is complete.

REFERENCES

[1] Baillon, J.B. and Clément, P., Ergodic theorems for non-linear Volterra equations in Hilbert space, Nonlinear Analysis 5 (1981) 789-801.

[2] Bénilan, Ph., Solution intégrales d'équations d'évolution dans un espace de Banach, C. R. Acad. Sci. 274 (1972) 47-50.

[3] Browder, F. E., Nonlinear operators and nonlinear equations of evolution in Banach spaces, in Proc. Symp. Pure Math. 18, part 2 (Amer. Math. Soc., Providence, 1976).

[4] Bruck, R. E., A simple proof of the mean ergodic theorem for nonlinear contractions in Banach spaces, Israel J. Math. 32 (1979) 107-116.

[5] Bruck, R. E. and Passty, G. B., Almost convergence of the infinite product of resolvents in Banach spaces, Nonlinear Analysis 3 (1979) 279-282.

[6] Clément, Ph. and Nohel, J. A., Asymptotic behavior of solutions of nonlinear Volterra equations with completely positive kernels, SIAM J. Math. Anal. 12 (1981) 514-535.

[7] Clément, Ph., MacCamy, R. C. and Nohel, J. A., Asymptotic properties of solutions of nonlinear abstract Volterra equations, J. Integral Eq. 3 (1981) 185-216.

[8] Crandall, M. G. and Nohel, J. A., An abstract functional differential equation and a related nonlinear Volterra equation, Israel J. Math. 29 (1978) 313-328.

[9] Hirano, N., Asymptotic behavior of solutions of nonlinear Volterra equations, J. Differential Eq. 47 (1983) 163-179.

[10] Hulbert, D. S. and Reich, S., Ergodic theorems for nonlinear Volterra integral equations, in: Lakshmikantham, V. (ed.), Trends in theory and practice of nonlinear differential equations (Marcel Dekker, New York and Basel, 1984).

[11] Kato, N., Kobayasi, K. and Miyadera I., On the asymptotic behavior of solutions of evolution equations associated with nonlinear Volterra equations, to appear in Nonlinear Analysis.

[12] Tanabe, H., Note on nonlinear Volterra integral equation in Hilbert space, Proc. Japan Acad. 56 (1980) 9-11.

This paper is in final form and no version of it will be submitted for publication elsewhere.

Trends in the Theory and Practice of Non-Linear Analysis
V. Lakshmikantham (Editor)

RANDOM DIFFERENCE INEQUALITIES*

G. S. Ladde
Department of Mathematics
University of Texas
Arlington, TX 76019

M. Sambandham
Department of Mathematics
and Computer Science
Atlanta University
Atlanta, GA 30314

Several random difference inequalities are derived for a sequence of random variables and for stochastic difference systems.

§1. Introduction

Several mathematical models of dynamical systems in Engineering and Physics are usually modeled by stochastic difference equations. However, if one is interested in numerical solutions of such models, it is essential to study stochastic difference equations. That is, a difference system in which some of the variables can change stochastically at discrete time. Though the solutions of such stochastic difference systems can be analyzed in several ways, the qualitative properties are known through stability of such systems. To study the qualitative properties one needs stochastic difference inequalities. In this article we prove a few results concerning stochastic difference inequalities. These results provide a powerful tool to analyze stability properties, error estimates, sampled data control systems, etc. These qualitative properties will be reported elsewhere.

In Section 2, we prove a few stochastic difference inequalities for a sequence of random variables. In Section 3, comparison theorems are proved which are useful to study the solution properties of random difference systems. In Section 4, we demonstrate the scope of the comparison principle and analyze some unknown random difference inequalities in the context of the presented framework.

Most of the results in this article would be discrete versions of results in Ladde and Lakshmikantham [1] and random version in Sugiyama [3]. Related results can be found in Fai Ma and Caughey [2].

---

*Research partially supported by U.S. Army Research Contract: DAAG29-83-G-0116 and DAAG29-81-G-0008.

§2. Random Difference Inequalities

In this section, we prove a few basic comparison theorems for a sequence of random variables. We assume in the following that all the equalities and inequalities are true with probability one $(\omega \cdot p \cdot 1)$.

Theorem 2.1. Let $\{\alpha_k(\omega)\}$, $\{\beta_k(\omega)\}$ be sequences of random variables defined on a complete probability space (Ω, F, P) into R^n and $k \in I(0) = \{0,1,2,\ldots\}$. Let $F(k,x,\omega)$ be a sequence of Borel measurable functions in Ω and staisfying the following:

$$F(k,x,\omega) - F(k,y,\omega) \geq -(x-y), \quad \text{for } x \geq y. \tag{2.1}$$

Assume that

$$\begin{aligned} \Delta\alpha_k(\omega) &\leq F(k,\alpha_{k-1}(\omega),\omega), \\ \Delta\beta_k(\omega) &\geq F(k,\beta_{k-1}(\omega),\omega). \end{aligned} \tag{2.2}$$

where $\Delta\eta_k(\omega) = \eta_k(\omega) - \eta_{k-1}(\omega)$ and η_k can be either α_k or β_k. Then $\alpha_k(\omega) \leq \beta_k(\omega)$ $\omega \cdot p \cdot 1$ for $k \in I(0)$, provided $\alpha_0(\omega) \leq \beta_0(\omega)$ $\omega \cdot p \cdot 1$.

Proof: We prove the theorem by induction. Set

$$m(k,\omega) = \beta_k(\omega) - \alpha_k(\omega).$$

We note that

$$m(0,\omega) = \beta_0(\omega) - \alpha_0(\omega) \geq 0.$$

From (2.1) and (2.2), we get

$$\begin{aligned} m(1,\omega) &= \beta_1(\omega) - \alpha_1(\omega) \\ &\geq \beta_0(\omega) + F(1,\beta_0,\omega) - \alpha_0(\omega) - F(1,\alpha_0,\omega) \\ &\geq (\beta_0(\omega) - \alpha_0(\omega)) - (\beta_0(\omega) - \alpha_0(\omega)) = 0. \end{aligned}$$

Therefore,

$$m(1,\omega) \geq 0.$$

Let $m(i,\omega) \geq 0$, $i = 0,1,2,\ldots,n$. Then for $i = n+1$

$$\begin{aligned} m(n+1,\omega) &= m(n) + F(n+1,\beta_n,\omega) - F(n+1,\alpha_n,\omega) \\ &\geq m(n) - (\beta_n(\omega) - \alpha_n(\omega)) \\ &= m(n) - m(n) = 0. \end{aligned}$$

Therefore, we get $m(n+1,\omega) \geq 0$.

From the principle of mathematical induction, we conclude that

$$m(k,\omega) \geq 0, \quad \text{for } k \in I(0).$$

Therefore

$$\alpha_k(\omega) \le \beta_k(\omega), \quad \omega\cdot p\cdot 1$$

provided $\alpha_0(\omega) \le \beta_0(\omega)$, $\omega\cdot p\cdot 1$.

Corollary 2.1. If we replace (2.2) by

$$\begin{aligned} \alpha_{k+1}(\omega) &\le F(k,\alpha_k(\omega),\omega), \\ \beta_{k+1}(\omega) &\ge F(k,\beta_k(\omega),\omega), \end{aligned} \tag{2.3}$$

and replace (2.1) by

$$F(k,y,\omega) - F(k,x,\omega) \ge 0 \quad \omega\cdot p\cdot 1 \text{ for } x \le y, \tag{2.4}$$

where $F(k,x,\omega)$ is nondecreasing for each $k \in I(0)$ $\omega\cdot p\cdot 1$. Then

$$\alpha_k(\omega) \le \beta_k(\omega), \quad \omega\cdot p\cdot 1,$$

provided $\alpha_0(\omega) \le \beta_0(\omega), \quad \omega\cdot p\cdot 1.$

Proof: Suppose that

$$m(k,\omega) = \beta_k(\omega) - \alpha_k(\omega).$$

Then

$$m(0,\omega) = \beta_0(\omega) - \alpha_0(\omega) \ge 0.$$

$$\begin{aligned} m(1,\omega) &= \beta_1(\omega) - \alpha_1(\omega) \\ &\ge F(k,\beta_0(\omega),\omega) - F(k,\alpha_0(\omega),\omega) \ge 0. \end{aligned}$$

Similarly, we can prove by mathematical induction that

$$m(k,\omega) \ge 0 \quad \text{for } k \in I(0);$$

that is

$$\alpha_k(\omega) \le \beta_k(\omega), \quad \omega\cdot p\cdot 1 \text{ for } k \in I(0),$$

provided $\alpha_0(\omega) \le \beta_0(\omega), \quad \omega\cdot p\cdot 1.$

Theorem 2.2. Let $m_k(\omega)$ be a sequence of random variables and $m_k(\omega)$ satisfies

$$\Delta m_k(\omega) \le g(k,m_{k-1}(\omega),\omega). \tag{2.5}$$

Let $r_k(\omega)$ be the maximal solution process of

$$\Delta u_k(\omega) = g(k,u_{k-1}(\omega),\omega), \tag{2.6}$$

where $g(k,r,\omega)$ is a sequence of Borel measurable functions in Ω, satisfying the following relation

$$g(k,r,\omega) - g(k,u,\omega) \ge -(r-u) \quad \text{for } u \le r. \tag{2.7}$$

Then

$$m_k(\omega) \le r_k(\omega), \quad \omega\cdot p\cdot 1,\ k \in I(0),$$

provided

$$m_0(\omega) \leq u_0(\omega), \quad \omega\cdot p\cdot 1.$$

Proof of this theorem is similar to the proof of Theorem 2.1.

Corollary 2.2. Suppose that

$$m_{k+1}(\omega) \leq g(k, m_k(\omega), \omega). \tag{2.8}$$

Let $r_k(\omega)$ be the maximal solution process of

$$u_{k+1}(\omega) = g(k, u_k(\omega), \omega), \tag{2.9}$$

where $g(k,r,\omega)$ is a sequence of Borel measurable functions in Ω, satisfying the following relation.

$$g(k,r,\omega) - g(k,u,\omega) \geq 0 \quad \omega\cdot p\cdot 1 \text{ for } u \leq r. \tag{2.10}$$

Then

$$m_k(\omega) \leq r_k(\omega), \quad \omega\cdot p\cdot 1,\ k \in I(0)$$

provided $\quad m_0(\omega) \leq u_0(\omega), \quad \omega\cdot p\cdot 1.$

Corollary 2.3. If $m_k(\omega)$, $k_k(\omega)$, $p_k(\omega)$ are ≥ 0 and $m_k(\omega)$ satisfies an inequality

$$m_k(\omega) \leq k_{k-1}(\omega)m_{k-1}(\omega) + p_{k-1}(\omega), \quad \omega\cdot p\cdot 1 \tag{2.11}$$

then

$$m_k(\omega) \leq r_{k_0}(\omega) \prod_{s=k_0}^{k-1} k_s(\omega) + \sum_{s=k_0}^{k-1} p_s(\omega) \sum_{\tau=s+1}^{k-1} k_\tau(\omega), \quad \omega\cdot p\cdot 1 \tag{2.12}$$

and moreover,

$$\begin{aligned} m_k(\omega) \leq\ & r_{k_0}(\omega) \exp\left[\sum_{s=k_0}^{k-1} (k_s(\omega) - 1)\right] \\ & + \sum_{s=k_0}^{k-1} p_s(\omega)\exp\left[\sum_{\tau=s+1}^{k-1} (k_\tau(\omega) - 1)\right] \quad \omega\cdot p\cdot 1. \end{aligned} \tag{2.13}$$

Proof: Consider

$$u_k(\omega) = k_{k-1}(\omega)u_{k-1}(\omega) + p_{k-1}(\omega), \quad u_0(\omega) = k_0(\omega)m_0(\omega) + p_0(\omega). \tag{2.14}$$

We note that the maximal solution of (2.14) is given by

$$r_k(\omega) = u_0(\omega) \prod_{s=0}^{k-1} k_s(\omega) + \sum_{s=0}^{k-1} p_s(\omega) \prod_{\tau=s+1}^{k-1} k_\tau(\omega).$$

From (2.11), (2.14) and an application of Corollary 2.2 yields (2.12). The verification of (2.13) follows from (2.12) since for any x, $1+x \leq e^x$.

§3. Comparison Theorem

In this section, we state and prove a main comparison theorem. Consider a stochastic difference system

$$\Delta y(k,\omega) = F(k,y(k-1,\omega),\omega), \qquad y(k_0,\omega) = y_0(\omega) \quad k > k_0, \tag{3.1}$$

where $F(k,y,\omega)$ is a sequence of Borel measurable functions on $R^n \times \Omega$ into R^n, $y_0 \in R[\Omega,R^n]$. We further assume that F satisfies a suitable regularity condition so that (3.1) has a sample solution process existing for $k \geq k_0$.

We assume that there exists a sequence of Borel measurable function V defined on $I(k_0)\times R^n \times \Omega$ into $R[\Omega,R_+]$. Define

$$\Delta V(k,x,\omega) = V(k,x+F(k,x,\omega),\omega) - V(k-1,x,\omega) \tag{3.2}$$

for all $k > k_0$, $(x,\omega) \in R^n \times \Omega$.

Theorem 3.1. Assume that there exists a sequence of Borel measurable functions V defined on $I(k_0) \times R^n \times \Omega$ into $R[\Omega,R_+]$. Further assume that V satisfies

$$\Delta V(k,x,\omega) \leq g(k,V(k-1,x,\omega)) \tag{3.3}$$

for $k > k_0$, $(x,\omega) \in R^n \times \Omega$, where g satisfies the hypothesis of Theorem 2.2. Then

$$V(k,y(k,\omega),\omega) \leq r(k,\omega), \tag{3.4}$$

for all $k \geq k_0$, whenever

$$V(k_0,y_0,\omega) \leq r_0(\omega), \tag{3.5}$$

where $y(k,\omega) = y(k,k_0,y_0,\omega)$ is a solution of (3.1) existing for all $k \geq k_0$ and $r(k,\omega) = r(k,k_0,r_0,\omega)$ is the maximal solution process of (2.6).

Proof: Set

$$m_k(\omega) = V(k,y(k,k_0,y_0,\omega),\omega).$$

Note that $m_{k_0}(\omega) \leq r_0(\omega)$. For (3.2) the definitions of $m_k(\omega)$ and $\Delta V(k,x,\omega)$, we have

$$\begin{aligned}\Delta m_k(\omega) &= V(k,y(k,\omega),\omega) - V(k-1,y(k-1,\omega),\omega)\\ &= V(k,y(k-1,\omega) + F(k,y(k-1,\omega),\omega),\omega)\\ &\qquad - V(k-1,y(k-1,\omega),\omega)\\ &\leq g(k,m_k(\omega),\omega).\end{aligned}$$

This together with the hypothesis of theorem verifies the hypothesis of Theorem 2.2. Hence by applying Theorem 2.2., we conclude that

$$m_k(\omega) \leq r_k(\omega) \quad \text{for } k \geq k_0.$$

This together with the definition of $m_k(\omega)$, implies the relation (3.4). Hence the proof of the theorem is complete.

In the following, we present another comparison theorem for the difference equation of the type

$$y(k,\omega) = F(k,y(k-1,\omega),\omega), \quad y(k_0,\omega) = y_0(\omega) \tag{3.6}$$

for $k > k_0$, where F satisfies the suitable regularity conditions to guarantee the existence of solutions of (3.6).

Theorem 3.2. Assume that

(i) $V \in M[I(k_0) \times R^n \times \Omega, R[\Omega,R_+]]$ and it satisfies the relation

$$V(k,x,\omega) \leq g(K,V(k-1,x,\omega),\omega) \quad \text{for all } k > k_0 \text{ and } x,y \in R^n; \tag{3.7}$$

(ii) g satisfies the hypothesis of Corollary 2.2;

(iii) $V(k_0,y_0(\omega),\omega) \leq r_0(\omega)$.

Then

$$V(k,y(k,\omega),\omega) \leq r(k,\omega) \tag{3.8}$$

where $y(k,\omega) = y(k,k_0,y_0(\omega),\omega)$ is a solution process of (3.6) and $r(k,\omega) = r(k,k_0,r_0(\omega),\omega)$ is the maximal solution process of (2.9).

Proof: Set $m_k(\omega) = V(k,y(k,\omega),\omega)$. Note that $m_{k_0}(\omega) \leq r_0(\omega)$. By following the argument that is used in the proof of Theorem 3.1 and applying Corollary 2.2, the proof of the theorem follows immediately.

§4. Scope of Comparison Principle

In this section we demonstrate the usefulness of Theorem 3.1.

Let $A(\omega) = (a_{ij}(\omega))$ be an $n \times n$ random matrix whose elements are defined random variables. Define

$$a(\omega) = \|I + A(\omega)\| - 1, \tag{4.1}$$

where $\|\cdot\|$ is a matrix norm and I is an identity matrix. We note that $a(\omega)$ is a random function defined on R_+ into $R[\Omega, R]$. Further using the properties of norm we get

$$-\|A(\omega)\| \leq a(\omega) \leq \|A(\omega)\| \quad \omega\cdot p\cdot 1,$$

and this proves that $a(\omega)$ is bounded for fixed $\omega \in \Omega$.

Definition 4.1. For any $n \times n$ random matrix

$$\mu(A(\omega)) = \|I + A(\omega)\| - 1, \quad \omega\cdot p\cdot 1 \tag{4.2}$$

is called the logarithmic norm of $A(\omega)$.

Definition 3.1 is the discrete version of Definition 2.9.1 of [1]. We note that the logarithmic norm $\mu(A(\omega))$ of an $n \times n$ random matrix is a random variable defined on (Ω, F, P) with values in R.

Without proof we state the following results.

Lemma 4.1. The logarithmic norm $\mu(A(\omega))$ of an $n \times n$ matrix $A(\omega)$ possesses the following properties:

(i) $|\mu(A(\omega))| \leq \|A(\omega)\|$,

(ii) $|\mu(A(\omega)) - \mu(B(\omega))| \leq \|A(\omega) - B(\omega)\|$.

For our further discussion we rewrite (3.1) in a suitable form. Suppose that the sample derivatives $\frac{\partial}{\partial x} F(k,x(k-1,\omega),\omega)$ of $F(k,x(k-1,\omega),\omega)$ exists and $F(k,0,\omega) \equiv 0$. Then

$$\Delta y(k,\omega) = A(k,y(k-1,\omega),\omega)y(k-1,\omega), \quad y(k_0,\omega) = y_0(\omega) \tag{4.3}$$

where

$$A(k,y(k-1,\omega),\omega) = \int_0^1 \frac{\partial}{\partial x} F(k,sy,\omega)ds,$$

and $A(k,y(k-1,\omega),\omega) = (a_{ij}(k,y,\omega))$ is an $n \times n$ matrix such that $a_{ij}: I(k_0+1) \times R \to R[\Omega,R]$. Similarly the linear version of (4.3) can be derived as follows:

$$\Delta y(k,\omega) = A(k,\omega)y(k-1,\omega), \quad y(k_0,\omega) = y_0(\omega) \tag{4.4}$$

We demonstrate the scope and usefulness of (4.2) in the following.

Theorem 4.1. The solution $y(k,\omega)$ of (4.3) satisfies the relation

$$\|y(k,\omega)\|^p \leq \|y_0(\omega)\|^p \exp \sum_{s=k_0}^{k-1} \Lambda_p(A(k,y(s,\omega),\omega)), \tag{4.5}$$

where $\Lambda_p(A(k,y(s,\omega),\omega)) = \|I + A(k,y(s,\omega),\omega)\|^p - 1$, $p \geq 1$.

Proof: Let $V(k,y,\omega) = \|y\|^p$. Then

$$\begin{aligned} \Delta V(k,y,\omega) &= \|y + A(k,y,\omega)y\|^p - \|y\|^p \\ &\leq \|I + A(k,y,\omega)\|^p \|y\|^p - \|y\|^p \\ &= (\|I + A(k,y,\omega)\|^p - 1)\|y\|^p \\ &= \Lambda_p(A(k,y,\omega))V(k-1,y,\omega). \end{aligned} \tag{4.6}$$

The maximal solution of

$$\Delta u(k,y,\omega) = \Lambda_p(A(k,y,\omega)u(k-1,y,\omega), \tag{4.7}$$
$$y(k_0,\omega) = u(k_0,\omega) = r_0(\omega)$$

is given by

$$r(k,\omega) = r_0(\omega) \prod_{s=k_0}^{k-1} [\Lambda_p(A(k,y,\omega)) + 1]. \tag{4.8}$$

Hence by an application of Theorem 3.1, and (4.6)-(4.8), we obtain

$$V(k,y(k,\omega),\omega) \leq u_0(\omega) \prod_{s=k_0}^{k-1} [\Lambda_p(A(k,y(s,\omega),\omega) + 1]$$

$$\leq u_0(\omega)\exp \sum_{s=k_0}^{k-1} \Lambda_p(A(k,y(s,\omega),\omega).$$

This together with the definition of V gives

$$||y(k,\omega)||^p \leq ||y_0(\omega)||^p \exp\left[\sum_{s=k_0}^{k-1} \Lambda_p(A(k,y(s,\omega),\omega))\right] \quad \omega\cdot p\cdot 1.$$

This completes the proof of the theorem.

Remark 4.1. We remark that in Theorem 4.1 if $p = 1$, then $\Lambda_1(A(k,y(s,\omega),\omega) = \mu(A(k,y(s,\omega),\omega)$. Therefore (4.5) reduces as follows

$$||y(k,\omega)|| \leq ||y_0(\omega)|| \exp\left[\sum_{s=k_0}^{k-1} \mu(A(k,y(s,\omega),\omega))\right] \quad \omega\cdot p\cdot 1. \tag{4.9}$$

In the next theorem we state the random difference inequality for (4.4).

Theorem 4.2. The solution $y(k,\omega)$ of (4.4) satisfies the relation

$$||y(k,\omega)||^p \leq ||y_0(\omega)||^p \exp\left[\sum_{s=k_0}^{k-1} \Lambda_p(A(s,\omega))\right] \omega\cdot p\cdot 1, \tag{4.10}$$

where $\Lambda_p(A(k,\omega)) = ||I + A(k,\omega)||^p - 1$, $p \geq 1$.

Proof of this theorem is similar to the proof of Theorem 4.1.

Remark 4.2. We remark that in Theorem 4.2 if $p = 1$, then $\Lambda_1(A(k,\omega)) = \mu(A(k,\omega))$. Therefore, (4.10) reduces to

$$||y(k,\omega)|| \leq ||y_0(\omega)|| \exp\left[\sum_{s=k_0}^{k-1} \mu(A(s,\omega))\right] \omega\cdot p\cdot 1. \tag{4.11}$$

Suppose that we rewrite (3.6) as follows:

$$y(k,\omega) = A(k,y(k-1,\omega),\omega)y(k-1,\omega), \quad y(k_0,\omega) = y_0(\omega). \tag{4.12}$$

Let the linear version of (4.12) be

$$y(k,\omega) = A(k,\omega)y(k-1,\omega), \quad y(k_0,\omega) = y_0(\omega) \tag{4.13}$$

In the next theorem we prove the random difference inequality for (4.12).

<u>Theorem 4.3</u>. The solution $y(k,\omega)$ of (4.13) satisfies the relation

$$\|y(k,\omega)\|^p \leq \|y_0(\omega)\|^p \exp \sum_{s=k_0}^{k-1} \Lambda_p(A(k,y(s,\omega),\omega)), \quad \omega\cdot p\cdot 1. \tag{4.14}$$

$p \geq 1$.

<u>Proof</u>: Let $V(k,y,\omega) = \|y\|^p$. Then

$$\begin{aligned} V(k,y,\omega) &= \|y + A(k,y,\omega)y\|^p \\ &\leq \|1 + A(k,y,\omega)\|^p \ \|y\|^p \\ &= [\Lambda_p(A(k,y,\omega) + 1] \ \|y\|^p . \end{aligned} \tag{4.15}$$

The maximal solution of

$$u(k,y,\omega) = (\Lambda_p(A(k,y,\omega) + 1)u(k-1,y,\omega), \tag{4.16}$$

$$y(k_0,\omega) = u(k_0,\omega) = r_0(\omega),$$

is given by

$$r(k,\omega) = r_0(\omega) \prod_{s=k_0}^{k-1} [\Lambda_p(A(k,y,\omega)) + 1]. \tag{4.17}$$

Hence by an application of Theorem 3.2 and (4.15)-(4.17), we obtain

$$\begin{aligned} V(k,y(k,\omega),\omega) &\leq u_0(\omega) \prod_{s=k_0}^{k-1} [\Lambda_p(A(k,y(s,\omega),\omega) + 1] \\ &\leq u_0(\omega) \exp\left[\sum_{s=k_0}^{k-1} \Lambda_p(A(k,y(s,\omega),\omega)]\right] . \end{aligned}$$

This together with the definition of V gives

$$\|y(k,\omega)\|^p \leq \|y_0(\omega)\|^p \exp\left[\sum_{s=k_0}^{k-1} \Lambda_p(A(k,y(s,\omega),\omega)\right], \quad \omega\cdot p\cdot 1.$$

This completes the proof of the theorem.

<u>Remark 4.3</u>. We remark that in Theorem 4.3, when $p = 1$, $\Lambda_1(A(k,y(s,\omega),\omega) = \mu(A(k,y(s,\omega),\omega)$. Therefore (4.14) reduces to

$$\|y(k,\omega)\| \leq \|y_0(\omega)\| \exp\left[\sum_{s=k_0}^{k-1} \mu(A(k,y(s,\omega),\omega))\right], \quad \omega\cdot p\cdot 1. \tag{4.18}$$

In the next theorem we state the random difference inequality for (4.13).

Theorem 4.4. The solution $y(k,\omega)$ of (4.13) satisfies the relation

$$\|y(k,\omega)\|^p \leq \|y_0(\omega)\|^p \exp\left[\sum_{s=k_0}^{k-1} \Lambda_p(A(k,\omega))\right], \quad \omega\cdot p\cdot 1 \qquad (4.19)$$

where $\Lambda_p(A(k,\omega)) = \|I + A(k,\omega)\|^p - 1$, $p \geq 1$.

Proof of this theorem is similar to Theorem 4.3.

Remark 4.4. We remark that in Theorem 4.4 if $p = 1$, $\Lambda_1(A(k,\omega)) = \mu(A(k,\omega))$. Therefore (4.19) reduces to

$$\|y(k,\omega)\| \leq \|y_0(\omega)\| \exp\left[\sum_{s=k_0}^{k-1} \mu(A(s,\omega))\right], \quad \omega\cdot p\cdot 1. \qquad (4.20)$$

The random difference inequalities (4.5), (4.9), (4.10), (4.11), (4.14), (4.18), (4.19), and (4.20) for the solutions of (4.3), (4.4), (4.12), and (4.13) provide sample estimates. These estimates can be used for studying the qualitative properties of the random solutions.

References

[1] Ladde, G. S., and Lakshmikantham, V., Random differential inequalities. Academic Press (1980).

[2] Ma, Fai and Caughey, T. K., On the stability of linear and nonlinear stochastic transformations, Int. J. Control. 34 (1981), 501-511.

[3] Sugiyama, S., Difference inequalities and their applications to stability problems. In Lecture Notes in Mathematics, Vol. 243 (1971), 1-15, Springer-Verlag.

This paper is in final form and no version of it will be submitted for publication elsewhere.

Trends in the Theory and Practice of Non-Linear Analysis
V. Lakshmikantham (Editor)

SINGULARLY PERTURBED STOCHASTIC DIFFERNETIAL SYSTEMS*

G.S. Ladde and O. Sirisaengtaksin

Department of Mathematics
University of Texas at Arlington
Arlington, Texas 76019

By employing generalized variation of constants formula and theory of differential inequalities, the mean square convergence of solutions process of a singularly perturbed stochastic differential system of Itô-type is investigated. Moreover, the solution process of such a system is approximated by the solution processes of corresponding reduced and boundary layer systems.

INTRODUCTION

A mathematical model of dynamical systems consisting of fast and slow phenomena is described by a system of singularly perturbed differential equations. The interaction of fast and slow phenomena in high-order systems results in "stiff" numerical problems which require expensive integration routines. The singular perturbation approach [4] alleviates both dimensionality and stiffness difficulties. The approach consists of (i) the lowering of the model order by first neglecting the fast phenomena, and (ii) the improvement of the approximation by reintroducing boundary layer corrections calculated in separate time scales. This approach [3,7,8,10] is used, successfully, to investigate singularly perturbed problems. In this paper, we extend the above mentioned singular perturbation approach to a system of singularly perturbed stochastic differential equations of Itô-type. Recently [11,12], an attempt was made to study the weak convergence of the slow modes in the space of continuous functions. Very recently [1], singular perturbation problems arising in stochastic control were investigated. Furthermore, by using asymptotic expansion method, the convergence of covariant of solution process of a system of singularly perturbed stochastic differential equation of Itô-type with respect to the covariant of solution processes of reduced system is analyzed in [2]. The main contribution of the present work is to investigate the mean square convergence of solutions process of a system of singularly perturbed stochastic differential equations with the solution processes of corresponding reduced and boundary layer systems. The paper is organized as follows: The problem formulation and the similarity transformation is outlined in section 2. Certain results in [3,7,8] are summarized in section 3. Section 4 consists of the main results of the present work.

STATEMENT OF THE PROBLEM

Consider the system

$$(2.1)\quad \begin{cases} dx = [A_{11}(t)x + A_{12}(t)y]dt + \sqrt{\sigma_1}\, g_1(t,\omega)dW_1(t,\omega), x(t_0) = x_0, \\ \varepsilon dy = [A_{21}(t)x + A_{22}(t)y]dt + \sqrt{\sigma_2}\, g_2(t,\omega)dW_2(t,\omega), y(t_0) = y_0, \end{cases}$$

*Research partially supported by U.S. Army Research Grants No. DAAG 29-81-G-0008 and DAAG 29-84-0060.

where $x \in \mathbb{R}^{n_1}$, $y \in \mathbb{R}^{n_2}$, $w_1 \in \mathbb{R}^{m_1}$, $w_2 \in \mathbb{R}^{m_2}$ and the dimension of the entire system (2.1) is $n = n_1 + n_2$; $g_i(t,\cdot)$ are F_t - measurable for all $t \in \mathbb{R}$, $E||g_i(t,\omega)||^2 \le k_i$, $i = 1,2$; all coefficient matrix functions in (2.1) are continuous and have appropriate dimensions; σ_1 and σ_2 are positive numbers.

Assumption 2.2. $A_{22}(t)$ is nonsingular for all $t \ge t_0$.

Assumption 2.3. $\lim_{\varepsilon\to 0} \frac{\sqrt{\sigma_2(\varepsilon)}}{\varepsilon} = \ell$; ℓ is constant.

The reduced system can be found by setting $\varepsilon = 0$ in (2.1). In fact the reduced system is described by

$$
(2.4) \qquad \begin{cases} d\bar{x} = A_R(t)\bar{x}dt + \sqrt{\sigma_1}g_1(t,\omega)dW_1(t,\omega), \ \bar{x}(t_0) = x_0, \\ \bar{y} = -A_{22}^{-1}(t)A_{21}(t)\bar{x}, \ \bar{y}(t_0) = -A_{22}^{-1}(t_0)A_{21}(t_0)x_0, \end{cases}
$$

where $A_R(t) = A_{11}(t) - A_{12}(t)A_{22}^{-1}(t)A_{21}(t)$.

The boundary layer system can be obtained by applying the ε-stretched time scale, i.e., let $\tau = \frac{t-t_0}{\varepsilon}$ and set $\varepsilon = 0$,

$$
(2.5) \qquad \begin{cases} d\hat{x} = \sqrt{\sigma_1}g_1(t,\omega)d\hat{W}_1(\tau,\omega), \hat{x}(0) = x_0, \\ d\hat{y} = [A_{21}(t_0)\hat{x} + A_{22}(t_0)\hat{y}]d\tau + \frac{\sqrt{\sigma_2}}{\varepsilon} g_2(t_0,\omega)d\hat{W}_2(\tau,\omega), \hat{y}(0) = y_0, \end{cases}
$$

where $\hat{W}_i(\tau,\omega) = W_i(\varepsilon\tau + t_0,\omega)$; $i = 1,2$.

By letting $\tilde{y} = \hat{y} + A_{22}^{-1}(t_0)A_{21}(t_0)\hat{x}$ in (2.5), the boundary layer system (2.5) is reduced to

$$
(2.6) \qquad \begin{cases} d\tilde{y} = A_{22}(t_0)\tilde{y}d\tau + \sqrt{\sigma_1}A_{22}^{-1}(t_0)A_{21}(t_0)g_1(t_0,\omega)dW_1(\tau,\omega) \\ \qquad\qquad + \frac{\sqrt{\sigma_2}}{\varepsilon} g_2(t_0,\omega)d\hat{W}_2(\tau,\omega), \\ \tilde{y}(0) = y_0 - \bar{y}(t_0). \end{cases}
$$

Our objective is to approximate the solution process of (2.1) with respect to the solution processes of reduced and boundary layer systems in the sense of mean square. This is achieved by using the transformation in [3] which transforms the deterministic parts of the original system (2.1) into a diagonal form. The deterministic parts of the system are totally decoupled. This enables us to obtain the approximation of the solutions of system (2.1) in a systematic way.

By rewriting system (2.1) as

$$
(2.7) \qquad \begin{bmatrix} dx \\ \varepsilon dy \end{bmatrix} = \begin{bmatrix} A_{11}(t) & A_{12}(t) \\ A_{21}(t) & A_{22}(t) \end{bmatrix} \begin{bmatrix} x \\ y \end{bmatrix} dt + \begin{bmatrix} \sqrt{\sigma_1}g_1(t,\omega) & 0 \\ 0 & \sqrt{\sigma_2}g_2(t,\omega) \end{bmatrix} \begin{bmatrix} dW_1(t,\omega) \\ dW_2(t,\omega) \end{bmatrix}
$$

and using the transformation

$$(2.8)\qquad \begin{bmatrix} u \\ v \end{bmatrix} = \begin{bmatrix} I-\varepsilon ML & -\varepsilon M \\ L & I \end{bmatrix} \begin{bmatrix} x \\ y \end{bmatrix}$$

as given in [3], the transformed system is described by

$$(2.9)\qquad \begin{bmatrix} du \\ \varepsilon dv \end{bmatrix} = \begin{bmatrix} A_{11}-A_{12}L & 0 \\ 0 & A_{22}+\varepsilon LA_{12} \end{bmatrix} \begin{bmatrix} u \\ v \end{bmatrix} dt + \begin{bmatrix} \sqrt{\sigma_1}g_1-\sqrt{\sigma_1}MLg_1-\sqrt{\sigma_2}Mg_2 \\ \varepsilon\sqrt{\sigma_1}Lg_1 & \sqrt{\sigma_2}g_2 \end{bmatrix} \begin{bmatrix} dW_1 \\ dW_2 \end{bmatrix}$$

where L and M are submatrix functions that are determined by

$$(2.10)\qquad \begin{cases} \varepsilon\dot{L} = A_{22}L-A_{21}-\varepsilon LA_{11}+\varepsilon LA_{12}L, \\ \varepsilon\dot{M} = -MA_{22}+A_{21}-\varepsilon MLA_{12}+\varepsilon A_{11}M-\varepsilon A_{12}LM, \end{cases}$$

with initial condition

$$(2.11)\qquad \begin{cases} L(t_0) = A_{22}^{-1}(t_0)A_{21}(t_0), \\ M(t_0) = A_{12}(t_0)A_{22}^{-1}(t_0). \end{cases}$$

Notice that the systems (2.10) and (2.11) are the same as in [3].

PRELIMINARY RESULTS

Before proving the main results, some assumptions are needed to establish certain preliminary results. These results will be used, subsequently.

Assumption 3.1. There exists a positive number α_{22} such that

$$L[A_{22}(t)] \leq -\alpha_{22}, \quad \text{for all} \quad t \geq t_0,$$

where $L[\cdot]$ is the logarithmic norm defined by

$$L[M] = \limsup_{h\to 0^+} \frac{||I+hM||-1}{h}.$$

Assumption 3.2. The matrix A_{22} is Lipschitzian on $\mathbb{R}$, that is, there exists a positive number γ such that

$$||A_{22}(t)-A_{22}(s)|| \leq \gamma|t-s|, \quad \text{for all} \quad t,s \in \mathbb{R}.$$

Assumption 3.3. $\Gamma(t,\varepsilon)$ is continuous in t and satisfies

$$\limsup_{t\to+\infty} \left\{ \frac{1}{t-t_0} \int_{t_0}^{t} ||\Gamma(t,\varepsilon)||d\tau \right\} < +\infty$$

where $\Gamma(t,\varepsilon) = L(t)A_{12}(t)$ and $L(t)$ is defined by (2.10) and (2.11). From

Lemmas 4.12 and 4.17 in [8], we have

Lemma 3.4. Under Assumption 3.1,

$$||\phi_{22}(t,s,\varepsilon)|| \le \exp[-\frac{\alpha_{22}}{\varepsilon}(t-s)], \quad t \ge s, \quad \alpha > 0,$$

where $\phi_{22}(t,s,\varepsilon)$ is the fundamental matrix solution of

(3.5) $$\varepsilon\dot{z} = A_{22}(t)z.$$

Lemma 3.6. Under Assumptions 3.1 and 3.3, there exists a positive number α_2 such that

$$||\phi_2(t,s,\varepsilon)|| \le \exp[-\frac{\alpha_2}{\varepsilon}(t-s)], \quad t \ge s,$$

where $\alpha_2 = \frac{\alpha_{22}}{\varepsilon}$ and $\phi_2(t,s,\varepsilon)$ is the fundamental matrix solution of

(3.7) $$\varepsilon\dot{z} = (A_{22} + \varepsilon\Gamma(t,\varepsilon))z.$$

Assumption 3.8. There exists a positive number α_R such that

$$\limsup_{t\to\infty} \left[\frac{1}{t-t_0} \int_{t_0}^{t} L[A_R(t)]\,d\tau \right] \le -\alpha_R.$$

From results 6.3 in [7], we have

Lemma 3.9. Under Assumption 3.8, there exist positive numbers $\hat{\varepsilon} \ge \varepsilon$ and $k > 0$ such that

$$||\phi(t,s,\varepsilon)|| \le k \exp[-\beta(t-s)], \quad t \ge s,$$

where $\phi(t,s,\varepsilon)$ is the fundamental matrix solution of

(3.10) $$\dot{p} = (A_R(t) + 0(\varepsilon))p;$$

$$\beta = \alpha_R - \varepsilon\hat{k}, \quad \hat{k} > 0, \quad \text{and}$$

$$k = \max_{t\in[S,S+T]} \exp\left\{ \int_s^t [L[A_R(\tau)]+\alpha_R]d\tau \right\}.$$

MAIN RESULTS

The transformed system (2.9) is used to find the approximate solutions to (2.7) in terms of reduced (2.4) and boundary layer (2.5) systems. The following result provides a basis for such an approximation.

Theorem 4.1. Let the assumption 2.2, 2.3, 3.1, 3.2, 3.3, and 3.8 hold. Then the solutions of (2.9), can be approximated by means of the solutions of reduced and boundary layer systmes in the mean square sense, that is

(4.2) $$u(t,t_0,u_0) = \bar{x}(t,t_0,x_0) + 0(\varepsilon),$$

and

(4.3) $$v(t,t_0,v_0) = \tilde{y}(\tau,0,\tilde{y}_0) + 0(\varepsilon),$$

both in the mean square sense, where u, v, $\bar{x}$ and $\tilde{y}$ are solutions of (2.9), (2.4) and (2.5), respectively.

Proof. First, we show that, as $\varepsilon \to 0$, $u - \bar{x} \to 0$ in the mean square sense.

From (2.9) and (2.4), if we let $p = u - \bar{x}$, then

(4.4)
$$dp = [A_{11}(t)-A_{12}(t)L(t)]pdt-A_{12}(t)0(\varepsilon)\bar{x}(t)dt$$
$$-\varepsilon\sqrt{\sigma_1}M(t)L(t)g_1(t,\omega)dW_1(t,\omega)$$
$$-\sqrt{\sigma_2}M(t)g_2(t,\omega)\ W_2(t,\omega),\ p(0) = 0(\varepsilon).$$

By considering (4.4) as a stochastic perturbed system of (3.10) and applying generalized variation of constant formula in [6], we obtain

$$E[\| p(t)\|^2] = \| 0(\varepsilon)\|^2 - 2\int_{t_0}^{t} E(p^T(s))\phi^T(t,s,\varepsilon)\phi(t,s,\varepsilon)A_{12}(s)0(\varepsilon)\phi_R(s,t_0)x_0 ds$$
$$+ \int_{t_0}^{t} E[\varepsilon^2\sigma_1 tr(g_1^T L^T M^T \phi^T(t,s,\varepsilon)\phi(t,s,\varepsilon)MLg_1)$$
$$+ \sigma_2 tr(g_2^T M^T \phi^T(t,s,\varepsilon)Mg_2)]ds,$$

where $\phi_R(s,t_0)$ is the fundamental matrix solution of

$$\dot{\bar{x}} = A_R(t)\bar{x}.$$

By using $2ab \le a^2 + b^2$, the above relation reduced to

$$E[\|p(t)\|^2] \le \|0(\varepsilon)\|^2$$
$$+ \|A_{12}\|\ \|0(\varepsilon)\|\ \|x_0\| \int_{t_0}^{t} (1+\|E(p(s))\|^2)\|\phi(t,s,\varepsilon)\|^2\ \|\phi_R(s,t_0)\|ds$$
$$+ \int_{t_0}^{t} (\varepsilon^2\sigma_1\|L\|^2\|M\|^2 E\|g_1\|^2+\sigma_2\|M\|^2 E\|g_2\|^2)\|\phi(t,s,\varepsilon)\|^2 ds.$$

This together with Lemma 3.9, assumptions on g_i's and Jensen's inequality, yields

$$E[\|p(t)\|^2] \le \|0(\varepsilon)\|^2$$
$$+\|A_{12}\|\ \|0(\varepsilon)\|\ \|x_0\|k^2\int_{t_0}^{t}(1+ E[\|p(s)\|^2])\exp[-2\beta(t-s)-\alpha_R(s-t_0)]ds$$
$$+ (\varepsilon^2\sigma_1\|L\|^2\|M\|^2k_1+\sigma_2\|M\|^2k_2)k^2\int_{t_0}^{t}\exp[-2\beta(t-s)]ds,$$
$$\le \bar{0}(\varepsilon)+\bar{k}(\varepsilon)\int_{t_0}^{t}\exp[-2\beta(t-s)-\alpha_R(s-t_0)]E[\|p(s)\|^2]ds$$

where

$$\bar{k}(\varepsilon) = \|A_{12}\|\ \|0(\varepsilon)\|\ \|x_0\|k^2,$$
$$\tilde{k}(\varepsilon) = (\varepsilon^2\sigma_1\|L\|^2\|M\|^2k_1+\sigma_2\|M\|^2k_2)k^2$$
$$= \varepsilon^2(\sigma_1\|L\|^2\|M\|^2k_1+\frac{\sigma_2}{\varepsilon^2}\|M\|^2k_2)k^2, \quad \text{and}$$

$$\bar{0}(\varepsilon) = \|0(\varepsilon)\|^2 + \bar{k}(\varepsilon)/(2\beta-\alpha_R) + \tilde{k}(\varepsilon)/2\beta.$$

Then

$$\exp[2\beta t]E[\|p(t)\|^2] \leq \exp[2\beta t]\bar{0}(\varepsilon) + \bar{k}(\varepsilon)\exp[\alpha_R t_0]\int_{t_0}^{t} \exp[-\alpha_R s]\exp[2\beta s]E[\|p(s)\|^2]ds.$$

By corollary 1.9.1 p. 38 in [9], we have

$$E[\|p(t)\|^2] \leq \bar{0}(\varepsilon)+\bar{0}(\varepsilon)\bar{k}(\varepsilon)\exp[-2\beta t+\alpha_R t_0]\exp[\tilde{0}(\varepsilon)]\int_{t_0}^{t} \exp[(2\beta-\alpha_R)s]ds,$$

where $\tilde{0}(\varepsilon) = \bar{k}(\varepsilon)/\alpha_R$. Thus

$$E[\|p(t,t_0,p_0)\|^2] \leq 0(\varepsilon), \quad \text{i.e.,}$$

$$u(t,t_0,u_0) = \bar{x}(t,t_0,x_0) + 0(\varepsilon)$$

in the mean square sense. This verifies our assertion that $u - \bar{x} \to 0$ as $\varepsilon \to 0$ in the mean square sense.

Now we will show that, as $\varepsilon \to 0$, $v \to \tilde{y}$ in the mean square sense. By following ideas in the above proof, assumption 3.2, the definition of α_2, and applying Lemmas 3.4 and 3.6, we have

$$E[\|z(t)\|^2] \leq \|\tilde{y}_0\|\int_{t_0}^{t} [\varepsilon^{-1}\gamma(s-t_0)+B]\exp[-\alpha_{22}(t-t_0)/\varepsilon](1+E\|z(s)\|^2)ds$$
$$+ (\sigma_1\|L\|^2k_1+\frac{\sigma_2}{\varepsilon^2}k_2+\sigma_1\|L(t_0)\|^2k_1+\frac{\sigma_2}{\varepsilon^2}k_2)\int_{t_0}^{t} \exp[-\alpha_{22}(t-s)/\varepsilon]ds$$

where $z = v - \tilde{y}$,

$$dz = (\varepsilon^{-1}A_{22}+LA_{12})zdt+[\varepsilon^{-1}(A_{22}-A_{22}(t_0))+LA_{12}\tilde{y}dt \tag{4.5}$$
$$+ \sqrt{\sigma_1}Lg_1dW_1 + \frac{\sqrt{\sigma_2}}{\varepsilon}g_2dW_2-\sqrt{\sigma_1}L(t_0)g_1(t_0)dW_1 + \frac{\sqrt{\sigma_2}}{\varepsilon}g_2(t_0)dW_2$$

with $z(t_0) = 0$, and

$$B = \|LA_{12}\|.$$

After certain computation, we obtain

$$E[\|z(t)\|^2] \leq 0_1(\varepsilon)$$
$$+ \varepsilon\|\tilde{y}_0\|\max(\frac{\gamma}{2\alpha_{22}^2}, \frac{B}{\alpha_{22}})[\frac{\alpha_{22}^2}{\varepsilon^2}(t-t_0)^2 + \frac{\alpha_{22}}{\varepsilon}(t-t_0)]\exp[-\alpha_{22}(t-t_0)/\varepsilon]$$
$$+ \exp[-\alpha_{22}(t-t_0)/\varepsilon]\|\tilde{y}_0\|\int_{t_0}^{t} [\varepsilon^{-1}\gamma(s-t_0)+B]E[\|z(s)\|^2]ds$$

where

$$0_1(\varepsilon) = (\sigma_1\|L\|^2 k_1 + \frac{\sigma_2}{\varepsilon^2} k_2 + \sigma_1\|L(t_0)\|^2 k_1 + \frac{\sigma_2}{\varepsilon^2} k_2) \frac{\varepsilon}{\alpha_{22}},$$

and since the 2nd term is bounded then

$$E[\|z(t)\|^2] \leq 0_2(\varepsilon) + \|\tilde{y}_0\| \exp[-\alpha_{22}(t-t_0)/\varepsilon] \int_{t_0}^{t} [\varepsilon^{-1}\gamma(s-t_0) + B]E\|z(s)\|^2 ds.$$

Hence, apply Corollary 1.9.1, p. 38 in [9], we have

$$E\|z(t,t_0,z_0)\|^2 \leq 0(\varepsilon), \text{ i.e.,}$$

$$v(t,t_0,v_0) = \tilde{y}(\tau,0,\tilde{y}_0) + 0(\varepsilon)$$

in the mean square sense. This completes the proof of the theorem.

Finally, we are ready to present the main result.

Theorem 4.6. Let the assumptions of Theorem 4.1 hold. There exists a positive number ε^* such that for all $t \geq t_0$, $t_0 \in R_+$ and all $0 < \varepsilon \leq \varepsilon^*$, the solution processes of (2.1) is approximated by the solution of the reduced problem (2.4) and the boundary layer problem (2.6), that is

$$\text{(4.7)} \qquad \begin{cases} x(t) = \bar{x}(t) + 0(\varepsilon), \\ y(t) = -A_{22}^{-1}(t)A_{21}(t)\bar{x}(t) + \tilde{y}(\tau) + 0(\varepsilon) \end{cases}$$

in the mean square sense.

Proof. From Theorem 4.1, we have

$$\text{(4.8)} \qquad \begin{cases} u(t) = \bar{x}(t) + 0(\varepsilon) \\ v(t) = \tilde{y}(\tau) + 0(\varepsilon) \end{cases}$$

in the sense of mean square for all $t \geq t_0$ and $0 < \varepsilon \leq \varepsilon^*$ and ε^* is determined in Lemma 3.9. From (2.8) we get

$$x = u + \varepsilon Mv$$

$$y = -Lu + (I-\varepsilon LM)v$$

which implies that

$$x(t) = u(t) + 0(\varepsilon)$$

$$y(t) = -A_{22}^{-1}(t)A_{21}(t)u(t) + v(t) + 0(\varepsilon).$$

This together with (4.8) gives

$$x(t) = \bar{x}(t) + O(\varepsilon)$$

$$y(t) = -A_{22}^{-1}(t)A_{21}(t)\bar{x}(t) + \tilde{y}(t) + O(\varepsilon)$$

in the mean square sense. This proves the theorem.

REFERENCES

[1] Bensoussan, A. "On some singular perturbation problems arising in stochastic control", Stochastic Anal. and Appl. 2 (1984), p. 13.

[2] Hong, L. D. "Stochastic perturbations of almost periodic solution for singularly perturbed systems", Rev. Roum Math. Pures et Appl. XXVIII (1983), p. 56.

[3] Khalil, H. K. and Kokotovic, P. V. "D-stability and multi-parameter singular perturbation", SIAM J. Control and Optimization 17 (1979), p. 56.

[4] Kokotovic, P. V., O'Malley, Jr., R. E. and Sannuti, P. "Singular perturbations and order reduction in control theory - an overview", Automatica 12 (1976), p. 123.

[5] Ladde, G. S. and Lakshmikantham, V. Random Differential Inequalities, Academic Press, 1981.

[6] Ladde, G. S. and Kulkarni, R. M. "Stochastic perturbations of nonlinear systems of differential equations", J. Math. Ply. Sci. 10 (1976), p. 33.

[7] Ladde, G. S. and Rajalakshmi, S. G. "Diagonalization and stability of multi-time scale singularly perturbed linear systems", J. of Appl. Math. and Comput. (in press).

[8] Ladde, G. S. and Siljak, D. D. "Multi-parameter singular perturbations of linear systems with multiple time scales", Automatica 19 (1982), p. 385.

[9] Lakshmikantham, V. and Leela, S. Differential and Integral Inequalities, Vol 1, Academic Press, 1969.

[10] O'Malley, Jr., R. E. Introduction to Singular Perturbations, Academic Press, 1974.

[11] Papanicolaov, G. C., Strook, D. and Varadhan, S. R. S. "Martingale approach to some limit theorems", in Statistical Mechanics, Dynamical Systems, and the Duke Turbulence Conference, ed. D. Ruelle, Duke University Math. Series, 3, Durham, N. C., 1977.

[12] Sastry, S. and Hijab, O. "Bifurcation in the presence of small noise", Technical Report #LIDS-P-1089, Laboratory for Information and Decision Systems, MIT, 1981.

This paper is in final form and no version of it will be submitted for publication elsewhere.

Trends in the Theory and Practice of Non-Linear Analysis
V. Lakshmikantham (Editor)

SYSTEMS OF FIRST ORDER PARTIAL DIFFERENTIAL EQUATIONS AND MONOTONE ITERATIVE TECHNIQUE*

G. S. Ladde† and A. S. Vatsala‡

†Department of Mathematics
University of Texas at Arlington
Arlington, Texas 76019

‡Department of Mathematics
Bishop College
Dallas, Texas 75241
U.S.A.

1. INTRODUCTION

Because of the fact first order partial differential equations arise naturally in modelling growth population of cells which constantly change in their properties, a study of existence, uniqueness and stability properties was initiated in [3,4]. Furthermore, in [3] monotone iterative technique was employed to obtain improvable upper and lower bounds for solutions.

In this paper, we wish to extend such results for systems of first order partial differential equations. If the coefficients of the gradient terms are different, proving existence results for the system by the method of characteristics seems to be difficult. However, if we employ monotone iterative technique, this difficulty can be eliminated, since in this case, we can reduce the study of the given system to the study of linear uncoupled systems. For this purpose, we first investigate comparison results and then develop monotone technique in the context of quasi-solutions and mixed monotone operators. One of comparison results proved provides bounds for solutions in terms of solutions of ordinary differential equations, which in turn contains as a very special case, the well known Haar's lemma [1].

2. COMPARISON RESULTS

Consider the initial value problem for a system of first order partial differential equations

(2.1) $$u_{i,t} + f_i(t,x)u_{i,x} = g_i(t,x,u), \quad u_i(0,x) = \phi_i(x),$$

where $f_i \in C[\Omega,R^k]$, $g_i \in C[\Omega\times R^n,R]$, $\Omega = [(t,x): 0 \leq t \leq T$ and $a \leq x \leq b$, $a,b,x \in R^k]$, $f_i(t,x)u_{i,x} = \sum_{j=1}^{k} f_i(t,x)u_{i,x_j}$ and $\phi_i \in C^1[[a,b],R]$, for $i \in I = \{1,2,\ldots,n\}$.

To define a mixed quasi-monotonicity of the function g, let p_i, q_i be two nonnegative integers such that $p_i + q_i = n - 1$ and we split $u \in R^n$ into the form $u = (u_i,[u]_{p_i},[u]_{q_i})$. Also let $[u,v]_i$ denote an element of R^n with the description $[u,v]_i = (u_i,[u]_{p_i},[u]_{q_i})$. Without further mention, we assume that $i \in I$ and all inequalities between vectors are componentwise.

Definition 2.1. The function g is said to possess a mixed quasi-monotone property (mqmp for short) if for each $i \in I$, $g_i(t,x,u_i,[u]_{p_i},]u]_{q_i})$ is monotone

*Research partially supported by U.S. Army Research Grant No. DAAG 29-84-G-0060

nondecreasing in $[u]_{p_i}$ and monotone nonincreasing in $[u]_{q_i}$.

We need the following comparison result for our discussion.

Theorem 2.1. Assume that

(A_0) $$\alpha_i,\beta_i \in C^1[\Omega,R], \quad \alpha_{i,t} + f_i(t,x)\alpha_{i,x} \le g_i(t,x,[\alpha,\beta]_i)$$
$$\beta_{i,t} + f_i(t,x)\beta_{i,x} \ge g_i(t,x,[\beta,\alpha]_i)$$

for $(t,x) \in \Omega$ and $i \in I$;

(A_1) For each $i \in I$, $f_i(t,x)$ is quasimonotone nonincreasing in x for each t, $0 \le f_i(t,a)$, and $0 \le f_i(t,b)$;

(A_2) For each $i \in I$,

$$g_i(t,x,u_i,[u]_{p_i},[v]_{q_i})-g_i(t,x,\bar{u}_i,[u]_{p_i},[v]_{q_i}) \le L_i(u_i-\bar{u}_i)$$

for some $L_i \ge 0$, whenever $\alpha_{0,i} \le \bar{u}_i \le u_i \le \beta_{0,i}$, where

$$\alpha_{0,i} = \min\{\min_{(t,x)\in\Omega} \alpha(t,x), \min_{(t,x)\in\Omega} \beta(t,x)\}, \text{ and}$$

$$\beta_{0,i} = \max\{\max_{(t,x)\in\Omega} \beta(t,x), \max_{(t,x)\in\Omega} \alpha(t,x)\}.$$

(A_3) $g(t,x,u)$ possesses a mixed monotone property. Then $\alpha(t,x) \le \beta(t,x)$ on Ω, provided $\alpha(0,x) \le \beta(0,x)$, $a \le x \le b$.

Proof. We first prove the theorem for strict inequalities. If the conclusion is not true, then there exists a (t_0,x_0) and an index j, with $t_0 > 0$, $x_0 \in [a,b]$ such that

$$\alpha_j(t_0,x_0) = \beta_j(t_0,x_0), \quad \alpha_i(t_0,x_0) \le \beta_i(t_0,x_0) \quad \text{for } i \ne j,$$

and $\alpha_j(t_0-h,x_0) < \beta_j(t_0-h,x_0)$ for sufficiently small $h > 0$. Also $\alpha_j(t_0,x) \le \beta_j(t_0,x)$ for $a \le x \le b$. If $a < x_0 < b$, we also have $\alpha_{j,t}(t_0,x_0) \ge \beta_{j,t}(t_0,x_0)$ and $\alpha_{j,x_i}(t_0,x_0) = \beta_{j,x_i}(t_0,x_0)$ for $i \in I$. In this case, using (A_0) and (A_3) we are led to the following contradiction

$$\begin{aligned} g_j(t_0,x_0,\alpha_j(t_0,x_0),[\alpha(t_0,x_0)]_{p_j},[\beta(t_0,x_0)]_{q_j}) &\ge \alpha_{j,t}(t_0,x_0)+f_j(t_0,x_0)\alpha_{j,x}(t_0,x_0) \\ &\ge \beta_{j,t}(t_0,x_0)+f_j(t_0,x_0)\beta_{j,x}(t_0,x_0) \\ &> g_j(t_0,x_0,\beta_j(t_0,x_0),[\beta(t_0,x_0)]_{p_j},[\alpha(t_0,x_0)]_{q_j}) \\ &\ge g_j(t_0,x_0,\alpha_j(t_0,x_0),[\alpha(t_0,x_0)]_{p_j},[\beta(t_0,x_0)]_{q_j}). \end{aligned}$$

If, on the other hand, for some k, $x_{0,k} = b_k$ and $x_{0,\mu}$, $\mu \neq k$ and $a < x_0$, then we have

$$\alpha_{j,x_\mu}(t_0,x_0) = \beta_{j,x_\mu}(t_0,x_0), \quad \mu \ne j \text{ and } \alpha_{j,x_k}(t_0,x_0) \ge \beta_{j,x_k}(t_0,x_0).$$

Hence using assumption (A_1) we obtain $f_j^k(t_0,x_0) \ge f_j^k(t_0,b) \ge 0$, which implies

$f_j^k(t_0,x_0)\alpha_{j,x_k}(t_0,x_0) \geq f_j^k(t_0,x_0)\beta_{j,x_k}(t_0,x_0)$. Consequently, we arrive at a similar contradiction. Similar arguments hold if $x_{0,k} = a_k$ and so on. This proves $\alpha(t,x) < \beta(t,x)$ on Ω.

If one of the inequalities in (A_0) is not strict, we set $\tilde{\alpha}_j(t,x) = \alpha_j(t,x) - \varepsilon e^{2L_j t}$ for $j \in I$, and note that $\tilde{\alpha}_j < \alpha_j$. Then using (A_0), (A_2) and (A_3), we see that

$$\begin{aligned}\tilde{\alpha}_{j,t} + f_j(t,x)\tilde{\alpha}_{j,x} &= \alpha_{j,t} + f_j(t,x)\alpha_{j,x} - 2L_j\varepsilon e^{2L_j t}\\ &\leq g_j(t,x,\alpha_j,[\alpha]_{p_j},[\beta]_{q_j}) - 2L_j\varepsilon e^{2L_j t}\\ &\leq g_j(t,x,\tilde{\alpha}_j,[\alpha]_{p_j},[\beta]_{q_j}) + L_j\varepsilon e^{2L_j t} - 2L_j\varepsilon e^{2L_j t}\\ &< g_j(t,x,\tilde{\alpha}_j,[\tilde{\alpha}]_{p_j},[\beta]_{q_j}),\end{aligned}$$

and $\tilde{\alpha}_j(0,x) < \phi_j(x)$ on Ω. Arguing as before, we get $\tilde{\alpha}_j(t,x) < \beta_j(t,x)$ on Ω. Taking limit as $\varepsilon \to 0$, we then obtain $\alpha(t,x) \leq \beta(t,x)$ on Ω, and the proof is complete.

Next we prove another comparison theorem which is useful in obtaining bounds for solutions of (2.1) in terms of solutions of systems of ordinary differential equations.

Theorem 2.2. Assume that

(B_0) $$u_i \in C^1[\Omega,R^n],$$

(2.2) $$|u_{i,t}(t,x)| \leq h_i(t,x,|u|) + F_i(t,x)|u_{i,x}|$$

for $(t,x) \in \Omega$, $i \in I$, where $h_i \in C[\Omega\times R_+^n,R_+]$ and $F_i \in C[\Omega,R_+^k]$;

(B_1) For each $i \in I$, $F_i(t,x)$ is quasimonotone nondecreasing in x for each t, with $0 \leq F_i(t,a)$, $F_i(t,b) = 0$,

(B_2) For each $i \in I$, $h_i(t,x,|u|)) \leq H_i(t,u)$, for $u \geq 0$, where $H \in C[[0,T],R_+^n]$ such that $H(t,0) \equiv 0$, and $H(t,u)$ is quasimonotone nondecreasing in u for each t.

Then $-y_0 \leq u(0,x) \leq y_0$ implies $-r(t,0,y_0) \leq u(t,x) \leq r(t,0,y_0)$ for all $(t,x) \in \Omega$, where $r(t,0,y_0)$ is the maximal solution of the IVP

(2.3) $$y' = H(t,y),\ y(0) = y_0$$

existing for $t \in [0,T]$.

Proof. Let $r(t,\varepsilon) = r(t,0,y_0,\varepsilon)$ be any solution of the IVP

$$r_i' = H_i(t,r) + \varepsilon,\quad r_i(0) = y_i(0) + \varepsilon,$$

for sufficiently small $\varepsilon > 0$. Since it is known that $\lim_{\varepsilon\to 0} r(t,0,y_0,\varepsilon) = r(t,0,y_0)$, uniformly, where $r(t,0,y_0)$ is the maximal solution of (2.3), it is sufficient to show that $-r_i(t,\varepsilon) < u_i(t,x) < r_i(t,\varepsilon)$. If the conclusion is not true, then there exists an index j and a $(t_1,x_1)\in\Omega$ such that either $u_j(t_1,x_1) = r_j(t_1,\varepsilon)$

or $u_j(t_1,x_1) = -r_j(t_1,\varepsilon)$ and $-r_i(t_1,\varepsilon) \le u_i(t_1,x_1) \le r_i(t_1,\varepsilon)$ for $i \ne j$. We shall consider the first case. The proof is similar for the latter. It is clear that $t_1 > 0$ and $u_j(t_1-h,x) < r_j(t_1-h,\varepsilon)$ for sufficiently small $h > 0$ for $a \le x \le b$. If $a < x_1 < b$, we have $u_{j,t}(t_1,x_1) \ge r_j'(t_1,\varepsilon)$ and $u_{j,x}(t_1,x_1) = 0$. Hence we arrive at

$$\begin{aligned} H_j(t_1,r(t_1,\varepsilon)) &< H_j(t_1,r(t_1,\varepsilon)) + \varepsilon = r_j'(t_1,\varepsilon) \le u_{j,t}(t_1,x_1) \\ &\le h_j(t_1,x_1,|u_j(t_1,x_1)|) \le h_j(t_1,x_1,r(t_1,\varepsilon)) \\ &\le H_j(t_1,r(t_1,\varepsilon)), \end{aligned}$$

using (B_2). This is a contradiction. If $x_{1,\mu} > b$ for some μ, and $a_i < x_{1,i} < b_i$, for $i \ne \mu$, then $u_{j,x_{1,\mu}}(t_1,x_1) \ge 0$. Using the quasimonotone character of F, we have

$$\begin{aligned} F_j^{\mu}(t_1,x_{1,1},x_{1,2},\ldots,x_{1,\mu},\ldots,x_{1,k}) &= F_j^{\mu}(t_1,x_{1,1},\ldots,b_{\mu},\ldots,x_{1,k}) \\ &\le F_j^{\mu}(t_1,b_1,b_2,\ldots,b_{\mu},\ldots,b_n) \le 0 \end{aligned}$$

which leads to a contradiction as before. Similarly if on the other hand, we have $x_{1,\mu} = a_{\mu}$ for some μ, a similar argument will lead to a contradiction. Hence the proof is complete.

Let us consider the example,

$$(2.4)\qquad \begin{aligned} u_{1,t} - F_1(t,x)u_{1,x} &= a_{11}(t,x)u_1 + a_{12}(t,x)u_2 + b_1(t,x), \quad u_1(0,x) = \phi_1(x) \\ u_{2,t} - F_2(t,x)u_{2,x} &= a_{2,1}(t,x)u_1 + a_{22}(t,x)u_2 + b_2(t,x), \quad u_2(0,x) = \phi_2(x) \end{aligned}$$

where F_1, F_2 satisfy the assumptions of Theorem 2.2, and the functions a_{ij} and b_i, $i,j = 1,2$ are continuous on Ω. Let $u = (u_1,u_2)$ be any solution of (2.4). Then we have

$$\begin{aligned} |u_{1,t}| &\le F_1(t,x)|u_{1,x}| + a(|u_1|+|u_2|) + b \\ |u_{2,t}| &\le F_2(t,x)|u_{2,x}| + a(|u_1|+|u_2|) + b \end{aligned}$$

where $a = \max|a_{ij}(t,x)|$, $b = \max|b_i(t,x)|$. It is clear that all the hypotheses of Theorem 2,2 are verified. Consequently, by Theorem 2.2, we obtain the following estimates

$$|u_1(t,x)| \le \phi_{1,0}\left(\frac{e^{2at}-1}{2}\right) + \phi_2(0)\left(\frac{e^{2at}+1}{2}\right) - b/2a$$

$$|u_2(t,x)| \le \phi_{1,0}\left(\frac{e^{2at}+1}{2}\right) + \phi_{2,0}\left(\frac{e^{2at}-1}{2}\right) - b/2a,$$

where $\phi_{1,0} = \max_{x\in\Omega}|\phi_1(x)|$, $\phi_{2,0} = \max_{x\in\Omega}|\phi_2(x)|$.

Remark. If in Theorem 2.2, we suppose that $x \in R^n$ on Ω, then we see that the assumption (B_1) is no longer needed. Corresponding to this situation if we also suppose that $F_i = \lambda_i \ge 0$ the conclusion of the above example is precisely Haar's lemma. See [1].

3. MONOTONE ITERATIVE TECHNIQUE

Now we are in a position to describe the monotone iterative technique which yields monotone sequences that converge uniformly to the unique solution of (2.1). This is precisely the following result.

Theorem 3.1. Assume that (A_0), (A_1) and (A_3) hold with $\alpha \leq \beta$ on Ω. Suppose further that

(A_4) for each $i \in I$, there exists a unique solution $x(t,t_0,x_0)$ of

(3.1) $$x^1 = f_i(t,x), \quad x(t_0) = x_0^i,$$

on $0 \leq t \leq T$, where $x(t,t_0,x_0^i)$ is continuously differentiable with respect to (t_0,x_0^i) and satisfies the relation

$$\frac{\partial x}{\partial t_0}(t,t_0,x_0^i) + \frac{\partial x}{\partial x_0}(t,t_0,x_0^i)f_i(t_0,x_0^i) = 0;$$

and

(A_5) $g_{i,x}$, $g_{i,u}$ exist and are continuous on $\Omega \times R$ such that for some $M_i > 0$

$$g_i(t,x,u_i,[v]_{p_i},[v]_{q_i}) - g_i(t,x,\bar{u}_i,[v]_{p_i},[v]_{q_i}) \geq -M_i(u_i-\bar{u}_i)$$

whenever $\alpha_i \leq \bar{u}_i \leq u_i \leq \beta_i$ and $\alpha \leq v \leq \beta$.

Then there exists monotone sequences $\{\alpha_n(t,x)\}$, $\{\beta_n(t,x)\}$ which converge uniformly to the unique solution u of (2.1) such that

$$\alpha \leq \alpha_1 \leq \dots \leq \alpha_n \leq u \leq \beta_n \leq \dots \leq \beta_1 \leq \beta \text{ on } \bar{\Omega}.$$

Proof. Consider the uncoupled linear system of IVP

(3.2) $$u_{i,t} + f_i(t,x)u_{i,x} = G_i(t,x,u;\eta,\mu), \quad u_i(0,x) = \phi_i(x)$$

where $G_i(t,x,u;\eta,\mu) = g_i(t,x,\eta_i,[\eta]_{p_i},[\mu]_{q_i}) - M_i(u_i-\eta_i)$ and $\eta,\mu \in C[\Omega,R]$ such that $\alpha \leq \eta$, $\mu \leq \beta$. We first prove that there exists a unique solutions to the IVP (3.2). Observe that, by (A_4) and (A_5), it follows that $x(t,0,x_0^i)$, $y_i(t,0,\phi_i(x_0^i);x_0^i)$ are unique solutions of (3.1) and

(3.3) $$y_i' = G_i(t,x(t,0,x_0^i),y_i), \quad y_i(0) = \phi_i(x_0^i)$$

respectively. If $x = x(t,0,x_0^i)$, then because of uniqueness $x_0^i = x(0,t,x)$. Also, the solution $(x(t,0,x_0^i),y(t,0,\phi_i(x_0^i);x_0^i))$ of (3.1) and (3.3) is a characteristic equation of (3.2) for each $i \in I$. Hence, for each solution of (3.1) and (3.3), we set

$$u(t,x(t,0,x_0^i)) = y_i(t,0,\phi_i(x_0^i);x_0^i)$$

then we have

(3.5) $$u_i(t,x) = y_i(t,0,\phi_i(x(0,t,x));\ x(0,t,x)), \quad (t,x) \in \Omega.$$

Using assumptions (A_4) and (A_5), it is easy to show that $u_i(t,x)$ defined by (3.5) satisfies (3.2), which proves the existence of a solution of (3.2). In order to establish uniqueness of solutions of (3.2), suppose, if possible, that $u_{1,i}(t,x)$

and $u_{2,i}(t,x)$ are two solutions of (3.2) on Ω. Then setting $\alpha_i = u_{1,i}$, $\beta_i = u_{2,i}$ and applying Theorem 2.1, with $n = 1$, it follows $u_{1,i}(t,x) \le u_{2,i}(t,x)$ on Ω. Similarly we can show that $u_{2,i}(t,x) \le u_{1,i}(t,x)$ on Ω proving uniqueness of solutions of (3.2).

As a result there exists a unique solution $u_i(t,x)$ of (3.2) on Ω for every $\eta,\mu \in C[\Omega,R^n]$ such that $\alpha \le \eta$, $\mu \le \beta$.

We define a mapping A by $A[\eta,\mu] = u$, where u is the unique solution of (3.2) corresponding to η, μ such that $\alpha \le \eta$, $\mu \le \beta$. Concerning this mapping, we shall show that (i) $\alpha \le A[\alpha,\beta]$, $\beta \ge A[\beta,\alpha]$; (ii) A satisfies a mixed monotone property on the sector $[\alpha,\beta]$. That is for fixed μ, if $\eta_1 < \eta_2$, $A[\eta_1,\mu] \le A[\eta_2,\mu]$ and for fixed η, if $\mu_1 \le \mu_2$, then $A[\eta,\mu_1] \ge A[\eta,\mu_2]$.

Let $\eta = \alpha$, $\mu = \beta$ and let $A[\alpha,\beta] = \alpha_1$ where α_1 is the unique solution of (3.2). Then we have

$$\alpha_{i,t} + f_i(t,x)\alpha_{i,x} \le G_i(t,x,\alpha;\alpha,\beta), \quad \alpha_i(0,x) \le \phi_i(x)$$

and $\alpha_{1,i,t} + f_i(t,x)\alpha_{1,i,x} = G_i(t,x,\alpha;\alpha,\beta)$, $\alpha_i(0,x) = \phi_i(x)$ on Ω. By Theorem 2.1, with $n = 1$, this implies that $\alpha_i \le \alpha_{1,i}$ for eacj $i \in I$. Hence $\alpha \le \alpha_1 = A[\alpha,\beta]$. Similarly we can show that $\beta \ge A[\beta,\alpha]$ proving (i). To prove (ii), let $\eta_1,\eta_2,\mu \in C[\Omega,R^n]$ be such that $\alpha \le \eta_1 \le \eta_2 \le \beta$ and $\alpha \le \mu \le \beta$. Let $A[\eta_1,\mu] = u_1$ and $A[\eta_2,\mu] = u_2$, where u_1,u_2 are unique solutions of (3.2) relative to $\eta = \eta_1$, $\mu = \mu$ and $\eta = \eta_2$, $\mu = \mu$ respectively. Then

$$\begin{aligned} u_{1,i,t}+f_i(t,x)u_{1,i,x} &= G_i(t,x,u_1;\eta_1,\mu) \\ &= g_i(t,x,\eta_{1,t},[\eta_1]_{p_1},[\mu]_{q_i})-M_i(u_{1,i}-\eta_{1,i}) \\ &\le g_i(t,x,\eta_{1,i},[\eta_2]_{p_i},[\mu]_{q_i})-M_i(u_{1,i}-\eta_{1,i}) \\ &\le g_i(t,x,\eta_{2,i},[\eta_2]_{p_i},[\mu]_{q_i})-M_i(u_{1,i}-\eta_{1,i})+M_i(\eta_{2,i}-\eta_{1,i}) \\ &= g_i(t,x,\eta_{2,i},[\eta_2]_{p_i},[\mu]_{q_i})-M_i(u_{1,i}-\eta_{2,i}) = G_i(t,x,u_1;\eta_2,\mu) \end{aligned}$$

using (A_3) and (3.1). Also $u_{2,i,t} + f_i(t,x)u_{2,i,x} = G_i(t,x,u_2;\eta_1,\mu)$. Hence by Theorem 2.1, with $n = 1$, we have $u_1 \le u_2$ on Ω. This proves $A[\eta_1,\mu] \le A[\eta_2,\mu]$. Similarly one can show that $A[\eta,\mu_1] \ge A[\eta,u_2]$ whenever $\mu_1 \le \mu_2$. Consequently it follows $A[\eta,\mu] \le A[\eta,\eta] \le A[\mu,\eta]$ whenever $\eta \le \mu$.

We now define the sequences $\{\alpha_n(t,x)\}$, $\{\beta_n(t,x)\}$ with $\alpha = \alpha_0$, $\beta = \beta_0$ such that $\alpha_{n+1} = A[\alpha_n,\beta_n]$, $\beta_{n+1} = A[\beta_n,\alpha_n]$. Because of the properties (i), (ii) of A, it is easy to concluded that $\alpha_0 \le \alpha_1 \le \dots \le \alpha_n \le \beta_n \le \dots \le \beta_1 \le \beta_0$ on Ω.

Consider the sequences $\{\alpha_n(t,x)\}$ and note that $\alpha_{n,i}(t,x) = y_{n,i}(t,0,\phi_i(x(0,t,x));x(0,t,x)]$ on Ω where $y_{n,i} = y_{n,i}(t,0,{}_i(x_0^i);x_0^i)$ is the unique solution of (3.3). Thus $\alpha_{n,i}(t,x(t,0,x_0^i) = y_{n,i}(t,0,\phi_i(x_0^i);x_0^i)$ and $\alpha \le y_n \le \beta$. Similarly $\beta_{n,i}(t,x(t,0,x_0^i) = y_{n,i}(t,0,\phi_i(x_0^i);x_0^i)$ on Ω and $\alpha \le y_n \le \beta$. Since the sequences $\{y_n\}$ and $\{Y_n\}$ are monotone sequences, it is easy to conclude $y_{n,i}(t,0,\phi_i(x_0^i);x_0^i)$ and $Y_{n,i}(t,0,\phi_i(x_0^i);x_0^i)$ converges uniformly as $n \to \infty$. Suppose that $\lim_{n\to\infty} y_{n,i}(t,0,\phi_i(x_0^i);x_0^i) = y_i^*(t,0,\phi_i(x_0^i);x_0^i)$ on $0 \le t \le T$ and $\lim_{n\to\infty} Y_{n,i}(t,0,\phi_i(x_0^i);x_0^i) = Y_i^*(t,0,\phi_i(x_0^i);x_0^i)$. Then we define $\rho_i(t,x) = y_i^*((t,0,\phi_i(x_i(0,t,x));x(0,t,x))$ on Ω and $r_i(t,x) = Y_i^*(t,0,\phi_i(x(0,t,x));x(0,t,x))$ on Ω. It is easy to see that ρ_i and r_i satisfy, in view of the assumptions (A_4) and (A_5) the relations

$$\rho_{i,t} + f_i(t,x)\rho_{i,x} = g_i(t,x,\rho_i,[\rho]_{p_i},[r]_{q_i}), \quad \rho_i(0,x) = \phi_i(x) \tag{3.4}$$

$$r_{i,t} + f_i(t,x)r_{i,x} = g_i(t,x,r_i,[r]_{p_i},[\rho]_{q_i}), \quad r_i(0,x) = \phi_i(x) \tag{3.5}$$

showing that ρ, r are coupled solutions. It is clear that $\rho \le r$, we shall show that $r \le \rho$ so that $u = r = \rho$ will be the unique solution of (2.1). For this purpose, it is easy to check that the assumptions of Theorem 2.1 are satisfied with $\alpha = r$, $\beta = \rho$. This implies $r \le \rho$ and the proof of the Theorem is complete.

Consider the following example

$$u_t + f_1(t,x)u_x = au - uv, \quad u(0,x) = \phi_1(x),$$

$$v_t + f_2(t,x)v_x = bv + uv, \quad v(0,x) = \phi_2(x).$$

where $a, b > 0$. Suppose that f_1, f_2 satisfy the assumptions (A_1) and (A_4). Letting $g_1 = au - uv$, $g_2 = bv + uv$, it is clear that g_1, g_2 satisfy (A_3).

Let $\alpha_1 = 0$, $\alpha_2 = 0$, $\beta_1 = \phi_0 e^{at}$, and $\beta_2 = \phi_0 e^{bt+\frac{\phi_0}{a}e^{at}}$, where $\phi_0 = \max(\phi_1,\phi_2)$. Furthermore since g_i are differentiable, it is not difficult to show g_i satisfies a Lipschitz condition with constant M_i.

Clearly, α, β are coupled quasi-lower and upper solutions. Consequently, by Theorem 3.1, there exists a unique solution u which is the uniform limit of monotone sequences $\{\alpha_n\}$ and $\{\beta_n\}$.

REFERENCES

[1] Hartman, P. Ordinary Differential Equations (John Wiley and Sons, Inc., New York, 1964).

[2] Lakshmikantham, V. and Leela, S. Differential and Integral Inequalities, Vol. II (Academic Press, New York, 1969).

[3] Lakshmikantham, V., Oguztoreli, M. N. and Vatsala, A. S. Monotone iterative technique for partial differential equations of first order, Jour. Math. Anal. and Appl. (to appear).

[4] Lasoto, A. Stable and chaotic solutions of a first order partial differential equation.

This paper is in final form and no version of it will be submitted for publication elsewhere.

Trends in the Theory and Practice of Non-Linear Analysis
V. Lakshmikantham (Editor)

IMPROVED A POSTERIORI ERROR BOUNDS FOR QUASILINEAR BOUNDARY-VALUE PROBLEMS BY THE METHOD OF PSEUDOLINEAR EQUATIONS

John E. Lavery

Department of Mathematics and Statistics
Case Western Reserve University
Cleveland, Ohio 44106

Abstract. The standard a posteriori error bounds based on complementary energy principles for approximate solutions of a certain class of quasilinear elliptic boundary-value problems are discussed. It is shown that these a posteriori error bounds can be significantly improved by using the method of pseudolinear equations to solve the given problem and its conjugate (dual) problem instead of solving these problems separately by other methods. Extensions to more general classes of quasilinear elliptic boundary-value problems, two-point boundary-value problems and their numerical analogues are indicated.

1. INTRODUCTION

A posteriori error bounds based on complementary energy principles [1,2,3,4, 5,6] are important computationally as tools for reliable error measurement and for stopping iterative calculations. In the present paper, we consider whether the a posteriori error bounds for quasilinear elliptic boundary-value problems and quasilinear two-point boundary-value problems can be improved by using the method of pseudolinear equations instead of standard strategies to solve the given problem. A second-order elliptic boundary-value problem in two dimensions is used in Sections 2-4 to introduce the concept. Analogous results for any problem that has a convex energy functional and a conjugate (dual) problem with a complementary energy principle and that admits using the method of pseudolinear equations are mentioned in Section 5.

2. CONJUGATE PROBLEMS AND COMPLEMENTARY ENERGIES

Consider the quasilinear elliptic boundary-value problem that consists in finding U such that

$$-[p_1(U_x,U_y,x,y)]_x - [p_2(U_x,U_y,x,y)]_y = 0 \quad \text{in } D\subset R^2, \quad U|_{\partial D} = g. \tag{2.1}$$

The weak formulation of the Dirichlet problem (2.1) that will be used in this paper consists in finding $U\in H^1_g(D) := \{u \in H^1(D)|\ u|_{\partial D} = g\}$ such that

$$\iint_D [p_1(U_x,U_y,x,y)h_x + p_2(U_x,U_y,x,y)h_y]dD = 0 \tag{2.2}$$

for all h in $H^1_0(D)$. We assume conditions on the domain D (Jordan domain with Lipschitz-continuous boundary), the functions p_1 and p_2 (Jacobian matrix symmetric and uniformly positive definite from above and below) and g (an element of $H^{1/2}(\partial D)$) that make problem (2.2) a well-posed variational problem with a convex energy functional given by

$$E(u) = \int_0^1\iint_D [p_1(tu_x,tu_y,x,y)u_x + p_2(tu_x,tu_y,x,y)u_y]dD\, dt \tag{2.3}$$

over $H^1_g(D)$. The unique solution U of problem (2.2) occurs at the minimum of E. The conditions on D, p_1, p_2 and g that are stated above in abbreviated form as well as the theory presented in this section and the next have been developed in [2,3].

Define from the solution U a function V by the relations

$$V_y = -p_1(U_x,U_y,x,y), \qquad V_x = p_2(U_x,U_y,x,y). \tag{2.4a}$$

The necessary and sufficient condition that such a V exist in $L^1_2(D):=H^1(D)$ modulo constants is that $(V_x)_y = (V_y)_x$, which is nothing more than equality (2.1). Let $\hat{p}_1$, $\hat{p}_2$ be the inverse functions for p_1, p_2 with respect to the first two arguments. Equalities (2.4a) can then be written in the equivalent form

$$U_x = \hat{p}_1(-V_y,V_x,x,y), \qquad U_y = \hat{p}_2(-V_y,V_x,x,y) \tag{2.4b}$$

(notation altered from the notations in [2] and [3]). From relations (2.4b), one finds that

$$[\hat{p}_1(-V_y,V_x,x,y)]_y - [\hat{p}_2(-V_y,V_x,x,y)]_x = [U_x]_y - [U_y]_x = 0 \text{ in } D, \tag{2.5a}$$

$$-n_2\hat{p}_1(-V_y,V_x,x,y) + n_1\hat{p}_2(-V_y,V_x,x,y) = -n_2U_x + n_1U_y = \frac{dg}{ds} \text{ on } \partial D \tag{2.5b}$$

((n_1,n_2) is the outward normal, dg/ds is the arclength derivative of g in a counterclockwise direction). Equalities (2.5) are a quasilinear elliptic equation for V with Neumann boundary conditions. The weak formulation of this problem, namely,

$$\iint_D[\hat{p}_1(-V_y,V_x,x,y)(-h_y) + \hat{p}_2(-V_y,V_x,x,y)h_x]dD - (\tfrac{dg}{ds},h|_{\partial D}) = 0 \tag{2.6}$$

for all h in $L^1_2(D)$, is the problem conjugate (dual) to problem (2.2). The conditions assumed on p_1, p_2 imply that the $\hat{p}_1$, $\hat{p}_2$ are such that problem (2.6) is a well posed variational problem with a convex energy functional

$$F(v) = \int_0^1\iint_D[\hat{p}_1(-tV_y,tV_x,x,y)(-V_y) + \hat{p}_2(-tV_y,tV_x,x,y)V_x]dD - (\tfrac{dg}{ds},V|_{\partial D}). \tag{2.7}$$

The unique solution V of problem (2.6) occurs at the minimum of F. This V satisfies relations (2.4).

It is known (see Theorem 1 of [2], Theorem 1 of [3]) that the energy functional E of problem (2.2) and the complementary energy functional F of the conjugate problem (2.6) satisfy the relation

$$E(U) + F(V) + c = 0 , \tag{2.8}$$

where c is an a priori known constant. It is on equality (2.8) that the a posteriori error bounds described in the next section are based.

3. A POSTERIORI ERROR BOUNDS

A posteriori error bounds for approximate solutions u of problem (2.2) and approximate solutions v of problem (2.6) are derived in the following manner. Since the minima of E and F occur at U and V, respectively,

$$E(u) + F(v) + c \geq E(u) + F(V) + c = E(u) - E(U) , \tag{3.1a}$$

$$E(u) + F(v) + c \geq F(v) + E(U) + c = F(v) - F(V) \tag{3.1b}$$

(equality (2.8) is used here). Now, by the Taylor formula applied to the functions $E(U+t(u-U))$ and $F(V+t(v-V))$ as functions of a real variable t, one obtains

$$E(u) - E(U) = E'(U,u-U) + \tfrac{1}{2}E''(U+t_1(u-U),u-U,u-U) , \tag{3.2a}$$

$$F(v) - F(V) = F'(V,v-V) + \tfrac{1}{2}F''(V+t_2(v-V),v-V,v-V) , \tag{3.2b}$$

where $E'(u,h)$, $E''(u,k,h)$ and $F'(v,h)$, $F''(v,k,h)$ are the first and second differentials of $E(u)$ and $F(v)$ and t_1 and t_2 are some numbers between 0 and 1. Since the minima of E and F occur at U and V,

$$E'(U,h) = 0, \qquad F'(V,h) = 0 \tag{3.3}$$

for all h in the appropriate spaces. Let μ and $\hat{\mu}$ denote the smallest eigenvalues of the Jacobian matrices of $\{p_1,p_2\}$ and $\{\hat{p}_1,\hat{p}_2\}$, respectively. Then

$$E''(\tilde{u},h,h) \geq \mu \|h\|^2 := \mu\iint_D [h_x^2 + h_y^2]\,dD, \qquad F''(\tilde{v},h,h) \geq \hat{\mu}\|h\|^2 \tag{3.4}$$

for all $\tilde{u}$, $\tilde{v}$ and h in the appropriate spaces. Relations (3.1)-(3.4) yield the a posteriori error bounds

$$E(u)+F(v)+c \geq \frac{\mu}{2}\|u - U\|^2 , \qquad E(u)+F(v)+c \geq \frac{\hat{\mu}}{2}\|v - V\|^2 \tag{3.5}$$

or, equivalently,

$$\|u - U\| \leq \sqrt{\frac{2}{\mu}\,[E(u)+F(v)+c]} , \qquad \|v - V\| \leq \sqrt{\frac{2}{\hat{\mu}}\,[E(u)+F(v)+c]} . \tag{3.6}$$

4. IMPROVED A POSTERIORI ERROR BOUNDS

Perusing the relations from which inequalities (3.6) were obtained, one observes that all of these relations are sharp except the initial inequalities (3.1). In (3.1), something has been "given away" by replacing $F(v)$ by $F(V)$ and $E(u)$ by $E(U)$. We propose now to redo the logic in the following manner. By (2.8),

$$E(u) + F(v) + c = E(u) - E(U) + F(v) - F(V). \tag{4.1}$$

By Taylor expansion (equalities (3.2)) and use of relations (3.3) and (3.4), one obtains that

$$E(u) + F(v) + c \geq \frac{\mu}{2}\|u - U\|^2 + \frac{\hat{\mu}}{2}\|v - V\|^2 . \tag{4.2}$$

Inequality (4.2) is an improvement over inequalities (3.5). However, (4.2) cannot be used to obtain a posteriori error bounds that are better than (3.6) unless lower bounds for $\|v-V\|$ and $\|u-U\|$ can be given. When one solves problems (2.2) and (2.6) separately, as is usually the case, such lower bounds are not available and the a posteriori error bounds in (3.6) are the best that can be obtained.

Let us now consider whether the method of pseudolinear equations can be of help in obtaining improved a posteriori error bounds. Let there be given two symmetric 2x2 matrix-valued functions $Q(x,y)$ and $\hat{Q}(x,y)$ the elements of which are in $L_\infty(D)$ and that are positive definite for almost all (x,y) in D. We assume that the smallest eigenvalues of Q and $\hat{Q}$ are essentially uniformly bounded below by positive constants κ and $\hat{\kappa}$, respectively. Let there be given an approximate solution

$$u^{(i)} \in H_g^1(D)$$

of problem (2.2). From this $u^{(i)}$, define an approximate solution $v^{(i)}$ of problem (2.6) as the element of $L_2^1(D)$ that minimizes the $L_2(D)$ error with weight $\hat{Q}$ in the conjugation relations (2.4a) in which U and V are replaced by $u^{(i)}$ and v, that is, minimizes the quadratic functional

$$\iint_D \begin{pmatrix} v_y + p_1(u_x^{(i)},u_y^{(i)},x,y) \\ v_x - p_2(u_x^{(i)},u_y^{(i)},x,y) \end{pmatrix}^T \hat{Q} \begin{pmatrix} v_y + p_1(u_x^{(i)},u_y^{(i)},x,y) \\ v_x - p_2(u_x^{(i)},u_y^{(i)},x,y) \end{pmatrix} dD \tag{4.3}$$

over v in $L_2^1(D)$. This minimization problem is equivalent to a linear elliptic boundary-value problem with Neumann boundary conditions.

From the $v^{(i)}$ just obtained, define the next approximation $u^{(i+1)}$ of U to be the element of $H_g^1(D)$ that minimizes the $L_2(D)$ error with weight Q in the equalities (2.4b) in which U and V are replaced by u and $v^{(i)}$, respectively, that is, minimizes the quadratic functional

$$\iint_D \begin{pmatrix} u_x - \hat{p}_1(-v_y^{(i)},v_x^{(i)},x,y) \\ u_y - \hat{p}_2(-v_y^{(i)},v_x^{(i)},x,y) \end{pmatrix}^T Q \begin{pmatrix} u_x - \hat{p}_1(-v_y^{(i)},v_x^{(i)},x,y) \\ u_y - \hat{p}_2(-v_y^{(i)},v_x^{(i)},x,y) \end{pmatrix} dD \tag{4.4}$$

over u in $H_g^1(D)$. This minimization problem is equivalent to a linear elliptic boundary-value problem with Dirichlet boundary conditions.

The method of pseudolinear equations consists in computing the sequence

$$u^{(0)} \longrightarrow v^{(0)} \longrightarrow u^{(1)} \longrightarrow v^{(1)} \longrightarrow u^{(2)} \longrightarrow \dots \tag{4.5}$$

starting from any $u^{(0)}$ in $H_g^1(D)$. The elements of sequence (4.5) satisfy the inequalities

$$\|u^{(i)} - U\| \leq r\|v^{(i-1)} - V\|, \quad \|v^{(i)} - V\| \leq \hat{r}\,\|u^{(i)} - U\|, \tag{4.6}$$

where r and $\hat{r}$ are constants that depend on the functions p_1, p_2, $\hat{p}_1$ and $\hat{p}_2$ and on the coefficient matrices Q and $\hat{Q}$ (see [2,3], keeping in mind that different notation is used there). The method of pseudolinear equations converges as long as

$$r\hat{r} < 1 . \tag{4.7}$$

Inequalities (4.6) yield the lower bounds needed to utilize inequality (4.2) effectively. Inequalities (4.2) and (4.6) imply

$$E(u^{(i)}) + F(v^{(i-1)}) + c \geq \frac{\mu}{2}\|u^{(i)} - U\|^2 + \frac{\hat{\mu}}{2r^2}\|u^{(i)} - U\|^2 , \tag{4.8a}$$

$$E(u^{(i)}) + F(v^{(i)}) + c \geq \frac{\mu}{2\hat{r}^2}\|v^{(i)} - V\|^2 + \frac{\hat{\mu}}{2}\|v^{(i)} - V\|^2 \tag{4.8b}$$

and hence the a posteriori error bounds

$$\|u^{(i)} - U\| \leq \left\{ \frac{2r^2}{\mu r^2+\hat{\mu}} [E(u^{(i)}) + F(v^{(i-1)}) + c] \right\}^{\frac{1}{2}} , \tag{4.9a}$$

$$\|v^{(i)} - V\| \leq \left\{ \frac{2\hat{r}^2}{\mu+\hat{\mu}\hat{r}^2} [E(u^{(i)}) + F(v^{(i)}) + c] \right\}^{\frac{1}{2}} . \tag{4.9b}$$

The bounds in inequalities (4.9) are smaller than those in (3.6) by factors of

$$\sqrt{\frac{\mu r^2}{\mu r^2 + \hat{\mu}}} \quad \text{and} \quad \sqrt{\frac{\hat{\mu}\hat{r}^2}{\mu + \hat{\mu}\hat{r}^2}} , \tag{4.10}$$

respectively. No matter what r and $\hat{r}$ are, the a posteriori error bounds in (4.9) obtained when using the method of pseudolinear equations are an improvement over the standard a posteriori error bounds in (3.6). The improvement is, moreover, for practical problems, often quite significant. If $\mu = \hat{\mu}$ and $r = \hat{r} = 0.90$, for example, we see an improvement in the bounds for both u and v by a factor of $\sqrt{0.81/1.81} = 0.669$. The smaller r and $\hat{r}$ are, that is, the faster the method of pseudolinear equations converges, the greater the amount of improvement of the bounds in (4.9) over those in (3.6).

5. CONCLUDING REMARKS

We have focused attention in Sections 2-4 on problem (2.2). The improvement in the a posteriori error bounds by factors (4.10) is, however, valid for any system of quasilinear elliptic boundary-value problems or quasilinear two-point boundary-value problems that has a convex energy functional and a conjugate (dual) problem and is such that the method of pseudolinear equations converges. Classes of such problems are treated in [2,3,4]. Moreover, the concepts of Sections 2-4 (energy functionals, conjugate problems, method of pseudolinear equations) carry over to the numerical analogues of the differential equations. Indeed, one need merely interpret the differential equation as a symbolic representation of a high-dimensional nonlinear vector equation. The results developed in this paper are thus immediately applicable to a computational environment.

REFERENCES

[1] Ekeland, I. and Temam, R., Convex Analysis and Variational Problems (American Elsevier, New York, 1976).

[2] Lavery, J.E., Solution of inhomogeneous quasilinear Dirichlet and Neumann problems by reduction to the Poisson equation and a posteriori error bounds, J. reine angew. Math. 299/300 (1978) 73-79.

[3] Lavery, J., Iterative solution of quasilinear elliptic systems with a posteriori error bounds, in: Marchuk, G.I. (ed.), Metody Rešenija Sistem Variatsionno-Raznostnykh Uravnenii̇̆, Vyp. 5 (Computing Center SOAN SSSR, Novosibirsk, 1979) 60-81 [in Russian].

[4] Lavery, J.E., Solution of quasilinear two-point boundary-value problems by the method of pseudolinear equations, Nonlinear Analysis 8 (1984) 193-207.

[5] Sewell, M.J., On dual approximation principles and optimization in continuum mechanics, Phil. Trans. Royal Soc. London 265 (1969) 319-351.

[6] Velte, W., On complementary variational inequalities, in: New Variational Techniques in Mathematical Physics (Edizioni Cremonese, Rome, 1974) 407-420.

This paper is in final form and no version of it will be submitted for publication elsewhere.

Trends in the Theory and Practice of Non-Linear Analysis
V. Lakshmikantham (Editor)

PERIODIC OR UNBOUNDED SOLUTIONS FOR A CLASS OF THREE-DIMENSIONAL ODE SYSTEMS WITH BIOLOGICAL APPLICATIONS

Daniel S. Levine

Department of Mathematics
University of Texas at Arlington
Arlington, TX 76019
U.S.A.

A class of autonomous systems of three ordinary differential equations is studied with certain sign constraints on the elements of its Jacobian matrix. Examples of systems with these conditions arise in population dynamics and chemical kinetics. If there is a unique equilibrium with a two-dimensional unstable manifold, essentially two cases can occur. Either all non-trivial solutions are unbounded, or solutions on the unstable manifold approach a stable or semi-stable periodic orbit.

INTRODUCTION

The purpose of this article is to study the qualitative theory of a general class of nonlinear autonomous systems of three ordinary differential equations that arises frequently in biological and chemical applications. The general system to be studied is

$$\begin{aligned} \dot{x} &= f(x,y,z) \\ \dot{y} &= g(x,y,z) \\ \dot{z} &= h(y,z) \end{aligned} \tag{1}$$

with certain conditions on the signs of the partial derivatives of f, g, and h. These conditions are

$$\left.\begin{array}{l} f_x \leq 0,\ f_y \geq 0,\ f_z < 0 \\ g_x > 0,\ g_y \leq 0,\ g_z \geq 0 \\ \quad h_y > 0,\ h_z \leq 0 \\ \text{on some set } B \subset \mathbb{R}^3 \text{ which is} \\ \text{positively invariant for (1)} \end{array}\right] \tag{2}$$

It will also be assumed that the closure of B contains a unique equilibrium point (x_0,y_0,z_0) of (1), that is

$$\left.\begin{array}{c} f(x_0,y_0,z_0) = g(x_0,y_0,z_0) = h(y_0,z_0) = 0 \\ (x_0,y_0,z_0) \in B \\ ((x,y,z) \in \bar{B},\ f(x,y,z) = g(x,y,z) = h(y,z) = 0) \Rightarrow \\ (x,y,z) = (x_0,y_0,z_0) \end{array}\right] \tag{3}$$

It is assumed also that the equilibrium is unstable and has a two-dimensional unstable manifold, that is,

$$\left[\begin{matrix} f_x & f_y & f_z \\ g_x & g_y & g_z \\ h_x & h_y & h_z \end{matrix}\right]\Bigg|_{(x_0,y_0,z_0)} \text{ has eigenvalues } -\lambda,\ \mu + i\rho,\ \nu - i\rho,\ \lambda > 0,\ \mu > 0,\ \nu > 0,\quad \mu = \nu \text{ unless } \rho = 0. \tag{4}$$

In many examples studied, the equilibrium is stable for some parameter values and has a two-dimensional unstable manifold for others, leading to a bifurcation of periodic solutions. Finally, a technical condition will be imposed on g, namely

$$\lim_{x\to-\infty} g(x,y,z) < 0 < \lim_{x\to\infty} g(x,y,z) \text{ for each } y,z. \tag{5}$$

The mathematical results to be sketched here are proved in more detail in Cohen and Levine [3] and Levine [11]. Cohen and Levine [3] studied the subcase of (1), where f depends only on x and z, g only on x, and h only on y, and (2) - (5) are obeyed. For this system, they showed that all solutions not on the one-dimensional stable manifold of the equilibrium become unbounded for large times, spiraling outward from the equilibrium. Examples of such a system were illustrated in several models of interacting populations, including age-dependent predation (Gurtin and Levine [6];Levine [10]); predation with a time lag between discovery of prey and their being eaten (MacDonald [12]); and three-species competition (Coste et al [4]). (The variables x.y, and z are not actual population numbers but are various functions or functionals of population numbers.) The proof of unboundedness extends to all examples of (1) obeying (2) - (5) with the additional assumption that $g_y \equiv h_z \equiv 0$ (that is, that two of the three "self-limitation" terms are absent from the equations.)

Other examples of (1) - (5), however, yield not unbounded solutions but periodic orbits that are either stable or, in some sense, semi-stable. Such examples include many systems where the existence of a periodic orbit has been previously shown by familiar methods (Pliss [14]) involving invariant boxes, Poincaré maps, and the Brouwer fixed point theorem. These systems have arisen in numerous applications, such as end-product inhibition in genetics (Tyson [15]); the Belousov-Zhabotinskii chemical reaction (Hastings and Murray [8], Noszticzius et al [13]); Michaelis-Menten enzyme kinetics (Dai [5]); and a population cannibalizing its own young (Gurtin and Levine [6]). The systems leading to periodic orbits typically include "self-limitation" in all variables, and for these systems the invariant set B of (2) is bounded. The methods used in [3] can be applied to this case to show that all solutions of (1) on the unstable manifold of the equilibrium approach the periodic orbit.

GENERAL RESULTS

The following theorem describes the dynamics of system (1) with assumptions (2) - (5), including the possibility of either unboundedness or periodicity.

Theorem 1. Suppose the functions f, g, and h in system (1) obey conditions (2) - (5). Let U be the global unstable manifold of (x_0,y_0,z_0) as $t\to-\infty$. Then

(a) Either all non-trivial solutions of (1) on U approach the same periodic orbit K as $t\to\infty$ (the periodic case) or all non-trivial solutions of (1) on U become unbounded as $t\to\infty$ (the unbounded case);

(b) Let $\hat{U}$ be the open cylinder $\{(x,y,z)\,|\,(\hat{x},y,z) \in U \text{ for some } \hat{x} \in \mathbb{R}\}$. Then in the periodic case, if the two eigenvalues with positive real part are not real, all trajectories of (1) that remain in $\hat{U} \cap B$ except those on the stable manifold of (x_0,y_0,z_0) approach K. In some subcases of the unbounded case, all trajectories in $\hat{U} \cap B$ except those on the stable manifold become unbounded as $t\to\infty$.

A sketch of the proof of Theorem 1 follows. Without loss of generality, we can assume that the unstable equilibrium (x_0,y_0,z_0) is the origin, so $f(0,0,0) = g(0,0,0) = h(0,0) = 0$.

First, the invariant set B can be divided into the eight octants

$$\begin{aligned} I &= B \cap \{x \geq 0,\ y \geq 0,\ z \geq 0\} & V &= B \cap \{x \leq 0,\ y \leq 0,\ z \geq 0\} \\ II &= B \cap \{x \geq 0,\ y \geq 0,\ z \leq 0\} & VI &= B \cap \{x \leq 0,\ y \geq 0,\ z \geq 0\} \\ III &= B \cap \{x \geq 0,\ y \leq 0,\ z \leq 0\} & VII &= B \cap \{x \geq 0,\ y \leq 0,\ z \geq 0\} \\ IV &= B \cap \{x \leq 0,\ y \leq 0,\ z \leq 0\} & VIII &= B \cap \{x \leq 0,\ y \geq 0,\ z \leq 0\} \end{aligned}$$

The conditions (2) can easily be shown to constrain the allowable transitions between octants as time increases. These transitions are

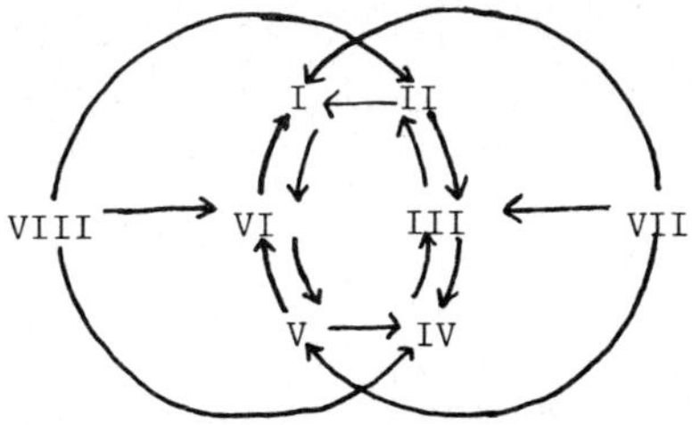

Also, any trajectory of (1) in B that remains in VII or VIII for large t must approach (0,0,0). For if $\gamma = (\hat{x}(t), \hat{y}(t), \hat{z}(t))$ is a trajectory that stays in VII, $\hat{x} \geq 0$, $\hat{y} \leq 0$, $\hat{z} \geq 0$, so by (2), $\dot{\hat{x}}(t) = f(\hat{x},\hat{y},\hat{z}) \leq f(0,\hat{y},\hat{z}) \leq f(0,0,z) \leq f(0,0,0) = 0$. Similarly, $\dot{\hat{y}} \geq 0$ and $\dot{\hat{z}} \leq 0$. Hence $\hat{x},\hat{y},\hat{z}$ are monotone and bounded, which means γ must approach a limit as $t\to\infty$. This limit must be (0,0,0) which is the only equilibrium in $\bar{B}$. The demonstration for VIII is analogous.

Thus any trajectory γ of (1) not on the stable manifold of (0,0,0) must remain in the union of octants I through VI for large times. It will now be shown, by an argument used in [6], that if γ is bounded, it must cross the face between II and I infinitely often or else touch the x-axis infinitely often.

Suppose the transition II→I is not made infinitely often. Then γ stays in $II \cup III \cup IV$ or $I \cup V \cup VI$ for large t, and so z keeps the same sign for large t.

The constraints (3) yield the following inequalities on each of the variables x,y,z at maxima or minima of the other variables:

$$\begin{aligned} \dot{z} &= 0,\ \ddot{z} \leq 0 \Rightarrow \dot{y} \leq 0 \\ \dot{z} &= 0,\ \ddot{z} \geq 0 \Rightarrow \dot{y} \geq 0 \\ \dot{y} &= 0,\ \ddot{y} \leq 0 \Rightarrow \dot{x} \leq 0 \text{ or } \dot{z} < 0 \\ \dot{y} &= 0,\ \ddot{y} \geq 0 \Rightarrow \dot{x} \geq 0 \text{ or } \dot{z} > 0 \\ \dot{x} &= 0,\ \ddot{x} \leq 0 \Rightarrow \dot{z} \geq 0 \text{ or } \dot{y} < 0 \\ \dot{x} &= 0,\ \ddot{x} \geq 0 \Rightarrow \dot{z} \leq 0 \text{ or } \dot{y} > 0 \end{aligned} \tag{6}$$

Inequalities (6) will be used to show that maxima of z can occur only when $z \geq 0$ and minima only when $z \leq 0$.

Let τ_1 be a time at which z has a minimum, and let τ_2 be the first time after τ_1 at which z has a maximum. Suppose $\dot{x}(\tau_2) > 0$. By $(6)_{1,2}$, $\dot{y}(\tau_1) \geq 0$ and $\dot{y}(\tau_2) \leq 0$. Hence for some $\sigma_1 \in [\tau_1,\tau_2]$, $\dot{y}(\sigma_1) = 0$, and $\ddot{y}(\sigma_1) \leq 0$. Since $\tau_1 \leq \sigma_1 \leq \tau_2$, $\dot{z}(\sigma_1) \geq 0$, so by $(6)_3$, $\dot{x}(\sigma_1) \leq 0$. Since $\dot{x}(\tau_2) > 0$, $\sigma_1 < \tau_2$ and x has a minimum $\sigma_2 \in [\sigma_1,\tau_2)$, so $\ddot{x}(\sigma_2) \geq 0$. Clearly, $\dot{z}(\sigma_2) \geq 0$. The argument used to show $(6)_6$ yields $\dot{z} < 0$ or $\dot{y} > 0$ if $f_y > 0$, and $\dot{z} \leq 0$ if $f_y = 0$. In the latter case, at $t = \sigma_2$, $\dot{z} = 0$ and $\ddot{z} = 0$, yielding $\dot{y} = 0$, an impossibility because that means $\dot{x}$, $\dot{y}$, and $\dot{z}$ are all 0. In the former case, $\dot{y}(\sigma_2) > 0$. Since $\dot{y}(\tau_2) \leq 0$ by $(6)_1$ this means there is a $\sigma_3 \in (\sigma_2,\tau_2]$ such that $\dot{y}(\sigma_3) = 0$, $\ddot{y}(\sigma_3) \leq 0$. Again, since $\dot{z}(\sigma_3) \geq 0$, $(6)_3$ yields $\dot{x}(\sigma_3) \leq 0$. By iteration of this process, a sequence $\{\sigma_n\}_{n=1}^{\infty}$ exists such that $\tau_1 \leq \sigma_1 \leq \sigma_2 \leq \sigma_3 \leq \sigma_4 \leq \ldots \leq \tau_2$ and $\dot{y}(\sigma_{2m-1}) = 0$, $\ddot{y}(\sigma_{2m-1}) \leq 0$, $\dot{x}(\sigma_{2m}) = 0$, $\ddot{x}(\sigma_{2m}) \geq 0$. The sequences $\{\sigma_{2m-1}\}_{m=1}^{\infty}$ and $\{\sigma_{2m}\}_{m=1}^{\infty}$ converge to the same limit σ. By continuity, $\dot{x}(\sigma) = \dot{y}(\sigma) = 0$ and $\ddot{x}(\sigma) \geq 0$, $\ddot{y}(\sigma) \leq 0$. But this means that at $t = 0$, $0 \leq x = f_z\dot{z}$, so that $\dot{z} \leq 0$ since $f_z < 0$. Since $\dot{z}(\sigma) \geq 0$, this again leads to the contradiction $\dot{x} = \dot{y} = \dot{z} = 0$.

The above demonstration and $(6)_1$ show that at every maximum of z past the first,

$$\dot{x} = f(x,y,z) \leq 0, \; \dot{y} = g(x,y,z) \leq 0 \tag{7}$$

The opposite signs hold at any minimum of z. Suppose that the sign of z remains the same for large t; for definiteness, say $z \geq 0$ for large t. It will now be shown that under that condition, z is eventually monotone or else γ touches the x-axis infinitely often.

Suppose z is not eventually monotone. Then after an arbitrarily large time, z has a maximum at $z = z_2 > 0$ followed by a minimum at $z = z_1$, $0 \leq z_1 < z_2$. Since $h_y > 0$, we can define the function $k(z)$ uniquely, on some real interval including 0, by $h(k(z),z) = 0$, and $k' \geq 0$ by (2). Since $h(y,z) = 0$ at a maximum or minimum of z, $k(z_2)$ and $k(z_1)$ exist, and y takes on those values at the corresponding points.

Define the function $s(y,z)$ by $g(s(y,z),y,z) = 0$. The function s is defined everywhere by (5), and is unique because $g_x > 0$. By (7) applied to the maximum where $z = z_2$, and its opposite applied to the mimimum where $z = z_1$, there exists values x_1, x_2 (the co-ordinates of those points) where

$$\begin{aligned} f(x_1,k(z_1),z_1) \geq 0, \; g(x_1,k(z_1),z_1) \geq 0 \\ f(x_2,k(z_2),z_2) \leq 0, \; g(x_2,k(z_2),z_2) \leq 0 \end{aligned} \tag{8}$$

By (8), since $g_x \geq 0$, $x_1 \geq s(k(z_1),z_1)$ and $x_2 \geq s(k(z_2),z_2)$. These inequalities combined with $f_x \leq 0$ and (8) yield $f(s(k(z_2),z_2),k(z_2),z_2)$ $\leq f(x_2,k(z_2),z_2) \leq 0 \leq f(x_1,k(z_1),z_1) \leq f(s(k(z_1),z_1),k(z_1),z_1)$. By continuity of the functions k and s, there is a value $\hat{z} \in [z_1,z_2]$ such that

$$f(s(k(\hat{z}),\hat{z}),k(\hat{z}),\hat{z}) = 0 \tag{9}$$

But by the definitions of k and s, $g(s(k(\hat{z}),\hat{z}),k(\hat{z}),\hat{z}) = h(k(\hat{z}),\hat{z}) = 0$, so that $s(k(\hat{z}),k(\hat{z}),\hat{z})$ is an equilibrium of (1). By (3), $\hat{z} = 0$ and so $z_1 = 0$. Hence also $k(z_1) = k(0) = 0$, so if z is not eventually monotone, all if its minima must lie on the x-axis.

Consider the case where $z > 0$ monotone and bounded for large time. Then $z \to z_0 < \infty$ and, since $\ddot{z}$ is bounded, $\dot{z} \to 0$. But also $\dot{z} = h(y,z) \to h(y,z_0)$, hence y must approach a value y_0 where $h(y_0,z_0) = 0$. Thus $\dot{x} \to f(x,y_0,z_0)$. This means x must also approach some limit x_0, and so γ approaches the equilibrium (x_0,y_0,z_0). This violates the assumption that γ is not on the stable manifold of (0,0,0), the only equilibrium in $\overline{B}$.

Let W be the local two-dimensional unstable manifold of (0,0,0). Then W can be represented ([2], pp. 330-335) by a C^2 functional relationship $\overline{x} = \overline{u}(\overline{y},\overline{z})$ where $(\overline{x},\overline{y},\overline{z}) = P[(x,y,z)]$ are the co-ordinates of the stable and unstable manifolds of the linearization of (1) at (0,0,0), and P is a change-of-basis matrix.

Near (0,0,0), W is tangent to the unstable manifold W' of the linearization of (1). If A is the Jacobian matrix of (1) at (0,0,0), then W' is spanned by the two (real or complex) eigenvalues of A corresponding to eigenvalues with positive real part. The normal to W' is in the direction of an eigenvector to A^T corresponding to the unique negative eigenvalue $-\lambda$. By (2),

$$A^T = \begin{bmatrix} -p & s & 0 \\ q & -t & v \\ -r & u & -w \end{bmatrix}, \qquad \begin{array}{l} p,q,t,u,w \geq 0 \\ \\ r,s,v > 0. \end{array}$$

If the polynomial $\pi(\omega)$ is defined as $\det(\omega I - A^T)$, then $\pi(0) = -\det A > 0$ by the signs of the eigenvalues of A. Also, it is easy to show that $\pi(-\max(p,w)) > 0$. Since $-\lambda$ is the unique negative root of λ, this means that $\lambda - w > 0$ and $\lambda - p > 0$. Hence, if $(a,b,c)^T$ is an eigenvector corresponding to $-\lambda$, $(\lambda-p)a + sb = 0$ so a and b have opposite sign. Since $-ra + ub + (\lambda-w)c = 0$, a and c have the same sign. Hence the equation for W' is $x = -\frac{b}{a}y - \frac{c}{a}z$, with $-\frac{b}{a} > 0$, $-\frac{c}{a} < 0$.

Now let π_s be the projection on the $\overline{x}$ co-ordinate, $\pi_{W'}$ the projection on the $(\overline{y},\overline{z})$ co-ordinates. Then $W = \{(x,y,z) \mid r(x,y,z) = \pi_s(p) - \overline{u}(\pi_{W'}(p)) = 0,\ p = (x,y,z)\}$. If $r_x(0,0,0) \neq 0$, the implicit function theorem defines u near the origin by $r(u(y,z),y,z) = 0$. The normal to W' is of the form $\vec{r}_1 = (a_1,-b_1,c_1)^T$, $a_1,b_1,c_1 > 0$, and is spanned by a vector $\vec{r}_2 = (a_2,-b_2,c_2)$, $a_2,b_2,c_2 > 0$. Since $(\overline{u}_{\overline{y}},\overline{u}_{\overline{z}})|_{(0,0)} = (0,0)$ because W is tangent to W' at (0,0,0), $r_x(0,0,0) = \frac{\partial}{\partial x}(\pi_s(p)) = \frac{a_2}{\vec{r}_1 \cdot \vec{r}_2} \neq 0$.

At (0,0), $u_y = -\frac{r_y}{r_x} = \frac{\frac{\partial}{\partial y}(\pi_s)}{\frac{\partial}{\partial x}(\pi_s)} = -\frac{\frac{-b_2}{\vec{r}_1\cdot\vec{r}_2}}{\frac{a_2}{\vec{r}_1\cdot\vec{r}_2}} = \frac{b_2}{a_2} > 0$, and

$$u_z = -\frac{r_z}{r_x} = \frac{\frac{\partial}{\partial z}(\pi_s)}{\frac{\partial}{\partial x}(\pi_s)} = -\frac{\frac{c_2}{\vec{r}_1\cdot\vec{r}_2}}{\frac{a_2}{\vec{r}_1\cdot\vec{r}_2}} = -\frac{c_2}{a_2} < 0.$$

By continuity, $u_y > 0$ and $u_z < 0$ in a neighborhood of (0,0).

Let U be the global unstable manifold of (1), that is, $\bigcup_{0 \le t < \infty} (\phi_t(W))$, where ϕ_t denotes the flow of (1). Then it can be shown that the appropriate two-dimensional projection of any of these flow maps is locally invertible, which leads to a local representation of U around any point (x_0,y_0,z_0) in U as $\{(\hat{u}(y,z),y,z) \mid (y,z) \in V_0\}$ for some neighborhood V_0 of (x_0,y_0) and a C^2 function $\hat{u}$ defined on V_0.

Since $x = \hat{u}(y,z)$ is locally invariant under the flow of (1), $u = \hat{u}$ satisfies the partial differential equation

$$g(u,y,z)\, u_y + h(y,z)\, u_z = f(u,y,z) \tag{10}$$

A given trajectory γ of (1) may go through different functions u (we will later prove that in fact u is globally single-valued, but we do not need that yet!), but u, u_y, and u_z are defined uniquely at each point of γ and obey (10). By (1), along any trajectory,

$$\begin{aligned} \dot{u}_y &= u_{yy}\dot{y} + u_{yz}\dot{z} = g(u,y,z)\, u_{yy} + h(y,z)\, u_{yz} \\ \dot{u}_z &= u_{yz}\dot{y} + u_{zz}\dot{z} = g(u,y,z)\, u_{yz} + h(y,z)\, u_{zz} \end{aligned} \tag{11}$$

Differentiating (10) with respect to y and z respectively and combining with (11) yields

$$\begin{aligned} \dot{u}_y &= f_x u_y + f_y - g_x (u_y)^2 - g_y u_y - h_y u_z \\ \dot{u}_z &= f_x u_z + f_z - g_x (u_y u_z) - g_z u_y - h_z u_z \end{aligned} \tag{12}$$

The set $\{u_y > 0,\ u_z < 0\}$ is positively invariant for (12), by (2), and $u_y > 0$, $u_z < 0$ on W, so $u_y > 0$, $u_z < 0$ on U.

Now the global single-valuedness of u on U follows by a theorem from [1], p.27. The theorem (CH) is that if π is any map of metric spaces that is proper (the inverse image of a compact set is compact), the cardinality of $\pi^{-1}(\{p\})$ is finite and constant over all p in each connected component of (range π) − $\pi(W)$, where W is the set of all critical points of π.

Let π be defined on U by projection on the yz-plane. Then since U_y and U_z

are never 0, π has no critical points. If π is restricted to a compact set of U, it is proper. Also, $\pi^{-1}(\{(0,0)\})$ has cardinality 1. For if $(x,0,0) \in U$, $x > 0$, then at $(x,0,0)$, $\dot{x} = f(x,0,0) \leq f(0,0,0) = 0$, $\dot{y} = g(x,0,0) > g(0,0,0) = 0$, and $\dot{z} = h(0,0) = 0$. But near this point, $x = u(y,z)$ for some local function u with $u_y > 0$, $u_z < 0$, so $\dot{x} = u_y\,\dot{y} + u_z\,\dot{z} = u_y\,\dot{y} > 0$ at $(x,0,0)$. A similar contradiction occurs if

$x < 0$. Hence $n(\pi^{-1}(\{(y_0,z_0)\}) \cap U)$ is at most 1 for any point (y_0,z_0). Thus $U = \{(u(y,z),y,z) \mid (y,z) \in P\}$ for some single-valued function u and a subset P of $\mathbb{R}^2$. (The exact conditions needed for (CH) to be used here have yet to be precisely determined, but are certainly obeyed in the periodic case where eigenvalues are complex because then trajectories spiral infinitely often through all octants in both directions. This allows for the correct choice of compact subset.)

The above demonstration also shows that no non-trivial trajectory on U touches the x-axis. Hence each non-trivial bounded trajectory on U must pass infinitely often through each of the octants I through VI. Consider such a trajectory γ. Its yz-projection (which is a homeomorphism since u is single-valued) must pass through successive points $(y_1,0)$, $(-w_1,0)$, $(y_2,0)$, $(-w_2,0)$, as shown below:

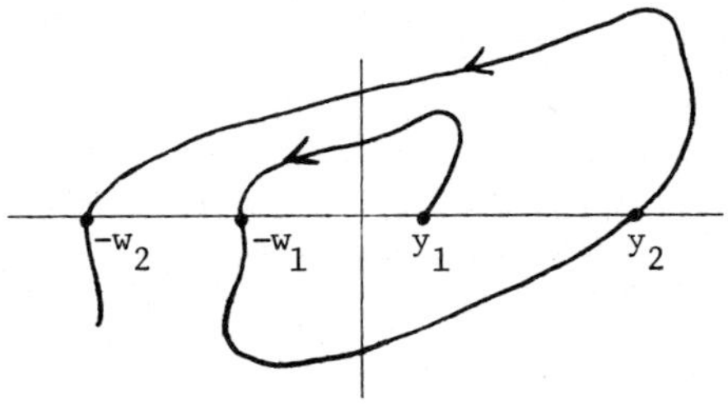

As the picture makes clear, since self-crossing cannot occur, either $y_{n+1} > y_n$ for all n, or $y_{n+1} < y_n$ for all n. The latter is impossible because $\gamma \to (0,0,0)$ as $t \to -\infty$.

Hence y_n is monotone increasing and bounded, so y_n approaches some $\bar{y}$ as $n \to \infty$. Consider the Poincaré map p of (1) on whatever subset of the face $F = \{x \geq 0,\ y \geq 0,\ z = 0\}$ it is defined on. Then $p(u(y_n,0),y_n,0) = (u(y_{n+1}(0),y_{n+1},0)$. Since γ is a projection of a bounded trajectory in U, $\{u(y_n,0)\}$ is bounded and so has a limit point $\bar{u}$. By continuity, if p is defined at $(\bar{u},\bar{y},0)$, then $(\bar{u},\bar{y},0)$ is a fixed point of p.

Suppose $p((\bar{u},\bar{y},0))$ is undefined. Then if the trajectory through $(\bar{u},\bar{y},0)$ is bounded, it cannot go through all octants, and so by Lemma 1, it must touch the x-axis. But that is impossible because $\pi(K)$ is disjoint from $\pi(U) = P$, which includes $(0,0)$.

Thus if $p((\bar{u},\bar{y},0))$ is undefined, the trajectory of (1) through $(\bar{u},\bar{y},0)$ is unbounded. But that trajectory consists of limit points of the trajectory $\pi^{-1}(\gamma)$, which is a bounded set and therefore inside the sphere of radius M around $(0,0,0)$ for some M. Hence the trajectory through $(\bar{u},\bar{y},0)$ is also inside the M-sphere. By this contradiction, $(\bar{u},\bar{y},0)$ is a fixed point of p, so the trajectory of (1) through that point is a periodic orbit.

Thus we have shown part of assertion (a): either there is a periodic orbit K of (1) which bounds the unstable manifold U, or all non-trivial solutions on U are unbounded. Suppose K exists. Then it remains to be shown that all non-trivial solutions on U approach K.

Let γ be a non-trivial trajectory of (1) on U. Then $(0,0) \in \pi(\gamma)$, and we have shown that $\pi(\gamma)$ cannot cross $\pi(K)$. Since $\pi(K)$ is a Jordan curve, with $(0,0)$ inside, this means $\pi(\gamma)$ is bounded. Since $\pi(\gamma)$ is a closed bounded set and u is continuous, the x-co-ordinates of γ are also bounded, so γ itself is bounded. Also, γ meets the x-axis only at $(0,0,0)$.

By the proof of Lemma 1, γ crosses the plane $z = 0$ infinitely often. Therefore, by the same argument used to establish the existence of K, the ω-limit set of γ must be <u>some</u> periodic trajectory K' on ∂U. By the same argument, $\pi(K')$ goes through some (unique) $(x_0',0)$, $x_0' > 0$. If K goes through $(x_0,0)$, then $x_0' = x_0$. For suppose $x_0' < x_0$. Then $(x_0,0)$ is on the ω-limit of some trajectory γ on U, so $(x_0-\varepsilon,0) \in \pi(\gamma)$, where ε can be chosen small enough that $x_0-\varepsilon > x_0'$. But the projection of the Jordan curve K' crosses the z-axis exactly twice, at $(x_0',0)$ and at some $(-u_0,0)$, $u_0 > 0$. Thus $(0,0) \in \pi(\gamma)$ is inside $\pi(K)$ and $(x_0-\varepsilon,0) \in \pi(\gamma)$ is outside $\pi(K)$. Hence $\pi(\gamma) \subset \pi(P)$ crosses $\pi(K)$, which was previously shown to be impossible. A similar contradiction arises if $x_0' > x_0$.

Hence $x_0' = x_0$ so $K = K'$. This proves that every non-trivial trajectory of (1) on U approaches the same periodic orbit K. This completes the proof of (a).

To prove (b), define the function

$$V(x,y,z) = x - u(y,z) \tag{13}$$

for $(x,y,z) \in \hat{U} \cap B$. Then by (1) and (13), along any trajectory,

$$\frac{d}{dt} V(x,y,z) = \frac{dV}{dx}\dot{x} + \frac{dV}{dy}\dot{y} + \frac{dV}{dz}\dot{z} = f(x,y,z) - g(x,y,z)\, u_y - h(y,z)\, u_z \tag{14}$$

Combining (14) with (10) yields $\frac{d}{dt} V = -u_y(g(x,y,z) - g(u(y,z),y,z))$, so that

$$\frac{d}{dt} V^2 = -u_y(x - u(y,z))(g(x,y,z) - g(u(y,z),y,z)) \tag{15}$$

Since g_x is strictly positive, $g(x,y,z) - g(u(y,z),y,z)$ has the same sign as $x - u(y,z)$. Hence (15) yields that $\frac{d}{dt} V^2 \leq 0$ along trajectories of (1), and $\frac{d}{dt} V^2 < 0$ except when $x = u(y,z)$, that is, except on U.
Hence the distance in the x-direction of any trajectory in $B \cap (\hat{U} - U)$ from U is strictly decreasing. In the case where a periodic orbit K exists, we can invoke the LaSalle invariance principle ([7], p. 296, Theorem 1.3). For any given trajectory γ, the ω-limit set of γ, $\omega(\gamma)$, is contained in the largest positively invariant set $\hat{M}$ in $\{(x,y,z) \in \hat{\bar{U}} \frac{d}{dt} V^2(x,y,z) = 0\}$. Now the definition of u extends continuously to $\bar{U}$ by setting $x = u(y,z)$ for $(x,y,z) \in K$. Since $\{(x,y,z) \in \hat{\bar{U}} | \frac{d}{dt} V^2 = 0\}$ is the subset of $\bar{U} = U \cup K$ on which $\dot{V}$ is defined, the invariant set $\ddot{M}$ is $\{(0,0,0)\} \cup K$, proving the first assertion of Theorem 1(b).

Now in the unbounded case, the invariance principle can again be invoked if the trajectory γ is bounded. In that case, $\gamma \subset B_M$, where B_m is the closed ball of radius M about the origin for some $M > 0$, hence $\omega(\gamma)$ is contained in the largest set in $U \cap B_M$. Since each non-trivial trajectory on U is unbounded, this invariant set must be $\{(0,0,0)\}$. Hence $\omega(\gamma) = (0,0,0)$, so the only bounded trajectories are on the stable manifold. The proof of Theorem 1 is complete.

THE UNBOUNDED AND PERIODIC CASES CONTRASTED

The following result for the unbounded case is a slight generalization of the main theorem of [3].

Theorem 2. If f,g,h obey conditions (2) - (5) and, in addition, $g_y \equiv h_z \equiv 0$, then every non-trivial trajectory is unbounded, subject to other conditions.

(The exact conditions needed for this theorem to be true are those which are needed for the function u to be single-valued.)

The proof involves a demonstration that the unstable manifold U cannot have a bounding periodic orbit K. It is easy to show that if K exists, then $\pi_{yz}(U) = P$ is exactly the inside of the Jordan curve $\pi_{yz}(K)$. Since $\dot{z} = h(y)$, $h(0) = 0$, and $h_y > 0$, then along K, z is increasing for $y > 0$ and decreasing for $y < 0$, as shown below:

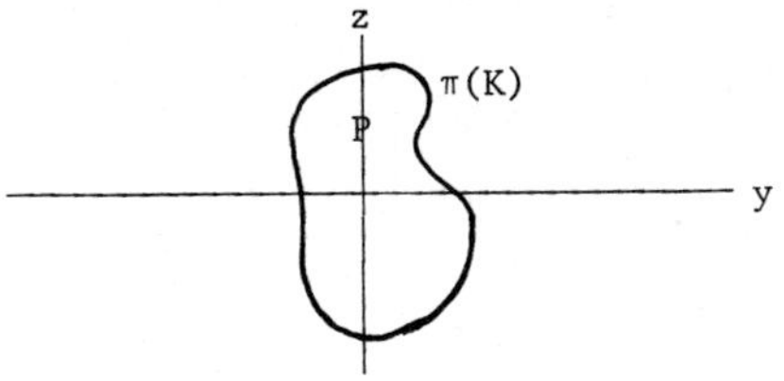

It is clear from the above figure that if $(y_0, z_0) \in P$, then $(y, z_0) \in P$/for all y between 0 and y_0, and $(0,z) \in P$/for all z between 0 and z_0. Hence we can define on P the Liapunov function

$$V(y,z) = -\int_0^z g(u(0,r),r)dr + \int_0^y h(s)ds \qquad (16),$$

g treated as a function of x and z and h of y.

By (1), (2), and (16), along trajectories of (1) which lie on U,

$\frac{d}{dt} V(y,z) = -g(u(0,z),0)\dot{z} + h(y)\dot{y} = h(y)(g(x,z) - g(u(0,z),z))$

$= h(y)(g(u(y,z),z) - g(u(0,z),z)) \geq 0$, and > 0 except when $y = 0$. Hence $V(y,z)$ is monotone increasing in time and $\dot{V}$ cannot approach 0. Since V and V are both bounded inside $\pi_{yz}(K)$, this is impossible, so K cannot exist for functions of this type.

A complete criterion for unboundedness versus periodicity has not been established. If a bounded invariant set can be constructed, periodicity will of course ensue. The following theorem describes the behavior of the periodic case.

<u>Theorem 3</u>. Suppose (2), (3), and (4) hold and trajectories on the unstable manifold U approach a periodic orbit K, and the functions f,g,h of (1) are analytic. Then the orbit K is either stable or semi-stable, that is, all of the Floquet multipliers of K have absolute value less than or equal to 1.

<u>Remark</u>. The intuitive picture of the periodic case is shown in Figure 1. The orbit K is approached on the cylinder $\hat{U}$ and from inside on the unstable manifold U. Stability follows if K is also an attractor from outside the cylinder.

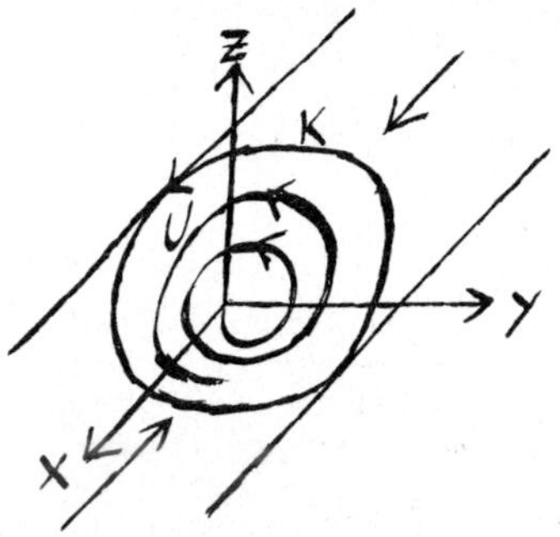

Figure 1

<u>Proof</u>. Let μ_1, μ_2, μ_3 be the Floquet multipliers of K, with $\mu_1 = 1$. It is a classical theorem (for example, [7], p.120) that

$\mu_1\mu_2\mu_3 = \exp\left(\int_K (f_x + g_y + h_z)\, dt\right)$. But by (2), $f_x + g_y + h_z \leq 0$ everywhere, so $0 < \mu_1\mu_2\mu_3 = \mu_2\mu_3 \leq 1$. Hence there is at most a unique Floquet multiplier, say μ_3, which is larger that 1 in absolute value, and if such a multiplier exists it must be real and larger than 1.

Suppose $\mu_3 > 1$. Then $\mu_2 < 1$ and Theorem 2.2 of [7], p. 218 can be invoked to show that K has a stable manifold S_K and an unstable manifold U_K in $\mathbb{R}^3$. By the construction of the stable manifold S_K, it includes $U \cup K$ and is a two-dimensional manifold, so it must extend $U \cup K$. Also, S_K is analytic because f, g, and h are.

Since S_K is an analytic manifold, if the function u is as defined in Lemma 4, u_y and u_z can be defined on K. By Lemma 3 and continuity, $u_y \geq 0$, $u_z \leq 0$ on K. Using equations (12) along K, those inequalities must be strict on K as they are on U. Hence, by continuity, there exists a neighborhood L of $\bar{U}$ in the relative topology of S_K such that $u_y > 0$, $u_z < 0$ on L. The arguments of Theorem 1 can be repeated to show that u is single-valued on L.

If $\hat{L}$ is the open cylinder $\pi_{yz}^{-1}(L)$, then by arguments previously used on $\hat{U}$, the function $(x - u(y,z))^2$ is strictly decreasing on any trajectory while it remains in L. Consider a trajectory $\gamma = \{\gamma(t) : -\infty < t < \infty\}$ on the unstable manifold U_K of K. Since $\hat{L}$ is an open neighborhood of K in $\mathbb{R}^3$ and the α-limit set of γ is K, there exists a time $-t_0$ such that for $t < -t_0$,

$\gamma(t) \in \hat{L}$. But if $\gamma(t) = (x(t),y(t),z(t))$, since $x = u(y,z)$ on K, $\lim_{t\to-\infty} (x(t) - u(y(t),z(t)))^2 = 0$, so that $(x(t) - u(y(t),z(t))))^2 < 0$ for $-\infty < t < -t_0$, which is impossible. By this contradiction, the unstable manifold U_K cannot exist. This proves Theorem 3.

Remark. The proofs of Theorems 1 and 3 rely on the structure of $\mathbb{R}^3$, but generalizations are likely to exist to dimensions higher than 3. Hastings et al [9] prove the existence of a periodic solution for one class of n-dimensional systems with specified sign constraints on the partial derivatives. The possible n-dimensional generalizations of the unbounded case are discussed in the introductory section of Cohen and Levine [3].

REFERENCES

[1] Chow, S.N. and J.K. Hale, Methods of Bifurcation Theory (Springer-Verlag, New York, 1982).

[2] Coddington, E.A. and N. Levinson, Theory of Ordinary Differential Equations (McGraw-Hill, New York, 1955).

[3] Cohen, M.A. and D.S. Levine , Unbounded oscillatory solutions for a system of interacting populations, to appear in SIAM J. Appl. Math.

[4] Coste, J., J. Peyraud, and R. Coullet, Asymptotic behaviors in the dynamics of competing species, SIAM J. Appl. Math. 36 (1979) 516-543.

[5] Dai, L.S., On the existence, uniqueness, and global asymptotic stability of the periodic solution of the modified Michaelis Menten mechanism, J. Diff. Eq. 31 (1979) 392-417.

[6] Gurtin, M.E. and D.S. Levine , On populations that cannibalize their young, SIAM J. Appl. Math. 42 (1982) 94-108.

[7] Hale, J.K., Ordinary Differential Equations (Wiley Interscience, New York, 1969).

[8] Hastings, S.P. and J.D. Murray, The existence of oscillatory solutions in the Field-Noyes model for the Belousov-Zhabotinskii reaction, SIAM J. Appl. Math. 28 (1975) 678-688.

[9] Hastings, S.P., J.J. Tyson, and D. Webster, Existence of periodic solutions for negative feedback cellular control systems, J. Diff, Eq. 25 (1977) 39-64.

[10] Levine, D.S., On the stability of predator-prey interactions with egg-eating predators, Math. Biosci. 56 (1981) 27-46.

[11] Levine, D.S., Qualitative theory of a third-order non-linear system with examples in population dynamics and chemical kinetics, to be submitted to Math. Biosci.

[12] MacDonald, N., Time delays in predator-prey models, Math Biosci. 28 (1976) 321-330.

[13] Noszticzius, Z., H. Farkas, and Z.A. Schelly, Explodator: a new skeleton mechanism for the halate driven chemical oscillators, J. Chem. Phys. 80 (1984) 6062-6070.

[14] Pliss, V.A., Nonlocal Problems in the Theory of Oscillations (Academic Press, New York, 1966).

[15] Tyson, J.J., On the existence of oscillatory solutions in negative feedback cellular control processes, J. Math. Biol. 1 (1975) 311-315.

The final (detailed) version of this paper will be submitted for publication elsewhere.

Trends in the Theory and Practice of Non-Linear Analysis
V. Lakshmikantham (Editor)

THE PHENOMENON OF QUENCHING: A SURVEY

Howard A. Levine

Department of Mathematics
Iowa State University
Ames, Iowa 50011
U.S.A.

In this paper we survey some recent results concerning the behavior of solutions of nonlinear evolutionary equations when the nonlinearity has a pole at a finite value of the solution. Simple examples of such equations are

$$u_t = u_{xx} + 1/(1-u)$$

or

$$u_{tt} = u_{xx} + 1/(1-u).$$

Results for these, for higher dimensional problems and for related problems in which the nonlinearity occurs in the boundary condition are also discussed.

§0. **Introduction:** In this talk, we present, without proofs, some results obtained over the last several years by several authors, concerning the phenomenon of quenching. Loosely speaking, we say that a solution of some evolutionary problem quenches in time T (finite or infinite) if some norm of the solution itself remains bounded while some norm of one of its derivatives becomes unbounded on the interval $[0,T)$.

This can occur in very simple problems. For example, on $[0, 1/2)$, the solution of

$$\frac{du}{dt} = \frac{1}{1-u}, \qquad u(0) = 0$$

is given by $u(t) = 1 - \sqrt{1-2t}$ so that $0 \leq u \leq 1$ on $[0, 1/2)$ but $u'(t)$ becomes infinite there.

A well known, less trivial example, is afforded by the initial value problem

$$u_t + uu_x = 0, \qquad x \in R^1, \quad t > 0$$

$$u(x,0) = f(x)$$

where f is a bounded, continuously differentiable function for which $f'(x)$ is negative at points of R^1. The classical solution is given implicitly by

$$u(x,t) = f\big(x - tu(x,t)\big)$$

which clearly remains bounded. Yet, if one calculates u_x or u_t from this equation, it is easy to see that they become unbounded in finite time T where

$$-1/T = \inf\{f'(\xi)\ ;\ \xi \in R^1\}\ .$$

If the infimum is infinite, there the classical initial value problem has no solution. See [5,7] for example.

Consider the following classical (unsolved) question from fluid mechanics: Does the vorticity of an inviscid, incompressible flow with nontrivial initial vorticity become unbounded in finite time ([4])? This may be viewed as a quenching problem because the velocity components remain bounded (the kinetic energy is uniformly bounded) while the curl of the velocity (the vorticity) presumably becomes unbounded. Numerical experiments (Chorin [4]) indicate that the answer is probably yes.

The term "quenching" has its origins in the study of electric current transients in polarized ionic conductors (Kawarda [6] and see also Scientific American, October 1980.)

Throughout this talk, we let $\phi:(-\infty,1) \to (0,\infty)$ denote an increasing, C^1 function with $\phi(0) = 1$ and

$$\lim_{x \to 1^-} \phi(x) = +\infty\ .$$

We define

$$\Phi(x) = \int_0^x \phi(\eta)d\eta\ .$$

We present some results and open questions concerning the following four semilinear initial boundary value problems of which two involve the heat operator and two involve the wave operator (in one space dimension). The problems are scaled so that the spacial domain is the reference interval [0,1]. Proofs are found in the references.

Problem Ia.

$$\frac{\partial u}{\partial t} = \frac{\partial^2 u}{\partial x^2} + L^2\phi\big(u(x,t)\big); \qquad 0 < x < 1,\quad t > 0.$$

$$u(x,0) = u(0,t) = u(1,t) = 0, \qquad 0 \leq x \leq 1,\quad t \geq 0.$$

Problem Ib.

$$\frac{\partial u}{\partial t} = \frac{\partial^2 u}{\partial x^2}\ ; \qquad 0 < x < 1,\quad t > 0,$$

$$u(x,0) = u(0,t) = 0 \qquad 0 \leq x \leq 1,\quad t \geq 0,$$

$$\frac{\partial u}{\partial t}(1,t) = L\phi\big(u(1,t)\big) \qquad t > 0\ .$$

Problem IIa.

$$\frac{\partial^2 u}{\partial t^2} = \frac{\partial^2 u}{\partial x^2} + L^2\phi\big(u(x,t)\big) \; ; \qquad 0 < x < 1, \quad t > 0,$$

$$u(x,0) = \frac{\partial u}{\partial t}(x,0) = 0 \qquad 0 \leq x \leq 1,$$

$$u(0,t) = u(1,t) = 0 \qquad t \geq 0 .$$

Problem IIb.

$$\frac{\partial^2 u}{\partial t^2} = \frac{\partial^2 u}{\partial x^2} \; ; \qquad 0 < x < 1, \quad t > 0$$

$$u(x,0) = \frac{\partial u}{\partial t}(x,0) = 0 \qquad 0 < x < 1$$

$$u(0,t) = 0$$

$$\frac{\partial u}{\partial x}(1,t) = L\phi\big(u(1,t)\big) \qquad t > 0 .$$

Three comments are in order. First, problems with nonlinearities in boundary conditions are not equivalent to problems with nonlinearities as source terms in the p.d.e.'s. Secondly, all data was chosen to be homogeneous in order to isolate the effects of the driving term, ϕ. Finally, for all of these problems, the existence of a unique, classical solution on $[0,1] \times [0,T]$ for some $T > 0$ has been established in the various references (local existence and uniqueness).

For each problem, we present the principal results and mention a few open problems in the hope that others might become interested in them.

§1. Quenching in parabolic equations. For (Ia), let f denote a solution of

$$\text{(BVP-1)} \qquad f''(x) + L^2\phi\big(f(x)\big) = 0, \qquad 0 < x < 1$$

$$f(0) = f(1) = 0,$$

It is not difficult to show by elementary means that if $L \leq L_0$ where

$$L_0 = 2\sqrt{2}\,\sup\Big\{ \int_0^{\Phi^{-1}(c)} \big(c - \Phi(\eta)\big)^{-1/2}\, d\eta, \quad c \in \text{Range}\ \Phi \Big\}$$

Then (BVP-1) has a unique smallest positive solution $f(x)$, while if $L > L_0$, there is no solution of (BVP-1). In the former case, there is a number $\delta \in (0,1)$ such that

$$0 \leq f(x) \leq 1 - \delta$$

for $x \in [0,1]$ and $L \in [0,L_0]$.

Acker and Walter [1,2] and, independently, Levine and Montgomery [11], proved the following:

Theorem. Let $u(x,t)$ denote the local solution of (Ia).

A. If $L \leq L_0$ then u exists for all $t > 0$ and

$$\lim_{t \to +\infty} u(x,t) = f(x)$$

from below. Moreover, u_t, u_{xx} remain bounded (no quenching).

B. If $L > L_0$, then there is $T = T(L) < \infty$ such that

$$\lim_{t \to T^-} u(\tfrac{1}{2},t) = 1 .$$

Moreover $u(x,t) < u(\tfrac{1}{2},t)$ for $x \neq \tfrac{1}{2}$. Therfore one of $u_t(\tfrac{1}{2},t)$, $u_{xx}(\tfrac{1}{2},t)$ becomes unbounded in finite time (u quenches).

Remarks.

1. Kawarada [6] showed that if $\phi(u) = 1/(1-u)$, then $L_0 \leq 2\sqrt{2}$. In fact, for this ϕ, $L_0 \approx 1.5307\ldots$. For any ϕ here considered, the elementary estimate

$$\sup\{x/\phi(x),\ x \in (-\infty,1)\} \leq L_0/(2\sqrt{2}) \leq 1 .$$

is shown in [11].

2. For $\phi(u) = (1-u)^{-1}$, Kawarada showed that whenever $u(\tfrac{1}{2},t)$ reaches one in finite time, then $u_t(\tfrac{1}{2},t)$ is unbounded in that time. This is not easy to show. Although it has not been done, Kawarada's proof should extend to more general ϕ's.

3. Quenching for (Ia) cannot occur in infinite time. (That is, Theorem 3b of [11] is vacuous in one dimension.)

4. Acker and Walter [1], considered quenching problems in higher dimensions. They considered, for example

$$u_t = \Delta u + \phi(u) \qquad (x,t) \in D_a \times [0,T),$$

$$u(x,0) = 0 \qquad x \in D_a$$

$$u(x,t) = 0 \qquad (x,t) \in \partial D_a \times (0,T)$$

where D_a, for $a > 0$ is a dilation of a reference domain $D \equiv D_1$. They showed the existence of a number $a_0 > 0$ such that if $a < a_0$, no quenching occured while if $a > a_0$, quenching occured in finite time. Unlike the one dimensional case, nothing was asserted at the critical case $a = a_0$.

5. We are thus led to three questions. (1) Can quenching occur in infinite time when $a = a_0$ in more than one dimension? (2) What happens to $\partial u/\partial t$ when $a > a_0$? (3) How does u_{xx}(or Δu) behave in the quenching regimes?

6. The proofs of the results here, as well as those for (Ib) below rest on various comparison theorems which follow from the Maximum Principle and the Boundary Point Lemma. That these theorems are not available for (IIa,b) was our original motivation for considering these problems.

7. Payne and Sperb have some work pending concerning what happens beyond quenching. That is, if $a > a_0$ $(L > L_0)$, let $w(x)$ solve

$$\Delta w + \phi(w) = 0 \qquad \text{if } w(x) < 1$$

$$w(x) = 1 \qquad \text{on } \Omega \subset D_a$$

$$w(x) = 0 \qquad \text{on } \partial D_a\ .$$

Presumably $\bar{\Omega} = \overline{\cup\{\Omega(t) \mid t > 0\}}$ where $\Omega(t) = \{x \in D_a \mid u(x,t) = 1\}$ and where we expect that $\{\Omega(t)\}$ is an increasing family of closed sets. We can view the problem of finding (w,Ω) as a free boundary value problem. The set Ω is sometimes called the <u>dead core</u>. See [22] and references.

For Problem (Ib), the situation is similar. Clearly the problem

(BVP-2) $$f''(x) = 0 \qquad 0 < x < 1$$

$$f(0) = 0$$

$$f'(1) = L\phi(f(1))$$

has solutions of the form $f(x) = ax$ (all solutions are necessarily of this form), if and only if there is $a \in (0,1)$ such that $a = L\phi(a)$. Thus, there is a number L_0 such that if $L \leq L_0$, (BVP-2) has a smallest solution while if $L > L_0$, there is no solution of this problem, where L_0 is given by

$$L_0 = \sup\{\delta/\phi(\delta),\ 0 < \delta < 1\} .$$

In [9], the following was established:

Theorem: Let $u(x,t)$ solve (Ib) on $[0,1] \times [0,T)$.

A. If $L \leq L_0$, we have $T = +\infty$ and $\lim_{t \to +\infty} u(x,t) = ax$, where a is the smallest root of $a = L\phi(a)$. There is no quenching here.

B. If $L > L_0$, then, for some $T = T(L)$, $\lim_{t \to T^-} u(1,t) = +1$ while

$$\lim_{t \to T^-} u_x(1,t) = \lim_{t \to T^-} u_t(1,t) = +\infty .$$

(Thus u quenches in finite time.)

Remarks.

1. It is possible to extend these results to higher dimensional problems. For example, let $D \subset R^n$ be bounded with $\partial D = \Sigma \cup \sigma$, $\Sigma \cap \sigma = \emptyset$ and Σ smooth. Let L_0 be as before and define

$$L_1 = \int_0^1 \frac{ds}{\phi(s)} .$$

Then $L_1 > L_0$. Let $w(x)$, $x \in \bar{D}$ solve $\Delta w = 0$, $w = 0$ on σ, $\frac{\partial w}{\partial n} = 1$ on Σ and let $w_0 = \max\{w(x) \mid x \in \bar{D}\}$. Let, in $D \times (0,T)$, $u(x,t)$ solve

$$\frac{\partial u}{\partial t} = \Delta u \qquad \text{in } D \times (0,T)$$

$$u = 0 \qquad \text{on } \sigma \times (0,T) \text{ and } D \times \{0\}$$

$$\frac{\partial u}{\partial n} = \phi(u) \qquad \text{on } \Sigma \times (0,T) .$$

It was shown in [10] that if $w_0 \leqslant L_0$, then $u \leqslant a$ where a is the smallest root of $s = L_0\phi(s)$ in $(0,1)$. Consequently, u exists for all t and no quenching is possible. If $w_0 > L_1$, then u reaches one in finite time on Σ so $\nabla_x u$ becomes unbounded in finite time. If $w_0 > L_1$ or $w_0 = L_1$ and u quenches in infinite time, then

$$\lim_{t \to \infty} u(x,t) = g(x)$$

where $g(x) \geqslant 0$ solves $\Delta g = 0$ in D, $g = 0$ on σ, $\partial g/\partial n = \phi(g)$ on Σ. In the latter case $g = 1$ on Σ whenever $w = L_1$. If this boundary value problem has no solution, then u must quench in finite time. Also, if $g_0 = \max\{g(x) \mid x \in D\} < 1$, then $u \leqslant g_0$ and no quenching can occur.

2. Concerning infinite time quenching, in [10] it was shown that there is a domain D and a partition of ∂D as well as initial values $u(x,0) > 0$ in D such that infinite time quenching actually occurs. Here $u(x,0) > g(x)$. For this domain $g(x) = 1$ at some points on Σ. It is not known whether or not this g is unique.

3. In the case of finite time quenching, it is not known what, if anything, happens to u_t on $\Sigma \times (0,T)$ in more than one dimension.

4. It has been suggested that initial boundary value problems of the form (1b) may be useful in the study of avalanches.*

§2. **Hyperbolic Problems.** As remarked earlier, we were lead to consider Problems IIa,b because of the lack of a maximum principle for them. Here the situation is less well understood than in the parabolic case, even in one space dimension.

In addition to the earlier requirements on ϕ, we shall assume that ϕ is convex. Define

$$H(x) = -\frac{1}{2}\pi^2 x^2 + L^2\Phi(x)$$

for $L > 0$ and $x \in (-\infty,1)$. Let $x_0(L)$ (if it exists) be a point in $(0,1)$ at which $H(x)$ has a positive minimum. For large L such points exist. Let

$$L_1 = \inf\{L \mid x_0(L) \text{ exists}\}.$$

---

* K. Hutter, Versuchsanstalt für Wasserbau, Hydrologie und Glaziologie, E.T.H. Zurich, Private Communication.

Define

$$L_0 = \pi \sup_{0 < \delta < 1} \left[\frac{\delta\pi}{\pi+\delta M(\delta)}\right]^{1/2}$$

where

$$M(\delta) = \sup_{|z| \leqslant \delta} \{2z^{-2}[\Phi(z) - z]\} .$$

($M(\delta) \geqslant \phi'(0)$.)

Chang and Levine [3] then showed the following:

Theorem. *Let* u *solve* (IIa) *on* $[0,1] \times [0,T)$.

A. *If* $L \leqslant L_0$ *then* $T = \infty$ *and there is* $\delta = \delta(L)$ (0,1) *such that* $|u(x,t)| \leqslant 1-\delta$ *on* $[0,1] \times [0,\infty)$. (*Quenching cannot occur in this case*.)

B. *If* $L \geqslant L_1$, *then* $T < \infty$ and

$$\lim_{t \to T^-} \left(\sup_{0 < x < 1} u(x,t)\right) = 1$$

Thus one of u_{tt}, u_{xx} *must become infinite on* $[0,1] \times [0,T)$.

Remarks.

1. The convexity of ϕ is needed only for part (B) of the Theorem. The idea of proof (discovered by Kaplan for parabolic problems) is to show that

$$F(t) \equiv \frac{2}{\pi} \int_0^1 u(x,t)\sin(\pi x)dx$$

must reach one in finite time. (See [8] for a discussion of various extensions of Kaplan's ideas.)

2. For part (A) of the Theorem one can give an indirect argument based on the conservation law

$$E(t) = \frac{1}{2} \int_0^1 u_t^2(x,t)dx + \frac{1}{2} \int_0^1 u_x^2(x,t)dx - \int_0^1 \Phi\big(u(x,t)\big)dx$$

$$= E(0) \quad (= 0 \text{ here})$$

and the sharp inequality

$$(E) \qquad 2|u(x,t)| \leq \left(\int_0^1 |u_x(y,t)|^2 dy \right)^{1/2} .$$

3. The Theorem (in the case of quenching) says nothing about which derivative in the equation becomes unbounded. This contrasts with Kawarada's result.

4. In the case $\phi(u) = 1/(1-u)$, L_1 is smaller than L_0 for problem 1a. That is, quenching occurs over a wider range for the hyperbolic problem than the parabolic problem.

5. Computations in the case $\phi(u) = 1/(1-u)$ suggest that there is a number $\bar{L}$, $L_0 < \bar{L} < L_1$ such that for $L < \bar{L}$ there is no quenching; for $L = \bar{L}$, there is infinite time quenching; while for $L > \bar{L}$, there is finite time quenching. Since the stationary solution does not play a role here, we strongly suspect this to be true.

6. In several space dimensions, results of type (B) are easy to obtain for problems where the dynamics take the form

$$u_{tt} = Au + \phi(u)$$

where A is an operator which possesses an eigenfunction (relative to the appropriate boundary conditions) which does not change sign. However, the method is restricted to such operators. In order to obtain results of the type (A), for such problems, one needs an embedding inequality of the form

$$|u(x,t)|^2 \leq c \int_D (u \cdot Au) dx$$

where $D \subset R^n$ is bounded and the boundary conditions (and A!!) are such that the weak conservation law

$$E(t) \equiv \frac{1}{2} \int_D u_t^2 \, dx + \frac{1}{2} \int_D (u \cdot Au) dx \leq E(0) \; . \; ,$$

holds. In [12], Smiley and Levine discuss several such problems in great generality.*

7. The phase plane portrait for the ordinary differential equation

$$y''(t) = - k^2 \, y(t) + \frac{k^2 \lambda}{(a-y(t))}$$

* In [9], p. 1142, the desired inequality is stated incorrectly. It should read

$$|u(x,t)|^2 \leq \text{const} \times \left(\int_\Omega |\Delta^p u(y,t)|^2 \, dy \right).$$

where λ, a are positive constants is discussed in [21]. The equation's physical origins are discussed in [13], also. Roughly, the equation describes the motion of a rigid current carrying wire composed of magnetic material and fixed at both ends by springs with string constants k^2 in the presence of a second, parallel, current carrying (rigid) wire of infinite length. Here λ is proportional to the current carried by the second wire. Our Problem (IIa) then can be given an analogous physical interpretation.

Finally, we turn to Problem (IIb). Again let

$$L_0 = \frac{1}{2} \sup_{0 \leq \delta \leq 1} \delta^2 \left(\int_0^{\delta} \phi(\sigma) d\sigma \right)^{-1}$$

and

$$L_1 = \sup_{\sigma \leq \delta \leq 1} \delta/\phi(\delta) .$$

An easy argument shows that $L_1 \geq L_0 \geq \frac{1}{2} L_1$. In fact, $L_1 > L_0$ unless there is $\delta_0 \in (0,1)$ such that both

$$\Phi(\delta_0) = \frac{1}{2} \delta_0 \ \phi(\delta_0) ,$$

$$\phi(\delta_0) = \delta_0 \ \phi'(\delta_0)$$

and this cannot happen if $\phi'' > 0$ on $[0,1)$.

We have established the following results ([9]):

Theorem: Let $u(x,t)$ solve (IIb) on $[0,1] \times [0,T)$.

A. If $L < L_0$, then $T = +\infty$ and there is $\delta_1 \in (0,1]$ such that $|u(1,t)| \leq 1 - \delta_1$ on $[0,\infty)$ while

$$u^2(x,t) \leq 2L \ \Phi(1 - \delta_1)$$

on $[0,1] \times [0,\infty)$. Quenching does not occur here.

B1. If $L > L_1$ and $\int_0^1 \phi(\eta) d\eta < \infty$, then $T < \infty$ and $\lim_{t \to T^{-1}} u(1,t) = 1$.

B2. If $L > L_1$ and $\int_0^1 \phi(\eta) d\eta = \infty$, then $T \leq \infty$ and $\lim_{t \to T^-} u(1,t) = 1$.

B3. *If* $u(1,t)$ *reaches one in finite time* T, *then*

$$\lim_{t \to T^-} u_x(1,t) = \lim_{t \to T^-} u_t(1,t) = +\infty ,$$

while if $u(1,t)$ *reaches one in infinite time, then at least*

$$\limsup_{t \to +\infty} \max_{0 \leq x \leq 1} u(x,t) = \limsup_{t \to \infty} \int_0^1 u_x^2(x,t)dx = \infty .$$

Remarks.

1. Unlike the parabolic problem (Ib), we know of no analogous results in higher dimensions.

References

1. A. Acker and W. Walter, **On the global existence of solutions of parabolic differential equations with a singular nonlinear term**, Nonlinear Analysis, 2(1978), 449-505.

2. __________, **The quenching problem for nonlinear parabolic equations**, Lecture Notes in Mathematics 564, Springer-Verlag, New York, 1976.

3. P. H. Chang and H. A. Levine, **The quenching of solutions of semilinear hyperbolic equations**, SIAM J. Math. Anal., 12(1981), 893-903.

4. A. J. Chorin, **The evolution of a turbulent vortex**, Comm. Math. Phys., 83(1982), 517-535.

5. P. R.Garabedian, **Partial Differential Equations**, John Wiley, New York, 1964.

6. H. Kawarada, **On the solutions of initial boundary value problems for** $u_t = u_{xx} + 1/(1-u)$, Publ. RIMS, Kyoto Univ., 10(1975), 729-736.

7. P. D. Lax, **Nonlinear hyperbolic equations**, Comm. Pure Appl. Math., 6(1953), 231-258.

8. H. A. Levine, **On the nonexistence of global weak solutions of some properly and improperly posed problems of mathematical physics: the method of unbounded Fourier coefficints**, Math. Ann., 214(1975), 205-220.

9. H. A. Levine, **The quenching of solutions of linear parabolic and hyperbolic equations with nonlinear boundary conditions**, SIAM J. Math. Anal., 14(1983), 1139-1153.

10. H. A. Levine and G. M. Lieberman, **Quenching of solutions of parabolic euqations with nonlinear boundary conditions in several dimensions**, J. reine ange. Math., 345(1983), 23-38.

11. H. A. Levine and J. T. Montgomery, **The quenching of solutions of some nonlinear parabolic equations**, this Journal, 11(1980), 842-847.

12. H. A. Levine and M. W. Smiley, **Abstract wave equations with a singular nonlinear forcing term**, J. Math. Anal. Appl. (in press).

13. N. Minorsky, **Introduction to Nonlinear Mechanics**, J. W. Edwards, Ann Arbor, MI, 1947.

14. L. E. Payne, **Improperly Posed Problems in Partial Differential Equations**, CBMS Regional Conference Series in Applied Mathematics 22, Society for Industrial and Applied Mathematics, Philadelphia, 1975.

15. M. H. Protter and H. F. Weinberger, **Maximum Principles in Differential Equations**, Prentice-Hall, Englewood Cliffs, NJ. 1967.

16. J. Rauch, **Singularities of solutions to semilinear wave equations**, J. Math. Pures Appl., 58(1979), 299-308.

17. J. Rauch and M. Reed, **Propagation of singularities for semilinear hyperbolic equations in one space variable**, Ann. Math., 111(1980), 531-552.

18. M. Reed, **Abstract Nonlinear Wave Equations**, Lecture Notes on Mathematics 507, Springer Verlag, New York, 1976.

19. W. Walter, **Differential and Integral Inequalities**, Ergib. der Math. Band 55, Springer-Verlag, Berlin, 1970.

20. W. Walter, **Parabolic differential equations with a singular nonlinear term**, Funkcial. Ekvac., 19(1976), 271-277.

21. J. J. Stoker, **Nonlinear Vibrations**, Interscience, New York, 1950.

22. J. I. Dias and J. Hernandez, **On the existence of a free boundary for a class of reaction-Diffusion systems**, SIAM J. Math. Anal., 15(1984), 670-685.

This paper is in final form and no version of it will be submitted for publication elsewhere.

Trends in the Theory and Practice of Non-Linear Analysis
V. Lakshmikantham (Editor)

INTERPOLATION BETWEEN SPACES OF CONTINUOUS FUNCTIONS

Alessandra Lunardi
Università di Pisa, Dipartimento di Matematica
via Buonarroti 2, 56100 PISA
ITALY

We characterize some interpolation spaces between spaces of continuous functions in a bounded open set $\Omega \subset \mathbb{R}^n$ and domains of elliptic operators with Dirichlet or mixed boundary condition.

INTRODUCTION

Let X be a Banach space with norm $\|\cdot\|$ and let $A:D(A) \subset X \to X$ be a linear operator such that:

(1) the resolvent set $\rho(A)$ of A contains a sector $S_\theta = \{z \in \mathbb{C};\ z \neq 0,\ |\arg z| < \theta\}$ with $\theta \in]\frac{\pi}{2},\pi[$ and there exists $M > 0$ such that the resolvent operator $R(z,A) = (z-A)^{-1}$ satisfies

$$\|R(z,A)\|_{L(X)} \leq \frac{M}{|z|} \qquad \forall z \in S_\theta$$

Then a semigroup of linear operators e^{tA} can be defined by the usual Dunford integral; e^{tA} is a strongly continuous at $t = 0$ if and only if $D(A)$ is dense in X. Even if this condition fails, e^{tA} has most of the properties of the usual analytic semigroups (see [15]) and A is said to generate the semigroup e^{tA}.

If A satisfies (1), for $\alpha \in]0,1[$ the interpolation spaces $D_A(\alpha,\infty)$ and $D_A(\alpha)$ are defined by (see [5], [7], [15]):

$$D_A(\alpha,\infty) = \{x \in X;\ \sup_{t \geq 0} \|t^{-\alpha}(e^{tA}x-x)\| < +\infty\}$$

$$D_A(\alpha) = \{x \in X;\ \lim_{t \to 0^+} t^{-\alpha}(e^{tA}x-x) = 0\}$$

$$\|x\|_\alpha = \|x\| + \sup_{t>0} \|t^{-\alpha}(e^{tA}x-x)\|$$

Then $D_A(\alpha,\infty)$ and $D_A(\alpha)$ are Banach spaces with the norm $\|\cdot\|_\alpha$, $D_A(\alpha)$ is the completion of $D(A)$ and of $D_A(\alpha+\varepsilon,\infty)$ (for each $\varepsilon \in]0,1-\alpha[$) in the norm $\|\cdot\|_\alpha$.

In the sequel A will be an elliptic second order differential operator in a bounded open set $\Omega \subset \mathbb{R}^n$, with Dirichlet or mixed boundary condition. Under some regularity assumptions on the coefficients and on the boundary of Ω, estimate (1) holds if $X = C(\bar{\Omega})$ is the space of the continuous functions in $\bar{\Omega}$.

Our aim is the characterization of $D_A(\alpha,\infty)$ and $D_A(\alpha)$ in this case. Such a characterization reveals useful in the study of existence and regularity for the solutions of some nonlinear evolution equations (see [5], [10], [11], [12], [14]).

ASSUMPTIONS AND RESULTS

Throughout this paper it will be assumed that:

(2) $\Omega \subset \mathbb{R}^n$ is a bounded open set with boundary $\partial\Omega$ of class C^3

(3) $a_{ij}, b_i, c : \bar{\Omega} \longrightarrow \mathbb{C}$ are continuous $(i,j = 1,..,n)$

(4) $\mathrm{Re} \sum_{i,j=1}^{n} a_{ij}(x)\xi_i\xi_j > 0 \quad \forall x \in \bar{\Omega},\ \forall \xi = (\xi_1,..,\xi_n) \in \mathbb{R}^n \backslash \{0\}$

(5) $d_i, e : \partial\Omega \longrightarrow \mathbb{C}$ are continuously differentiable $(i = 1,..,n)$

(6) $\sum_{i=1}^{n} d_i(x)\upsilon_i(x) \neq 0 \quad \forall x \in \partial\Omega$

where $\upsilon(x) = (\upsilon_1(x),..,\upsilon_n(x))$ is the unit outward normal vector to $\partial\Omega$ at x.

Set:

$$\Lambda = \sum_{i,j=1}^{n} a_{ij}(\cdot)D_{ij} + \sum_{i=1}^{n} b_i(\cdot)\, D_i + c(\cdot)$$

$$B = \sum_{i=1}^{n} d_i(\cdot)D_i + e$$

Then the following results hold:

<u>Theorem 1</u> ([16], [17]) Let (2),..,(6) hold and set:

$X = C(\bar{\Omega})$ (X is endowed with the sup norm $\|\cdot\|_\infty$)

$$(7)\quad \begin{cases} D(A_1) = \{\phi \in \bigcap_{q\geq 1} W^{2,q}(\Omega);\ \Lambda\phi \in C(\bar{\Omega}),\ \ \phi(x) = 0 \ \ \forall x \in \partial\Omega\} \\ \\ D(A_2) = \{\phi \in \bigcap_{q\geq 1} W^{2,q}(\Omega);\ \Lambda\phi \in C(\bar{\Omega});\ \ B\phi(x) = 0 \ \ \forall x \in \partial\Omega\} \end{cases}$$

Then there exists $\omega \in \mathbb{R}$ such that the operators:

$$(8)\quad \begin{cases} A_1 : D(A_1) \longrightarrow X;\ A_1\phi = \Lambda\phi - \omega\phi \\ \\ A_2 : D(A_2) \longrightarrow X;\ A_2\phi = \Lambda\phi - \omega\phi \end{cases}$$

satisfy (1). Moreover there exist $M > 0$, $\theta \in]\frac{\pi}{2},\pi[$ such that for each $z \in S_\theta$ and $\phi \in X$ we have:

$$(9) \qquad |z|\,\|R(z,A_i)\phi\|_\infty + |z|^{\frac{1}{2}} \sum_{j=1}^{n} \|D_j(R(z,A_i)\phi\|_\infty \leq M\|\phi\|_\infty$$

Before stating our main result, we recall the definitions of some functional spaces. For $\theta \in]0,1[$, $C^\theta(\bar{\Omega})$ is the space of all Hölder continuous functions of order θ in $\bar{\Omega}$, endowed with the usual norm. $h^\theta(\bar{\Omega})$ is the closure of $C^1(\bar{\Omega})$ in the C^θ-norm; it consists of all little-Hölder continuous functions of order θ, i.e. of all $\phi : \bar{\Omega} \longrightarrow \mathbb{C}$ such that

$$\lim_{\tau \to 0^+} \sup_{0<|x-y|<\tau} (|x-y|^{-\theta}|\phi(x)-\phi(y)|) = 0$$

If $k \in \mathbb{N}$ and $\theta \in]0,1[$, $C^{k+\theta}(\bar{\Omega})$ is the set of all differentiable functions on $\bar{\Omega}$ whose k-th derivatives belong to $C^\theta(\bar{\Omega})$, it is endowed with the usual norm. $h^{k+\theta}(\bar{\Omega})$ is the closure of $C^{k+1}(\bar{\Omega})$ in the $C^{k+\theta}$-norm; it consists of all $\phi \in C^{k+\theta}(\bar{\Omega})$ whose k-th derivatives belong to $h^\theta(\bar{\Omega})$.

If $Y(\bar{\Omega})$ is a subset of $C(\bar{\Omega})$ we put $Y_0(\bar{\Omega}) = \{\phi \in Y(\bar{\Omega}),\ \phi(x) = 0\ \forall x \in \partial\Omega\}$ and, if $Y(\bar{\Omega}) \subset C^1(\bar{\Omega})$, we put $Y_B(\bar{\Omega}) = \{\phi \in Y(\bar{\Omega});\ B\phi(x) = 0\ \forall x \in \partial\Omega\}$. It can be shown that for each $\theta \in]0,1[$, $h_0^\theta(\bar{\Omega})$ is the closure of $C_0^1(\bar{\Omega})$ in the C^θ-norm, and that $h_0^{1+\theta}(\bar{\Omega})$ (resp. $h_B^{1+\theta}(\bar{\Omega})$) is the closure of $C_0^2(\bar{\Omega})$ (resp. $C_B^2(\bar{\Omega})$) in the $C^{1+\theta}$-norm.

Finally, if Z and Y are Banach spaces, $Y \hookrightarrow Z$ means that Y is continuously embedded in Z, $Y \cong Z$ means that $Y \hookrightarrow Z$ and $Z \hookrightarrow Y$.

<u>Theorem 2</u> Let (2),..,(6) hold, let A_1 and A_2 be given by (8). Then we have, for each $\alpha \in]0,\frac{1}{2}[$:

$$(10) \qquad D_{A_1}(\alpha,\infty) \cong C_0^{2\alpha}(\bar{\Omega})$$

$$(11) \qquad D_{A_1}(\alpha) = h_0^{2\alpha}(\bar{\Omega})$$

If $a_{ij} \in C^\varepsilon(\bar{\Omega})$ for some $\varepsilon > 0$ $(i,j = 1,..,n)$, then (10) and (11) hold also for each $\alpha \in]\frac{1}{2},1[$.

If $a_{ij} \in C^1(\bar{\Omega}), d_i, e \in C^2(\partial\Omega)$ $(i,j = 1,..,n)$ we have, for each $\alpha \in]0,\frac{1}{2}[$:

$$(12) \qquad D_{A_2}(\alpha,\infty) \cong C^{2\alpha}(\bar{\Omega})$$

$$(13) \qquad D_{A_2}(\alpha) = h^{2\alpha}(\bar{\Omega})$$

Finally, if $a_{ij} \in C^2(\bar{\Omega})$, $d_i, e \in C^3(\partial\Omega)$ $(i,j = 1,..,n)$ and $\partial\Omega \in C^4$ we have, for all $\alpha \in]\frac{1}{2},1[$:

$$(14) \qquad D_{A_2}(\alpha,\infty) \cong C_B^{2\alpha}(\bar{\Omega})$$

(15) $D_{A_2}(\alpha) \cong h_B^{2\alpha}(\bar{\Omega})$

PROOF OF THEOREM 2

Here we shall give a brief proof which utilizes, togheter with theorem 1, also some general results of the interpolation theory and some classical regularity results about linear parabolic equations. A more direct approach can be found in [1], where also the case $\alpha = \frac{1}{2}$ is studied and the assumptions on the regularity of a_{ij} are weaker.

The results that will be used here are the following:

Proposition 3 ([7], [8], [12]) Let $A:D(A) \subset X \to X$ satisfy (1). Then for all $\alpha \in]0,1[$ we have:

(16) $D_A(\alpha,\infty) \cong (D(A),X)_{1-\alpha,\infty}$

where $(D(A),X)_{1-\alpha,\infty}$ is the real interpolation space between $D(A)$ and X (see [6], [7]). Moreover:

(17) $D_A(\alpha,\infty) = \{x \in X;\ \sup_{t>0} \|t^{\alpha}AR(t,A)x\| < +\infty\}$

and there exists $k \geq 1$ such that:

(18) $\frac{1}{k}\|x\|_{\alpha} \leq \|x\| + \sup_{t>0} \|t^{\alpha}AR(t,A)x\| \leq k\|x\|_{\alpha} \quad \forall x \in D_A(\alpha,\infty)$

PROPOSITION 4 ([5], [12], [15], [18 p. 60]). Let $A:D(A) \subset X \to X$ satisfy (1). For $\beta \in]0,1[$ set:

$$D_A(\beta+1,\infty) = \{x \in D(A); Ax \in D_A(\beta,\infty)\}; D_A(\beta+1) = \{x \in D(A); Ax \in D_A(\beta)\}$$

$$\|x\|_{\beta+1} = \|x\| + \|Ax\|_{\beta} \qquad \forall x \in D_A(\beta+1,\infty).$$

Then the operators;

$$A_{\beta}:D_A(\beta+1,\infty) \to D_A(\beta,\infty) \ ; \ A_{\beta}x = Ax$$
$$\bar{A}_{\beta}:D_A(\beta+1) \to D_A(\beta) \ ; \ \bar{A}_{\beta}x = Ax$$

satisfy (1) with X replaced by $D_A(\beta,\infty)$ (resp. $D_A(\beta)$). For each $\gamma \in]0,1[$, $\gamma \neq 1-\beta$, we have also:

(19) $D_{A_{\beta}}(\gamma,\infty) \cong D_A(\beta+\gamma,\infty)$

Theorem 5 [6 p. 320]. Let (2),..,(6) hold and let $\alpha \in]0,1[$, $\alpha \neq \frac{1}{2}$, $a_{ij} \in C^{2\alpha}(\bar{\Omega})$, $\partial\Omega \in C^{2+2\alpha}$, $d_i, e \in C^{1+2\alpha}(\partial\Omega)$ $(i,j = 1,..,n)$, $g \in C^{2+2\alpha}(\bar{\Omega})$. If $\alpha < \frac{1}{2}$, assume that the compatibility condition

(20) $Bg = \sum_{i=1}^{n} d_i(\cdot)g_{x_i}(\cdot) + e(\cdot)g(\cdot) = 0$ on $\partial\Omega$

holds; if $\alpha > \frac{1}{2}$ assume also that

(21) $B(\sum_{i,j=1}^{n} a_{ij}(\cdot)g_{x_ix_j}(\cdot)) = 0$ on $\partial\Omega$.

Then for all $w \in \mathbb{R}$ the solution of the problem

$$(22)\quad \begin{cases} u_t(t,x) = \sum_{i,j=1}^{n} a_{ij}(t,x)u_{x_ix_j}(t,x) + wu(t,x) & t \in [0,1], x \in \bar{\Omega} \\ u(0,x) = g(x) & x \in \bar{\Omega} \\ Bu(t,x) = 0 & t \in [0,1], x \in \partial\Omega \end{cases}$$

is such that the function $v(t) = u_t(t,\cdot)$ belongs $C^{\alpha}([0,1];C(\bar{\Omega}))$, and there exists $C > 0$ not depending on g, such that:

$$(23)\quad \|v\|_{C^{\alpha}([0,1];C(\bar{\Omega}))} \leq C\|g\|_{C^{2+2\alpha}(\bar{\Omega})}$$

Proof of theorem 2 First we shall prove the inclusions $\hookrightarrow$ of (10) and (12), using the results of theorem 1 and the characterization (17).

Let $\alpha \in]0,\frac{1}{2}[$ and let $f \in D_{A_i}(\alpha,\infty)$ (i may be 1 or 2). For each $t \geq 1$ we have:

$$(24)\quad tR(t,A_i)f = R(1,A_i)f - \int_1^t R(s,A_i)A_iR(s,A_i)f\,ds$$

Then for each $x,y \in \bar{\Omega}$ with $|x-y| \leq 1$ and for each $t \geq 1$ we have:

$$|f(x)-f(y)| \leq |f(x)-[tR(t,A_i)f](x)| + |[tR(t,A_i)f](x)-[tR(t,A_i)f](y)| + \\ + |[tR(t,A_i)f](y)-f(y)|$$

(24), (9) and (18) imply that:

$$|[tR(t,A_i)f](x)-[tR(t,A_i)f](y)| \leq \sum_{j=1}^{n} (\|D_j(R(1,A_i)f)\|_\infty +$$

$$+ \int_1^t \|D_jR(s,A_i)A_iR(s,A_i)f\|_\infty ds)|x-y| \leq M(\|f\|_\infty + k\int_1^t s^{-\frac{1}{2}-\alpha}ds\|f\|_{D_{A_i}(\alpha,\infty)})|x-y|$$

and then, recalling that $f - tR(t,A)f = -AR(t,A)f$ and using again (18), we get, for $t = |x-y|^{-\frac{1}{2}}$:

$$(25)\quad |f(x)-f(y)| \leq (2k+M+\frac{k}{\frac{1}{2}-\alpha})\|f\|_{D_{A_i}(\alpha,\infty)}|x-y|^{2\alpha}$$

so that f is Hölder continuous with exponent 2α. Moreover, as $D_{A_i}(\alpha,\infty) \subset \overline{D(A_i)}$, if $f \in D_{A_1}(\alpha,\infty)$ then f vanishes on $\partial\Omega$.

Assuming now that $a_{ij} \in C^{\varepsilon}(\bar{\Omega})$ we shall prove the inclusions $\hookrightarrow$ of (10) and (14) for $\alpha \in]\frac{1}{2},1[$. Set:

$$(26)\quad D(\Lambda_1) = D(A_1);\ \Lambda_1\phi = \sum_{i,j=1}^{n} a_{ij}(\cdot)\phi_{x_ix_j}(\cdot) - w\phi(\cdot)$$

$$(27)\quad D(\Lambda_2) = D(A_2);\ \Lambda_2\phi = \sum_{i,j=1}^{n} a_{ij}(\cdot)\phi_{x_ix_j}(\cdot) - w\phi(\cdot)$$

where $w \in \mathbb{R}$ is such that Λ_1 and Λ_2 satisfy (1.1) and are invertible. Then, as the graph normy of Λ_i and A_i are equivalent, by (16) we have $D_{A_i}(\alpha,\infty) \cong D_{\Lambda_i}(\alpha,\infty)$. By Schauder's theorem and Agmon-Douglis-Nirenberg results (see [2], [13]) we have, for $\beta = \frac{\varepsilon}{2}$:

$$D_{\Lambda_1}(\beta+1,\infty) \cong \{\phi \in C^{2+2\beta}(\bar{\Omega}); \phi(x) = \Lambda_1\phi(x) = 0 \quad \forall x \in \partial\Omega\}$$

$$D_{\Lambda_2}(\beta+1,\infty) \cong \{\phi \in C^{2+2\beta}(\bar{\Omega}); \phi(x) = 0 \quad \forall x \in \partial\Omega\}$$

Take $\gamma = \alpha-\beta$. Using the notations and the results of proposition 4 and the characterization (16) we get:

$$D_{\Lambda_i}(\alpha,\infty) = D_{\Lambda_i}(\alpha,\infty) \cong (D_{\Lambda_{i\beta}}(\alpha,\infty); D_{\Lambda_i}(\alpha,\infty))_{1-\gamma,\infty}$$

But $D_{\Lambda_i}(\beta+1,\infty) \hookrightarrow C^{2+2\beta}(\bar{\Omega})$ and $D_{\Lambda_i}(\alpha,\infty) \hookrightarrow C^{2\alpha}(\bar{\Omega})$ imply that $D_{\Lambda_i}(\alpha,\infty) \hookrightarrow (C^{2+2\beta}(\bar{\Omega}), C^{2\beta}(\bar{\Omega}))_{1-\gamma,\infty}$ which is isomorphic to $C^{2\beta+2\gamma}(\bar{\Omega}) = C^{2\alpha}(\bar{\Omega})$ (see [18]). Moreover, using (24) it is easy to see that, if f belongs to $D_{\Lambda_i}(\alpha,\infty)$ $(\alpha \in]\frac{1}{2},1[)$ then $tR(t,\Lambda_i)f \to f$ in $C^1(\bar{\Omega})$, so that $f \in D_{\Lambda_i}(\alpha,\infty)$ implies $f(x) = 0$ $\forall x \in \partial\Omega$ and $f \in D_{\Lambda_2}(\alpha,\infty)$ implies $Bf(x) = 0$ $\forall x \in \partial\Omega$.

Let us prove now the inclusions $\hookleftarrow$. The inclusion $\hookleftarrow$ of (10) is obvious for all $\alpha \neq \frac{1}{2}$ because $C_0^2(\bar{\Omega}) \hookrightarrow D(A_1)$, and hence, by (16):

$$C_0^{2\alpha}(\bar{\Omega}) \cong (C_0^2(\bar{\Omega}), C_0(\bar{\Omega}))_{1-\alpha,\infty} \hookrightarrow (D(A_1),X)_{1-\alpha,\infty} \cong D_{A_1}(\alpha,\infty)$$

(for the first equivalence see [12]).

To prove the inclusion $\hookleftarrow$ of (12) and (14) it will be used theorem 5.

Let $\partial\Omega \in C^3$, $a_{ij} \in C^1(\bar{\Omega}), d_i, e \in C^2(\bar{\Omega})$ and $f \in C^{2\alpha}(\bar{\Omega})$, $\alpha \in]0,\frac{1}{2}[$ (resp. $\partial\Omega \in C^4$, $a_{ij} \in C^2(\bar{\Omega}), d_i, e \in C^3(\partial\Omega)$ and $f \in C_B^{2\alpha}(\bar{\Omega})$, $\alpha \in]\frac{1}{2},1[$) and set $g = \Lambda_2^{-1}f$, where Λ_2 is defined in (27). Then g belongs to $C_{t\Lambda_2}^{2+2\alpha}(\bar{\Omega})$ and satisfies (20) (resp. (20) and (21), and $e^{t\Lambda_2}f = \Lambda_2 e^{t\Lambda_2} g = u_t(t,\cdot)$, where u is the solution of (22). Then we have, using (23) and the estimate $M_0 = \sup_{t>0} \|e^{t\Lambda_2}\|_{L(X)} < +\infty$ (which is a consequence of (1), see [15]):

$$\sup_{t>0} \|t^{-\alpha}(e^{t\Lambda_2}f-f)\|_\infty \le \sup_{0<t\le 1} \|t^{-\alpha}(e^{t\Lambda_2}f-f)\|_\infty + \sup_{t\ge 1} \|t^{-\alpha}(e^{t\Lambda_2}f-f)\|_\infty \le$$

$$\le C\|g\|_{C^{2+2\alpha}(\bar{\Omega})} + (M_0+1)\|f\|_\infty \le C\|\Lambda_2^{-1}\|_{L(C^{2\alpha}(\bar{\Omega}), C^{2+2\alpha}(\bar{\Omega}))} \|f\|_{C^{2\alpha}(\bar{\Omega})}$$

so that $C^{2\alpha}(\bar\Omega) \hookrightarrow D_{\Lambda_2}(\alpha,\infty) \cong D_{\Lambda_2}(\alpha,\infty)$ (resp. $C_B^{2\alpha}(\bar\Omega) \hookrightarrow D_{\Lambda_2}(\alpha,\infty) \cong$ $\cong D_{A_2}(\alpha,\infty)$)

Let us prove now (11). Let $f \in D_{A_1}(\alpha)$ and $\varepsilon \in\,]0,2-2\alpha[$, then there exist a sequence $\{f_n\} \subset D(A_1) \hookrightarrow C^{2\alpha+\varepsilon}(\bar\Omega)$ which converges to f in $D_{A_1}(\alpha,\infty)$, and hence in $C^{2\alpha}(\bar\Omega)$. Then f belongs to $h_0^{2\alpha}(\bar\Omega)$. Let now $f \in h_0^{2\alpha}(\bar\Omega)$ and $\varepsilon \in\,]0,2-2\alpha[$: then there exists a sequence $\{f_n\} \subset C_0^{2\alpha+\varepsilon}(\bar\Omega) = D_{A_1}(\alpha+\varepsilon/2,\infty)$ which converges to f in $C^{2\alpha}(\bar\Omega)$ and hence in $D_{A_1}(\alpha,\infty)$: then f belongs to $D_{A_1}(\alpha)$. The proof of (1.13) and (1.16) is analogous.

<u>Remark 6</u> From theorem 2 and proposition 4 it follows that for each $\beta \in\,]0,1[$ the operators:

$$A_{1|D_\beta} : D_\beta \to C_0^\beta(\bar\Omega);\ D_\beta = \{\phi \in C_0^{2+\beta}(\bar\Omega);\ A_1\phi \in C_0^\beta(\bar\Omega)\}$$

$$A_{2|C_B^{2+\beta}(\bar\Omega)} : C_B^{2+\beta}(\bar\Omega) \to C^\beta(\bar\Omega)$$

satisfy estimate (1) and hence generate analytic semigroups in the spaces $C_0^\beta(\bar\Omega)$ and $C^\beta(\bar\Omega)$ respectively. In the variational case, with Dirichlet or Neumann boundary condition, this result was proved also in [3] and [4] with different methods.

REFERENCES

[1] Acquistapace, P., Terreni, B.: Characterization of some interpolation spaces, Pubb. Dip. Mat ., Univ. Pisa (June 1984).

[2] Agmon, S., Douglis, A., Nirenberg, L., Estimates near the boundary for solutions of elliptic partial differential equations satisfying general boundary conditions (II), Comm. Pure Appl. Math. 17 (1964) 35-92.

[3] Campanato, S., Generation of analytic semigroups by elliptic operators of second order in Hölder spaces, Ann. S.N.S. Pisa 8 (1981) 495-512.

[4] Campanato, S., Generation of analytic semigroups in the Hölder topology by elliptic operators of second order with Neumann boundary condition, to appear in Le Matematiche.

[5] Da Prato, G., Grisvard, P., Equations d'evolution abstraites non linéaires de type parabolique, Ann. Mat. Pura Appl. 120 (1979) 329-396.

[6] Ladyzhenskaja, D.A., Solonnikov, V.A., Ural'čeva, N.N., Linear and quasilinear equations of parabolic type (Am . Math. Soc., Providence 1968).

[7] Lions, J.L., Théorèmes de trace et d'interpolation (I), Ann. S.N.S. Pisa 13 (1959) 389-403.

[8] Lions, J.L., Peetre, J.: Sur une classe d'espaces d'interpolation,

Publ. I.H.E.S. Paris (1964) 5-68.
[9] Lunardi, A., Analyticity of the maximal solution of an abstract nonlinear parabolic equation, Nonl. An. 6 (1982) 503-521.
[10] Lunardi, A., Abstract quasilinear parabolic equations, Math. Ann. 267 (1984) 395-415.
[11] Lunardi, A., Asymptotic exponential behavior in quasilinear parabolic equations, to appear in Nonl. An.
[12] Lunardi, A., Characterization of interpolation spaces between domains of elliptic operators and spaces of continuous functions with applications to nonlinear parabolic equations, to appear in Math. Nachr.
[13] Schauder, J., Über lineare elliptische Differentialgleichungen zweiter Ordnung, Math. Z. 38 (1934) 257-282.
[14] Sinestrari, E., Continuous interpolation spaces and spatial regularity in nonlinear Volterra integrodifferential equations, J. Int. Eq. 5 (1983) 287-308.
[15] Sinestrari, E., On the abstract Cauchy problem in spaces of continuous functions, to appear in J. Math. An. Appl.
[16] Stewart, H.B., Generation of analytic semigroups by strongly elliptic operators, Trans. Am. Math. Soc. 199 (1974) 141-162.
[17] Stewart, H.B., Generation of analytic semigroups by strongly elliptic operators under general boundary conditions, Trans. Am. Math. Soc. 259 (1980) 299-310.
[18] Triebel, H., Interpolation theory, function spaces, differential operators (North Holland, Amsterdam, 1978).

This paper is in final form and no version of it will be submitted for publication elsewhere.

Trends in the Theory and Practice of Non-Linear Analysis
V. Lakshmikantham (Editor)

CONSTRUCTIVE EXISTENCE OF SOLUTION FOR NEGATIVE EXPONENT GENERALIZED EMDEN-FOWLER NODAL PROBLEMS

C. D. Luning

Department of Mathematics
Sam Houston State University,
Huntsville, TX 77340

W. L. Perry

Department of Mathematics
Texas A&M University,
College Station, TX 77843

A constructive proof is given for the existence of solutions to the nodal problem for generalized Emden-Fowler equations $y''(x) + a(x)y(x)|y(x)|^{\alpha} = 0$, $0 < x < 1$, $\alpha > -2$ with $a(x)$ continuous on $[0,1]$ and positive on $(0,1)$. Numerical results are also included.

1. INTRODUCTION.

The nodal problem for the generalized Emden-Fowler equation is: For each positive integer n, find a solution $y_n(x)$ of

$$y''(x) + a(x)y(x)|y(x)|^{\alpha} = 0, \quad 0 < x < 1 \tag{1}$$

$$y(0) = y(1) = 0, \quad a(x) \in C^{+}[0,1]$$

such that $y_n(x)$ has exactly n zeroes interior to the interval $(0,1)$.

For $\alpha > 0$, the superlinear case, existence of solution has been shown by variational methods [3], degree theoretic methods [5], and fixed point methods [1]. The fixed point method in [1] is also applicable to the particular sublinear case of $-1 < \alpha < 0$. In this paper we prove existence of solution of the nodal problem for the sublinear case $-2 < \alpha < -1$. The method of proof is different from those in [1], [3] or [5] in that our solutions are obtained as uniform limits of monotone sequences of Picard type iterations (see [2]). Implementation of the procedure is easily achieved by using numerical integration techniques for the Picard iterations and a bisection technique for placing the nodes.

Instead of solving the nodal problem for (1) directly, we solve the following nonlinear eigenvalue problem: For $a(x)$ as in (1) and for each positive integer n, find a function $u_n(x)$ in $C^2(0,1)$ with exactly n zeros interior to $(0,1)$ and find $\lambda_n > 0$ such that

$$u_n''(x) + \lambda_n a(x)u_n(x)|u_n(x)|^{\alpha} = 0, \quad 0 < x < 1$$

$$u_n(0) = u_n(1) = 0, \quad u_n'(0^{+}) = 1$$

A solution $y_n(x)$ to the nodal problem for (1) is obtained from a solution pair $(u_n(x), \lambda_n)$ of (2) by letting $y_n(x) = \lambda_n^{1/\alpha} u_n(x)$.

2. EXISTENCE OF SOLUTION FOR THE SINGLE NODE PROBLEM

We first establish the existence of the solution for the single node problem for (2). In section 3, the technique is extended to solve multinode problems. We use the following results proved in [2] and [4].

In [2] we considered the problem

$$\begin{aligned} &v'' + \lambda A(x)v^{\alpha+1} = 0 \qquad 0 < x < 1 \\ &v(0) = v(1) = 0, \quad v'(0^+) = b > 0, \quad A \in C^+[0,1] \end{aligned} \tag{3}$$

Let $G(x,\xi)$ be the Green's function for $u'' = 0$, $u(0) = u(1) = 0$. Define the sequences $\{v_k(x)\}_{k=0}^{\infty}$ and $\{\lambda_k\}_{k=1}^{\infty}$ by $v_0(x) = bx$ and for $k = 0,1,2,...$

$$\begin{aligned} \lambda_{k+1} &= b\left(\int_0^1 (1-\xi)A(\xi)v_k^{\alpha+1}(\xi)d\xi\right)^{-1} \\ v_{k+1}(x) &= \lambda_{k+1} \int_0^1 G(x,\xi)A(\xi)v_k^{\alpha+1}(\xi)d\xi. \end{aligned} \tag{4}$$

Theorem 1 (see [2]): For $-2 < \alpha < -1$, there exists a solution pair $(v(x),\lambda)$ of (3) with $v(x) > 0$, $0 < x < 1$ and $\lambda > 0$. Moreover $v(x)$ is the uniform limit of the sequence $\{v_k(x)\}$ and λ is the limit of the sequence $\{\lambda_k\}$ defined by (4). Also the sequences $\{v_k(x)\}$ and $\{\lambda_k\}$ are alternating monotone in that for $k = 1,2,3,...$ and $0 < x < 1$,

$$\begin{aligned} &0 < v_{2k-1}(x) < v_{2k+1}(x) < v_{2k}(x) < v_{2k-2}(x) \\ &0 < \lambda_{2k} < \lambda_{2k+2} < \lambda_{2k+1} < \lambda_{2k-1}\,. \end{aligned} \tag{5}$$

We note that for the solution pair $(v(x),\lambda)$ we have

$$\begin{aligned} \lambda &= b\left(\int_0^1 (1-\xi)A(\xi)v^{\alpha+1}(\xi)d\xi\right)^{-1} \\ v(x) &= \lambda\int_0^1 G(x,\xi)A(\xi)v^{\alpha+1}(\xi)d\xi \end{aligned} \tag{6}$$

In [4], Taliaferro proved the following result.

<u>Theorem 2</u>: For $A(x)$ positive and continuous on the open interval (0,1) and for $\alpha < -1$, the boundary value problem

$$\begin{aligned} &v''(x) + A(x)v^{\alpha+1}(x) = 0 \qquad 0 < x < 1 \\ &v(0) = v(1) = 0 \end{aligned} \tag{7}$$

has a solution if and only if $\int_0^1 x(1-x)A(x)dx < \infty$, in which case the solution is unique. Furthermore $v'(1^-)$ is finite if and only if $\int_{1/2}^1 A(x)(1-x)^{\alpha+1}dx < \infty$ and $v'(0^+)$ is finite if and only if $\int_0^{1/2} A(x)x^{\alpha+1}dx < \infty$. Moreover there exists β such that as $x \to 1^-$,

$$v(x) = \beta(1-x) + \beta^{\alpha+1}(1 + o(1)) \int_x^1 (x-\xi)A(\xi)\xi^{\alpha+1}d\xi \tag{8}$$

and $\gamma > 0$ such that as $x \to 0^+$

$$v(x) = \gamma x - \gamma^{\alpha+1}(1 + o(1)) \int_0^x (x-\xi)A(\xi)\xi^{\alpha+1}d\xi. \tag{9}$$

We now generate a solution to the single node problem of (2) for $-2 < \alpha < -1$ by splicing solutions of problems of type (6). Let $a(x)$ be the coefficient function in (2) and for each c, $0 < c < 1$, let $(w_c(t), \mu_c)$ be the positive solution pair of

$$w_c''(t) + \mu_c a(ct)w_c^{\alpha+1}(t) = 0 \qquad 0 < t < 1 \tag{10}$$

$$w_c(0) = w_c(1) = 0, \quad w_c'(0^+) = 1$$

and let $(z_c(t), \rho_c)$ be the positive solution pair of

$$\begin{gathered} z_c''(t) + \rho_c a((1-c)t + c)z_c^{\alpha+1}(t) = 0 \qquad 0 < t < 1 \\ z_c(0) = z_c(1) = 0, \quad z_c'(0) = w_c'(1). \end{gathered} \tag{11}$$

The existence and uniqueness of $w_c(t)$, μ_c, $z_c(t)$ and ρ_c follows from theorem 2. Also since $a(x)$ is continuous on $[0,1]$ and $-2 < \alpha < -1$, it follows that

$$\int_{1/2}^1 a(ct)(1-t)^{\alpha+1}\, dt < \infty$$

and thus $w_c'(1^-) < \infty$. From theorem 1 it now follows that $w_c(t)$, μ_c, $z_c(t)$ and ρ_c are limits of appropriate alternating monotone sequences. Define the function $u_c(x)$ and the values $\hat{\lambda}_c$ and $\check{\lambda}_c$ respectively by

$$u_c(x) = \begin{cases} cw_c(x/c) & 0 \le x \le c \\ (c-1)z_c((x-c)/(1-c)) & c \le x \le 1 \end{cases} \tag{12}$$

$$\hat{\lambda}_c = \mu_c/c^{\alpha+2}, \quad \check{\lambda}_c = \rho_c/(1-c)^{\alpha+2}.$$

It is easily verified that $u_c(x)$ satisfies the equation

$$u_c''(x) + \hat{\lambda}_c a(x) u_c^{\alpha+1}(x) = 0$$

$$u_c(0) = u_c(c^-) = 0$$

and

$$u_c''(x) + \check{\lambda}_c a(x) u_c^{\alpha+1}(x) = 0 \qquad c < x < 1$$

$$u_c(c^+) = u_c(1) = 0$$

Moreover

$$u_c'(c^+) = u_c'(c^-).$$

The final step to the proof of the existence of a solution pair $(u(x),\lambda)$ to the single node problem for (2) is to show there exists a c such that $\hat{\lambda}_c = \check{\lambda}_c$. Then for this value of c, the function $u_c(x)$ is the desired solution $u(x)$ and the eigenvalue λ is $\lambda = \hat{\lambda}_c = \check{\lambda}_c$.

Lemma 1: The eigenvalues $\hat{\lambda}_c$ and $\check{\lambda}_c$ have the properties

a) $\lim_{c\to 0^+} \hat{\lambda}_c = +\infty$, b) $\lim_{c\to 1^-} \hat{\lambda}_c < \infty$, c) $\lim_{c\to 1^-} \check{\lambda}_c = +\infty$, d) $\lim_{c\to 1^-} \check{\lambda}_c = +\infty$

e) $\hat{\lambda}_c$ and $\check{\lambda}_c$ depend continuously on c.

Proof: From (6) we have

$$\mu_c = \left(\int_0^1 (1-\xi)a(c\xi)w_c^{\alpha+1}(\xi)d\xi\right)^{-1}$$

and (13)

$$\rho_c = w_c'(1)\left(\int_0^1 (1-\xi)a(c\xi)z^{\alpha+1}(\xi)d\xi\right)^{-1}$$

Using the asymptotic properties (8) and (9), we have the integrals in μ_c and ρ_c remaining bounded as $c \to 0^+$. Thus properties a) and d) follow from the definitions of $\hat{\lambda}_c$ and $\check{\lambda}_c$ in (12). Properties b) and c) are a direct result from Theorem 2. Finally property (e) is a consequence of the formulas for μ_c and ρ_c along with the continuity of $a(x)$.

From lemma 1, it now follows that there exists a value of c such that $\check{\lambda}_c = \hat{\lambda}_c$. Thus we have

Theorem 3: For $-2 < \alpha < -1$, there exists a solution to the single node problem for (2) and therefore for (1).

3. EXISTENCE OF SOLUTION FOR THE MULTI-NODE PROBLEM

For the two node problem (2), the procedure is to take a solution $u_{1,c}(x)$ to the single node problem (2) on $(0,c)$, which can of course be obtained from a single node solution on $(0,1)$ by the change of variables $t = x/c$ as in (12), and adjoin a solution to

$$u''_{1,r}(x) + \check{\lambda}_c a(x) u_{1,r}^{\alpha+1}(x) = 0 \qquad c < x < 1$$

$$u_{1,r}(c^+) = u_{1,r}(1) = 0 \tag{14}$$

$$u'_{1,r}(c) = u'_{1,c}(c) .$$

<u>Lemma 2</u>: The solution $(u_{1,c}(x), \lambda_{1,c})$ of the single node problem on $(0,c)$ for

$$u''_{1,c}(x) + \lambda_{1,c} a(x) u_{1,c}^{\alpha+1}(x) = 0 \qquad 0 < x < c$$

$$u_{1,c}(0) = u_{1,c}(c) = 0$$

$$u'_{1,c}(0) = 1$$

has the properties: a) $\lim_{c \to 0^+} \lambda_{1,c} = +\infty$, b) $\lim_{c \to 1^-} \lambda_{1,c} < \infty$, c) $u'_{1,c}(c)$ is bounded for all $c > 0$.

<u>Proof</u>: As $c \to 0$, the first node c_1, approaches zero. Since $u_{1,c}(x)$ is a positive solution on $(0,c_1)$ property a) is a direct result of Lemma 1. Property b) follows from Theorem 3 and continuity of $\lambda_{1,c}$ in c. Property c) follows from Theorem 2.

From lemma 1 we have $\lim_{c \to 0^+} \lambda_c < \infty$ and $\lim_{c \to 1^-} \lambda_c = +\infty$. Since $\lambda_{1,c}$ and λ_c depend continuously on c it follows there is a value c_2 such that $\lambda_{1,c_2} = \rho_{c_2}$. Let this common value be denoted by λ_2 and define the function $u_2(x)$ by

$$u_2(x) = \begin{cases} u_{1,c_2}(x) & 0 \le x \le c_2 \\ u_{1,r}(x) & c_2 \le x \le 1 . \end{cases}$$

The pair $(u_2(x), \lambda_2)$ is then a solution to the two node problem for (2).

For the n-node problem proceed inductively by matching a solution $(u_{n-1,c}, \lambda_{n-1,c})$ of the (n-1)-node problem on $(0,c)$ to a solution of

$$u'' + \check{\lambda}_c a(x) u^{1+\alpha} = 0 \qquad 0 < x < 1$$

$$u(c^+) = u(1) = 0$$

$$u'(c) = (-1)^n u'_{n-1}(c).$$

The existence of c such that $\lambda_{c_n} = \check{\lambda}_{n-1,c}$ is accomplished using results similar to lemma 1 and lemma 2.

We have thus shown

Theorem 4: For each positive integer n, the nodal problem for the generalized Emden-Fowler equation (1) has a solution for $-2 < \alpha < -1$.

We note that the results in (2) are also valid for $\alpha > -1$ and thus theorem 4 also holds for $\alpha > -1$. Of course the result of theorem 4 is already known for $\alpha > -1$ (see [1]), however our constructive method is new.

4. Example

Consider the single node problem for

$$u''(x) + \lambda x u(x)|u(x)|^2 = 0, \quad 0 < x < 1$$

$$u(0^+) = u(1^-) = 0, \quad u'(0) = 1.$$

The problem was solved by solving (8) and (9) numerically utilizing the iterative method in [2]. The value c_1 of c was found by bisection of a bracketing interval.

We obtained a node of .54767 with $\lambda = 2513.78$. As noted before the solution of the corresponding problem (1) is $y_1(x) = (2513.78)^{1/2} u_1(x)$.

As a computational note, we mention the sensitivity of λ_c to changes in c. For example, as c: $1 \to .56 \to .55$, λ_c: $95.16 \to 2248.93 \to 2460.96$. Similar behavior occurs for ρ_c. As the number of nodes increases and the first node is pushed toward the origin this sensitivity will limit the use of our method for efficient computational use.

REFERENCES

[1] Gustafson, G., Fixed point methods for nodal problems in differential equations, in: Swaminathan, S. (ed.) Fixed Point Theory and its Applications (Academic Press, New York, 1976).

[2] Luning, C. D., and Perry, W. L., Positive solutions of negative exponent generalized Emden-Fowler boundary value problems, SIAM J. Math. Anal. 12 (1981) 874-879.

[3] Nehari, Z., On a class of second order differential equations, Trans. Amer. Math. Soc. 93 (1959) 101-123.

[4] Taliaferro, S., A nonlinear singular boundary value problem, Nonlinear Analysis 3 (1979) 897-904.

[5] Turner, R. E. L., Superlinear Sturm-Liouville problems, J. Diff. Eq. 13 (1973) 157-171.

This paper is in final form and no version of it will be submitted for publication elsewhere.

Trends in the Theory and Practice of Non-Linear Analysis
V. Lakshmikantham (Editor)

EXISTENCE RESULTS FOR THE INVERSE PROBLEM OF THE VOLUME POTENTIAL

Carla Maderna (*), Carlo D.Pagani (**), Sandro Salsa (*)

(*) Dipartimento di Matematica "F.Enriques", Via C.Saldini 50 - Milano (Italy)

(**) Dipartimento di Matematica del Politecnico
P.za Leonardo da Vinci, 32 - Milano (Italy)

One proves the existence of a body G, whose density is known, which creates on its surface a given potential.

INTRODUCTION

Inverse problems in potential theory arise in several fields of applied sciences, as in geophysics, biology, solid state physics, etc.. They can be roughly divided into two groups: i) one group of problems consists in finding the density δ of a given material (or charged) body G, or the surface density of a single layer supported by a given surface Γ, or the intensity of a double layer, by measuring the potentials (the volume potential, the single layer or the double layer potential respectively); these problems are linear and one of the main questions to be studied is the question of the stability (cfr., e.g., [1]); ii) in the second group of problems we assume that the distribution of mass on G or on Γ (or the intensity distributed on Γ) is known (at least, we assume a model distribution) but the figure of G or of Γ is unknown and we want to determine it by measuring the potentials; the problems, of course, are highly non linear, and usually one is interested in seeking a local solution. In this case a quite natural tool is offered by an implicit function theorem; but, as we clarify later, a phenomenon arises in all such problems which prevents the use of the ordinary implicit function theorem. The so called "hard implicit function theorem" then provides a tool to handle the problem.
Here we will concern only about the problem for the volume potential and will limit ourselves to present the main result (about the existence of a local solution) with some comments. The proofs and the illustrations of the adopted techniques are contained in the paper [2]. Results about stability are given in [3].

POSITION OF THE PROBLEM

Let us consider a class of bodies G_u parametrized by smooth functions u:

$$u: S^2 \to R \qquad |u| \leqslant \text{constant} < r_o$$

as follows (here $S^2 = \{x \in R^3 : x_1^2 + x_2^2 + x_3^2 = 1\}$ and r_o is a positive arbitrary constant). Let $\phi_u : S^2 \to R^3$ be a differentiable embedding

$$S^2 \ni \omega \to \phi_u(\omega) = \omega\,(r_o + u(\omega))$$

Define $\Gamma_u = \phi_u(S^2)$ and G_u as the bounded domain whose boundary is Γ_o

(thus Γ_o is a reference sphere, centered at 0 with radius r_o). The data of the problem are: i) the assumed model for the density of G_u, namely a given smooth strictly positive function $\delta: R^3 \to R$, and ii) the measured potential on Γ_u (pull-back to S^2), i.e. a function $v: S^2 \to R$. Denoting by V_u the potential created by a mass of density δ distributed over G_u,

$$V_u(x) = \int_{S^2} d\omega \int_o^{|\phi_u(\omega)|} |x - t\,\omega|^{-1}\, t^2\, \delta(t\omega)\, dt$$

($d\omega$ is the surface element on S^2), we impose that such a potential on Γ_u ($A(u)$ say) must equal v:

$$(1) \qquad A(u) = V_u \circ \phi_u = v$$

The problem is then to solve equation (1) for u for a given v (the density δ is assumed to be fixed, and we ignore the dependence of the relevant quantities on it).
Consider now the first differential of the map $A(u)$, $A'(u)\rho$ say, which is composed of two terms

$$(2) \qquad A'(u)\rho = M_u\rho + f_u\cdot\rho$$

one coming from differentiating V_u:

$$M_u\rho(\omega) = V'_u\rho\circ\phi_u(\omega) = \int_{S^2} \frac{\delta\circ\phi_u(\omega')\,\rho(\omega')\,|\phi_u(\omega')|^2}{|\phi_u(\omega)-\phi_u(\omega')|}\, d\omega'$$

and the other from differentiating ϕ_u:

$$f_u(\omega) = \langle \operatorname{grad} V_u \circ \phi_u(\omega),\ \omega \rangle$$

Now we must estimate the norm of this differential (as long as the norm of the map A itself) as a linear operator acting in some Sobolev space $H^s(S^2)$. Consider first M_u. We can prove the following result [3] : *M_u acts continuously from $H^s(S^2)$ onto $H^{s+1}(S2)$ $(s>o)$ if $u \in C^{2+s}(S^2)$ and the best estimate available for $M_u\rho$ is ($\|\cdot\|_s$ is the norm in $H^s(S^2)$):*

$$\| M_u\rho \|_{s+1} \leq c\, \|\rho\|_s$$

where the constant depends on the norm of u in $C^{2+s}(S^2)$. One cannot obtain a better estimate, i.e. an estimate involving fewer derivatives of u. The complete differential $A'(u)$ exhibits similar behaviour. This is the mechanism by which derivations are lost in problems of this kind, and this is why we cannot work at a fixed step of the class of Banach spaces $H^s(S^2)$ (or $C^s(S^2)$ but we are forced to work with the complete family of the spaces $H^s(S^2)$, equipped with the structure of a Fréchet space. The hard implicit function theorem, (or Nash-Moser theorem) is precisely an extension of the ordinary implicit function theorem to (a special class of) Fréchet spaces.

EXISTENCE OF A SOLUTION

Let $u_o: S^2 \to R$, $\delta: R^3 \to R$ be smooth functions such that the linearized equation

$$(3) \qquad M_u\rho + f_u\rho = h$$

is satisfied by a unique smooth function $\rho: S^2 \to R$ for every h smooth on S^2. In [2] sufficient conditions are given on u_o and δ that guarantee the fulfillment of these hypotheses. Our purpose is to show that if v is sufficiently close to $v_o = A(u_o)$, then there is a unique function u close to u_o such that equation (1) is satisfied; proximity will be defined in terms of Sobolev norms. As we said, we solve (1) by an appeal to the Nash-Moser procedure. This technique is extensively described in [4]. Two main kinds of hypotheses must be verified to work with this technique. First one must estimate very carefully the norm of A, of its differential A' and of the nonlinear part of A. This is tedious and not always straightforward. If $\| u \|_4 \leqslant 1$, we prove, for $s \geqslant 0$, the following estimates [2]:

$$\| A(u) - A(u_o) \|_s \leqslant c \, \| u - u_o \|_{s+3}$$

$$\| A'(u)\rho \|_s \leqslant c \, \{ \| \rho \|_2 \, \| u \|_{s+3} + \| \rho \|_{s+2} \}$$

$$\| A'(u+\sigma)\rho - A'(u)\rho \|_s \leqslant c \, \{ \| \rho \|_{s+2} \| \sigma \|_3 + \| \rho \|_2 \| \sigma \|_{s+3} \}$$

Second, one must show that A' is invertible and obtain estimates for the inverse similar to those obtained for A' itself. This is not obvious at all, because we need to invert A' uniformly on a range of H^s spaces; to achieve this result, we require that the solution of the linearized equation with $u = u_o$ satisfies an inequality of the form:

$$\| \rho \|_s \leqslant c \, \| h \|_s \quad (s \geqslant 0)$$

with c independent on s (this is certainly true, e g., when u_o is constant, for suitable density models: see [2]). Then we get (we call $\Lambda(u)$ the inverse of $A'(u)$:

$$\| \Lambda(u)h \|_s \leqslant c \, \{ \| h \|_2 \| u \|_{s+3} + \| h \|_{s+2} \}$$

if $\| u - u_o \|_3$ is small.
The crucial fact in the estimates above is that, when the order of differentiability s grows at the left member, it grows in *only one factor* of the terms at the right member, and there it grows exactly as s. From these estimates it is not hard to deduce a uniqueness theorem for equation (1). But moreover we can prove the following result of existence.
THEOREM *Let* u_o *satisfy the hypotheses stated above and assume that* $\| v - A(u_o) \|_q$, *with* $q > 9$, *is sufficiently small; then there exists a (unique)* u *satisfying equation* (1) *and*

$$\| u_o - u \|_{n-a} \leqslant c \, \| v - A(u_o) \|_n$$

for any $n > 9$ *and* $a > 2$.
As we see, there is an initial gap of 9 derivatives; this loss could be lowered at cost of additional work (by working in spaces C^s instead of H^s); but the spirit of this result is essentially to show that smoothness of data implies smoothness of the solution, what is not affected assuming the control of some derivatives more than necessary.

References

[1] A.Lorenzi - C.D.Pagani, An Inverse problem in potential theory, Ann. Mat. Pura e Appl. 129 (1981) pp. 281-303

[2] C.Maderna - C.D.Pagani - S.Salsa, Existence results in an inverse problem of potential theory, to appear

[3] C.D.Pagani, Stability of a surface determined from measures of potential, to appear on: SIAM J. on Math. Anal.

[4] R.S.Hamilton, The inverse function theorem of Nash and Moser, Bull. Amer. Math. Soc. 7 (1) (1982) pp. 65-222.

The detailed version of this paper has been submitted for publication elsewhere.

Trends in the Theory and Practice of Non-Linear Analysis
V. Lakshmikantham (Editor)

VARIATIONAL PROBLEMS GOVERNED BY A MULTI-VALUED EVOLUTION EQUATION

Toru Maruyama

Department of Economics
Keio University
2-15-45 Mita, Minato-ku
Tokyo, JAPAN

Throughout this paper, $\mathbb{H}$ stands for a real Hilbert space of finite dimension and let a correspondence (=multi-valued mapping) $\Gamma : [0, T] \times \mathbb{H} \twoheadrightarrow \mathbb{H}$ be given. Define $\Delta(a)$ as a set of all elements x of $W^{1,2}([0, T], \mathbb{H})$ that satisfy

$$\dot{x}(t) \in \Gamma(t, x(t)) \quad \text{a.e.}$$
$$x(0) = a \qquad (*)$$

We shall examine a couple of variational problems governed by a multi-valued differential equation of the form (*), and establish sufficient conditions which assure the existence of optimal solutions for them.

EXISTENCE OF SOLUTION FOR (*).

Assumption 1. Γ is a compact-convex-valued; i.e. $\Gamma(t, x)$ is a non-empty, compact and convex subset of $\mathbb{H}$ for all $t \in [0, T]$ and all $x \in \mathbb{H}$.

Assumption 2. The correspondence $x \twoheadrightarrow \Gamma(t, x)$ is upper hemi-continuous (abbreviated as u.h.c.) for each fixed $t \in [0, T]$.

Assumption 3. The correspondence $t \twoheadrightarrow \Gamma(t, x)$ is measurable for each fixed $x \in \mathbb{H}$. For the concept of "measurability" of a correspondence, see Maruyama [4] Chap. 6.

Assumption 4. There exists $\psi \in L^2([0, T], R_+)$ such that

$$\Gamma(t, x) \subset S_{\psi(t)} \qquad \text{for every } (t, x) \in [0, T] \times \mathbb{H},$$

where $S_{\psi(t)}$ is the closed ball in $\mathbb{H}$ with the center 0 and the radius $\psi(t)$.

EXISTENCE THEOREM (Maruyama [2]). *Suppose that Γ satisfies Assumption 1-4, and let A be a non-empty, convex and compact subset of $\mathbb{H}$. Then*

(i) $\Delta(a^*) \neq \phi$ *for any* $a^* \in$ A, *and*

(ii) *the correspondence* $\Delta : A \twoheadrightarrow W^{1,2}$ *is compact-valued and upper hemi-continuous (abbreviated as u.h.c.) on* A *in the weak topology for* $W^{1,2}$.

VARIATIONAL PROBLEM (1).

Let $u : [0, T] \times \mathbb{H} \times \mathbb{H} \to R$ be a given function and consider the following variational problem:

$$\text{(P-1)} \qquad \underset{x \in \Delta(a)}{\text{Maximize}}\ J(x) = \int_0^T u(t, x(t), \dot{x}(t))dt$$

for a given $a \in A$.
We shall begin by specifying the properties of u.

Assumption 5. $-u$ is a normal integrand; i.e. a) $u(t, x, y)$ is measurable in (t, x, y) and b) the function $(x, y) \to u(t, x, y)$ is upper semi-continuous (abbreviated as u.s.c.) for every fixed $t \in [0, T]$.

Assumption 6. The function $y \to u(t, x, y)$ is concave for every fixed $(t, x) \in [0, T] \times H$.

Assumption 7. There exists a couple of functions, $\theta \in L^1([0, T], R)$ and $b \in L^2([0, T], H)$, such that $u(t, x, y) - \langle b(t), y\rangle \leqq \theta(t)$ for every $(t, x, y) \in [0, T] \times H \times H$.

Assumption 8. There exists some $x \in W^{1,2}$ such that $|J(x)| < \infty$

REMARK. 1) $\Delta(A) = \bigcup_{a\in A}\Delta(a)$ is a weakly compact set in $W^{1,2}$ because it is an image of a compact set $A \subset H$ under the compact-valued u.h.c. correspondence Δ.
2) Under the Assumption 7, it is clear that

$$\sup_{x\in\Delta(a)} J(x) < \infty.$$

If a sequence $\{x_n\}$ in $W^{1,2}$ weakly converges to x^*, then (i) $\{x_n\}$ strongly converges to x^* in L^1 and (ii) $\{\dot{x}_n\}$ weakly converges to $\dot{x}^*$ in L^2. Hence by the Ioffe's theorem [1], the integral functional J is sequentially u.s.c. on $W^{1,2}$ endowed with the weak topology. Since the weak topology on $\Delta(A)$ is metrizable, we obtain the following lemma.

LEMMA. *Under the Assumption* 1-8, *the integral functional J is u.s.c. on* $\Delta(A)$ *in the weak topology for* $\Delta(A)$.

Combining the Existence Theorem for the differential equation (*) together with the above Lemma, we obtain the following theorem, which is a revised version of the previous result in Maruyama [3].

THEOREM 1. *Under the Assumption* 1-8, *the problem* (P-1) *has a solution.*

VARIATIONAL PROBLEM (2).

Finally we consider a different kind of variational problem, in which the initial value is also variable:

$$\text{(P-2)} \qquad \underset{a\in A,\ x\in\Delta(a)}{\text{Maximize}} \int_0^T u(t, x(t), \dot{x}(t))dt .$$

In order to solve such a two-stage maximization problems as (P-2), the celebrated Berge's Maximum Theorem plays a crucial role.

BERGE'S THEOREM. *Let* Y *and* Z *be any topological spaces. And assume that the function* $f : Y \times Z \to R$ *is u.s.c. and the correspondence* $\theta : Y \twoheadrightarrow Z$ *is compact-valued and u.h.c. The the function* $f^* : Y \to R$ *defined by*

$$f^*(y) = \text{Max}\,\{f(y, z) \mid z \in \theta(y)\}$$

is u.s.c. on Y. (c.f. Maruyama [4] Chap2, §4.)

Applying Berge's theorem to our problem (P-2), we can assert that the function

$$a \to \underset{x \in \Delta(a)}{\text{Max}} \int_0^T u(t, x(t), \dot{x}(t))dt$$

is u.s.c. on A. Thus we get the following.

THEOREM 2. *Under the Assumption* 1-8, *the problem* (P-2) *has a solution.*

REFERENCES

[1] Ioffe, A.E., On Lower Semicontinuity of Integral Functionals, I, SIAM J. Control and Optimization, 15 (1977), 521-538.

[2] Maruyama, T., On a Multi-valued Differential Equation: An Existence Theorem, Proc. Japan Acad., 60A (1984), 161-164.

[3] ———, Variational Problems Governed by a Multi-valued Differential Equation, Proc. Japan Acad., 60A (1984), 212-214.

[4] ———, Functional Analysis, (Keio Tsushin, Tokyo, 1980), (in Japanese).

This paper is in final form and no version of it will be submitted for publication elsewhere.

Trends in the Theory and Practice of Non-Linear Analysis
V. Lakshmikantham (Editor)
Elsevier Science Publishers B.V. (North-Holland), 1985

ON THE STABILITY OF EQUILIBRIUM FOR PERIODIC MECHANICAL SYSTEMS

Vinicio Moauro

Dipartimento di Matematica
Università di Trento
Povo, Trento
ITALY

Some results, obtained in [1,2] , relative to the stability properties of an isolated equilibrium position of a holonomic mechanical system are presented. Here the proofs will be only sketched.

Let $\mathcal{S}$ be a holonomic mechanical system with $n \geq 1$ degrees of freedom and let $q_1, \ldots, q_n$ be a system of Lagrangian coordinates for $\mathcal{S}$. Denoting by q the n-vector $(q_1, \ldots, q_n)$, we will suppose that

(h_1) the kinetic energy of $\mathcal{S}$ has the form $\mathcal{T} = \frac{1}{2} \sum_1^n{}_h \dot{q}_h^2$
(h_2) the forces acting on $\mathcal{S}$ are the following:

1) a force depending on a potential function $U(q,t) = f(t)U^*(q)$, where $f(t)$ is a periodic function of the time t, with period $T > 0$, not identically equal to zero, and $U^*(q)$ is a C^∞-function defined in a neighbourhood of $q = 0$ which is not flat at $q = 0$. Further we assume that the minimum order of the non zero derivatives of U^* at $q = 0$ is $m+1$ with $m \geq 2$ and we will denote by U^*_{m+1} the homogeneous polynomial in q of degree $m+1$ such that $U^*(q) = U^*_{m+1} + o(|q|^{m+1})$, with $|\cdot|$ any Euclidean norm in R^n;
2) a linear complete dissipative force with Lagrangian components $Q_h = -\dot{q}_h$;

(h_3) $q = 0$ is an isolated equilibrium position for $\mathcal{S}$.

Remark. The stability problem of the equilibrium position $q = 0$ of $\mathcal{S}$ is completely solved in the case $f(t) \equiv 1$. In fact, in such a case, $q = 0$ is asymptotically stable if U^* has a maximum at $q = 0$ and it is unstable if U^* has not a maximum at $q = 0$. This result is true also when the kinetic energy is any positive definite quadratic form in $\dot{q}$ with coefficients depending on q, U^* is flat and the complete dissipative force is not linear.

The equations of motion of $\mathcal{S}$ are the following:

$$(1) \qquad \begin{aligned} \frac{d\dot{q}_h}{dt} &= f(t)\frac{\partial U^*}{\partial q_h} - \dot{q}_h \\ \frac{dq_h}{dt} &= \dot{q}_h \end{aligned} \qquad h = 1\ldots,n.$$

CASE n = 1.

In [1] the stability problem of the null solution of (1) (that is of the equilibrium position q = 0 of $\mathcal{S}$) has been considered, in connection to the problem of generalized Hopf bifurcation for periodic systems, in the case n = 1. In this case the linear approximation of (1) has 0 as simple characteristic exponent. Therefore, by using also a procedure due to Liapunov [3], it is possible to get the following results:

(A) If f(t) has mean value equal to zero, then the null solution of (1) is (2m-1)-asymptotically stable;

(B) If f(t) has mean value $M \neq 0$, then the null solution of (1) is m-asymptotically stable or m-unstable whether MU^*_{m+1} is negative definite or not.

Remarks. For any integer $k \geq 1$, saying that the null solution of(1) is k-asymptotically stable or k-unstable means that its asymptotic stability or its instability is determined only by the terms of order less than or equal to k in the expansion of the right hand sides of (1) and k is the minimum integer for which this happens. Result (B) shows that, if $f(t) \equiv 1$, the null solution of (1) is m-asymptotically stable or m-unstable whether U^*_{m+1} has or has not a maximum at q = 0. Thus, our hypotheses (h_1), (h_2) assure that the stability properties can be recognized at the order m.

Result (A) shows that the introduction in the potentialfunction of a periodic factor with mean value equal to zero can change drastically the stability properties of the equilibrium position.

In the one degree of freedom case, the property of the equilibrium position to be isolated is assured by the non flatness of U* and it is recognizable by inspecting U^*_{m+1} only.

The procedure to get results (A) and (B) can be schematized in the following two steps:

STEP 1. Look for a change of coordinates which trasforms system (1) into a system of the form

$$\begin{aligned} \dot{x} &= gx^k + X(x,y,t) \\ \dot{y} &= -y + Y(x,y,t) \end{aligned} \qquad x,y \in R, \tag{2}$$

where k is an integer greater than or equal to 2, $g \neq 0$ is a constant, X,Y are functions of order greater than or equal to 2 in x,y at x = y = 0, T-periodic in t and such that X(x,0,t), Y(x,0,t) are of order greater than k in x at x = 0. This change of coordinates is required to preserve the stability properties of the null solution and the order at which such properties are recognized.

STEP 2. Construct a Liapunov function, by using only the terms up to the order k in the r.h.s. of (2), which assures the asymptotic stability or the instability of the null solution of (2).

We do not give the details of this procedure because in the following we will show that it can be generalized to the case $n \geq 1$. Therefore we will give the details in this general case.

CASE $n \geq 1$.

By following the scheme of the procedure used in the case $n = 1$, it is possible to show [2] that
- system (1) can be transformed, by using changes of coordinates which do not modify the stability properties of the null solution, into a system of the form

$$(3) \qquad \begin{aligned} \dot{\eta}_h &= g_h(\eta) + \Pi_h(\eta,\zeta,t) \\ \dot{\zeta}_h &= -\zeta_h + W_h(\eta,\zeta,t) \end{aligned} \qquad h = 1,\dots,n,$$

where $\eta = (\eta_1,\dots,\eta_n)$, $\zeta = (\zeta_1,\dots,\zeta_n)$, the functions $g_h(\eta)$ are homogeneous polynomials of same degree $k \geq 2$, Π_h, W_h are functions of order greater than or equal to 2 in η,ζ at $\eta = \zeta = 0$, T-periodic in t and such that $\Pi_h(\eta,0,t)$, $W_h(\eta,0,t)$ are infinitesimal of order greater than k in η at $\eta = 0$;
- under the hypoyhesis

(H) $|\text{grad}U^*_{m+1}|^2$ is positive definite, i.e. the property of the equilibrium position $q = 0$ to be isolated is recognizable by inspecting only U^*_{m+1},

a Liapunov function can be constructed, by using only the terms of order less than or equal to k in the r.h.s. of (3), so that asymptotic stability or instability of the null solution of (3) is assured.

To reduce (1) to a system of form (3), let us introduce first the new variables defined by

$$x_h = q_h + \dot{q}_h \qquad y_h = \dot{q}_h \qquad h = 1,\dots,n.$$

System (1) is transformed into

$$(4) \qquad \begin{aligned} \dot{x}_h &= f(t)\left(\frac{\partial U^*}{\partial q_h}\right)_{q=x-y} \\ \dot{y}_h &= -y_h + f(t)\left(\frac{\partial U^*}{\partial q_h}\right)_{q=x-y} \end{aligned} \qquad \begin{aligned} h &= 1,\dots,n, \\ x &= (x_1,\dots,x_n), \\ y &= (y_1,\dots,y_n). \end{aligned}$$

We can construct as in [1,4] n polynomials Φ_h, $h = 1,\dots,n$, in x of degree 2m-1, whose coefficients are T-periodic functions of t, such that along the solutions of (4) we have

$$\left\{\frac{d}{dt}[y_h - \Phi_h(x,t)]\right\}_{y=\Phi(x,t)} = o(|x|^{2m-1}) \qquad \begin{aligned} h &= 1,\dots,n, \\ \Phi &= (\Phi_1,\dots,\Phi_n). \end{aligned}$$

We have

$$\Phi_h(x,t) = \Phi_{h,m}(x,t) + o(|x|^m) \qquad h = 1,\dots,n,$$

with $\Phi_{h,m}$ homogeneous polynomials in x of degree m satisfying the condition

$$(5) \qquad -\Phi_{h,m} + f(t)\left(\frac{\partial U^*_{m+1}}{\partial q_h}\right)_{q=x} - \frac{\partial \Phi_{h,m}}{\partial t} = 0 .$$

By means of the polynomials Φ_h's, we define the new variables

$$x_h = x_h \qquad \zeta_h = y_h - \Phi_h(x,t) \qquad h = 1,\dots,n,$$

and (4) is transformed into

$$(6)\quad \begin{aligned} \dot{x}_h &= f(t)\left[\left(\frac{\partial U^*}{\partial q_h}\right)_{q=x} - \sum_1^n {}_k\left(\frac{\partial^2 U^*_{m+1}}{\partial q_h \partial q_k}\right)_{q=x} \Phi_{k,m}\right] + X_h(x,\zeta,t) \\ & \qquad h = 1,\dots,n,\ \zeta=(\zeta_1,\dots,\zeta_n), \\ \dot{\zeta}_h &= -\zeta_h + \bar{W}_h(x,\zeta,t) \end{aligned}$$

where for $h = 1,\dots,n$, $X_h, \bar{W}_h$ are functions of order greater than or equal to m in (x,ζ) at $x = \zeta = 0$, T-periodic in t and $X_h(x,0,t)$, $\bar{W}_h(x,0,t)$ are of order greater than 2m-1 in x at $x = 0$. We will distinguish now the two situations:

(I) f(t) has mean value equal to zero;

(II) f(t) has mean value M different from zero.

(I) By means of changes of coordinates, used in the averaging method [5,6], we can transform system (6) into the following system:

$$(7)\quad \begin{aligned} \dot{\eta}_h &= -\frac{1}{T} \sum_1^n {}_k\left(\frac{\partial^2 U^*_{m+1}}{\partial q_h \partial q_k}\right)_{q=\eta} \int_0^T f(s)(\Phi_{k,m})_{\substack{x=\eta \\ t=s}} ds + \Pi_h(\eta,\zeta,t) \\ & \qquad h = 1,\dots,n,\ \zeta = (\zeta_1,\dots,\zeta_n), \\ \dot{\zeta}_h &= -\zeta_h + W_h(\eta,\zeta,t) \end{aligned}$$

which has the form (3) with the g_h's homogeneous polynomials of degree 2m-1.

(II) By averaging directly system (6), we get the system

$$(8)\quad \begin{aligned} \dot{\eta}_h &= \frac{M}{T}\left(\frac{\partial U^*_{m+1}}{\partial q_h}\right)_{q=\eta} + \Pi_h(\eta,\zeta,t) \\ & \qquad h = 1,\dots,n, \\ \dot{\zeta}_h &= -\zeta_h + W_h(\eta,\zeta,t) \end{aligned}$$

which has also the form (3) with the g_h's homogeneous polynomials of degree m.

Now we will show how it is possible to construct Liapunov functions for systems (7) and (8) in such a way that we can prove the following

THEOREM. Under hypothesis (H), results (A) and (B) hold true in the case $n \geq 1$.

In situation (I) let us consider the function

$$V = \sum_1^n {}_h\left[\frac{1}{2}(\eta_h^2 + \zeta_h^2) + \zeta_h P_h(\eta,t)\right],$$

where the P_h's are polynomials in η of degree 2m-1 and infinitesimal of order greater than or equal to m at $\eta = 0$, whose coefficients are T-periodic functions of t. These coefficients can be chosen in such a way that the derivative of V along the solutions of system (7) is given by

$$\dot{V} = -\sum_1^n {}_h\left[\frac{m}{T}\left(\frac{\partial U^*_{m+1}}{\partial q_h}\right)_{q=\eta} \int_0^T f(s)(\Phi_{h,m})_{\substack{x=\eta \\ t=s}} ds + \zeta_h^2\right] + S(\eta,\zeta,t)$$

with

$$(9)\quad S = \sum_{i_1+\dots+i_n=2m} \mu_{i_1,\dots,i_n} \eta_1^{i_1}\cdots\eta_n^{i_n} + \sum_{s,\sigma=1}^n \nu_{s,\sigma}\zeta_s\zeta_\sigma$$

where $\mu_{i_1,\dots,i_n}$, $\nu_{s,\sigma}$ are continuous functions of η,ζ,t, T-periodic in t and infinitesimal in η,ζ at $\eta = \zeta = 0$. The function $\dot{V}$ is negative definite because, by (5) and hypothesis (H), we have for $\eta \neq 0$

$$\sum_1^n {}_h\left(\frac{\partial U^*_{m+1}}{\partial q_h}\right)_{q=\eta} \int_0^T f(s)(\Phi_{h,m})_{\substack{x=\eta \\ t=s}} ds = \sum_1^n {}_h \int_0^T (\Phi_{h,m})^2_{\substack{x=\eta \\ t=s}} ds > 0.$$

Thus, the null solution of (7) is asymptotically stable and it is (2m-1)-asymptotically stable because the determination of the polynomials P_h's in V depends only on the terms of order less than or equal to 2m-1 in the r.h.s. of (7).

In situation (II), let us suppose first that MU^*_{m+1} is not sign-constant in a neighbourhood of q = 0. In such a case, let us consider the function

$$V = -MU^*_{m+1}(\eta) + \sum_{1}^{n}{}_h[\frac{1}{2}\zeta_h^2 + \zeta_h P_h(\eta,t)],$$

where the P_h's are homogeneous polynomials of degree 2m-1 in η whose coefficients are T-periodic functions of t determined in such a way that along the solutions of system (8) we have

$$\dot{V} = -\sum_{1}^{n}{}_h[\frac{M^2}{T}(\frac{\partial U^*_{m+1}}{\partial q_h})^2_{q=\eta} + \zeta_h^2] + S(\eta,\zeta,t)$$

with S a function defined as in (9). Thus, by hypothesis (H), $\dot{V}$ is negative definite and the null solution of (8) is unstable. As the determination of the polynomials P_h's in V depends only on the terms of order less than or equal to m in the r.h.s. of (8), the null solution of (8) is m-unstable. Let us suppose now that MU^*_{m+1} is sign constant in a neighbourhood of q = 0. Because of hypothesis (H), MU^*_{m+1} will be sign definite. If it is negative definite, we consider the function

$$V = \sum_{1}^{n}{}_h[\frac{1}{2}(\eta_h^2 + \zeta_h^2) + \zeta_h\bar{P}_h(\eta,t)],$$

where the $\bar{P}_h$'s are homogeneous polynomials of degree m in η whose coefficients are T-periodic functions of t determined in such a way that along the solutions of (8)

$$\dot{V} = \frac{M}{T}(U^*_{m+1})_{q=\eta} - \sum_{1}^{n}{}_h\zeta_h^2 + \bar{S}(\eta,\zeta,t)$$

with

$$\bar{S} = \sum_{i_1+\dots+i_n=m}\bar{\mu}_{i_1,\dots,i_n}\eta_1^{i_1}\dots\eta_n^{i_n} + \sum_{s,\sigma=1}^{n}\bar{\nu}_{s,\sigma}\zeta_s\zeta_\sigma,$$

where $\bar{\mu}_{i_1,\dots,i_n}$, $\bar{\nu}_{s,\sigma}$ are continuous functions of η,ζ, t, T-periodic in t and infinitesimal in η,ζ at $\eta = \zeta = 0$. Therefore the null solution of (8) is asymptotically stable and the way of constructing the polynomials $\bar{P}_h$'s implies that such property is recognizable at the order m. Finally, if MU^*_{m+1} is positive definite, it is possible to prove that the null solution of (8) is m-unstable by using a function of the type

$$V = \sum_{1}^{n}{}_h[\frac{1}{2}(-\eta_h^2 + \zeta_h^2) + \zeta_h\bar{P}_h(\eta,t)]$$

with the $\bar{P}_h$'s suitable homogeneous polynomials in η of degree m.

<u>Remark</u>. As a consequence of result (B), the equilibrium position q= 0 of $\mathcal{S}$ is m-unstable if f(t) has mean value different from zero and m is even.

REFERENCES:

[1] Salvadori,L., Visentin,F., Sul problema della biforcazione generalizzata di Hopf per sistemi periodici, Rend. Sem. Mat. Univ. di Padova 68 (1982).

[2] Moauro,V., A stability problem for holonomic mechanical systems, Ann. Mat. Pura e Appl., to appear.

[3] Liapunov,A.M., Problème général de la stabilité du mouvement, Ann. of Math. Studies 17 (Princeton Univ. Press, New Jersey 1947).

[4] Bernfeld,S.R., Negrini,P., Salvadori,L., Quasi invariant manifolds, stability and generalized Hopf bifurcation, Ann. Mat. Pura e Appl. (IV) 130 (1982).

[5] Hale,J.K., Ordinary differential equations (Wiley Interscience, New York, 1969).

[6] Ladis,N.N., Asymptotic behaviour of solutions of quasi homogeneous differential systems, Differential Equations 9 (1973).

The detailed version of this paper has been submitted for publication elsewhere.

Trends in the Theory and Practice of Non-Linear Analysis
V. Lakshmikantham (Editor)

REMARKS ON TIME DISCRETIZATION OF CONTRACTION SEMIGROUPS

Olavi Nevanlinna

Institute of Mathematics
Helsinki University of Technology
SF-02150 Espoo 15
Finland

The paper surveys results concerning the preservation of contractivity in the numerical solution of initial value problems.

1. INTRODUCTION

In this paper we discuss numerical integration methods for initial value problems from the point of view whether they preserve the contractivity of semigroups.

In the following X denotes a Banach space, H a Hilbert space, f will be a (nonlinear) mapping and A a linear operator. The stability properties of the methods are first tested on a scalar test equation

(1.1) $\quad u' = \lambda u$

where λ is a complex scalar. We set

$$S = \left\{ h\lambda \;\middle|\; \begin{array}{l}\text{method produces only bounded solutions when} \\ \text{applied to (1.1) with step length } h\end{array} \right\}$$

and denote

$$D_r = \{z \in \mathbb{C} \mid |z+r| \le r\}, \qquad r < \infty$$

$$D_\infty = \{z \in \mathbb{C} \mid \operatorname{Re} z \le 0\}.$$

Correspondingly, for linear operators in X we set

$$D_r(X) = \{A \mid |A+r| \le r\}$$

$$D_\infty(X) = \{A \mid A \text{ is the infinitesimal generator of a contraction semigroup}\}.$$

Assume that $D_r \subset S$. Then we might ask whether the method produces bounded solutions also when applied to

(1.2) $\quad x' = Ax$

with $hA \in D_r(X)$. If we obtain a discrete contraction semigroup then we can try to treat also time dependent and nonlinear problems with the same discretization method.

We concentrate on a class of methods which we call multistage multistep formulas. When such a formula is applied to (1.1) it produces a difference equation

(1.3) $\quad U_{n+1} = C(h\lambda)U_n$

where $C(z)$ is a kxk-matrix with rational entries. On the shape and size of possible stability regions S see [9], [10], [11]. If one uses variable step integration then $C(z)$ will typically also depend on the step ratios.

2. NONLINEAR PROBLEMS IN BANACH SPACES

Let $-f$ be an m-accretive operator in X. Then the "solution" of the evolution equation

(2.1) $x' \in f(x)$

can be defined via the use of implicit Euler method

(2.2) $x_n - x_{n-1} \in hf(x_n)$,

see Crandall and Liggett [2]. Implicit Euler method has convergence order $p = 1$ on compact time intervals, provided that the solutions happen to be smooth enough. It is therefore natural that more accurate methods will have difficulties with some problems.

Results of the form: "preserving contractivity of the semigroups in Banach spaces implies $p \leq 1$" have been presented for some classes of methods, see [18], [20]. The same barrier is met in gas dynamics, where monotonicity of the scheme is needed to preserve the L^1-contractivity.

3. LINEAR PROBLEMS IN BANACH SPACES

We formulate and prove here a model result which, for example, can be used to prove results for variable step integration with multistage multistep methods.

Assume given a sequence $\{C_\nu\}$ of kxk-matrices whose elements are rational functions. Our basic stability assumption is:

(H1) there exists $K < \infty$ such that for all $0 \leq m \leq n$ and $z \in D_r$

$$\left| \prod_{\nu=m}^{n} C_\nu(z) \right| \leq K.$$

In particular, in the constant matrix case this simply means uniform power boundedness in D_r. In addition we need some uniformity over the index ν:

For $r < \infty$ we require

(H2) there exists $L < \infty$ such that for all ν and $z \in D_r$

$$\left|\frac{d}{dz} C_\nu(z)\right| \leq L.$$

For $r = \infty$ we require

(H2) there exists $L < \infty$ such that for all ν and $z \in D_\infty$

$$\left|\frac{d}{dz} C_\nu(z)\right| \leq \frac{L}{1+|z|^2}$$

and $$|C_\nu(z) - C_\nu(\infty)| \leq \frac{L}{1+|z|} .$$

Observe that in the constant matrix case (H2) follows from (H1).

While $C_\nu(z)$ operates in $\mathbb{C}^k$ the operator $C_\nu(A)$ operates in $X^k (= X\times X\times \ldots \times X)$ if A is a linear operator in X. We need to specify the norm in X^k. In $\mathbb{C}^k$ we use the Euclidean norm, and the matrix norms above and operator norms in general are the induced operator norms. In X^k we use the norm $\|\cdot\|$ defined for $x = (x_1,\ldots,x_k)^T \in X^k$ by $\|x\| = (\Sigma_1^k |x_j|^2)^{1/2}$.

Our aim is to bound $\|\prod_{\nu=1}^{n} C_\nu(A)\|$ for $A \in D_r(X)$. Examples using e.g. trapezoidal rule show that this product can grow as fast as $\sqrt{n}$, [1], [20].

<u>Theorem 3.1</u>

Assume $\{C_\nu\}$ satisfies (H1) and (H2) for some $r \le \infty$. Then there exists $M < \infty$ such that for all $A \in D_r(X)$

$$(3.1)\quad \|\prod_{\nu=1}^{n} C_\nu(A)\| \le M\sqrt{n}. \qquad \square$$

<u>Proof.</u> Take a fixed n and set $R_m(z) := \prod_{\nu=1}^{n} C_\nu(z)$. We consider the easier case $r < \infty$ first. By (H1) $R_n(z)$ is analytic in an open neighborhood of D_r. Expand

$$(3.2)\quad R_n(z) = \sum_{j=0}^{\infty} (1 + z/r)^j R_{nj}.$$

Since the matrices R_{nj} decay exponentially with j we can estimate

$$\sum_{j=0}^{\infty} |R_{nj}| \le |R_{n0}| + (\sum_1^{\infty} j^{-2})^{1/2} (\sum_1^{\infty} |R_{nj}|^2 j^2)^{1/2}.$$

By (H1) we have

$$|R_{n0}| \le K.$$

Furthermore

$$\sum_1^{\infty} |R_{nj}|^2 j^2 = \frac{1}{2\pi} \int_{-\pi}^{\pi} |\frac{d}{d\tau} R_n(re^{i\tau} - r)|^2 d\tau \le rnK^2L$$

since for $z \in D_r$ we have by (H1) and (H2)

$$|\frac{d}{dz} R_n(z)| \le \sum_{k=1}^{n} |\prod_{\nu=k+1}^{n} C_\nu(z)| \; |C_k'(z)| \; |\prod_{\nu=1}^{k-1} C_\nu(z)| \le \sum_{k=1}^{n} KLK.$$

Hence $\sum_{j=0}^{\infty} |R_{nj}| \le \text{Const}\,\sqrt{n}$.

Let now B be any $k\times k$-matrix. Then, e.g. using the max-norm in $\mathbb{C}^k$, one observes that there exists a constant c, only depending on k, such that for any bounded operator A in X one has $\|B \otimes A\| \le c|B|\;|A|$. Hence

$$\|R_n(A)\| \le \sum_{j=0}^{\infty} \|R_{nj} \otimes (I + \frac{1}{r} A)^j\| \le c \sum_{j=0}^{\infty} |R_{nj}|,$$

since, by assumption, $|I + \frac{1}{r} A| \le 1$.

Consider now $r = \infty$. In [1] Brenner and Thomée proved this for powers of scalar functions. They represented the functions as Laplace transforms of positive measures and combined Carlson's inequality with a partition of unity. All we do here is to point out that their essential estimates hold. These are lines (15) and (16) in [1], corresponding to

$$(3.3)\quad |R_n(i\xi) - R_n(\infty)| \le \text{Const}\,\min\{1, \frac{n}{1+|\xi|}\}$$

(3.4) $\left|\frac{d}{d\xi} R_n(i\xi)\right| \leq \text{Const}\ \frac{n}{1+|\xi|^2}$.

Let $\Delta_j := C_j(z) - C_j(\infty)$. Then we obtain (3.3) from

$$\prod_1^n C_j(z) = \prod_1^n C_j(\infty) + [\prod_2^n C_j(\infty)]\Delta_1 + \ldots +$$

$$+ [\prod_m^n C_j(\infty)]\Delta_{m-1}[\prod_1^{m-1} C_j(z)] + \ldots + \Delta_n \prod_1^{n-1} C_j(z)$$

using (H1) and (H2). The other estimate (3.4) follows immediately by differentiating the product $R_n(z)$. The proof can now be completed by following the proof of Theorem 1 in [1]. □

For related work see [20], [12], [21].

4. LINEAR PROBLEMS IN HILBERT SPACES

Here we shall focus on k×k-matrices $C(z)$ and on linear operators A in a Hilbert space H.

Theorem 4.1.

For all $A \in D_r(H)$ (with $r \leq \infty$) we have

(4.1) $\|C(A)\| \leq \sup_{z \in D_r} |C(z)|$. □

This is a simple generatization of a well known result of von Neumann; a proof can be found in [17]. Here $|\cdot|$ denotes any norm in $\mathbb{C}^k$ defined by an inner product $\xi^T G\bar{\eta}$ while $\|\cdot\|$ denotes the norm coming from the corresponding inner product in H^k: $((x,y)) := \Sigma_{ij}\, g_{ij}(x_i,y_j)$, where (,) denotes the inner product in H.

For scalar functions, i.e. for one-step methods $D_r \subset S$ trivially implies $|c(z)| \leq 1$ for $z \in D_r$. Furthermore, in [4] Dahlquist has shown that for the companion matrices corresponding to fixed step integration by one-leg methods $D_r \subset S$ implies the existence of a positive definite G so that in that "G-norm" one has $|C(z)| \leq 1$ for all $z \in D_r$.

Remark 4.1. Theorem 4.1 applies to variable step integration, too, by applying it to the matrices

$$R_n(z) = \prod_1^n C_\nu(z).$$

In fact, if (H1) holds, then

$$\|\prod_1^n C_\nu(A)\| \leq K \quad \text{for } A \in D_r(H).$$ □

5. NONLINEAR PROBLEMS IN HILBERT SPACES

When we want to extend some of the results on linear semigroups to nonlinear ones we face an obvious difficulty: knowing the matrix $C(z)$ does not indicate how we would apply our method to a nonlinear problem

(5.1) $x' \in f(x)$.

(Here, naturally, $-f$ is a maximally monotone operator in H.) Indeed, for Runge-

Kutta methods Hairer and Türke [8] have shown that if $|c(z)| \le 1$ for $z \in D_\infty$ then there exists an implicit Runge-Kutta method such that it reduces to $c(z)$ when applied to the test equation and preserves the contractivity when applied to (5.1). This result is in contrast with the situation for explicit Runge-Kutta methods. There one can have $D_r \subset S$ while no method exists with the same stability polynomial $c(z)$, and preserving contractivity for

(5.2) $u' = \lambda(t)u, \quad \text{with } \lambda(t) \in D_\rho$

and $\rho > 0$, see [5].

For one-leg methods, however, with constant time steps the situation is satisfactory, because of the existence of a "G-norm" for $C(z)$ whenever $D_r \subset S$, [4]. For further details on this we refer to [7].

If the nonlinear mapping f is smooth enough one can linearize it and require that $hf'(x) \in D_r(H)$ at all relevant points x. This can also be written as a "circle condition" for the nonlinear mapping f:

(5.3) $\operatorname{Re}(x-y,\ f(x)-f(y)) \le -\frac{h}{2r}\, |f(x) - f(y)|^2.$

Hence, conditions of the form (5.3), which do not assume differentiability on f, can be considered as nonlinear analogues of $D_r(H)$. In particular, f is (for $r<\infty$) Lipschitz-continuous (and single valued). Recall that one-leg methods on (5.1) yield difference equations of the form

(5.4) $\sum_0^k \alpha_j x_{n+j} \in hf(\sum_0^k \beta_j x_{n+j}), \qquad n \ge 0.$

<u>Theorem 5.1</u>. For a one-leg method assume that $D_r \subset S$ with $r \le \infty$. If $\{x_n\}$ and $\{y_n\}$ are two solutions of (5.4), and f satisfies (5.3), then

$$\|X_{n+1} - Y_{n+1}\| \le \|X_n - Y_n\| \qquad \text{for } n \ge 0. \qquad \square$$

Here we denote $X_n = (x_n, \ldots, x_{n+k-1})^T$ and $\|\cdot\|$ is the norm obtained from the "G-norm" and from the inner product of H, as explained after Theorem 4.1. The proof of this (for differentiable f and finite dimensional H) can be found e.g. in [4], [7].

6. ONE-LEG METHODS AND ANGLE-BOUNDED NONLINEARITIES

As long as we work with operators in $D_r(H)$ or with the nonlinear analogue (5.3) we fail to cover parabolic evolutions properly: we would need to assume A-stability, i.e. that $D_\infty \subset S$, and this sets the order of accuracy down to $p \le 2$, [3]. However, higer order one-leg methods do exist which are stable in sectors containing the negative half line, so that they should be suitable for parabolic equations. In [19] Nevanlinna and Odeh developed a multiplier technique to cover fixed time step integration of angle-bounded problems by one-leg methods. Similarly to the G-stability analysis this technique does not satisfactorily work for variable time steps, see [16]. Here we use the term angle-bounded in a non-technical sense; [19] contains a discussion of different ways to make it precise.

When the nonlinearity in the problem is angle-bounded, the solutions often show some nonoscillation properties for large time values. In [13], [14] some results are given for their discretized counterparts.

Söderlind [22] has considered another approach ("double-step contractivity") to cover cases where the method is not A-stable. Furthermore, Dahlquist and Söderlind [7] discuss the possibility of working simultaneously with different circular sets inside S and with correspondingly decomposed systems of equations.

7. TWO-STEP METHODS FOR VARIABLE TIME STEPS

Two-step one-leg methods for (5.1) can be written as

$$(7.1)\quad \sum_{j=0}^{2} \alpha_j(r_n)x_{n+j} \in h_{n+1} f\Big(\sum_{j=0}^{2} \beta_j(r_n)x_{n+j}\Big)$$

where h_{n+1} denotes the new time step and $r_n = h_{n+1}/h_n$ is the step ratio. The coefficients α_j and β_j can typically be chosen to be low order rational functions of the step ratio.

We can now require that our method produces bounded solutions whenever applied, with an arbitrary step sequence $\{h_n\}$, to a <u>time dependent</u> test problem

$$(7.2)\quad u' = \lambda(t)u, \qquad \mathrm{Re}\,\lambda(t) \le 0.$$

It turns out that among second-order accurate one-leg methods there is a one-parameter family of such methods and they are those for which the companion matrix $C(z)$ satisfies $|C(z)| \le 1$ for $\mathrm{Re}\, z \le 0$ in the matrix norm induced by the max-norm in $\mathbb{C}^2$, [15], [6]. In [6] these formulas are called A-contractive.

<u>Theorem 7.1.</u>

If $-f$ is a maximally monotone operator in H and we discretize (5.1) using an A-contractive one-leg method and an arbitrary step sequence, then any two solutions of the resulted difference equation satisfy

$$(7.3)\quad |x_{n+2}-y_{n+2}| \le \max\{|x_{n+1}-y_{n+1}|, |x_n-y_n|\}. \qquad \square$$

This result is included in Theorem 6.1, [17].

REFERENCES:

[1] P. Brenner, V. Thomée, On rational approximations of semigroups. SIAM J. Num. Anal. 16 (1979), 683-694.

[2] M. Crandall, T. Liggett, Generation of semi-groups of nonlinear transformations on general Banach spaces. Amer. J. Math. 93 (1971), 265-298.

[3] G. Dahlquist, A special stability problem for linear multistep methods, BIT 3 (1963), 27-43.

[4] G. Dahlquist, G-stability is equivalent to A-stability, BIT 18 (1978), 384-401.

[5] G. Dahlquist, R. Jeltsch, Generalized disks of contractivity for explicit and implicit Runge-Kutta methods, Report TRITA-NA-7906 (1979).

[6] G. Dahlquist, W. Liniger, O. Nevanlinna, Stability of two-step methods for variable integration steps. SIAM J. Num. Anal. 20 (1983), 1071-1085.

[7] G. Dahlquist, G. Söderlind, Some problems related to stiff nonlinear differential systems. Computing Methods in Applied Sciences and Engineering, V., Glowinski, Lions (Eds.), North-Holland Publ. Co., INRIA, 1982.

[8] E. Hairer, H. Türke, The equivalence of B-stability and A-stability, BIT 24 (1984).

[9] R. Jeltsch, O. Nevanlinna, Stability of explicit time discretizations for solving initial value problems, Numer. Math. 37 (1981), 61-91.

[10] R. Jeltsch, O. Nevanlinna, Stability and accuracy of time discretizations for initial value problems, Numer. Math. 40 (1982), 245-296.

[11] R. Jeltsch, O. Nevanlinna, Accuracy of multistage multistep formulas. Bericht Nr. 23 (1983), RWTH Aachen, Institut für Geometrie und Praktische Mathematik.

[12] J. Kraaijevanger, Absolute Monotonicity of Polynomials occuring in the Numerical Solution of Initial Value Problems. Report Nr. 84-01 (1984), Institute of Applied Mathematics and Computer Science, University of Leiden.

[13] O. Nevanlinna, On the numerical integration of nonlinear initial value problems by linear multistep methods. BIT 17 (1977), 58-71.

[14] O. Nevanlinna, On the behavior of global errors at infinity in the numerical integration of stable initial value problems. Numer. Math. 28 (1977),445-454.

[15] O. Nevanlinna, Stability of variable step integration. Report Mathematics 32 (1979), University of Oulu.

[16] O. Nevanlinna, Some Remarks on Variable Step Integration. ZAMM 64 (1984), 315-316.

[17] O. Nevanlinna, Matrix valued versions of a result of von Neumann with an application to time discretization. REPORT HTKK-MAT-A224 (1984).

[18] O. Nevanlinna, W. Liniger, Contractive methods for stiff differential equations. Part I, BIT 18, 457-474, Part II, BIT 19, 53-72 (1978).

[19] O. Nevanlinna, F. Odeh, Multiplier techniques for linear multistep methods. Numer. Funct. Anal. and Optim. 3(4), (1981), 377-423.

[20] M. Spijker, Contractivity in the numerical solution of initial value problems. Numer. Math. 42 (1983), 271-290.

[21] M. Spijker, Stepsize restrictions for stability of one-step methods in the numerical solution of initial value problems. Report Nr. 84-04 (1984). Institute of Applied Mathematics and Computer Science, University of Leiden.

[22] G. Söderlind, Multiple-step G-contractivity with applications to slowly varying linear systems. Report TRITA-NA-8107 (1981).

This paper is in final form and no version of it will be submitted for publication elsewhere.

Trends in the Theory and Practice of Non-Linear Analysis
V. Lakshmikantham (Editor)
Elsevier Science Publishers B.V. (North-Holland), 1985

LOW-DIMENSIONAL BEHAVIOR OF THE PATTERN FORMATION CAHN-HILLIARD EQUATION

Basil Nicolaenko

Theoretical Division, MS-B284
Center for Nonlinear Studies
Los Alamos National Laboratory
Los Alamos, NM 87545

Bruno Scheurer

Centre d'Etudes de Limeil
and Universite Paris-Sud (Orsay), France

We investigate the fourth-order Cahn-Hilliard parabolic partial differential equation which describes pattern formation in phase transition. Neumann and periodic boundary conditions are considered for a domain in R^n, $1 \le n \le 3$. This equation is characterized by a negative (backward) second order diffusion and multiple steady states for the appropriate range of parameters. We establish compactness of the orbits in $H^1(\Omega)$ and convergence to some steady state. We demonstrate that the Cahn-Hilliard equation admits an intrinsic low dimensional behavior: in R^1, the number of determining modes (in a Galerkin expansion) is proportional to $L^{3/2}$; where L, the diameter of the domain, is also proportional to the number of unstable modes for the linearized equation. Similar results hold for n = 2,3.

1. INTRODUCTION

We investigate the low dimensional behavior of the Cahn-Hilliard equation with a quartic homogeneous free energy, in R^n, $1 \le n \le 3$:

$$\frac{\partial u}{\partial t} = \operatorname{div}\,[M(u)\,\nabla\,(-\Delta u + \alpha u^3 - \beta u)]$$

$$\equiv \operatorname{div}\,[M(u)\,\nabla\,J(u)] \quad \text{in } \Omega \subset R^n \;,$$

$$u(0) = u_0 \in H^2(\Omega) \quad , \alpha > 0 \text{ and } \beta > 0 \;; \tag{1.1a}$$

the following hypotheses are made for the mobility coefficient M(u):

$$M(u) > 0 \quad , \text{monotone non-increasing in } |u|,\ C^1$$

$$\text{and } M(u) > M(0) \exp -\lambda|u| \quad , \lambda > 0 \;; \tag{1.1b}$$

the boundary conditions on $\partial\Omega$ (boundary of the pattern cell) are either of the Neumann type or periodic (periodic cell structure):

$$\left.\frac{\partial u}{\partial \nu}\right|_{\partial\Omega} = 0 \quad , \left.\frac{\partial J}{\partial \nu}\right|_{\partial\Omega} = 0 \quad , \tag{1.1c}$$

or

$$u(x + Le_i t) = u(x,t) \quad 1 \le i \le n \quad , \tag{1.1d}$$

L being the size of a typical pattern cell.

Eq. (1.1) is in fact a normalized form for the classical Cahn-Hilliard equation [2,5,9]:

$$\frac{\partial c}{\partial t} = \mathrm{div}\ [M(u)\ \nabla\ (-\Delta c + b_2 c + b_3 c^2 + b_4 c^3)] \quad ,$$

$$b_2 \text{ either } > 0 \text{ or } < 0, \quad b_3 < 0, \ b_4 > 0 \quad , \qquad (1.2)$$

with the same boundary conditions. As shown below (1.2) reduces to (1.1) by a simple translation $c(x,t) = u(x,t) + c^*$, c^* constant.

Eq. (1.2) is a continuum model for pattern formation resulting from phase transition. It is associated to a classical Landau-Ginzburg free energy [1]:

$$\hat{F} = \int_\Omega (\tfrac{1}{2}(\nabla\hat{c})^2 + f(\hat{c}))\ dx \quad , \quad \int_\Omega \hat{c}\ dx \equiv \int_\Omega c(x,0)\ dx = ct \quad , \qquad (1.3a)$$

where the homogeneous free energy $f(c)$ is a quartic polynomial whose derivative is:

$$\frac{\partial f}{\partial c} = b_2 c + b_3 c^2 + b_4 c^3 \quad , \ b_3 < 0 \quad , \ b_4 > 0 \ . \qquad (1.3b)$$

Steady-state solutions of (1.2) are given by critical points of the <u>non-convex</u> functional F. The corresponding Euler-Lagrange equation is:

$$-\Delta\hat{c} + b_2\hat{c} + b_3\hat{c}^2 + b_4\hat{c}^3 = ct \quad , \qquad (1.3c)$$

plus appropriate boundary conditions.

The influence of the homogeneous free energy function $f(c)$ appears in the sign of b_2 and the parameter B [9]:

$$B = \frac{b_3}{(|b_2|b_4)^{\frac{1}{2}}} \ . \qquad (1.4)$$

If $b_2 \leq 0$, there is a "negative viscosity" destabilizing mechanism somewhat similar to the one observed in the Kuramoto-Sivashinsky equation for unstable flame fronts [6-8]. The <u>zero solution is unstable</u> and this regime is referred to as "unstable subspinodal." The special limit case $b_2 = 0$ is called the "spinodal regime."

If $b_2 > 0$ and $B^2 > 3$, the cubic $\frac{\partial f}{\partial c}$ defined in (1.3b) possesses two distinct extrema. If $B^2 < 3$, $b_2 > 0$, it is well known that zero is a monotonically stable attractor [5,9]. A. Novick-Cohen and L. A. Segel [9] have extensively studied the case $3 \leq B^2 < \infty$ in a one-dimensional geometry. They have specified the full set of equilibrium solutions. They have also established that for $4.5 \leq B^2 < \infty$, the basin of attraction of zero is bounded, whereas there exists at least another nontrivial equilibrium with its own basin of attraction. $B^2 = 4.5$ is the distinguished "binodal" case.

We investigate some global dynamical properties of (1.2) when $b_2 > 0$ and $B^2 > 3$, or $b_2 \leq 0$. <u>Either case reduce to the normalized equation</u> (1.1); set:

$$u(x,t) = c(x,t) - c^* \quad , \qquad (1.5a)$$

where

$$c^* = -b_3/3b_4 > 0 \quad , \tag{1.5b}$$

and is such that

$$\left.\frac{\partial^3 f}{\partial c^3}\right|_{c=c^*} = 0 \quad ;$$

through the translation (1.5), the cubic $\frac{\partial f}{\partial c}$ is changed into:

$$\frac{\partial f}{\partial c} = c^* + [b_2 - \frac{1}{3}\frac{b_3^2}{b_4}]\, u + b_4\, u^3 \quad . \tag{1.6a}$$

We define

$$\alpha = b_4 > 0 \tag{1.6b}$$

$$\beta = -\ [b_2 - \frac{1}{3}\frac{b_3^2}{b_4}] \quad , \ \beta > 0 \quad ; \tag{1.6c}$$

indeed $B^2 > 3$, $b_2 > 0$ implies $\beta > 0$. Injecting (1.5) and (1.6) into the Cahn-Hilliard Eq. (1.2) yields the normalized form (1.1), with $M \equiv M(c^* + u)$, and $u_0 = c(x,0) - c^*$.

In Section 1, we verify boundedness of orbits in $H^1(\Omega)$ and the existence of Lyapunov functional. Although the above is implicit in the literature, compactness of orbits in $H^1(\Omega)$ has not previously been established, to our knowledge. This is done in Section 2, and enables the correct application of a classical topological dynamics theorem of Hale [4]: all orbits strongly converge in $H^1(\Omega)$ to critical points of the non-convex functional (1.3a).

However, the most important results are found in Section 4; we establish the intrinsically low-dimensional behavior of the Cahn-Hillard equation. Essentially, we project any orbit onto the linear manifold of the first m-eigenmodes of the biharmonic Δ^2. Suppose that the m-dimensional projected orbit converges to some m-dimensional fixed point; we will say that the first m-eigenmodes are determining if this implies convergence of the infinite dimensional orbit.

Following ideas developed in the Navier-Stokes context by Foias-Manley-Temam-Treve [3], we prove that for the one-dimensional Cahn-Hilliard equation:

$$m \geq ct\ L^{3/2} \quad ,$$

where L is the pattern size.

L is also proportional to the number of unstable modes of (1.1) linearized at $u = 0$; indeed the eigenvalue spectrum is:

$$\Lambda_k = \beta^2 \left(-\left(\frac{2\pi k}{\sqrt{\beta}L}\right)^4 + \left(\frac{2\pi k}{\sqrt{\beta}L}\right)^2\right) \quad , \ k = 0,1,2,\ldots.$$

and

$$\# \{\Lambda_k | \Lambda_k > 0\} = [\frac{\sqrt{\beta}}{2\pi} L] \quad ,$$

where [a] is the usual integer part of a. So for the determining modes:

$$m \geq ct\ (\#\ \text{unstable modes})^{3/2} \quad ;$$

in some heuristic sense, the impact of the nonlinearity is reflected only through the exponent ½. Similar results hold for n = 2 and n = 3, periodic boundary conditions.

To simplify the technical derivations, we restrict ourselves to M(u) = constant; the general case is easily disposed of, as soon as one obtains an estimate such as:

$$\overline{\lim_{t\to\infty}} ||u(x,t)||_{L^\infty(\Omega)} < K \ ;$$

then from (1.1b)

$$0 < M(0) \leq M(u) \leq M(K) \quad .$$

2. BOUNDEDNESS OF ORBITS IN $H^1(\Omega)$: THE LYAPUNOV FUNCTION

We consider the normalized problem:

$$\frac{\partial u}{\partial t} - \Delta J(u) = 0 \text{ in } \Omega \quad , \tag{2.1a}$$

$$J(u) = -\Delta u + \alpha u^3 - \beta u \quad , \ \alpha \text{ and } \beta > 0$$

$$u(0) = u_0 \in H^2(\Omega) \tag{2.1b}$$

with either

- periodic boundary conditions , $u(x + Le_i, t) = u(x,t)$, $1 \leq i \leq n$ (2.1c)

(L being the size of a typical pattern cell) or

$$\left.\frac{\partial u}{\partial \nu}\right|_{\partial\Omega} = \left.\frac{\partial J}{\partial \nu}\right|_{\partial\Omega} = 0 \quad . \tag{2.1d}$$

In this section, $\Omega \subset R^n$, $1 \leq n \leq 3$.

First we have the:

Lemma 2.1. $\bar{u}(t) \equiv \bar{u}(0)$, *where* $\bar{u}(t)$ *is the average* $\frac{1}{|\Omega|} \int u(x,t)\, dx$ *and* $|\Omega|$ = meas Ω.

Remark 2.2. The previous lemma implies that Poincaré-like inequalities hold, as u can be renormalized to a function of null mean value. From now on, we set

$$||u|| = (\int u^2 \, dx)^{\frac{1}{2}} \quad ,$$

unless specified otherwise.

We now look for a Lyapunov function associated with (2.1). Multiply (4.1) by J(u) and integrate by parts over Ω. With either set of boundary conditions:

$$\int_\Omega \frac{\partial u}{\partial t} J(u) \, dx + \int_\Omega (\nabla J(u))^2 \, dx = 0 \tag{2.2a}$$

and injecting the explicit form of J(u) into the first integral:

$$\frac{d}{dt}\left(\frac{1}{2} \int_\Omega (\nabla u)^2 \, dx - \frac{\beta}{2} \int_\Omega u^2 dx + \frac{\alpha}{4} \int_\Omega u^4 dx \right) + \int (\nabla J)^2 dx = 0 \quad . \tag{2.2b}$$

Let us define V(t) as:

$$V(t) = \frac{1}{2} \int_\Omega (\nabla u)^2 \, dx - \frac{\beta}{2} \int_\Omega u^2 dx + \frac{\alpha}{4} \int_\Omega u^4 dx \quad . \tag{2.3}$$

Then (2.2b) implies:

$$\frac{d}{dt} V(t) \leq 0 \quad . \tag{2.4}$$

To establish that V(t) is a Lyapunov function, we must show the boundedness of orbits in $H^1(\Omega)$ and that V(t) is <u>bounded from below</u> in $H^1(\Omega)$. Remark that:

$$V(t) \equiv \frac{1}{2} \int_\Omega (\nabla u)^2 dx + \int_\Omega \left(\frac{\sqrt{\alpha}}{2} u^2 - \frac{\beta}{2\sqrt{\alpha}}\right)^2 dx - \frac{\beta^2}{4\alpha} |\Omega| \quad ; \tag{2.5}$$

now

$$V(t) \leq V(0) \quad , \tag{2.6}$$

so

$$\frac{1}{2} \int_\Omega (\nabla u)^2 \, dx + \int_\Omega \left(\frac{\sqrt{\alpha}}{2} u^2 - \frac{\beta}{2\sqrt{\alpha}}\right)^2 dx \leq \frac{1}{2} \int_\Omega (\nabla u_0)^2 \, dx + \int_\Omega \left(\frac{\sqrt{\alpha}}{2} u_0^2 - \frac{\beta}{2\sqrt{\alpha}}\right)^2 dx \quad . \tag{2.7}$$

This proves the

Theorem 2.3. $\overline{\lim_{t\to\infty}} \, ||\nabla u(t)|| \leq F(u_0)$, *where*

$$F(u_0) = (||\nabla u_0^2|| + 2 \int_\Omega (\frac{\sqrt{\alpha}}{2} u_0^2 - \frac{\beta}{2\sqrt{\alpha}})^2 dx)^{\frac{1}{2}} . \tag{2.8}$$

Corollary 2.4. $\overline{\lim_{t\to\infty}} ||u||_{L^4}$ *is bounded.*

Proof. Use the continuous imbedding

$$H^1(\Omega) \hookrightarrow L^4(\Omega) \quad , n \leq 4$$

or specifically Eq. (2.7), together with Poincaré's inequality.

Corollary 2.5. V(t) is a continuous, bounded from below, Lyapunov functional on $H^1(\Omega)$.

Remark 2.6. All of the above results are valid if we consider the more general equation (1.1) with the coefficient of diffusion M(u) given as in (1.1b). Indeed:

$$\frac{\partial u}{\partial t} - \text{div } M(u) \nabla J(u) = 0 \quad ;$$

multiplying by J(u) and integrating over Ω:

$$\int_\Omega \frac{\partial u}{\partial t} J(u) \, dx + \int_\Omega M(u) (\nabla J)^2 \, dx = 0 \quad ,$$

and we still have

$$\frac{d}{dt} V(t) \leq 0 \quad ,$$

with V(t) same as in (2.3).

3. ASYMPTOTIC BEHAVIOR OF ORBITS.

We wish to establish some kind of convergence of the orbits u(x,t) to the critical manifold M of fixed points $\hat{u}(x)$ of:

$$-\Delta\hat{u} + \alpha \hat{u}^3 - \beta \hat{u} = \gamma \tag{3.1a}$$

$$\int_\Omega \hat{u} \, dx = |\Omega|\bar{u}(0) \tag{3.1b}$$

$$\frac{\partial u}{\partial \nu}\Big|_{\partial\Omega} = 0 \text{ or periodic boundary conditions} \quad . \tag{3.1c}$$

To apply classical topological dynamics results of Hale [4], we first need the relative compactness of orbits u(t) in $H^1(\Omega)$:

Theorem 3.1. $\overline{\lim_{t\to\infty}} ||D^2u||$ *is bounded*(1), *for either periodic boundary conditions (2.1c) or Neumann conditions (2.1d) if* $\Omega \subset R^1$; *and for periodic boundary conditions if* $\Omega \subset R^2$ *or* R^3.

The proof is technical and will be outlined below. Theorem 3.1 ensures the relative compactness of the orbit u(t) in $H^1(\Omega)$; hence, the ω-limit set associated to u_0 is nonempty, compact, invariant and connected. Using a classical theorem for such flows with Lyapunov functions [4], namely that V(t) is constant on $\omega(u_0)$, we deduce:

Corollary 3.2. As $t \to\infty$, $\lim \text{dist} |u(x,t) - M| = 0$ in $H^1(\Omega)$, *for either boundary conditions if* $\Omega \subset R^1$, *and for periodic boundary conditions if* $\Omega \subset R^2$ *or* R^3.

Remark 3.3. Problem (3.1) usually admits multiple solutions, whether one considers β or L = diam Ω as a bifurcation parameter [9].

Proof of Theorem 3.1. Multiply (2.1) by $\frac{\partial^4}{\partial x_1^{2\delta_1} \cdots \partial x_n^{2\delta_n}} u$, integrate by parts and take the sumation over all $\delta = (\delta_1, \ldots, \delta_n)$ such that $|\delta| = 2$; we get:

$$\frac{1}{2}\frac{d}{dt} ||D^2u||^2 + ||D^4u||^2 - \beta||D^3u||^2 = \sum_{|\delta|=2} \alpha \int \Delta u^3 \, D^{2\delta}u \, dx$$

$$= \sum_{|\delta|=2} (6\alpha \int u|\nabla u|^2 D^{2\delta}u \, dx + 3\alpha \int u^2 \Delta u \, D^{2\delta}u \, dx) \quad . \tag{3.2}$$

Apply Cauchy-Schwartz and Cauchy-Young's inequalities to the R.H.S. of (3.2):

$$\frac{1}{2}\frac{d}{dt} ||D^2u||^2 + (1-\varepsilon) ||D^4u||^2 \leq \beta ||D^3u||^2 + C(\varepsilon) \int u^2(\nabla u)^4 \, dx$$

$$+ C(\varepsilon) \int u^4 (\Delta u)^2 \, dx \quad ; \tag{3.3}$$

from now on C(ε) will be a generic symbol for any constant depending upon ε.

We will estimate:

$$J_1 = \int u^2 (\nabla u)^4 \, dx \quad , \tag{3.4}$$

$$J_2 = \int u^4 (\Delta u)^2 \, dx \quad . \tag{3.5}$$

(1) For brevity, we set $||D^ku||^2 = \sum_{|\alpha|=k} ||D^\alpha u||^2$.

We will need the Agmon inequalities (for functions periodic and/or with zero mean value):

$$||u(t)||_{L^\infty} \le \begin{cases} \gamma_1 ||u(t)||^{\frac{1}{2}} \; ||\nabla u(t)||^{\frac{1}{2}} & , \text{ if } n = 1 \; , \\ \gamma_2 ||u(t)||^{\frac{1}{2}} ||\Delta u(t)||^{\frac{1}{2}} & , \text{ if } n = 2 \\ \gamma_3 ||u(t)||^{\frac{1}{4}} \; ||\Delta u(t)||^{\frac{3}{4}} & , \text{ if } n = 3 \; . \end{cases} \tag{3.6}$$

We also need the following general interpolation inequalities:

$$||D^{k+1}u|| \le ||D^{k-1}u||^{1/3} \; |D^{k+2}u||^{2/3} \tag{3.7}$$

$$||D^{k}u|| \le ||D^{k-1}u||^{\frac{1}{2}}||D^{k+1}u||^{\frac{1}{2}} \tag{3.8}$$

Also, as $H^{\frac{1}{2}} \hookrightarrow L^4$ ($n = 2$) or $H^{\frac{3}{4}} \hookrightarrow L^4$ ($n = 3$), we will need:

$$||Du||^4_{L^4} \le ||Du||^3 \; ||D^3u|| \quad , \; n = 2 \quad ; \tag{3.9a}$$

$$||Du||^4_{L^4} \le ||Du||^{5/2} \; ||D^3u||^{3/2} \; , \; n = 3 \quad ; \tag{3.9b}$$

which are obtained by interpolation of $H^{\frac{1}{2}}$ (resp. $H^{\frac{3}{4}}$) between L^2 and H^2. We will give explicit technical details only for $n = 2$. The case $n = 1$ and $n = 3$ are similar.

In (3.3), we first consider the term $\beta||D^3u||^2$; from (3.7) and using Cauchy-Young's inequality with $p = 3/2$, $q = 3$:

$$\begin{aligned} ||D^3u||^2 \le ||D^4u||^{4/3} \; ||Du||^{2/3} \; || &\le \varepsilon \; ||D^4u||^2 + C(\varepsilon) \; ||Du||^2 \\ &\le \varepsilon \; ||D^4u||^2 + C(\varepsilon) \quad , \end{aligned} \tag{3.10}$$

since $\overline{\lim_{t\to\infty}} \; ||\nabla u|| \le F(u_0)$ (Theorem 2.3).

Now estimate J_1 in (3.4):

$$\int u^2(\nabla u)^4 dx < ||u||^2_{L^\infty} \; ||\nabla u||^4_{L^4} \quad ;$$

using Agmon's inequalities (3.6) and the interpolation inequality (3.9a):

$$J_1 < Ct \; ||u|| \; ||D^2u|| \; ||Du||^3 \; ||D^3u|| \quad ,$$

and from Theorem 2.3:

$$J_1 < Ct \; ||D^2u|| \; ||D^3u|| < Ct \; ||D^3u||^2$$

(using Poincaré's inequality) and

$$J_1 < \varepsilon \; ||D^4u||^2 + C(\varepsilon) \quad , \tag{3.11}$$

following (3.10).

Now estimate J_2 in (3.5):

$$\int u^4 (\Delta u)^2 \, dx \leq ||\Delta u||^2_{L^\infty} \; ||u^4||^4_{L^4} \quad ;$$

using Agmon's inequalities (3.6):

$$J_2 \leq Ct \; ||\Delta u|| \; ||D^4 u|| \; ||u^4||^4 \leq Ct \; ||D^2 u|| \; ||D^4 u|| \quad ,$$

(using Corollary 2.4); now using the interpolation inequality (3.8):

$$J_2 \leq Ct \; ||Du||^{\frac{1}{2}} \; ||D^3 u||^{\frac{1}{2}} \; ||D^4 u|| \leq Ct \; ||D^3 u||^{\frac{1}{2}} \; ||D^4 u|| \quad ;$$

but from the interpolation inequality (3.7):

$$||D^3 u|| \leq ||Du||^{1/3} \; ||D^4 u||^{2/3} \quad ;$$

so:

$$J_2 \leq Ct \; ||Du||^{1/6} \; ||D^4 u||^{4/3} \quad ,$$

and using Cauchy-Young's inequality with $p = 3/2$, $q = 3$:

$$J_2 \leq \varepsilon \; ||D^4 u||^2 + C(\varepsilon) \; ||Du||^{\frac{1}{2}} \quad ,$$

$$J_2 \leq \varepsilon \; ||D^4 u||^2 + C(\varepsilon) \quad . \tag{3.12}$$

We now collect all terms in Eq. (3.3), applying (3.10, 3.11, 3.12):

$$\frac{1}{2} \frac{d}{dt} ||D^2 u||^2 + (1 - 3\varepsilon - \beta\varepsilon) \; ||D^4 u||^2 < C(\varepsilon) \quad . \tag{3.13}$$

We conclude with the help of Poincaré's inequality and Gronwall's Lemma, that:

$$\overline{\lim_{t \to \infty}} \; ||D^2 u|| < \infty \quad . \qquad \square$$

4. NUMBER OF DETERMINING MODES

This section gives our main result, namely an upper bound of the number of determining modes for any solution of the Cahn-Hilliard equation (2.1) with <u>periodic boundary conditions</u>. This bound is formulated in terms of L. Although we give the detailed derivation for space dimension $n = 1$, analogue results can easily be derived for $n = 2$ and $n = 3$.

Consider u,v two solutions of (2.1), corresponding to two initial data (in $H^2(\Omega)$); set $w = u-v$. Due to the periodicity of u,v, we can use a Fourier mode decomposition of w and set:

$$P_m w(x,t) = \sum_{|k| \leq m} w_k(t) \exp\frac{2i\pi}{L} k.x \tag{4.1}$$

where $k \in Z^n$, and $w_k(t)$ is the kth Fourier coefficient of $w(x,t)$. We will also use:

$$Q_m \, w(x,t) = (I - P_m) w(x,t) \quad . \tag{4.2}$$

Definition 4.1. We say that the first m Fourier modes of $w = u-v$ are determining if:

$$\lim_{t\to\infty} ||P_m (u(t) - v(t))|| = 0 \;\rightarrow\; \lim_{t\to\infty} ||u(t) - v(t)|| = 0 \;. \tag{4.3a}$$

Remark 4.2. For Neumann boundary conditions (2.1d), we use the appropriate eigenfunctions of (Δ^2) as a Galerkin basis in (4.1 - 4.2).

Remark 4.3. If Ξ is a compact positive invariant set under the semi-flow defined in Section 3, then from (4.3) we deduce:

$$\lim_{t\to\infty} \text{dist} =(P_m u(t), P_m \Xi) = 0 \;\rightarrow\; \lim_{t\to\infty} \text{dist}(u(t), \Xi) = 0 \;,$$

since $v(t) \in \Xi$ for all times if $v(0) \in \Xi$.

In particular, if $u \equiv u^*$, where u^* is some equilibrium solution belonging to the set of M of fixed points (cf. Eq. (3.1), then:

$$\lim_{t\to\infty} ||P_m u(t) - P_m u^*|| = 0 \rightarrow \lim_{t\to\infty} ||u(t) - u^*|| = 0 \;; \tag{4.3b}$$

if the projection of the orbit converges to some (projected) fixed point, the same is true of the infinite-dimensional orbit.

The main result of this section is stated for space dimension $n = 1$; with $\Omega = [0,L]$ and periodic boundary conditions:

Theorem 4.4. The first m Fourier modes are determining if

$$m + 1 \geq K\, L^{3/2} \;, \tag{4.4}$$

where K is some constant depending on α,β and ζ_0, with initial values $||\nabla u(0)|| \leq \zeta_0$.

Proof of Theorem 4.4. For sake of brevity, in the sequel, we will denote $q_m \equiv Q_m w$, $p_m \equiv P_m w$. Now, if u,v are two solutions, w satisfy the following equation:

$$\frac{\partial w}{\partial t} + \Delta(\Delta w + \beta w - \alpha[u^2 + uv + v^2]w) = 0 \;. \tag{4.5a}$$

Multiplying by q_m and integrating:

$$\tfrac{1}{2} \frac{d}{dt} ||q_m||^2 + ||\Delta q_m||^2 - \beta\, ||\nabla q_m||^2 - \alpha \int [u^2 + uv + v^2]\, w\, \Delta q_m\, dx = 0 \tag{4.5b}$$

But $w = q_m + p_m$, and so by Hölder's inequality:

$$\int (u^2 + uv + v^2) w\, \Delta q_m\, dx$$

$$\leq ||u^2 + uv + v^2||_{L^\infty} (||p_m|| + ||q_m||)\, ||\Delta q_m|| \tag{4.6}$$

and

$$\frac{1}{2}\frac{d}{dt}\,||q_m||^2 + \frac{1}{||q_m||^2}\,\{||\Delta q_m||^2 - \beta\,||\nabla q_m||^2$$

$$- \alpha||u^2 + uv + u^2||_{L^\infty}\,||\Delta q_m||\,||q_m||\}\,||q_m||^2$$

$$\leq \alpha\,||u^2 + uv + v^2||_{L^\infty}\,||\Delta q_m||\,||p_m|| \quad . \tag{4.7}$$

<u>We must prove that</u> $||p_m|| \to 0$ <u>implies</u> $||q_m|| \to 0$. This will be completed by verifying the three assumptions of the generalized Gronwall's Lemma 4.1 of [3]. We recall this Lemma:

Let $\xi(t)$ be an absolutely continuous nonnegative function on $(0,\infty)$ such that

$$\frac{d\xi}{dt} + A(t)\xi \leq B(t) \quad \text{a.e. on } (0,\infty) \quad ,$$

where $A(t)$ is a locally integrable function on $(0,\infty)$ satisfying for some T, $0 < T < \infty$:

$$\liminf_{t\to\infty} \int_t^{t+T} A\,ds = \gamma > 0 \tag{H1}$$

$$\limsup_{t\to\infty} \int_t^{t+T} A^-\,ds = \Gamma < \infty \quad , \tag{H2}$$

where $A^- = \max(-A, 0)$ and $B(t)$ is a measurable function on $(0,\infty)$ such that

$$B(t) \to 0 \quad , \; t \to \infty \quad , \tag{H3}$$

then

$$\xi(t) \to 0 \text{ as } t \to \infty \quad .$$

(Here, we set $\xi(t) \equiv ||q_m(t)||^2$.) We define:

$$A_m(t) = 2\,\frac{||\Delta q_m|| - \beta\,||\nabla q_m||^2}{||q_m||^2} - 2\alpha\,\frac{||u^2 + uv + v^2||_{L^\infty}}{||q_m||}\,||\Delta q_m|| \tag{4.8}$$

$$B_m(t) = 2\alpha\,||u^2 + uv + v^2||_{L^\infty}\,||\Delta q_m||\,||p_m|| \quad , \tag{4.9}$$

$$\rho_m(t) = \frac{||\Delta q_m||^2}{||q_m||^2} \quad , \quad \tilde{\rho}_m(t) = \frac{1}{T}\int_t^{t+T} \rho_m(s)\,ds \quad , \tag{4.10}$$

$$R(u,v) = \alpha||u^2 + uv + v^2||_{L^\infty} \quad . \tag{4.11}$$

Inequality (4.7) now can be rewritten in a more compact way:

$$\frac{d}{dt} ||q_m||^2 + A_m(t) \, ||q_m||^2 \leq B_m(t) \quad . \tag{4.12}$$

We first verify Hypothesis (H1) from the generalized Gronwall's Lemma:

$$A_m(t) \geq \frac{2||\Delta q_m||^2}{||q_m||^2} - \frac{2\beta||\Delta q_m||}{||q_m||} - 2\,R(u,v)\,\frac{||\Delta q_m||}{||q_m||}$$

$$= 2\,\rho_m(t) - 2\,\beta\,\rho_m(t)^{\frac{1}{2}} - 2\,R(u,v)\,\rho_m(t)^{\frac{1}{2}} \quad . \tag{4.13}$$

From (4.13):

$$\frac{1}{T}\int_t^{t+T} A_m(s)\,ds \geq 2\,\tilde{\rho}_m(t) - 2\,\beta\,\tilde{\rho}_m(t)^{\frac{1}{2}} - \frac{2}{T}\int_t^{t+T} R(u,v)\,\rho_m(s)^{\frac{1}{2}}\,ds$$

$$\geq 2\,\tilde{\rho}_m(t) - 2\,\beta\,\tilde{\rho}_m(t)^{\frac{1}{2}} - 2\left(\frac{1}{T}\int_t^{t+T} R(u,v)^2\,ds\right)^{\frac{1}{2}} \tilde{\rho}_m(t)^{\frac{1}{2}}$$

$$= 2\,\tilde{\rho}_m(t)^{\frac{1}{2}}\,[\tilde{\rho}_m(t)^{\frac{1}{2}} - \beta - (\frac{1}{T}\int_t^{t+T} R(u,v)^2\,ds)^{\frac{1}{2}}] \quad , \tag{4.14}$$

where we use a classical interpolation inequality for $||\nabla q_m||^2$ and Jenssen's inequality. From (4.14), a sufficient condition for (H1) is:

$$\tilde{\rho}_m(t)^{\frac{1}{2}} \geq \beta + (\frac{1}{T}\int_t^{t+T} R(u,v)^2\,ds)^{\frac{1}{2}} \quad ; \tag{4.15}$$

but

$$\tilde{\rho}_m(t) \geq E_{m+1} \quad , \tag{4.16}$$

where E_{m+1} is the $(m+1)^{th}$ eigenvalue of the biharmonic; $E_{m+1} = (\frac{2\pi(m+1)}{L})^4$. Then a sufficient condition for hypothesis (H1) is:

$$\frac{4\pi^2(m+1)^2}{L^2} > \beta + 4\alpha\,[\frac{1}{T}\int_t^{t+T} \max\,(||u^2||^2_{L^\infty} \;,\; ||v^2||^2_{L^\infty}\,ds]^{\frac{1}{2}} \quad . \tag{4.17}$$

We will further elaborate on (4.17). But we first verify Hypothesis (H2) and (H3) from the generalized Gronwall's Lemma. To verify (H2), notice that (4.14) implies by the Cauchy-Young inequality:

$$\frac{1}{T}\int_t^{t+T} A_m(s)\,ds \geq 2\,\tilde{\rho}_m(t) - 2\,\beta\,\tilde{\rho}_m(t)^{\frac{1}{2}} - \tilde{\rho}_m(t) - \overline{\lim_{t\to\infty}}\,R(u,v)^2 \quad ; \tag{4.18}$$

(H2) is satisfied as soon as

$$\tilde{\rho}_m(t) \geq 4\,\beta^2 \tag{4.19}$$

which is implied by (4.16) and (4.17). To verify (H3), remember that $R(u,v)$ and $||\Delta q_m||$ are uniformly bounded in time (cf., Section 3); moreover, $||p_m(t)|| \to 0$ from the very hypothesis of theorem 4.4.

We now further explicit the remaining sufficient condition (4.17). Using (Lemma 2.1), namely that

$$\bar{u}(t) \equiv \bar{u}(0) \quad ,$$

the continuous injection of $H^1(\Omega)$ into $L^\infty(\Omega)$ can be sharpened as:

$$||u||_{L^\infty} \leq \sqrt{L}\ ||\nabla u||_{L^2} + \bar{u}(0) \quad . \tag{4.20}$$

Then:

$$\left(\frac{1}{T}\int_t^{t+T} \max\ (||u^2||^2_{L^\infty}\ ,\ ||v^2||^2_{L^\infty}\ ds\right)^{\frac{1}{2}}$$

$$\leq \max\ (\overline{\ell im}_{t\to\infty}\ ||u||^2_{L^\infty}\ ,\ \overline{\ell im}_{t\to\infty}\ ||v||^2_{L^\infty})$$

$$\leq \max\ ((\sqrt{L}\ F(u_0) + \bar{u}(0))^2\ ,\ (\sqrt{L}\ F(v_0) + \bar{v}(0))^2) \quad , \tag{4.21}$$

where we have used Theorem 2.3, i.e., $\overline{\ell im}_{t\to\infty}\ ||\nabla u(t)|| \leq F(u_0)$. Then for m and L large enough, (4.17) is equivalent to:

$$\frac{4\pi^2(m+1)^2}{L^2} \sim Ct(\alpha,\beta,u_0,v_0)\ L \quad , \tag{4.22a}$$

$$m + 1 \sim Ct(\alpha,\beta,\zeta_0)\ L^{3/2} \quad , \tag{4.22b}$$

where we have taken both $||\nabla u(0)||$ and $||\nabla v(0)|| < \zeta_0$. □

Acknowledgments. This work was completed while the second author visited the Center for Nonlinear Studies at Los Alamos. The authors wish to thank Darryl D. Holm for stimulating discussions. This research was supported by the Center for Nonlinear Studies, Los Alamos National Laboratory. Work also performed under the auspices of the U.S. Department of Energy under contract W-7405-ENG-36 and contract KC-04-02-01, Division of Basic and Engineering Sciences.

References

1. K. Binder, Z. Physik 267 (1974), 213.
2. J. W. Cahn and J. E. Hilliard, J. Chem. Phys. 28 (1958), 258.
3. C. Foias, O. P. Manley, R. Temam and Y. M. Treve, "Asymptotic analysis of the Navier-Stokes equations," Physica 9D (1983), 157-188.
4. J. K. Hale, "Dynamical Systems and Stability," J. Math. Anal. Appl. 26 (1969), 39-59.
5. J. S. Langer, Annals of Physics (N.Y.) 65 (1971), 53.

6. B. Nicolaenko and B. Scheurer, "Remarks on the Kuramoto-Sivashinsky Equation," Physica 12D (1984), 391-395.
7. B. Nicolaenko, B. Scheurer and R. Temam, "Some Global Dynamical Properties of the Kuramoto-Sivashinsky Equations: Nonlinear Stability and Attractors," Los Alamos National Laboratory report LA-UR-84-2326 (1984).
8. B. Nicolaenko, B. Scheurer and R. Temam, "Quelques proprietes des attracteurs pom l'équation de Kuramoto-Sivashinsky," C. R. Acad. Sc. Paris 298 (1984), 23-25.
9. A. Novick-Cohen and L. A. Segel, "Nonlinear Aspects of the Cahn-Hilliard Equation," Physica D (1984), to appear.

This paper is in final form and no version of it will be submitted for publication elsewhere.

Trends in the Theory and Practice of Non-Linear Analysis
V. Lakshmikantham (Editor)

STOCHASTIC CONTINUITY AND RANDOM DIFFERENTIAL INEQUALITIES

Juan J. Nieto

Departamento de Teoria de Funciones
Facultad de Matematicas
Universidad de Santiago
SPAIN

We introduce a new concept of continuity for stochastic processes, that is stronger than the sample continuity, but it preserves some properties of the deterministic functions. We show the relation between both concepts with several examples and counterexamples. Using this concept, we are able to prove random versions of some of the (deterministic) theorems involving differential inequalities.

NOTATION AND PRELIMINARIES

Let $(\Omega,\mathcal{F},P)$ be a complete probability space. The class of random variables, $u:\Omega \longrightarrow \mathbb{R}^n$ is denoted by $\mathcal{R}[\Omega,\mathbb{R}^n]$.
Let I denote an arbitrary index set. The class of random functions defined on I into $\mathcal{R}[\Omega,\mathbb{R}^n]$ is denoted by $\mathcal{R}[I,\mathcal{R}[\Omega,\mathbb{R}^n]]$. Without loss in generality, we shall assume that all random processes that are considered in this paper are separable processes.

Let $a,b \in \mathbb{R}$. A random process $u \in \mathcal{R}[[a,b],\mathcal{R}[\Omega,\mathbb{R}^n]]$ is said to be almost-surely continuous at $t \in (a,b)$ if

$$P\{\omega : \lim_{h\to 0} | u(t+h,\omega) - u(t,\omega) | \neq 0 \} = 0 .$$

The random process u is said to be sample continuous in $t \in (a,b)$ if

$$P\{\bigcup_{t\in(a,b)}\{\omega: \lim_{h\to 0} |u(t+h,\omega)-u(t,\omega)|\neq 0 \}\} = 0 .$$

If a process is sample continuous in $t \in (a,b)$ and has one-sided sample continuity at the end points a and b, then it is said to be sample continuous on $[a,b]$. Thus, $C[[a,b],\mathcal{R}[\Omega,\mathbb{R}^n]]$ is the set of all $\mathbb{R}^n$-valued separable and sample continuous random processes defined on $[a,b]$. Obviously, this set is a separable complete metric space.
We say that $u \in M[[a,b],\mathcal{R}[\Omega,\mathbb{R}^n]]$ if $u:[a,b]\times\Omega \longrightarrow \mathbb{R}^n$ is product-measurable on $([a,b]\times\Omega,\mathcal{F}'\times\mathcal{F},m\times P)$ where $([a,b],\mathcal{F}',m)$ is a sample Lebesgue-measurable space.
A random function $u \in C[[a,b],\mathcal{R}[\Omega,\mathbb{R}^n]]$ is said to posses an almost-sure derivative $u'(t,\omega)$ at $t \in (a,b)$ if

$$P\{\omega : \lim_{h\to 0} | \frac{u(t+h,\omega)-u(t,\omega)}{h} - u'(t,\omega)| \neq 0 \} = 0 .$$

The random process u is said to posses a sample derivative $u'(t,\omega)$ in $t \in (a,b)$ if

$$P\{\bigcup_{t\in(a,b)}\{\omega: \lim_{h\to 0} |\frac{u(t+h,\omega)-u(t,\omega)}{h} - u'(t,\omega) | \neq 0 \}\} = 0 .$$

If a process is differentiable at every $t \in (a,b)$ and it has one sided derivatives at the end points a and b, then it is said to be differentiable on $[a,b]$. We note that the sample derivative is a random variable.
For $u \in C[[a,b],\mathcal{R}[\Omega,\mathbb{R}^n]]$ and $t \in [a,b]$ we define the sets

$$A(u,t) = \{ \omega: u(t,\omega) > 0 \} \quad \text{and} \quad B(u,t) = \{ \omega: u(t,\omega) \geqslant 0 \} .$$

For $t \in [a,b]$ we write $u(t,\cdot) > 0$ if $P(A(u,t)) = 1$. Similarly, $u(t,\cdot) \geqslant 0$ means that $P(B(u,t)) = 1$. For $\varepsilon \in \mathbb{R}$ and $t \in [a,b]$, we say that $u(t,\cdot) > \varepsilon$ if $P\{\omega: u(t,\omega) > \varepsilon\} = 1$.
For $u \in C[[a,b],\mathcal{R}[\Omega,\mathbb{R}]]$ and $t \in (a,b]$ we define

$$D_{-}u(t,\omega) = \liminf_{h\to 0^{-}} \frac{u(t+h,\omega)-u(t,\omega)}{h} .$$

We note that $D_{-}u(t,\omega)$ exists with probability one (henceforth denoted by w.p. 1). Now, let $u:[a,\bar{b}] \longrightarrow \mathbb{R}$ be a continuous function. If $t \in [a,b]$ and $u(t) > 0$, then there exists $\delta > 0$ such that $u(\underline{t}) > 0$ for every $\underline{t} \in [a,b] \cap (t-\delta,t+\delta)$. This property is used in several areas of mathematics repetitively. For instance, in the theory of differential inequalities (see Apendix A) which is our main interest in this paper. For random functions, that property is not valid in general. Thus, we introduce a new concept of stochastic continuity and with this we shall prove random versions of deterministic theorems involving differential inequalities.

EXAMPLES AND COUNTEREXAMPLES

Most of the examples of this section are taken from [10]. Let $u \in C[[a,b],\mathcal{R}[\Omega,\mathbb{R}]]$ and $t_0 \in [a,b]$. Assume that $u(t_0,\cdot) > 0$. Does this imply that there exists $\delta > 0$ such that

(1) $$u(t,\cdot) > 0 \quad \text{for} \quad t \in [a,b] \cap (t_0-\delta,t_0+\delta) \quad ?$$

In general, the answer is no.

Example 1: Let $\Omega = (0,1) \subset \mathbb{R}$ and $u_1:[0,1]\times\Omega \longrightarrow \mathbb{R}$ defined by $u_1(t,\omega) = \omega - t$ Thus, $u_1(0,\omega) = \omega > 0$ and $P(A(u_1,0)) = 1$. Howewer, for $t > 0$ we have $A(u_1,t) = \{\omega: \omega > t\}$ and $P(A(u_1,t)) < 1$. Hence, there is no δ such that (1) is satisfied.

Note that Ω (as a topological space) is not compact. That is not a problem since taking $\Omega = [0,1]$ we get the same conclusion. The point is that $u(0,0) = 0$, that is $u(0,\cdot)$ is zero at some point $\omega \in \Omega$.
Now, assume that there exists $\varepsilon > 0$ such that $u(t_0,\cdot) > \varepsilon$. Even in this case (1) needs not to be true. The following example illustrates this.

Example 2: Let $\Omega = [1,\infty) \subset \mathbb{R}$ and define $u_2:[0,1]\times\Omega \longrightarrow \mathbb{R}$ by $u_2(t,\omega) = \omega - t\cdot\omega^2$. Thus, $u_2(0,\omega) = \omega \geqslant 1$ and we can take $\varepsilon = \frac{1}{2}$ so that $u_2(0,\cdot) > \varepsilon$. For $t > 0$ we have $u_2(t,\omega) > 0$ if and only if $\omega < 1/t$. Hence, $P(A(u_2,t)) < 1$ for $t > 0$.

Note that in this last example Ω is not bounded.
Now, let (Ω,τ) be a topological compact space and $\mathcal{B}$ the σ-algebra of the Borel sets (smallest σ-algebra containing τ) and P a probability on $(\Omega,\mathcal{B})$.

Theorem 1: Let (Ω,τ) be a compact topological space such that $P(M) > 0$ for every $M \in \tau$. Let $u:[a,b]\times\Omega \longrightarrow \mathbb{R}$ be continuous (in both variables). If $t_0 \in [a,b]$ is such that $u(t_0,\omega) > \varepsilon$ for some $\varepsilon > 0$, then there exists $\delta > 0$ such that $u(t,\omega) > 0$ for $t \in [a,b] \cap (t_0-\delta,t_0+\delta)$ and $\omega \in \Omega$. Consequently, (1) is verified

Proof: If not, for every $n \in N$, there exists $t_n \in [a,b]$ and $\omega_n \in \Omega$ such that $u(t_n,\omega_n) \leqslant 0$ where $|t_n-t_0| \leqslant 1/n$. Thus, $\{t_n\} \to t_0$. Since Ω is compact, there exists a subsequence (again $\{\omega_n\}$) such that $\{\omega_n\} \to \omega_0$. Hence $0 \geqslant \lim_n u(t_n,\omega_n) = u(t_0,\omega_0) > 0$, which is a contradiction.

If Ω is compact but $u(t_o,\omega) = 0$ for some $\omega \in \Omega$ then the result of the theorem is not valid.

Example 3: Let $\Omega = [-1,1]$, $u_3:[0,1]\times\Omega \longrightarrow \mathbb{R}$ defined by $u_3(t,\omega) = \omega^2 - t$. Then $P(A(u_3,0)) = 1$, but for $t > 0$ we have that $P(A(u_3,t)) < 1$.

We note that in the three examples, the process is sample differentiable and $u'(0,\cdot) < 0$. Indeed, $u_1'(t,\omega) = -1$, $u_2'(t,\omega) = -\omega^2$ and $u_3'(t,\omega) = -1$. Thus, we have $P\{\omega : u_i'(t,\omega) < 0\} = 1$ and $P\{\omega : u_i'(0,\omega) < -\frac{1}{2}\} = 1$ for $i = 1,2,3$. Then, it is natural to ask the following:

Do $u(t_o,\cdot) > 0$ and $u'(t_0,\cdot) > 0$ imply that (1) is satisfied ? As before, the answer is no.

Example 4: Let $\Omega = (-\infty,0)$, $u_4:[0,1]\times\Omega \longrightarrow \mathbb{R}$, $u_4(t,\omega) = -t^2\omega^2 - t\omega - \omega$. Thus, $u_4(0,\omega) = -\omega$ and $P(A(u_4,0)) = 1$. By the other hand, $u_4'(t,\omega) = -2\cdot t\cdot\omega^2 - \omega$ and $P(A(u_4',0)) = 1$. Howewer, there exists no δ such that $u_4(t,\cdot) > 0$ for $t \in [0,\delta)$ and (1) is not satisfied (see [10]).
Even if $u(t_0,\cdot) > \varepsilon$ and $u'(t_0,\cdot) > \varepsilon$ for some $\varepsilon > 0$, (1) needs not to hold [10].

If $u:[a,b] \longrightarrow \mathbb{R}$ is a continuous function (deterministic), $u(a) > 0$ and $D\_u(t) > 0$ for $t \in (a,b]$, then $u(t) > 0$ for every $t \in [a,b]$. A generalitation of this result is given in Apendix A. For random processes, we have similar results if g is continuous. If g is product measurable, the generalitation to the stochastic case is not possible (compare this with theorem 2.3.1 in [7]).

Example 5: Let $\Omega = (0,1)$ and define $g:[0,1]\times\mathbb{R}\times\Omega \longrightarrow \mathbb{R}$ as $g(t,u,\omega) = 0$ if $u = 0$ and $g(t,u,\omega) = -2$ if $u \neq 0$. Thus, $g \in M[[0,1]\times\mathbb{R},\mathcal{R}[\Omega,\mathbb{R}]]$. Let $u,v:[0,1]\times\Omega \longrightarrow \mathbb{R}$, $u(t,\omega) = 1-t-\omega$ and $v(t,\omega) = 0$. Then, $u(0,\omega) = 1-\omega > 0 = v(0,\omega)$ for every $\omega \in \Omega$. We have that $D\_u(t,\omega) = -1$ and $D\_v(t,\omega) = 0$. Thus, $D\_v(t,\omega) = 0 \leqslant g(t,v(t,\omega),\omega) = 0$. By the other hand $g(t,u(t,\omega),\omega) = g(t,1-t-\omega,\omega) < -1$ if $\omega \neq 1-t$.
Hence, $P\{\omega : D\_u(t,\omega) > g(t,u(t,\omega),\omega)\} = P\{\omega : \omega \neq 1-t\} = 1$.
Howewer, for $t > 0$, $P(A(u,t)) < 1$, that is $u(t,\cdot) > v(t,\cdot) = 0$ is not true for $t > 0$.

Remark: In this example $P\{\omega : D\_u(t,\omega) \leqslant g(t,u(t,\omega),\omega)\} = 0$, but

$$P\{\bigcup_{t\in(0,1)}\{\omega : D\_u(t,\omega) \leqslant g(t,u(t,\omega),\omega)\}\} = 1.$$

It is possible to modify example 5 (see [10]) and to get

$$P\{\bigcup_{t\in(0,1)}\{\omega : D\_u(t,\omega) \leqslant g(t,u(t,\omega),\omega)\}\} = 0.$$

COMPARISON RESULTS

Let $u \in C[[a,b],\mathcal{R}[\Omega,\mathbb{R}]]$, we introduce the following

Definition: We say that the process u is positively continuous at $t_0 \in [a,b]$ if $P(A(u,t_o)) = 1$ implies that there exists $\delta > 0$ such that $P(A(u,t)) = 1$ for every $t \in [a,b] \cap (t_0-\delta,t_0+\delta)$. Let $\varepsilon > 0$, we say that u is ε-positively continuous at t_0 if $u(t_0,\cdot) > \varepsilon$ implies that there exists $\delta > 0$ such that (1) is satisfied.
If u is positively continuous at every $t \in [a,b]$, we write $u \in C_p[[a,b],\mathcal{R}[\Omega,\mathbb{R}]]$.
We say that u is positively sample continuous at t_0 if $P(A(u,t_0)) = 1$ implies that there exists $\delta > 0$ such that $P(N_\delta) = 0$ where

$$N_\delta = \{\bigcup_{t\in\Delta}\{\omega : u(t,\omega) < 0\} \quad \text{and} \quad \Delta = [a,b] \cap (t_0-\delta,t_0+\delta) .$$

We note that if $P(A(u,t_0)) = 1$ and u is positively sample continuous at t_0 , then there exists $\delta > 0$ such that $P(A(u,t)) = 1$ for every $t \in [a,b] \cap (t_0-\delta,t_0+\delta)$.
In what follows $t_0 \in \mathbb{R}$ and $a > 0$.

Theorem 2: Let $g \in M[[t_0,t_0+a)\times\mathbb{R},\mathcal{R}[\Omega,\mathbb{R}]]$ and $u,v \in C[[t_0,t_0+a),\mathcal{R}[\Omega,\mathbb{R}]]$ and assume that for every $t \in (t_0,t_0+a)$ we have $P\{\omega : D\_v(t,\omega) \leqslant g(t,v(t,\omega),\omega)\} = P\{\omega : D\_u(t,\omega) > g(t,u(t,\omega),\omega)\} = 1$. In addition, suppose that

i) $u(t_0,\cdot) > v(t_0,\cdot)$ and $u-v$ is positively sample continuous at every $t \in [t_0,t_0+a)$.
ii) If $\Omega_0 \in \mathcal{F}$ is such that $P(\Omega_0) > 0$, then $v-u \in C_p[[t_0,t_0+a),\mathcal{R}[\Omega_0,\mathbb{R}]]$.

Then, $P\{\omega : v(t,\omega) < u(t,\omega)\} = 1$ for every $t \in [t_0,t_0+a)$.

Proof: If not, there exists $t^* \in [t_0,t_0+a)$ such that $P\{\omega : v(t^*,\omega) \geqslant u(t^*,\omega)\} > 0$. Define

$$t_1 = \operatorname{Sup}\{t \in [t_0,t_0+a) : P\{\bigcup_{t_0<\bar{t}<t}\{\omega : u(\bar{t},\omega) \leqslant v(\bar{t},\omega)\}\} = 0\} .$$

Thus, $t_0 < t_1$ since $u-v$ is positively continuous at t_0. Then,

$$P\{\omega : u(t_1,\omega) \leqslant v(t_1,\omega)\} > 0 . \tag{2}$$

If this is not true, then $P\{\omega : u(t_1,\omega) > v(t_1,\omega)\} = 1$ and by i), t_1 is not the sup. Hence, $P\{\omega : u(t_1,\omega) = v(t_1,\omega)\} > 0$. If not $P(\Omega_o) > 0$ where $\Omega_o = \{\omega : u(t_1,\omega) < v(t_1,\omega)\}$ and then by ii), t_1 is not the sup. Thus, we can conclude that for $t < t_1$, $u(t,\cdot) > v(t,\cdot)$. Now, for $n \in N$, let $t_n = t_1 - 1/n$ and define the set

$$\Omega_n = \bigcup_{t_o \leq t \leq t_n} \{\omega : u(t,\omega) < v(t,\omega)\} .$$

Clearly, $P(\Omega_n) = 0$. Thus, for $\omega \in \Omega - \Omega_n$ we have $u(t,\omega) > v(t,\omega)$, for every $t \in [t_o,t_n]$. Let $A = \bigcup_{n=1}^{\infty} \Omega_n$, $P(A) = 0$, and for $\omega \in \Omega - A$ and $t \in [t_o,t_1)$, we have that $u(t,\omega) > v(t,\omega)$.

Indeed, for $t < t_1$ there exists $n \in N$ such that $t < t_n < t_1$. Then, for $\omega \in \Omega - A$ we get that $\omega \in \Omega - \Omega_n$ and $u(t,\omega) > v(t,\omega)$. Thus, for $t < t_1$ and $\omega \in \Omega - A$, we have

$$\frac{u(t_1+h,\omega)-u(t_1,\omega)}{h} < \frac{v(t_1+h,\omega)-v(t_1,\omega)}{h}$$

where $h < 0$ and $t = t_1+h$. This implies that $D\_u(t_1,\omega) \leq D\_v(t_1,\omega)$ for $\omega \in \Omega - A$. By the other hand, there exists $N_1 \quad \Omega$ with $P(N_1) = 0$ such that

$$D\_v(t_1,\omega) \leq g(t_1,v(t_1,\omega),\omega) \quad \text{and} \quad D\_u(t_1,\omega) > g(t_1,u(t_1,\omega),\omega) \text{ for } \omega \in \Omega - N_1$$

Therefore, if $\omega \in \Omega_o-(A \quad N_1)$ we have:

$$D\_v(t_1,\omega) \leq g(t_1,v(t_1,\omega),\omega) = g(t_1,u(t_1,\omega),\omega) < D\_u(t_1,\omega) .$$

This is a contradiction since $P(\Omega_o-(A \quad N_1)) > 0$. This concludes the proof of the theorem.

If g is continuous, we have the following result

<u>Theorem 3</u>: Let $g \in C[[t_o,t_o+a)\times\mathbb{R},\mathcal{R}[\Omega,\mathbb{R}]]$ (that is: $g(\cdot,\cdot,\omega)$ is continuous w.p.1 and $g(t,u,\cdot)$ is a random variable for every $(t,u) \in [t_o,t_o+a)\times\mathbb{R}$).
Let $u,v \in C[[t_o,t_o+a),\mathcal{R}[\Omega,\mathbb{R}]]$ and assume that

a) $P\{\bigcup_{t_o<t<t_o+a} \{\omega : D\_v(t,\omega) > g(t,v(t,\omega),\omega)\}\} = 0$

b) $P\{\bigcup_{t_o<t<t_o+a} \{\omega : D\_u(t,\omega) \leq g(t,u(t,\omega),\omega)\}\} = 0$

c) $P\{\omega : v(t_o,\omega) < u(t_o,\omega)\} = 1$, that is, $v(t_o,\cdot) < u(t_o,\cdot)$.

Then, $P\{\bigcup_{t_o \leq t<t_o+a} \{\omega : v(t,\omega) \geq u(t,\omega)\}\} = 0$. In particular, $v(t,\cdot) < u(t,\cdot)$ for every $t \in [t_o,t_o+a)$.

Proof: By a) there exists $N_o \in \mathcal{F}$ such that $P(N_o) = 0$ and if $\omega \in \Omega - N_o$ we have

$$D\_v(t,\omega) \leq g(t,v(t,\omega),\omega) \quad \text{for every} \quad t \in [t_o,t_o+a) .$$

Similarly, there exists $N_1 \in \mathcal{F}$ such that $P(N_1) = 0$ and for $\omega \in \Omega - N_1$ we have

$$D\_u(t,\omega) > g(t,u(t,\omega),\omega) \quad \text{for every} \quad t \in [t_o,t_o+a) .$$

Since g,u and v are continuous, there exists $N_2 \in \mathcal{F}$ such that $P(N_2) = 0$ and $g(\cdot,\cdot,\omega) \in C[[t_o,t_o+a)\times\mathbb{R}]$ and $u(\cdot,\omega)$, $v(\cdot,\omega) \in C[[t_o,t_o+a)]$ for every $\omega \in \Omega - N_2$. Now if $\omega \in \Omega - N$ where $N = N_o \cup N_1 \cup N_2$ we can conclude that for every $t \in (t_o,t_o+a)$ we have

$$D\_u(t,\omega) > g(t,u(t,\omega),\omega) \quad \text{and} \quad D\_v(t,\omega) \leq g(t,v(t,\omega),\omega)$$

where $g(\cdot,\cdot,\omega) \in C[[t_o,t_o+a)\times\mathbb{R},\mathbb{R}]$.
Hence, by the deterministic comparison theorem given in Apendix A, we can conclude that for $\omega \in \Omega - N$, $u(t,\omega) > v(t,\omega)$ for every $t \in [to,t_o+a)$. Therefore,

$$P\{\bigcup_{t_o \leq t<t_o+a} \{\omega : u(t,\omega) \leq v(t,\omega)\}\} = 0$$

since

$$\bigcup_{t_o \leq t<t_o+a} \{\omega : u(t,\omega) \leq v(t,\omega)\} \subset N \quad \text{and} \quad P(N) = 0 .$$

This proves the theorem.

<u>Corollary</u>: Let $u \in C[[t_o,t_o+a),\mathcal{R}[\Omega,\mathbb{R}]]$ and assume that $P\{\omega : u(t_o,\omega) > 0\} = 1$. Then, $P\{\bigcup_{t_o<t<t_o+a} \{\omega : D\_u(t,\omega) \leq 0\}\} = 0$ implies that

$P\{\bigcup_{t_o\leq t<t_o+a}\{\omega: u(t,\omega) \leq 0\}\} = 0$. In particular, $P\{\omega : u(t,\omega) > 0\} = 1$ for every $t \in [t_0,t_0+a)$.

Proof: Take $g = 0$.

Let $L \in M[[t_0,t_0+a), [\Omega,\mathbb{R}^+]]$ and suppose that it is sample Lebesgue-integrable on $[t_0,t_0+a)$, (see [7]).

Theorem 4: Let $g \in M[[t_0,t_0+a)\times\mathbb{R},\mathcal{R}[\Omega,\mathbb{R}]]$, $u,v \in C[[t_0,t_0+a),\mathcal{R}[\Omega,\mathbb{R}]]$. Assume that for every $t \in (t_0,t_0+a)$ we have

$$P\{\omega : D_-v(t,\omega) \leq g(t,v(t,\omega),\omega)\} = P\{\omega : D_-u(t,\omega) \geq g(t,u(t,\omega),\omega)\} = 1$$

and

$$g(t,x,\omega) - g(t,y,\omega) \leq L(t,\omega)(x-y) \qquad \text{for } x \geq y.$$

Assume, in addition, that:

i) there exists $\delta > 0$ such that $P\{\bigcup_{t_0\leq t\leq t_0+\delta}\{\omega: v(t,\omega) > u(t,\omega)\}\} = 0$.

ii) for $\varepsilon > 0$, define

$$w_\varepsilon(t,\omega) = u(t,\omega) + \varepsilon\exp\{2\int_{t_0}^{t}[L(s,\omega)+\varepsilon]ds\},$$

and assume that $w_\varepsilon - v$ is positively sample continuous at every $t \in (t_0,t_0+a)$

iii) if $\Omega_0 \in \mathcal{F}$ and $P(\Omega_0) > 0$, then $w_\varepsilon - v \in C_p[[t_0+\delta,t_0+a),\mathcal{R}[\Omega_0,\mathbb{R}]]$.

Then, $P\{\omega : v(t,\omega) \leq u(t,\omega)\} = 1$ for every $t \in [t_0,t_0+a)$.

Proof: By the definition of w_ε we get

$$D_-w_\varepsilon(t,\omega) = D_-u(t,\omega) + \varepsilon[L(t,\omega)+\varepsilon]\exp\{2\int_{t_0}^{t}[L(s,\omega)+\varepsilon]ds\} \quad .$$

Thus, $D_-w_\varepsilon(t,\omega) > g(t,w_\varepsilon(t,\omega),\omega)$ w.p.1 for every $t \in (t_0,t_0+a)$, and we can conclude that:

a) $w_\varepsilon - v$ is positive at t_0 and positively sample continuous at every $t \in [t_0,t_0+a)$

b) for $\Omega_0 \in \mathcal{F}$ with $P(\Omega_0) > 0$, $w_\varepsilon - v \in C_p[[t_0,t_0+a),\mathcal{R}[\Omega_0,\mathbb{R}]]$.

Therefore, by theorem 2, we can write

$$P\{\omega : w_\varepsilon(t,\omega) > v(t,\omega)\} = 1 \quad \text{for } t \in [t_0,t_0+a) .$$

Now, fix $t \in [t_0,t_0+a)$ and for $n \in N$, let $A_n = \{\omega : w_{1/n}(t,\omega) > v(t,\omega)\}$, and consequently $P(A_n) = 1$. With this definition, we have

$$A = \bigcup_{n=1}^{\infty} A_n = \{\omega : u(t,\omega) > v(t,\omega) \quad .$$

Obviosly, $A_n \supset A_{n+1}$ and $P(A) = 1$. This completes the proof of the theorem.

Now, consider the random differential equation

(3) $\quad x' = f(t,x,\omega) \quad , \quad x(t_0,\omega) = x_0(\omega)$

where $f \in M[\mathbb{R}^+\times B(\rho),\mathcal{R}[\Omega,\mathbb{R}]]$ and $B(\rho) = \{x \in \mathbb{R} : |x| < \rho\}$.

Definition: We say ([7]) that a process x is a solution of the random differential equation (3) if there exists a δ such that

1) $x(t_0) = x_0$

2) $x(t)$ is sample continuous and differentiable

3) $x(t)$ is product measurable

4) $x'(t,\omega) = f(t,x(t,\omega),\omega)$ w.p.1 for every $t \in [t_0,t_0+\delta)$

Corollary: Let $g \in M[[t_0,t_0+a)\times\mathbb{R},\mathcal{R}[\Omega,\mathbb{R}]]$ where $g(t,x,\omega)-g(t,y,\omega) \leq L(t,\omega)(x-y)$ for $x \geq y$ and $L \in M[[t_0,t_0+a),[\Omega,\mathbb{R}^+]]$ is sample Lebesgue-integrable on $[t_0,t_0+a)$. Let u be a solution of the problem

$$u'(t,\omega) = g(t,u(t,\omega),\omega) \quad , \quad u(t_0,\omega) = u_0(\omega)$$

existing on $[t_0,t_0+a)$,(for sufficient conditions for the existence of solution of

a random differential equation, see [7]).
Let $m \in C[[t_o,t_o+a),\mathcal{R}[\Omega,\mathbb{R}]]$ be such that $P\{\omega : D_-m(t,\omega) \leq g(t,m(t,\omega),\omega)\} = 1$ for every $t \in [t_o,t_o+a)$. In addition, assume that there exists $\delta > 0$ such that

i) $P \bigcup_{t_o \leq t \leq t_o+\delta} \{\omega: m(t,\omega) > u(t,\omega)\}\} = 0$

ii) for any $\varepsilon > 0$, $w_\varepsilon - m$ is positively sample continuous at every $t \in [t_o,t_o+a)$

iii) if $\Omega_0 \in \mathcal{F}$ with $P(\Omega_0) > 0$, then $w_\varepsilon - m \in C_p[[t_o+\delta,t_o+a),\mathcal{R}[\Omega_o,\mathbb{R}]]$

Then, for every $t \in [t_o,t_o+a)$, we have $P\{\omega : m(t,\omega) \leq u(t,\omega)\} = 1$.

Proof: Take $v = m$ in the theorem.

We now consider the random differential equation (3). For $V \in C[\mathbb{R}^+ \times B(\rho),\mathcal{R}[\Omega,\mathbb{R}^+]]$ we define

$$D_-^f V(t,x,\omega) = \liminf_{h \to 0^-} \frac{V(t+h,x+hf(t,x,\omega),\omega) - V(t,x,\omega)}{h}$$

We note that for $(t,x) \in \mathbb{R}^+ \times B(\rho)$, that limit exists and $D_-^f V(t,x,\omega)$ is a product-measurable random process.
Using this definition and the concept of random Lyapunov functions we have the following

Theorem 5: Assume that

a) $g \in M[[\mathbb{R}^+ \times \mathbb{R},\mathcal{R}[\Omega,\mathbb{R}]]$ and g is sample continuous in u for fixed $t \in \mathbb{R}^+$ and there exists $L \in M[[t_o,\infty),\mathcal{R}[\Omega,\mathbb{R}^+]]$ such that L is sample Lebesgue-integrable on any compact interval of $\mathbb{R}^+$ and $g(t,x,\omega) - g(t,y,\omega) \leq L(t,\omega)(x-y)$ for $x \leq y$

b) u is a solution of $u'(t,\omega) = g(t,u(t,\omega),\omega)$, $u(t_o,\omega) = u_o(\omega)$ existing on $[t_o,\infty)$

c) $V \in C[\mathbb{R}^+ \times \mathbb{R},\mathcal{R}[\Omega,\mathbb{R}]]$ and $V(t,x,\omega)$ is locally Lipschitzian in x w.p.1 and for $(t,x) \in \mathbb{R}^+ \times \mathbb{R}$ we have
$$D_-^f V(t,x,\omega) \leq g(t,V(t,x,\omega),\omega)$$

d) there exists $\delta > 0$ such that $P\{\bigcup_{t_o \leq t \leq t_o+\delta} \{\omega: V(t,x(t,\omega),\omega) > u(t,\omega)\}\} = 0$ where $x(t,\omega)$ is any sample solution process of (3)

e) set $m(t,\omega) = V(t,x(t,\omega),\omega)$ and assume that for any $\varepsilon > 0$ we have that the process $w_\varepsilon - m$ is positively sample continuous at every $t \geq t_o$

f) if $\Omega_0 \in \mathcal{F}$ and $P(\Omega_0) > 0$, then $w_\varepsilon - m \in C_p[[t_o,\infty),\mathcal{F}[\Omega_o,\mathbb{R}]]$.

Then, $V(t,x(t,\omega),\omega) = m(t,\omega) \leq u(t,\omega)$ w.p.1 for $t \geq t_o$.

Proof: For small $h < 0$ we get: $m(t+h,\omega) - m(t,\omega) = V(t+h,x(t+h,\omega),\omega) - V(t,x(t,\omega),\omega) = V(t+h,x(t+h,\omega),\omega) - V(t+h,x(t,\omega)+hf(t,x(t,\omega),\omega) - V(t,x(t,\omega),\omega) + V(t+h,x(t,\omega)+hf(t,x(t,\omega),\omega) \geq -K[x(t+h,\omega) - x(t,\omega) - hf(t,x(t,\omega),\omega)] + V(t+h,x(t,\omega)+hf(t,x(t,\omega),\omega) - V(t,x(t,\omega),\omega)$ where K is the local Lipschitz constant relative to $V(t,x,\omega)$.
Now, dividing by $h < 0$ and using b) and c) we get

$$D_-m(t,\omega) \leq D_-V(t,x(t,\omega),\omega) \leq g(t,V(t,x(t,\omega),\omega),\omega) = g(t,m(t,\omega),\omega) .$$

Thus, we can conclude by the corollary of theorem 3 that

$$P\{\omega : m(t,\omega) \leq u(t,\omega)\} = 1 \quad \text{for every } t \geq t_o.$$

This completes the proof of the theorem.

Apendix A

For $u \in C[a,b]$ we define for $t \in (a,b]$

$$D_-u(t) = \liminf_{h \to 0^-} \frac{u(t+h) - u(t)}{h} .$$

We have the following comparison result

Theorem A: Let $t_o \in \mathbb{R}$, $a > 0$ and $u,v \in C[[t_o,t_o+a),\mathbb{R}]$ and $g \in C[[t_o,t_o+a) \times \mathbb{R},\mathbb{R}]$. Assume that for $t \in (t_o,t_o+a)$ we have

$$D_-v(t) \leq g(t,v(t)) \quad \text{and} \quad D_-u(t) > g(t,u(t)) \ .$$

then, $v(t_o) < u(t_o)$ implies that $v(t) < u(t)$ for every $t \in [t_o,t_o+a)$.

Proof: If not, there exists $t_1 \in (t_o,t_o+a)$ such that $u(t_1) \leq v(t_1)$. Let

$$t_2 = \text{Inf } \{ t \in [t_o,t_o+a) : u(t) \leq v(t) \} \ .$$

Thus, $t_o < t_2$ and $u(t_2) = v(t_2)$. Moreover, $u(t) > v(t)$ for $t \in [t_o,t_2)$. Now, take $h < 0$ and we get

$$\frac{u(t_2+h) - u(t_2)}{h} < \frac{v(t_2+h) - v(t_2)}{h} \ .$$

This implies that $D_-u(t_2) \leq D_-v(t_2)$. By the other hand,

$$D_-v(t_2) \leq g(t_2,v(t_2)) = g(t_2,u(t_2)) < D_-u(t_2)$$

which is a contadiction. This concludes the proof.

Apendix B

Theorem A can be improved if we assume that g satisfies a growth condition.

Theorem B: Let $g \in C[[t_o,t_o+a) \times \mathbb{R},\mathbb{R}]$, $u,v \in C[t_o,t_o+a)$. Assume that for every $t \in (t_o,t_o+a)$ we have $D_-v(t) \leq g(t,v(t))$ and $D_-u(t) \geq g(t,u(t))$. In addition, suppose that there exists $L:[t_o,t_o+a) \longrightarrow \mathbb{R}^+$ Lebesgue integrable such that

$$g(t,x)-g(t,y) \leq L(t)(x-y) \quad \text{for } x \geq y \ .$$

Then, $v(t_o) \leq u(t_o)$ implies that $v(t) \leq u(t)$ for $t \in [t_o,t_o+a)$.

Proof: For $\varepsilon > 0$, define

$$w_\varepsilon(t) = u(t) + \varepsilon \exp \{ 2 \int_{t_o}^{t} [L(s)+\varepsilon]ds \ .$$

Clearly $w_\varepsilon(t) > u(t)$ for any $t \in [t_o,t_o+a)$. Thus,

$$D_-w_\varepsilon(t) = D_-u(t) + 2\varepsilon[L(t)+\varepsilon] \exp \{ 2 \int_{t_o}^{t} [L(s)+\varepsilon]ds \geq$$

$$g(t,u(t)) + \varepsilon L(t) \exp \{ 2 \int_{t_o}^{t} [L(s)+\varepsilon]ds + \varepsilon[L(t)+2\varepsilon] \exp \{ 2 \int_{t_o}^{t} [L(s)+\varepsilon]ds$$

$$= g(t,u(t)) + L(t) [w_\varepsilon(t)-u(t)] + \varepsilon[L(t)+2\varepsilon] \exp \{ 2 \int_{t_o}^{t} [L(s)+\varepsilon]ds \geq$$

$$g(t,w_\varepsilon(t)) + \varepsilon[L(t)+2\varepsilon] \exp \{ 2 \int_{t_o}^{t} [L(s)+\varepsilon]ds > g(t,w_\varepsilon(t)) \ .$$

Therefore, $D_-w_\varepsilon(t) > g(t,w_\varepsilon(t))$ and $w_\varepsilon(t_o) > v(t_o)$. By theorem A we can conclude that

$$v(t) < w_\varepsilon(t) \quad \text{for } t \in [t_o,t_o+a) \ .$$

Now, taking inti account that $\lim_{\varepsilon \to o} w_\varepsilon(t) = u(t)$, we get the conclusion of the theorem.

REFERENCES

[1] Arnold,L. Stochastic Differential Equations: Theory and Applications (Wiley, New York,1974).

[2] Bartlett,M.S., Introduction to Stochastic Processes, 2nd ed. (Cambridge University Press, London, 1966).

[3] Chandra,J.,Ladde,G.S. and Lakshmikantham,V., On the fundamental Theory of nonlinear Second Order Stochastic Boundary Value Problems, Stochastic Anal. and Appl. 1 (1983), 1-19.

[4] Doob,J.L., Stochastic Processes (Wiley, New York, 1953).

[5] Gikhman,I.I. and Skorokhod,A.V., Stochastic Differential Equations, (Springer Verlag, Berlin and New York, 1972).

[6] Ladde,G.S., Systems of differential inequalities and stochastic differential equations II, J. Mathematical Phys. 16 (1975), 894-900.

[7] Ladde,G.S. and Lakshmikantham,V., Random Differential Inequalities (Academic Press, New York, 1980).

[8] Lakshmikantham,V, and Leela,S., Differential and Integral Inequalities, Vol I (Academic Press, New York, 1969).

[9] McShane,E.J., Stochastic Calculus and Stochastic Models (Academic Press, New York, 1974).

[10] Nieto,J.J., Stochastic Continuity: Some Counterexamples (to appear).

[11] Nieto,J.J., Stochastic Maximum Principles (to appear).

[12] Wong,E., Stochastic Processes in Information and Dynamical Systems (McGraw-Hill, New York, 1971).

This paper is in final form and no version of it will be submitted for publication elsewhere.

Trends in the Theory and Practice of Non-Linear Analysis
V. Lakshmikantham (Editor)

INFORMATION PROCESSING IN VERTEBRATE RETINA

M.N. OGUZTORELI, T.M. CAELLI, G. STEIL

Departments of Mathematics and Psychology
University of Alberta
Edmonton, Alberta, Canada T6G 2G1

The objective of this paper is to briefly describe the modelling, analysis, and simulation of visual information processing in vertebrate retina based on a nonlinear mathematical model studied in Oguztoreli (1979).

I. INTRODUCTION

The retina, considered as a subsystem of the brain, performs early processing of visual information which is then transmitted to higher centers through the optic nerve by compressing the extremely large range of light intensities of the external world into a narrower range of neural activities.

The retina in all vertebrates has the same five principal types of cells: receptor (rod or cone), bipolar, horizontal, amacrine, and ganglion cells. These cells form a complicated network particularly due to their complex interconnections.

When the receptor cells are exposed to light, ranging in wave length from violet, about 4000 A°, to red, 7500 A°, the photopigment molecules of the membranes of the receptor cells absorb the incoming photons, and initiate a complex chemical reaction causing discharge of a neural impulse. The amount of light energy which is necessary to affect a single receptor cell is very small. Note that rods function efficiently at low light intensities, cones at high.

The responses of the receptor cells to light inputs are further processed, filtered, and coded in the retina. Information received at the retinal ganglion cell layer is then transmitted along the optic nerve fibres to thalamocortical cells located in the lateral geniculate nuclei and in the visual cortex where visual perception occurs as the sensation of light in the conscious mind.

The synaptic organization of the retina has been studied extensively in the literature (cf. Cervetto and Fuortes, 1978; Dowling, 1970, 1972; Dowling and Ehinger, 1978; Dowling and Werblin, 1969; Michael, 1969; Shephard, 1979; Werblin, 1973). Some of the main characteristics of the neurons in the retina are described briefly and the form of the connectivities and interactions among the cells are outlined in Oguztoreli (1983), where further relevant references can be found. We schematically represent the main results of these past investigations in Figure 1, where circles represent single cells or groups of cells. Here, the basic circuitry for a retinal cell and its typical response profile to a rectangular input are also shown.

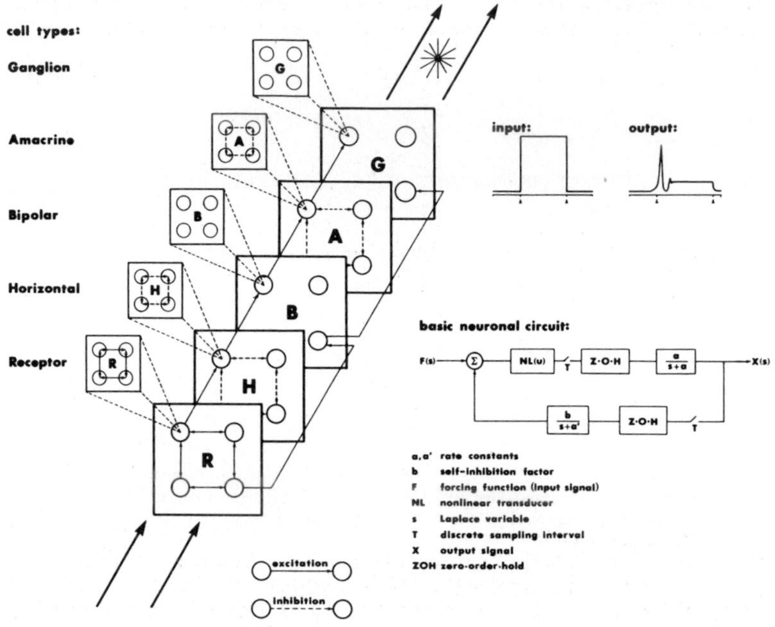

Figure 1

The retinal networks vary considerably from retinotopic location to location. Some of these circuitries are presented in Figure 2 which emphasizes the essential fact that such circuitries are distinguished by the degree and types of connections.

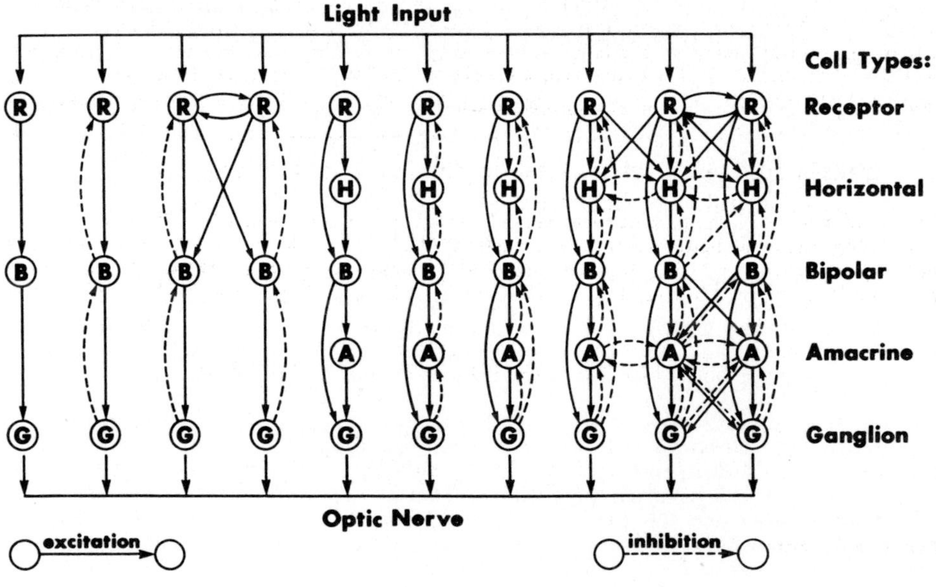

Figure 2

In Figure 3, the organization of the smallest functional unit containing all five types of retinal cells with all possible interactions is described.

Figure 3

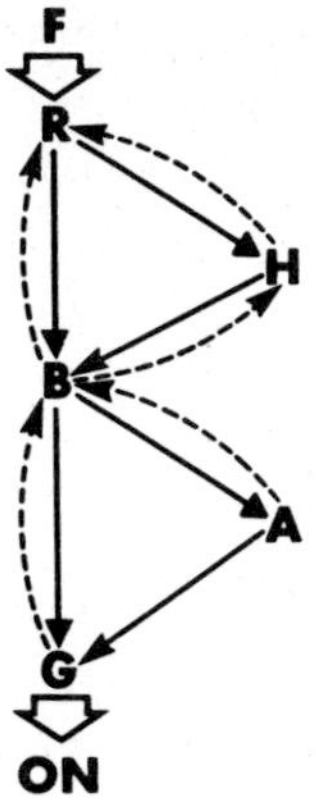

Since only the receptor cells contain photopigments and act as transducers, converting light energy into neuronal signals which are further processed by succeeding retinal cells, only the receptor R is subjected to an external force, denoted by F, which is a sensory input, and the receptor cell provides the input to the remainder of the network. Clearly this network can be adapted to special conditions by adding and/or deleting some of the connectivity pathways marked by arrows.

The objective of the present paper is to briefly describe the modelling, analysis, and simulation of neuronal processing in the retinal circuitries, particularly in the network shown in Figure 3, using the neural equations studied in Oguztoreli (1979). To simplify our presentation, the neural network in Figure 3 will be referred as the basic retinal network, or shortly BRN. Throughout this work we use the terminology, notation, and results of Oguztoreli (1979). The numerical algorithms described in this reference have been implemented in the University of Alberta in the ALGOL W and FORTRAN laguages.

II. MATHEMATICAL DESCRIPTION OF THE NEURAL ACTIVITIES IN BRN

In Oguztoreli (1979) a discrete neural model has been developed from a physiological, mathematical, and computational point of view. In this model a network containing finitely many neurons is described by a system of nonlinear ordinary integro-differential difference equations of the form

$$a_{10}^{-1}\frac{dx_1(t)}{dt} + x_1(t) = \mathcal{S}\{f_i + \sum_{j=1}^{n} c_{ij}x_j(t-\sigma_{ij}) + \sum_{j=1}^{m} b_{ik}\int_0^t x_i(\tau)e^{-a_{ik}(t-\tau)}dt\} \tag{1}$$

for $t > 0$, where

t time;
m number of the neurons in the network;
$m+1$ maximal order of the neurons;

$x_i(t)$ normalized firing rate, activity function, of the ith neuron at time t;
f_i $= f_i(t)$, external input to the ith neuron at the time t;
a_{i0} rate constant characterized by the fact that a step change in input to the ith neuron produces a exponential approach from the initial value $x_i(0)$ to a steady-state firing rate x_i^* with the rate constant a_{i0};
b_{ik} adaptation or self-inhibition factor for the ith neuron in the case $b_{ik} < 0$ and, self-excitation factor in the case $b_{ik} > 0$ with the rate constant a_{ik} (> 0);
c_{ij} interaction coefficient denoting the influence of the jth neuron on the ith neuron representing an inhibition in the case $c_{ij} < 0$ and an excitation in the case $c_{ij} > 0$; if the jth neuron is not connected directly to the ith neuron we have $c_{ij} = 0$; further, we always assume that $c_{ii} = 0$ for $i = 1,2,\cdots,n$ since the self-inhibition and self-excitation in the ith neuron are characterized by the parameters b_{ik} and a_{ik};
σ_{ij} (≥ 0), time lag occurring in the transfer of the activity of the jth neuron to the ith neuron; we assume $\sigma_{ii} = 0$ for $i = 1,2,\cdots,n$;

$$\mathcal{S}\{u\} = \frac{1}{1+e^{-u}} . \tag{2}$$

Put

$$\sigma = \max_{1\leq i,j\leq n} \sigma_{ij} , \quad \mathcal{I}_0 = [-\sigma,0] \tag{3}$$

Consider the system (1) with continuous input functions $f_i(t)$ for $t \geq 0$, $i = 1,\ldots,n$. Then, given the functions $\phi_i(t)$, $0 \leq \phi_i(t) \leq 1$, $i = 1,\ldots,n$, defined and continuous in the interval $\mathcal{I}_0$, the system (1) admits a unique continuously differentiable solution $\{x_1(t),x_2(t),\cdots,x_n(t)\}$ for $t \geq 0$ satisfying the initial conditions

$$x_i(t) = \phi_i(t) \quad \text{for} \quad t \in \mathcal{I}_0, \qquad\qquad x_i(+0) = \phi_i(0) , \quad i = 1,2,\cdots,n. \tag{4}$$

This solution depends analytically on all the a_{ik} , b_{ij} , c_{ik} , σ_{ij}, $f_i(t)$ and $\phi_i(t)$, and is such that

$$0 \leq x_i(t) \leq 1 \quad \text{for} \quad t \geq 0 \tag{5}$$

if $a_{i0} > 1$, $i,j = 1,2,\cdots,m$, $k = 1,2,\ldots,m$. For a general discussion of the functional differential equations of type (1) we refer to Oguztoreli (1966).

The physiological grounds of the neural equations (1) are discussed in Stein et al. (1974). The existence, uniqueness, stability, and numerical constructions of solutions and limit cycles of the system (1) are studied in Bingxi Li (1981), Deimling (1977), Heiden (1976), Leung et al (1974), Oguztoreli (1972, 1979, 1984), and Stein et al. (1974). Extensions of the system (1) to 1- and

2-dimensional continuous neural networks are given in Oguztoreli (1975, 1978). Certain questions concerning the case $n = \infty$ have been discussed in Deimling (1977) and Leung et al. (1974).

In a series of recent works the earlier results of Oguztoreli (1979) have been applied to the modelling, analysis, and simulation of information processing in the vertebrate retina and thalamocortical pathway (Oguztoreli, 1980, 1982, 1983; Oguztoreli and O'Mara, 1982). The present work can be considered as a continuation of these works.

In conformity with the neural equations (1), we also designate the retinal cells R, B, H, A, and G in Figure 3 by 1 , 2 , 3 , 4 , and 5 , respectively. The mathematical modelling of the BRN can then be easily achieved by considering specific characteristics of the neurons and network (such as the tendency to rhythmic oscillations, time constants, self adaptations, the form of the interactions and time delays) and by implementing the formation of the system (1) (cf. Oguztoreli, 1983). In this way we find the following functional differential equations for the neural activities in the basic retinal network:

$$
\begin{aligned}
\frac{1}{a_{10}}\frac{dx_1}{dt} + x_1 &= \mathcal{S}\{f_1 + c_{12}x_2(t-\sigma_{12}) + c_{13}x_3(t-\sigma_{13}) + b_{11}\int_0^t e^{-a_{11}(t-\tau)}x_1(\tau)dt\} \\
\frac{1}{a_{20}}\frac{dx_2}{dt} + x_2 &= \mathcal{S}\{c_{21}x_1(t-\sigma_{21}) + c_{23}x_3(t-\sigma_{23}) + c_{24}x_4(t-\sigma_{24}) \\
&\qquad + c_{25}x_5(t-\sigma_{25}) + b_{21}\int_0^t e^{-a_{21}(t-\tau)}x_2(\tau)d\tau\} \\
\frac{1}{a_{30}}\frac{dx_3}{dt} + x_3 &= \mathcal{S}\{c_{31}x_1(t-\sigma_{31}) + c_{32}x_2(t-\sigma_{32}) + b_{31}\int_0^t e^{-a_{31}(t-\tau)}x_3(\tau)d\tau\} \\
\frac{1}{a_{40}}\frac{dx_4}{dt} + x_4 &= \mathcal{S}\{c_{42}x_2(t-\sigma_{42}) + b_{41}\int_0^t e^{-a_{41}(t-\tau)}x_4(\tau)d\tau\} \\
\frac{1}{a_{50}}\frac{dx_5}{dt} + x_5 &= \mathcal{S}\{c_{52}x_2(t-\sigma_{52}) + c_{54}x_4(t-\sigma_{54}) + b_{51}\int_0^t e^{-a_{51}(t-\tau)}x_5(\tau)d\tau\},
\end{aligned}
\tag{6}
$$

subjected to the initial conditions

$$x_i(t) = \phi_i(t) \quad (t \in \mathcal{I}_0 \; ; \; i = 1,2,\cdots,5) \tag{7}$$

Hence, the neuronal processing in the basic retinal network depends on 43 parameters:

a_{i0}, $i = 1,2,\cdots,5$	(rate constants: positive)
a_{i1}, $i = 1,2,\cdots,5$	(rate constants: positive)
b_{i1}, $i = 1,2,\cdots,5$	(adaptation coefficients: negative)
c_{12}, c_{13}, c_{24}, c_{25}, c_{32}	(inhibitory coefficients: negative)
c_{21}, c_{23}, c_{31}, c_{42}, c_{52}, c_{54}	(excitatory coefficients: positive)
σ_{12}, c_{13}, σ_{21}, σ_{23}, σ_{24}, σ_{25}	(time-lags: nonnegative)
σ_{31}, σ_{32}, σ_{42}, σ_{52}, σ_{54}	
$\phi_i(t)$, $i = 1,2,\cdots,5$	(initial functions: $0 \leq \phi_i(t) \leq 1$)
$f_1(t)$	(external force (light intensity): nonnegative)

For the determination of the constant system parameters we refer to Oguztoreli (1980).

Although the existence and uniqueness of the solutions to the initial value problem (6)-(7) are assured under rather general conditions (cf. Oguztoreli, 1979), to express these solutions in the closed form is extremely difficult, if not impossible, because of the highly complex nonlinear structure of the system (6). Accordingly, in lieu of searching for analytical solutions in the closed form, we rather prefer to construct solutions numerically. In the next section we shall present some of our simulations. Here we set

$$
\begin{aligned}
& a_{i0} = 100, \quad a_{i1} = 30, \quad i = 1,2,\cdots,5 \\
& b_{11} = 2000, \; b_{21} = 1750, \; b_{31} = 1500, \; b_{41} = 1250, \; b_{51} = 1000 \\
& c_{12} = c_{13} = c_{24} = c_{25} = c_{32} = -10 \\
& c_{21} = c_{23} = c_{31} = c_{42} = c_{52} = c_{54} = 30 \\
& \sigma_{12} = \sigma_{13} = \sigma_{21} = \sigma_{23} = \sigma_{24} = \sigma_{25} \\
& \quad = \sigma_{31} = \sigma_{32} = \sigma_{42} = \sigma_{52} = \sigma_{54} = 0.003 \\
& \phi_i(t) \equiv 0, \; t \in \mathcal{J}_0 \equiv [-0.003,0], \; i = 1,2,\cdots,5
\end{aligned}
\tag{8}
$$

and considered the light inputs of the form

$$f_1(t) = w(t)\{F + \frac{A}{2}[1+\sin(2\pi\theta t)]\}, \quad 0 \leq t \leq 1.2 \tag{9}$$

where A, F, and θ are nonnegative constants, and

$$w(t) = \begin{cases} 1 & \text{if } 0.1 \leq t \leq 1.1 \\ 0 & \text{if } 0.0 \leq t < 0.1 \text{ or } 1.1 < t \leq 1.2 \end{cases} \tag{10}$$

Before closing the section we would like to note that the parametric configuration in (8) is different than those used in our earlier work (Oguztoreli, 1980, 1982, 1983; Oguztoreli and O'Mara, 1982).

III. SIMULATIONS AND DISCUSSIONS

The light input $f_1(t)$ in (9) is the superposition of a rectangular background luminance (the case $A = 0$ in (9)) and a sinusoidally varying luminance (the case $F = 0$ in (9)). Because of the nonlinearities in (6), the usual superposition principle is no longer valid for the neural processing in the BRN. In the following we shall briefly discuss the responses of the BRN to rectangular inputs, sinusoidal inputs, and superpositions of such inputs.

In Figure 4 and Figure 5 are shown the responses of the receptor cell R and the ganglion cell G to rectangular inputs with $F = 10, 25, 50, 100$, respectively. As it is expected, higher excitations yield higher activities in the cells. Note that higher activities in R produce higher activities in the bipolar and horizontal cells, B and H, which, in turn, inhibit R more and more, causing instabilities in the network responses. Under the same illuminations ,

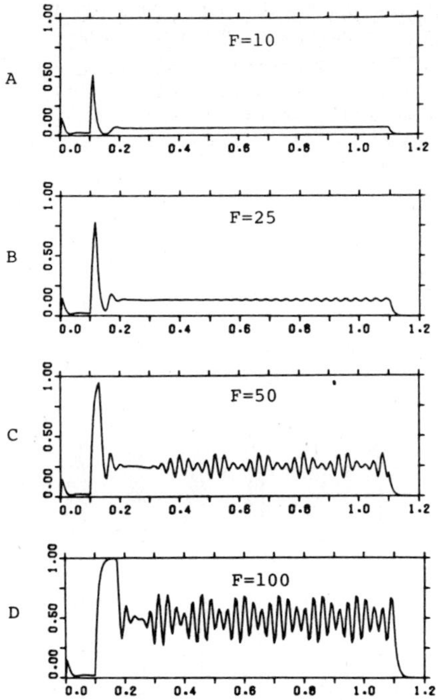

Figure 4 ←

Figure 5 →

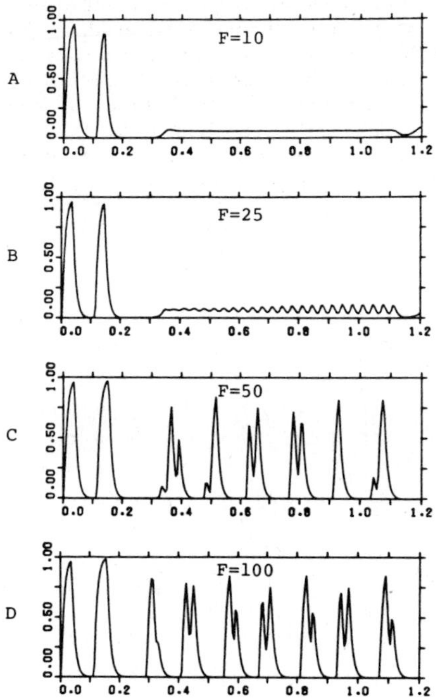

the R,B, and G cells, and, A and H cells, are exhibited in Figure 6 and Figure 7, respectively.

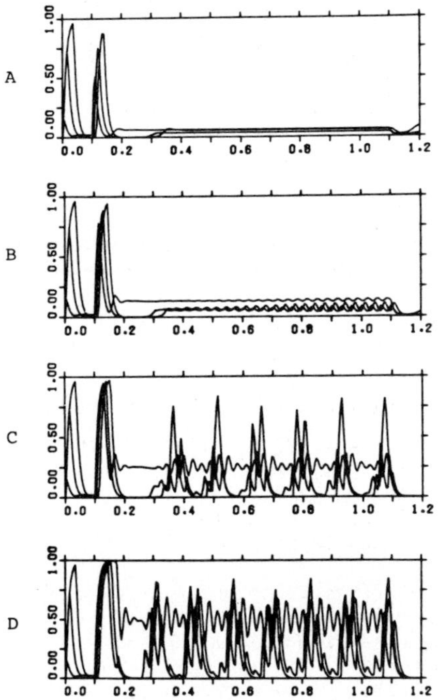

Figure 6

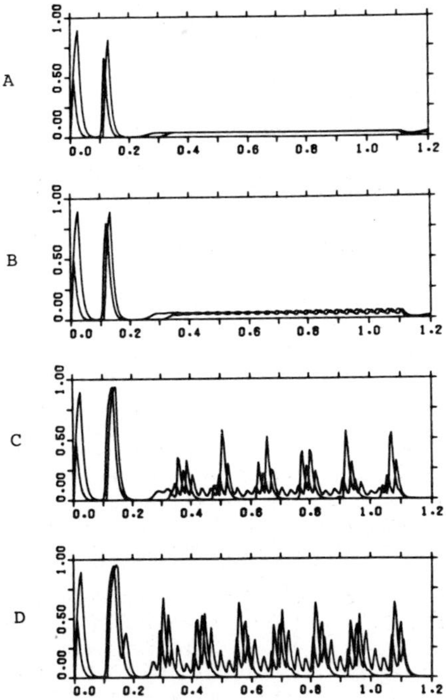

Figure 7

In Figure 8 (A) - (E) are shown the activities in the BRN in the order R, B, H, A, and G, respectively, when BRN is under the rectangular luminance with F = 150.

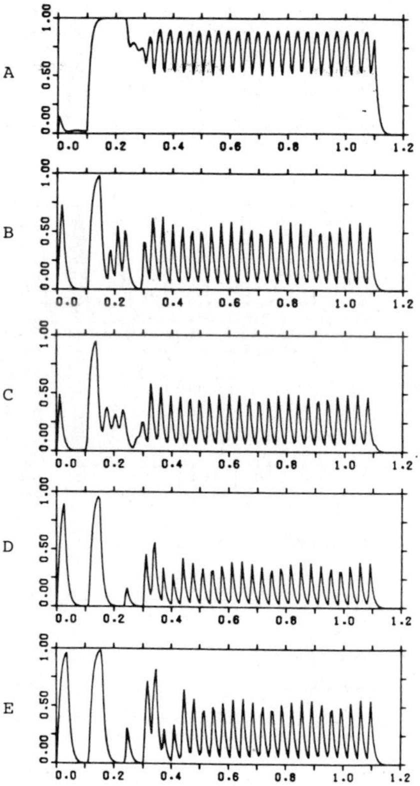

Figure 8

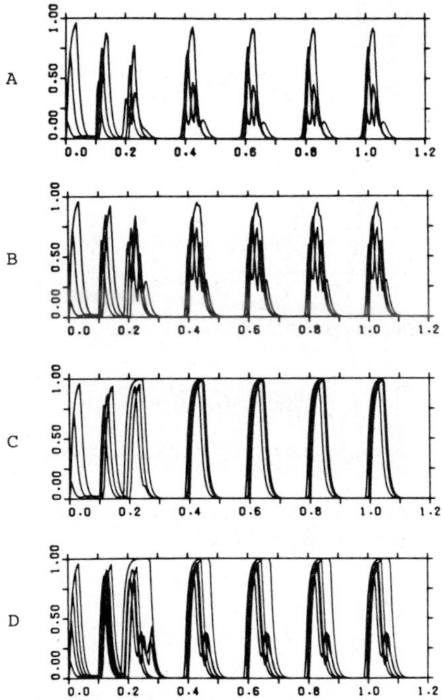

Figure 9

In Figure 9(A)-(C) are presented the responses of the cells R, B, and G to the sinusoidal illuminations with $\theta = 5$ and $A = 25, 50, 100$, respectively; the activities in the whole BRN are exhibited in Figure 9(D) in the case $\theta = 5$ and $A = 150$. The synchronous beatings in the activities with small delays are to be noted. The activities of the cells in the BRN under the sinusoidal illuminations described in Figure 10, 11 and 12 (A),(D), ($F = 0$, $A = 50$, $\theta = 1, 3, 5, 7, 10$ and 15, in (9)) are shown in Figure 10, 11 and 12 (B), (C), respectively. The close entrainment of the BRN by the inputs and the synthronous beatings in the activities are to be noted.

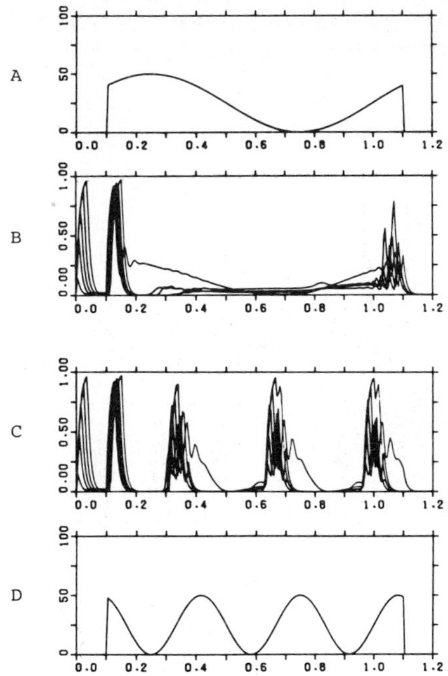

Figure 10

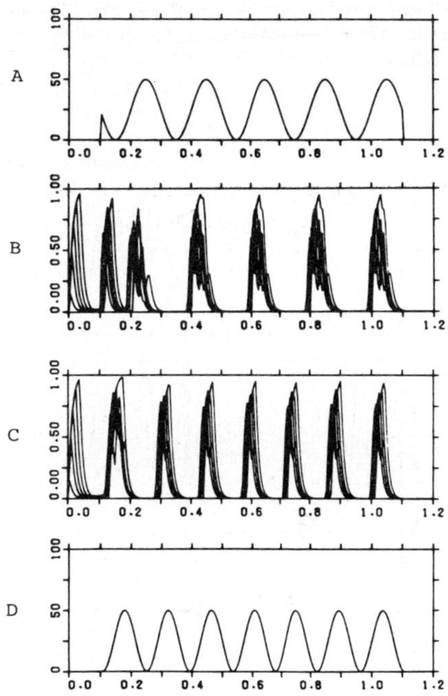

Figure 11

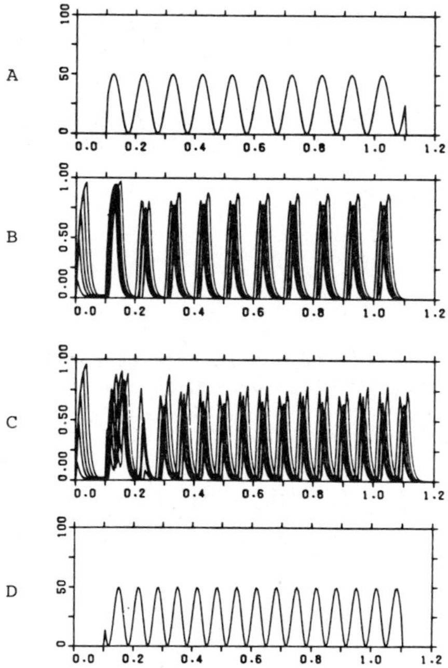

Figure 12

Let us note that when the frequency of the input exceeds 20, the responses of the

BRN lose their collective beating behavior particularly in the ganglion cell layer. This phenomenon has been illustrated in Figure 13, where the light input has the frequency $\theta = 25$ and amplitude $A = 50$, and A,B,D,C, and E correspond to R, B, H, A, and G cells, respectively.

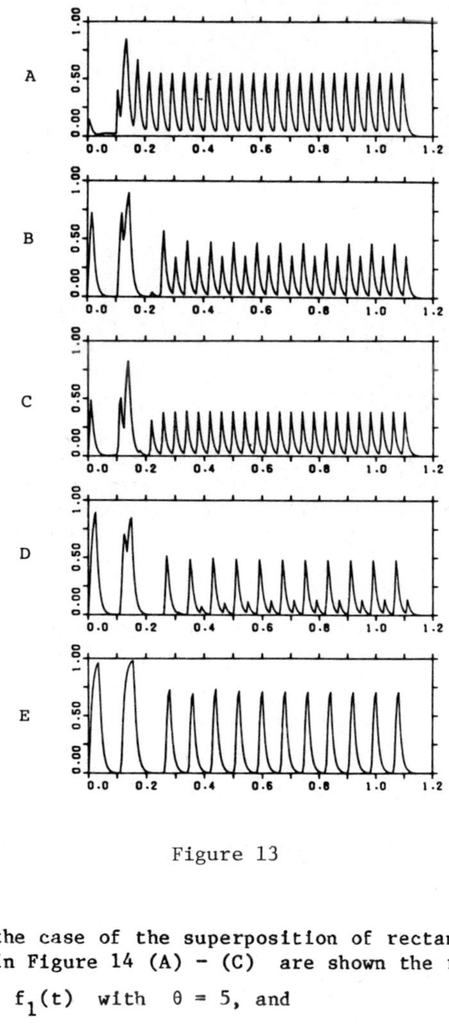

Figure 13

We now consider the case of the superposition of rectangular and sinusoidal illuminations. In Figure 14 (A) - (C) are shown the responses of the BRN to the light inputs $f_1(t)$ with $\theta = 5$, and

$$F = 25,\ A = 25\ ;\ F = 50\ ,\ A = 25\ ;\ F = 25,\ A = 50 \tag{11}$$

respectively. Comparing Figures 4, 5 (B), (C), Figure 9 (A), (B), and

A

B

C

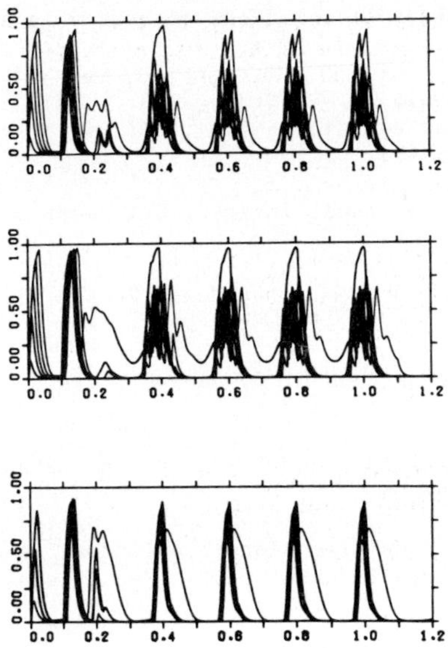

Figure 14

Figure 11(B), we conclude that the effects of the constant background illuminations are suppressed considerably by the sinusoidally varying illuminations with intensities (11). This loss in the superposition property is particularly significant in Figure 14 (C), and, in the activities of the ganglion cell G in Figure 14 (A) - (C).

ACKNOWLEDGEMENTS

This work partially supported by the Natural Sciences and Engineering Research Council of Canada udner Grant A-4345 to M.N.O. and Grant A2568 to T.M.C. through the University of Alberta.

REFERENCES

[1] Bingxi Li: Uniqueness and stability of a limit cycle for a third order dynamical system arising in neuron modelling. Nonlinear. Anal., Theor. Meth. Appl. 5(1981) 13-19.

[2] Cervetto, L. and Fuortes, M.G.F.: Excitation and interaction in the retina. Ann. Rev. Biophys. Bioeng., 7(1978) 229-251.

[3] Deimling, K.: Differential Equations in Banach Spaces. (Springer-Verlag, Berlin, Heidelberg, New York, 1977).

[4] Dowling, J.E.: Organization of vertebrate retina. Invest. Ophthal., 9(1970) 655-680.

[5] Dowling, J.E.: Functional organization of vertebrate retina. In: Retina Congress, R.C. Pruett and C.D.J. Regan (Eds.), (Appleton-Century-Crofts, 1972).

[6] Dowling, J.E. and Ehinger, B.: The interplexiform cell system. I. Synapses of the dopaminergic neurons of the goldfish retina. Proc. Royal Soc. (Lond.) B201(1978) 7-26.

[7] Dowling, J.E. and Werblin, F.S.: Organization of retina of the mudpuppy, necturus maculosus. I, II. J. Neurophsyiol. 32(1969) 315-338, 339-355.

[8] Heiden, U an der: Existence of periodic solutions of a nerve equation. Biol. Cybern. 21(1976) 37-39.

[9] Leung, K.V., Mangeron, D., Oguztoreli, M.N., Stein, R.B.: On a class of nonlinear integro-differential equations. III. Bull. Acad. R. Sci. Belgique, Ser. 5 59(1973) 492-499.

[10] Michael, C.R.: Retinal procesing of visual images. Sci. Amer., 220(1969) 104-114.

[11] Oguztoreli, M.N.: Time-lag Control System. (Academic Press, New York and London, 1966).

[12] Oguztoreli, M.N.: On the neural equations of Cowan and Stein . Utilitas Mathematica 2(1972) 305-317.

[13] Oguztoreli, M.N.: On the activities in a continuous neural network. Biol. Cybern. 21(1975) 41-48.

[14] Oguztoreli, M.N.: Activities in a one-dimensional continuous neural network. Arch. Math. 3, Scripta Fac. Sci. Nat. Ujep Brunensis 14(1968) 161-169.

[15] Oguztoreli, M.N.: Activity propagation in a continuous neural network. In: "Jubliee Volume" of the Institute of Mathematics of the Soviet Academy of Sciences, Moscow-USSR (1968) 450-458.

[16] Oguztoreli, M.N.: Activity analysis of neural networks. Biol. Cybern. 34(1979) 159-169.

[17] Oguztoreli, M.N.: Modelling and simulation of vertebrate retina. Biol. Cybern. 37(1980) 53-61.

[18] Oguztoreli, M.N., O'Mara, K.S.: Modelling and simulation of vertebrate retina: extended network. Biol. Cybern. 38(1980) 9-17.

[19] Oguztoreli, M.N.: Response analysis of vertebrate analysis. Biol. Cybern. 44(1982) 1-8.

[20] Oguztoreli, M.N.: Modelling and simulation of vertebrate primary visual system: Basic network. IEEE Trans. Syst. Man. Cybern. SMC-13,5(1983) 765-781.

[21] Oguztoreli, M.N.: Some problems concerning nonlinear oscillators in neural networks. In: Trends in Theory and Practice of Nonlinear Differential Equations, V. Lakshmikantham (Ed.), (Marcell Dekker, New York and Basel, 1984, 425-433).

[22] Shepherd, G.M.: The Synaptic of the Brain. (Oxford Univesity Press, London, Toronto, 1979).

[23] Stein, R.B., Leung, K.V., Oguztoreli, M.M., Williams, D.W.: Properties small networks. Kybernetik 14(1974) 223-230.

[24] Stein, R.B., Leung, K.V., Mangeron, D., Oguztoreli, M.N.: Improved neuronal models for studying neural networks. Kybernetik 15(1974) 1-9.

[25] Werblin, F.S.: The control of sensitivity in the retina. Sci. Amer. 228(1973) 70-79.

This paper is in final form and no version of it will be submitted for publication elsewhere.

Trends in the Theory and Practice of Non-Linear Analysis
V. Lakshmikantham (Editor)

NUMERICAL SOLUTION OF QUASILINEAR BOUNDARY VALUE PROBLEMS

M. C. Pandian

Department of Mathematics
University of Texas at Arlington
Arlington, Texas 76019
U.S.A.

1. INTRODUCTION

We consider the boundary value problem

$$(*)\qquad \begin{aligned} &Su \equiv -u_{xx} + f(x,u)u_x + g(x,u) = 0, \\ &u(0) = b_0,\ u(1) = b_1,\ 0 < x < 1, \end{aligned}$$

where f and $g \in C^2[I\times R,R]$, I being the interval $[0,1]$. Nonlinear convection-diffusion equations of the form (*) arise in many applications; for example, in gas lubricating films [4,10,14] and in porous flow (steady state) under the influence of gravity [7]. We do not restrict the number of zeros that the function f can have. The zeros of f are often called "turning points". The presence of turning points makes it more diffcult to obtain an accurate and stable numerical scheme.

The problem (*) is known for its numerical difficulty even in the linear case when the coefficient of the first order term is large. For example, consider a simple model problem

$$-u_{xx} + au_{xx} = 0,\ u(0) = 1,\ u(1) = 0,$$

where a is a nonzero constant. The exact solution is $u(x) = (e^{ax}-e^{a})/(1-e^{a})$. Let us consider the numerical solution through finite differences. Consider the grid: $x_i = ih$, with $(N+1)h = 1$, and let $u_i = u(x_i)$. Now if we use the centered difference approximation $u_x|_i \approx (u_{i+1}-u_{i-1})|2h$, and the standard three point difference approximation of u_{xx}, i.e.

$$u_{xx}|_i = (u_{i+1}-2u_i+u_{i-1})|h^2,$$

then the resulting difference equation is

$$(1-t)u_{i+1} - 2u_i + (1+t)u_{i-1} = 0,$$

where $t = \frac{ah}{2} \neq 1$. The general solution for this difference equation is

$$u_i = c_1 + c_2\left[\frac{1+t}{1-t}\right]^i .$$

Observe that this solution is oscillatory when $|t| > 1$, i.e., $h > \frac{|a|}{2}$. Since $|a|$ is large (say, 10^2-10^6) for most of the practical problems, the centered differencing of u_x gives rise to oscillations in the numerical solution at reasonable grid sizes. Oscillations can be avoided if one uses the following "upwind" form.

$$u_x\Big|_i \approx \begin{cases} (u_i-u_{i-1})|h & \text{for} \quad a > 0 \\ (u_{i+1}-u_i)|h & \text{for} \quad a < 0 \end{cases}$$

However, this has only first order $(O(h))$ accuracy while the centered scheme has second order $(O(h^2))$ accuracy. Notice that

$$(u_i-u_{i-1})|h = (u_{i+1}-u_{i-1})|2h - (u_{i+1}-2u_i+u_{i-1})|2h$$
$$= u_x(x_i) - \frac{h}{2} u_{xx}(x_i) + O(h^2).$$

Thus, the one sided scheme or upwind form (with second order accuracy) introduces an "artifical diffusion" of size $|a|h|2$. This artificial diffusion or numerical dispersion does stabilize the difference method, but the numerical error induced can be disastrous for some physical processes. An example [9, p. 21] is the numerical modeling of a burning front in the reservoir simulation. If the computed temperature distribution is artificially diffused to obtain temperatures below certain critical value, the numerical model will predict a cease in the burning when, in fact, combustion is still occurring. So the physics of the process will be lost.

In this paper, we present a weighted two-sided difference scheme which gives a balance between the centered form (accurate, unstable) and the one-sided upwind form (stable, excessively diffused). If the first order derivative term is not present, then the problem (*) becomes a mildly nonlinear problem for which the above difficulties do not appear. The standard three point difference approximation of u_{xx} can be used. For a mildly nonlinear problem, using the fact that the difference operator is inverse positive, monotonically converging iterative schemes have been studied in [8,11]. For more general nonlinear boundary value problems, based on the maximum principle, the monotone iterative method is used in theory to prove the existence of multiple solutions [2-6]. This theory is useful in order to obtain analytical bounds for the solution. However, the scheme is not numerical. Furthermore, the constant thay have used to obtain the monotonicity is too large and this, in turn, slows down the convergence rate drastically. Numerically the problem (*) becomes more difficult because of the above-mentioned inherent drawbacks with the difference approximations. In addition, we also study Newton (quasi-linearization) schemes. Analogous to [2-6], our iterative scemes are based on the inverse positivity of certain linearized difference operators. In our case, the iterative processes become very complicated especially when the problem has turning points. Because, the turning points give rise to different difference approximations in the opposite directions for the first order term. We have successfully circumvented these situations.

In the case when f is a function of u only, numerical solution of (*) is obtained recently in [1,12] using one-sided difference schemes. Their method is time evolving, that is, the solution is obtained as the limit of the solutions of $u_t = Su$ with the same time-independent boundary conditions and appropriate initial condition. The stability condition needed for the explicit method [12], in general, permits only small time steps. This results in a slow rate of convergence of the time-dependent solutions. In order to avoid the stability restriction, an implicit scheme is considered in [1]. But, this leads to a nonlinear system to be solved at each time step, for which, the Newton's method is proposed [1]. The great difficulty with the Newton's method is guessing a good starting point. It is noted in [1] that the starting point should be very close to the true solution for successful convergence at each time step. Further, in [1,12] it is assumed that $g_u>0$, which is not required for our method. Also, it is not known whether their scheme can be extended to the case in which f is a function of both u and x. Thus, our scheme covers a more general class of problems. Here, we present only some main results. Proofs and other details including the computer codes for a model quasilinear problem can be found in [13]. A complete report will be submitted elsewhere.

2. DISCRETE APPROXIMATION

We set up a grid: $x_i = ih$, $h = \Delta x$ with $(N+1)h = 1$. Unless otherwise stated, in what follows, we assume that $i \in \{1,2,\ldots,N\}$. Define $u_i = u(x_i)$,

$$\Delta u_i = \frac{1-\alpha}{2} u_{i+1} + \alpha u_i - \frac{(1+\alpha)}{2} u_{i-1}, \quad -1 \le \alpha \le 1$$

$$\Delta_+ u_i = \Delta u_i \quad \text{with} \quad \alpha = \alpha_1$$

$$\Delta_- u_i = \Delta u_i \quad \text{with} \quad \alpha = \alpha_2$$

where α_1, α_2 are constants such that $-1 \le \alpha_1 \le 0$ and $0 \le \alpha_2 \le 1$. For notational simplicity, we assume without loss of generality, that f is a function of u only. However, in the case f depends on x and u it is to be understood that $f(u_i)$ can be replaced by $f(x_i,u_i)$ and that f' represents the derivative of f w.r.t. u. Let

$$f_+(u_i) = \max(0,f(u_i)), \quad f_- = f - f_+.$$

At the ith grid, let $u_{xx} \approx u_{i+1} - 2u_i + u_{i-1}$, and the transport term is approximated as

$$f(u)u_x \approx \frac{1}{h}\,[f_+(u_i)\Delta_- u_i + f_-(u_i)\Delta_+ u_i]. \tag{2.1}$$

Notice that $\alpha = 0$ corresponds to the centered differencing and $\alpha = \pm 1$ leads to the full upwind form. The resulting nonlinear system can be written as

$$J(u) \equiv Au + F_+(u)D_- u + F_-(u)D_+ u + G(u) - b = 0, \tag{2.2}$$

where $u = (u_i)$, $F_\pm(u) = (f_\pm(u_i))$, $D_\pm = h\Delta_\pm$, $G(u) = (h^2 g(x_i,u_i))$, b is the given vector consisting of boundary conditions, and A is the $N\times N$ tridiagonal matrix with diagonal elements 2 and super and sub diagonal elements -1. For any two vectors v,w we denote $vw = (v_i w_i)$, and the inequalities are understood componentwise. We assume that there exists two vectors $\underline{u}$ and $\bar{u}$ auch that $\underline{u} \le \bar{u}$, and $J(\underline{u}) \le 0 \le J(\bar{u})$. The vectors $\underline{u}$ and $\bar{u}$ are called lower and upper solutions of (2.2), respectively.

Let $P = (p_{ij})$ be an $n\times n$ matrix. We write $P \ge 0$ if $p_{ij} \ge 0$ for all i and j.

3. MAIN RESULTS

In this section, we present a formula to choose the weights in the difference approximation of the first order term (2.1) and iterative schemes to solve (2.2).

We define the coefficients of upwind differencing in (2.1) as follows:

$$\alpha_{2i} = \max\{0,1-2|(ha_+)\},$$

$$\alpha_{1i} = \min\{0,-1+2|(ha_-)\},$$

where $a_\pm = \max\{|f_\pm(u_i)| : \underline{u}_i \le u_i \le \bar{u}_i\}$. For convenience, we delete the subscript i in α_1 and α_2.

Note that $G'(u) = h^2 \mathrm{diag}(g_u(x_i,u_i))$ and $F'(u) = \mathrm{diag}(f'(u_i))$. Choose $M = \mathrm{diag}(m_i)$ such that

$$G(u) - G(v) \leq M(u-v), \; u \geq v, \quad u,v \in [\underline{u},\bar{u}]. \tag{3.1}$$

Define, for $k = 1,2,\ldots,$

$$\begin{aligned} T^{(k-1)}u^{(k)} \equiv Au^{(k)} &+ F_+(u^{(k-1)})D_-u^{(k)} + F_-(u^{(k-1)})D_+u^{(k)} \\ &+ Q^{(k-1)}u^{(k)}, \end{aligned} \tag{3.2}$$

where $Q^{(k-1)}$ are defined in the following theorems.

We consider the following iterative scheme.

$$\begin{aligned} T^{(k-1)}u^{(k)} &= Q^{(k-1)}u^{(k-1)} - G(u^{(k-1)}) + b \\ u^{(0)} &= \underline{u}, \quad k = 1,2,\ldots . \end{aligned} \tag{3.3}$$

The notation $f'_{\pm}$ is used to refer $(f_{\pm})'$. Notice that $f'_{\pm}$ is defined, except possibly, at thezeros z of f. At z, we take $f'_{\pm}(z) = f'(z)$. Define

$$c_{\pm i} = \max\{|f'(u_i)| : \underline{u}_i \leq u_i \leq \bar{u}_i\}. \tag{3.4}$$

Theorem 3.1. Suppose that $F'(u) \geq 0$ in $[u,u]$. The sequence $\{u^{(k)}\}$, given by (3.3), with

$$\begin{aligned} Q^{(k-1)} &= \mathrm{diag}(q_i^{(k-1)}), \\ q_i^{(k-1)} &= \max\{0, c_{+i}d_{-i}^{(k-1)}, c_{-i}d_{+i}^{(k-1)}\} + m_i \\ d_{+i}^{(k-1)} &= h\left(\frac{(1-\alpha_1)}{2}\bar{u}_{i+1} + \alpha_1 u_i^{(k-1)} - \frac{(1+\alpha_1)}{2}u_{i-1}^{(k-1)}\right), \quad \text{and} \\ d_{-i}^{(k-1)} &= h\left(\frac{(1-\alpha_2)}{2}\bar{u}_{i+1} + \alpha_2 \bar{u}_i - \frac{(1+\alpha_2)}{2}u_{i-1}^{(k-1)}\right), \end{aligned} \tag{3.5}$$

converges monotonically from below to a solution u^* of (2.2).

Theorem 3.2. Suppose that $F'(u) \leq 0$ in $[\underline{u},\bar{u}]$. Now in the definition of $q_i^{(k-1)}$ in (3.5), replace $d_{\pm 1}^{(k-1)}$ by $e_{\pm i}^{(k-1)}$, where

$$\begin{aligned} e_{+i}^{(k-1)} &= h\left(\frac{(1+\alpha_1)}{2}\bar{u}_{i-1} - \alpha_1 \bar{u}_i - \frac{(1-\alpha_1)}{2}u_{i+1}^{(k-1)}\right), \quad \text{and} \\ e_{-i}^{(k-1)} &= h\left(\frac{(1+\alpha_2)}{2}\bar{u}_{i-1} - \alpha_2 u_i^{(k-1)} - \frac{(1-\alpha_2)}{2}u_{i+1}^{(k-1)}\right). \end{aligned}$$

Then the conclusion of Theorem 3.1 is valid.

We have the following general result which does not require any sign condition on f'.

Theorem 3.3. In the definition of $q_i^{(k-1)}$ in (3.5) replace $d_{\pm i}^{(k-1)}$ by $\gamma_{\pm i}^{(k-1)} = \max\{d_{\pm i}^{(k-1)}, e_{\pm i}^{(k-1)}\}$. Then the conclusion of Theorem 3.1 holds.

In general, the equation (2.2) can have more than one solution. We have the following result:

Corollary 3.1. The solution u^* obtained in the above theorems is a minimal solution of (2.2) in $[\underline{u},\bar{u}]$; i.e., if u is any other solution of (2.2) in $[\underline{u},\bar{u}]$ then $u^* \leq u$.

If f' satisfies the condition, for $i = 1,2,\ldots,n$,

$$|f'(u_i)| \leq |f'(w_i)|,\ u_i \geq w_i,\ u,w \in [\underline{u},\bar{u}].$$

then we can improve the rate of convergence in the iterative process (3.3) by updating the bound on f' at each step. That is, replace c_+ by $c_i^{(k-1)} = |f'(u_i^{(k-1)})|$ for $i = 1,2,\ldots,n$. In certain cases, the rate of convergence for the scheme (3.3) is quadratic.

Analogous results can easily be obtained for the iterative scheme (3.3) starting from an upper solution, i.e., $u^{(0)} = \bar{u}$. In this case, we have monotonically decreasing sequence $\{u^{(k)}\}$ which converges to a maximal solution in $[\underline{u},\bar{u}]$.

Using this scheme, numerical solution of the steady state Reynolds equation for gas lubricated slider bearings is obtained in [14]. The governing equation, under a transformation, becomes a quasilinear boundary value problem in two dimensions. The work [14] shows the extension of our scheme to higher dimensional problems.

REFERENCES

[1] Abrahamsson and Osher, S. Monotone difference schemes for singular perturbation problems, SIAM J. Nuemr. Anal. 19 (1982) 979-992.

[2] Bernfeld, S. R. and Chandra, J. Minimal and maximal solutions of nonlinear boudnary value problems, Pacific J. of Math. 71 (1977) 13-20.

[3] Bernfeld, S. R. and Lakshmikantham, V. Linear monotone method for nonlinear boudnary value problems in Banach spaces, Rocky Mountain J. of Math. 12 (1982) 807-815.

[4] Castelli, V. and Pirvics, J. Review of numerical methods in gas bearing film analysis, ASME J. of Lubrication Technology 90 (1968) 777-792.

[5] Chandra, J. and Davis, P. W. A monotone method for quasilinear boundary value problems, Arch. Rational Mech. Anal. 54 (1974) 257-266.

[6] Chandra, J., Lakshmikantham, V. and Leela, S. A monotone method for infinite systems of nonlinear boundary value problems, Arch. Rational Mech. Anal. 68 (1978) 179-190.

[7] Chavent, G. and Salzano G. A finite element method for the 1-D water flooding problem with gravity, J. Comp. Phys. 45 (1982) 307-344.

[8] Greenspan, D. and Parter, S. V. Mildly nonlinear elliptic partial differential equations and their numerical solution-II, Numer. Math. 7 (1965) 129-146.

[9] Ewing, R. E. Problems arising in the modeling of process for hydrocarbon recovery, in: Ewing, R. E. (ed.), The Mathematics of Reservoir Simulation (SIAM Frontiers, Philadelphia, 1983).

[10] Gross, W. A. (editor). Fluid Film Lubrication (John Wiley and Sons, Inc., New York, 1980).

[11] Ortega, J. M. and Rehinboldt, W. C. Iterative Solution of Nonlinear Equations in Several Variables (Academic Press, New York, 1970).

[12] Osher, S. Nonlinear singular perturbation problems and one sided difference schemes, SIAM J. Numer. Anal. 18 (1981) 129-144.

[13] Pandian, M. C. Numerical Studies of Nonlinear Systems and Quasilinear Boundary Value Problems With Applications to Gas Lubricating Films, Ph.D. Thesis, Dept. of Math., University of Texas at Arlington (July, 1984).

[14] Pandian, M. C. A new method for the numerical solution of the Reynolds equation for gas lubricated slider bearings, Journal of Engineering Mathematics (to appear).

The final (detailed) version of this paper will be submitted for publication elsewhere.

Trends in the Theory and Practice of Non-Linear Analysis
V. Lakshmikantham (Editor)

NUMERICAL METHODS FOR SIMULTANEOUSLY APPROACHING ROOTS OF POLYNOMIALS

L. Pasquini
Dipartimento di Metodi e Modelli Matematici per le Scienze Applicate, 1ª Università, via A. Scarpa 10, 00161 Roma, Italy

D. Trigiante
Dipartimento di Matematica
Università di Bari
via Nicolai 2, 70121 Bari, Italy

An algorithm for simultaneously finding roots of polynomials and dynamically determinating their multiplicities is outlined. It can work with polynomials f represented in the general form: $f = \sum_{h=0}^{N} \gamma_h P_{N-h}$, where the $P_i (i = 0,1,\ldots)$ are polynomials of degree i satisfying a three terms iterative relation $P_i(s) = \pi_i(s)P_{i-1}(s) + d_i P_{i-2}(s)$, $i = 1,2,\ldots (P_0(s) \equiv 1, P_{-1}(s) \equiv 0)$, with $\pi_i(s) = b_i s + c_i$ polynomial of degree 1 and d_i constant.

1. The first nonlinear problems the mathematicians had to face were the ones concerning polynomials. In particular, the problem of determining polynomial roots has been studied since the beginnings and almost all the most famous mathematicians have given contributions to it.

After the works done by Ruffini and Galois, it was clear that there could not exist a general direct method for solving algebraic equations and the direct methods had to give precedence to the iterative ones which had already been studied by Viete and Newton and have been used since the starting of mathematics. It is well known, for example, that the Sumerian mathematicians used the method $x_{n+1} = (x_n + A/x_n)/2$ for finding the square root of any positive number A.

Most of the iterative methods approximate only one or two roots and normally converge only if the starting points belong to suitable neighborhoods of the solutions. Recently, methods which simultaneously converge to all the roots of the polynomial and are almost-globally convergent have been published [1-6].

While new methods were getting on, a large theory of the geometry of the polynomial roots was developed [7], [9] and some of the more recent methods combine results derived from this theory with iterative procedures in order to get (almost) globally and simultaneously convergent algorithms [6], [10]. Unfortunately, these methods often exhibit a slow convergence especially in the case in which multiple roots occur.

The problem of multiple roots (their approximation and the determination of their multiplicities) is a very hard one to solve in order to get a robust software for algebraic equations. Besides the heavy loss of the convergence speed which almost all iterative methods suffer from when multiple roots, or even roots close to each other, are approximated, other troubles arise. The machine errors may radically change the situation and must be appropriately taken into account (for example the equation $Az^n = 0$ may be represented in the machine with $(A + \varepsilon_1)z^n + \varepsilon_2 = 0)$. Moreover, the numerical problem of solving algebraic equations with multiple roots may be very unstable and rounding errors may disturb the procedure, necessarily sophisticated, for determining the multiplicities.

Such problems have been faced in [12] by suitable tests. A set of subroutines has been designed in order to solve problems like the following: to individuate the last useful moment for getting the best output (obtainable with machine-precision) before the threshold noise arises; to determinate the multiplicities; to select a suitable starting point; etc. The basic method employed (subroutine BASMTH) is the one stated in [11], Section 5. It is an almost-globally convergent method (at least in the case of simple and real roots) for simultaneously finding all the roots of a polynomial f and exhibits, among the other properties, a quadratic-like convergence also in case multiple roots of the polynomial f are present.

In quite all the algorithms, as well as in [12], f is supposed to be given in the usual representation, i.e.: the Taylor's expansion at the origin. We present here an algorithm which has the same features as [12], but in addition it can work with polynomials f represented in a considerably more general form. This form has several advantages (see Section 2). The most important one is the possibility of obtaining better stability properties.

2. It is well known that the usual representation of a polynomial f:

$$f(s) = \sum_{h=0}^{N} a_h s^{N-h} \tag{1}$$

often is not the most useful. Sometimes it is better to use a different basis of polynomials $P_0, P_1, \ldots$ which verify a three terms recursive relation:

$$P_i(s) = \pi_i(s) P_{i-1}(s) + d_i P_{i-2}(s) \quad i = 1,2,\ldots \quad (P_0(s) \equiv 1, P_{-1}(s) \equiv 0), \tag{2}$$

where:

$$\pi_i(s) = b_i s + c_i \tag{3}$$

and $b_i \neq 0 \ \forall\, i$, c_i, d_i are constants, and put the polynomial in the form:

$$f(s) = \sum_{h=0}^{N} \gamma_h P_{N-h}(s) \quad (\gamma_h \text{ constants}). \tag{4}$$

The advantage of this more general representation with respect to the problem we are interested in (i.e.: the search of the roots of f) has been already discussed by Golub and Robertson [13]. We only quote some argumentations.

1) Sometimes the basis: $1, s, s^2, \ldots,$ used in (1), leads to ill-conditioning and instability (see, for example [14]); the generality of the representation (4) may allow one to get better conditioning and stability.

2) In some problems (for example, the one finding the eigenvalues of a three-diagonal matrix) representations like the (4) are more natural.

3) When f is defined in a not usual way (for example, as a determinant of a matrix, the elements of which are polynomials) the computation of the γ_h may be simpler than the computation of the a_h in (1) if the basis $\{P_i\}$ is chosen in a suitable way.

3. We present in this section two generalizations, which may be of interest in themselves, of the Horner's algorithms [8]. The Horner's rule is the less expensive algorithm (in terms of operations required) for computing the value (1) at a point s of a polynomial f given in the usual representation. An analogous algorithm can be given for the more general case of a polynomial represented in the form (4).

Lemma 1. The following representations of the polynomial f defined by (4) hold:

(5) $$f(s) = \sum_{h=\ell}^{N} \gamma_h^{(\ell)}(s) P_{N-h}(s) \quad \ell = 0,1,\dots,N ,$$

where the $\gamma_h^{(\ell)}(s)$ are the polynomials defined by the:

(6) $$\gamma_h^{(0)}(s) \equiv \gamma_h (h = 0,1,\dots,N) ,$$

(7) $$\gamma_{h+1}^{(h+1)}(s) = \pi_{N-h}(s)\gamma_h^{(h)}(s) + \gamma_{h+1}^{(h)}(s) ,$$

(8) $$\gamma_{h+2}^{(h+1)}(s) = d_{N-h}\gamma_h^{(h)}(s) + \gamma_{h+2}^{(h)}(s) ,$$

(9) $$\gamma_{h+j}^{(h+i)}(s) = \gamma_{h+j} \quad \text{if } j - i \geq 2.$$

Proof. If $\ell = 0$, the (5) becomes the (4) [see (6)] and is consequently true. We now assume that (5) is true with $\ell = i < N$ and prove that it holds also with $\ell = i + 1$. Since it results, in virtue of (2), $P_{N-i}(s) = \pi_{N-i}(s)P_{N-i-1}(s) + d_{N-i}P_{N-i-2}(s)$, one obtains from (5) written with $\ell = i$:

$$f(s) = \gamma_i^{(i)}(s)\left[\pi_{N-i}(s)P_{N-i-1}(s) + d_{N-i}P_{N-i-2}(s)\right] + \sum_{h=i+1}^{N} \gamma_h^{(i)}(s)P_{N-h}(s).$$

From this it follows, by using (7), (8) and (9),

$$f(s) = \sum_{h=i+1}^{N} \gamma_h^{(i+1)}(s)P_{N-h}(s) \blacksquare$$

Corollary 1. Let us simply denote by Γ_h, for the sake of brevity, the value $\gamma_h^{(h)}(s)$:

(10) $$\Gamma_h = \gamma_h^{(h)}(s).$$

Then the value of f at the point s is given by Γ_N:

(11) $$f(s) = \Gamma_N ,$$

and may be computed by the recursive relation:

(12) $$\Gamma_{h+1} = \pi_{N-h}(s)\Gamma_h + d_{N-h+1}\Gamma_{h-1} + \gamma_{h+1} \quad h = 0,1,\dots,N-1 \ (\Gamma_0 = \gamma_0, \Gamma_{-1} = 0).$$

Proof. The (11) follows from (5) by putting $\ell = N$; the recursive relation (12) can be obtained from (7) by writing $\gamma_{h+1}^{(h)}(s)$ according to (8) and noting that, in virtue of (9), one has $\gamma_{h+1}^{(h-1)}(s) = \gamma_{h+1}$ $\blacksquare$

Remark 1. The (12) can be regarded as a generalization of the Horner's rule. In fact, in the particular case: $b_i = 1$, $c_i = d_i = 0 (i = 0,1,\dots,N)$, one has $P_i(s) = s^i (i = 0,1,\dots,N)$, the (4) is the same as the (1) [i.e.: $a_h = \gamma_h (h = 0,1,\dots,N)$] and the (12) becomes:

(13) $$\Gamma_{h+1} = s\Gamma_h + a_{h+1} \quad h = 0,1,\dots,N-1 \ (\Gamma_0 = a_0, \Gamma_{-1} = 0),$$

which is just the recursive relation defining the Horner's rule for computing the value (1) (see again [8]).

Remark 2. The number of product operations required by (12) in the general case (i.e.: whatever the b_i, c_i, d_i may be) is 3N. If the representation (1) is known, the computation of $f(s)$ by (13), (1) requires N product operations. The difference of operations may be justified by the arguments quoted in Section 2.

The following result concerns the computation of the divided differences of f. Let us again simplify notations. We will briefly indicate by Γ_{hk} the value of the divided difference of Γ_h with respect to the arguments $x_1, x_2, \ldots, x_k$:

$$\Gamma_{hk} = \Gamma_h[x_1, x_2, \ldots, x_k] \quad h = 0,1,\ldots,N, \quad k \geq 1. \tag{14}$$

Since Γ_h is [see (6), (7),...,(10)] a polynomial of degree h having the coefficient of the term of degree h equal to $\gamma_0 \cdot b_N \cdot b_{N-1} \cdots b_{N-h+1}$ [equal to γ_0 if $h = 0$], one has:

$$\Gamma_{k-2,k} = 0 \quad k = 2,3,\ldots,N+2, \tag{15}$$

$$\Gamma_{k-1,k} = \gamma_0 \cdot b_N \cdot b_{N-1} \cdots b_{N-k+2} \quad k = 2,3,\ldots,N+1, \tag{16}$$

$$\Gamma_{0,1} = \gamma_0. \tag{17}$$

Finally, if we put

$$\Gamma_{h-1,0} = 0 \quad h = 1,2,\ldots,N, \quad \Gamma_{-1,1} = 0, \tag{18}$$

we are in a position to state the following

Theorem 1. The values of the divided differences $f[x_1, x_2, \ldots, x_k]$ are given by the Γ_{Nk}:

$$f[x_1, x_2, \ldots, x_k] = \Gamma_{Nk} \tag{19}$$

and can be computed by the following recursive relation:

$$\Gamma_{hk} = b_{N-h+1}\Gamma_{h-1,k-1} + \pi_{N-h+1}(x_k)\Gamma_{h-1,k} + d_{N-h+2}\Gamma_{h-2,k} + \delta_{1k}\gamma_h \quad h = k,k+1,\ldots,N, \quad k = 1,2,\ldots,N, \tag{20}$$

where δ_{hk} is the Kroneker symbol.

Proof. Eq. (19) is an obvious consequence of (11), (14). In order to prove (20) it is sufficient to observe that, in virtue of (12), (14), one has:

$$\Gamma_{hk} = (\pi_{N-h+1}\Gamma_{h-1})[x_1, x_2, \ldots, x_k] + d_{N-h+2}\Gamma_{h-2,k} + \delta_{1k}\gamma_h .$$

Then (20) easily follows by applying to the divided difference of the product $\pi_{N-h+1}\Gamma_{h-1}$ the Leibniz rule and bearing (3) in mind:

$$(\pi_{N-h+1}\Gamma_{h-1})[x_1,x_2,\dots,x_k] = \Gamma_{h-1}[x_1,x_2,\dots,x_{k-1}]\pi_{N-h+1}[x_{k-1},x_k] +$$
$$+ \Gamma_{h-1}[x_1,x_2,\dots,x_k]\pi_{N-h+1}[x_k] =$$
$$= \Gamma_{h-1,k-1}b_{N-h+1} + \Gamma_{h-1,k}\pi_{N-h+1}(x_k).$$

Finally, it is easy to check that the correct starting values of the $\Gamma_{h-1,0}, \Gamma_{0,1}$ and $\Gamma_{-1,1}$, for iterating $(h = 1,2,\dots,N)$ (20) with $k = 1$, and, afterwards, the correct starting values of the $\Gamma_{k-1,k}$ and $\Gamma_{k-2,k}$, for iterating $(h = k, k+1,\dots,N)$ (20) when $k \geq 2$, are respectively given by (18), (17) and by (16), (15)∎

<u>Remark 3</u>. Horner's rule can be generalized [8], [9] as follows:

$$(21) \qquad \tau_n^{(k)} = \tau_n^{(k-1)} + t\tau_{n-1}^{(k)} \quad n = 1,2,\dots,N-k \ (\tau_0^{(k)} = a_0)$$
$$k = 0,1,\dots,N-1 \ (\tau_n^{(-1)} = a_n)$$

which allows one to compute the coefficients $\tau_{N-k}^{(k)} = f^{(k)}(t)/k!$ of Taylor's expansion at the point t:

$$(22) \qquad f(s) = \sum_{k=0}^{N} \frac{f^{(k)}(t)}{k!}(s-t)^k$$

of a polynomial given in the usual representation (1). The algorithm stated in Theorem 1 allows computation of the coefficients of Newton's representation:

$$(23) \qquad f(s) = \sum_{k=1}^{N+1} f[x_1,x_2,\dots,x_k] \prod_{i=1}^{k-1}(s-x_i) = \sum_{k=1}^{N+1} \Gamma_{Nk} \prod_{i=1}^{k-1}(s-x_i)$$

of a polynomial f given in the form (4). Thus the algorithm defined by (15), (16),...,(20) is even more general than the one defined by (21) [extended Horner's algorithm]. In fact:

<u>Corollary 2</u>. <u>If</u>: $x_1 = x_2 = \dots = x_N = t$ <u>and</u>: $b_i = 1$, $c_i = d_i = 0 \ \forall \ i$, <u>the algorithm defined by (15, (16),...,(20) is exactly the same as the extended Horner's one (21)</u>.

<u>Proof</u>. In the case we are considering, (20) becomes:

$$\Gamma_{hk} = \Gamma_{h-1,k-1} + t\Gamma_{h-1,k} + \delta_{1k}\gamma_h \quad h = k,k+1,\dots,N$$
$$k = 1,2,\dots,N,$$

with $\Gamma_{h-1,0} = 0$ and $\Gamma_{k-1,k} = a_0$ [see (18), (16), (17)]. By putting now:

$$\tau_n^{(\ell)} = \Gamma_{hk}, \quad \text{with: } \ell = k-1, \quad n = h-k+1,$$

one obtains

$$(24) \qquad \tau_n^{(\ell)} = \tau_n^{(\ell-1)} + t\tau_{n-1}^{(\ell)} + \delta_{1,\ell+1}a_{n+\ell} \quad n = 1,2,\dots,N-\ell \ (\tau_0^{(\ell)} = a_0)$$
$$\ell = 0,1,\dots,N-1 \ (\tau_n^{(-1)} = 0)$$

which is just the algorithm (21). In fact, the last addend in (24) can be omitted by putting: $\tau_n^{(-1)} = a_n$ (instead of: $\tau_n^{(-1)} = 0$) ∎

Remark 4. The representation (23) is of the type (4) with: $\gamma_h = f[x_1,x_2,\dots,x_{N-h+1}]$, $P_{N-h}(s) = \prod_{i=1}^{N-h}(s - x_i)$ and $b_i = 1$, $c_i = x_i$, $d_i = 0 (i = 1,2,\dots)$ in (2), (3). The (15), (16),...,(20) can then be regarded as an algorithm to compute the coefficients of Newton's form (23) of a polynomial written in any form of the type (4). Taylor's form (22) can be obtained, as said above, by taking all the arguments x_i equal to t; for $t = 0$ one obtains the usual form (1).

4. Let us first introduce some notations. We denote by α_k, $k = 1,2,\dots,N$, the N zeros of the polynomial (4):

$$f(s) = \Gamma_{N,N+1} \prod_{k=1}^{N}(s - \alpha_k), \quad [\Gamma_{N,N+1} = \gamma_0 \prod_{i=1}^{N} b_i, \text{ see (16)}]$$

with β_k, $k = 1,2,\dots,M (M \le N)$, the distinct zeros of (4) and with μ_k the multiplicity of β_k.

Consider now the iterative algorithm for simultaneously finding roots of the polynomial (4) operatively defined by the following steps.

Algorithm 1.

1) Choose a suitable starting point $x^{(0)} = (x_1^{(0)},x_2^{(0)},\dots,x_N^{(0)})$, $x_k^{(0)}$ being an initial approximation for $\alpha_k (k = 1,2,\dots,N)$.

2) Starting with $x^{(0)}$ generate the sequence $\{x^{(n)}\}_{n=1,2,\dots}$, $x^{(n)} = (x_1^{(n)},x_2^{(n)},\dots,x_N^{(n)})$, by carrying into execution the following steps with $n = 1,2,\dots$.

2_1) Put $Q_i = 0$, $i = 0,1,\dots,N - 1$.

2_2) Effectuate the following operations consecutively with $k = 1,2,\dots,N$.

(25) $$R_h = b_{N-h+1}Q_{h-1} + \pi_{N-h+1}(x_k^{(n-1)})R_{h-1} + d_{N-h+2}R_{h-2} + \delta_{1k}\gamma_h \quad h = k,k+1,\dots,N$$

$$(R_{k-1} = \Gamma_{k-1,k} \text{ [see (17), (16)]}, \ R_{k-2} = \Gamma_{k-2,k} = 0 \text{ [see (18), (15)]}),$$

(26) $$\Delta_k = R_N$$

(27) $$Q_h = R_h, \quad h = k-1,k,\dots,N-1$$

(28) $$R_h = b_{N-h+1}Q_{h-1} + \pi_{N-h+1}(x_\ell^{(n-1)})R_{h-1} + d_{N-h+2}R_{h-2} \quad h = k,k+1,\dots,N$$

$$(R_{k-1} = \Gamma_{k-1,k+1} = 0 \text{ [see (15)]}, \ R_{k-2} = \Gamma_{k-2,k+1} = 0),$$

(29) $$\Delta_k = \begin{cases} \Delta_k - R_N\Delta_\ell & \text{if } \ell \le k-1 \\ \Delta_k/R_N & \text{if } \ell = k \end{cases}$$

(steps (28)–(29) for $\ell = 1,2,\dots,k$)

(30) $$x_k^{(n)} = x_k^{(n-1)} - \Delta_k$$

The convergence properties of the sequence $\{x^{(n)}\}$ generated by Algorithm 1 are described below. In Theorems 2,3,4, the case of the approximation of real roots is considered and the a_i,b_i,c_i,d_i,γ_h in (1), (2), (3), (4) are also supposed to be real; moreover $\alpha_{i_1 i_2 \dots i_N}$ denotes the point of $\mathbb{R}^N$:

(31) $$\alpha_{i_1 i_2 \ldots i_N} = (\alpha_{i_1}, \alpha_{i_2}, \ldots \alpha_{i_N})$$

where $i_1, i_2, \ldots, i_N$ is any permutation of the indexes $1,2,\ldots,N$. Theorems 2, 3, and 4 will be proved together after the statement of Theorem 4. They cover the following cases: i) all the α_k are real and simple; ii) all the α_k are real and at least one of them is a multiple root; iii) some of the α_k are complex.

Theorem 2. In the case i), the sequence $\{x^{(n)}\}$ converges towards one of the points (31) with almost every $x^{(0)}$ (i.e.: the set of the exceptional starting points is a closed set of $\mathbb{R}^N$ of measure zero) and each $x_k^{(n)}$ converges quadratically to the corresponding root α_{i_k} of (31).

The following theorems describe the local convergence properties of Algorithm 1 in the neighborhood of one of the points (31) (Theorem 3, case ii)), or in the neighborhood of a point of the type (31) but into a suitable subspace of $\mathbb{R}^N$ (Theorem 4, case iii)).

Theorem 3. In the case ii), let $I_h (h = 1,2,\ldots,M)$ be the set of indexes such that if $k \in I_h$ then $x_k^{(n)}$ converges to β_h and let $\hat{x}_h^{(n)}$ be the average of the $x_k^{(n)}, k \in I_h$:

(32) $$\hat{x}_h^{(n)} = \sum_{k \in I_h} x_k^{(n)} / \mu_h \, .$$

The average (32) approaches β_h with a convergence of quadratic type.

Theorem 4. In the case iii), let N_r be the number of real roots α_k. Then Algorithm 1 is at least locally convergent in $\mathbb{R}^{N_r}$ to a point having the coordinates equal to the N_r real roots α_k. The convergence properties are the same described in Theorem 2 if these real roots are simple or the ones described in Theorem 3.

Proofs of Theorems 2,3,4 (Sketch). It is easily checked that Algorithm 1 is an implementation for a polynomial given in the form (4) of the method defined in [11] for a sufficiently smooth function f:

(33a) $$x_k^{(n)} = x_k^{(n-1)} - \Delta_k(x^{(n-1)}) \quad k = 1,2,\ldots,N, \; n = 1,2,\ldots ,$$

where

(33b) $$\Delta_k(x) = \frac{f[x_1,x_2,\ldots,x_k] - \sum_{\ell=1}^{k-1} f[x_1,x_2,\ldots,x_\ell,x_\ell,\ldots,x_k] \cdot \Delta_\ell}{f[x_1,x_2,\ldots,x_k,x_k]} \quad \left(\sum_{\ell=1}^{0} = 0\right) .$$

The proofs follow then as in the corresponding cases therein ∎

Remark 5. A considerably more general routine than the one in [12] can be immediately obtained by simply substituting in [12] the subroutine BASMTH, designed for polynomials in the form (1), with another one, BASMT1, in which the basic method (30), (33) is implemented in the form stated before in Algorithm 1.

Finally, we want to observe that numerical evidence shows that the method (30), (33) works exactly in the same way for complex polynomials as well. The algorithms need only obvious changes to be adapted to the new ambient space.

REFERENCES

[1] I. O. Kerner, Ein Gesamtschrittverfahren zur Berechnung der Nullstellen von Polynomen, Num. Math. 8, (1966) 290-294.

[2] E. Durand, Solution Numérique des Equations Algébriques, Tome I. Paris: Massons (1968).

[3] L. W. Ehrlich, A Modified Newton Method for Polynomials, Comm. of the ACM 10, (1967) 107-109.

[4] O. Aberth, Iteration Methods for Finding all Zeros of a Polynomial Simultaneously, Math. of Comp. 27, (1973) 339-344.

[5] G. Alefeld and J. Herzberger, On the Convergence Speed of Some Algorithms for the Simultaneous Approximation of Polynomial Roots, SIAM JNA 11, (1974) 237-243.

[6] G. E. Collins, Infallible Calculation of Polynomial Zeros to Specified Precision, in J. R. Rice, Mathem. Software III (1982).

[7] M. Marden, Geometry of Polynomials, Amer. Math. Soc., Providence, Rhode Island (1966).

[8] P. Henrici, Elements of Numerical Analysis, J. Wiley, New York (1964).

[9] __________, Applied and Computational Complex Analysis, Vol I, J. Wiley, New York (1974).

[10] M. A. Jenkins and J. F. Traub, A Three-Stage Variable-Shift Iteration for Polynomial Zeros, SIAM JNA 7, (1970) 545-566.

[11] L. Pasquini and D. Trigiante, A Globally Convergent Method for Simultaneously Finding Polynomial Roots. To appear in Math. of Comp. (January 1985).

[12] M. L. Lo Cascio, L. Pasquini and D. Trigiante, Un polialgoritmo a convergenza rapida per la determinazione simultanea degli zeri reali di un polinomio e delle loro molteplicità, Pubbl. dell'IAC, Monografie di Software Matematico 30 (1984).

[13] G. H. Golub and T. N. Robertson, A Generalized Bairstow Algorithm, Comm. of the ACM 10, (1967) 371-373.

[14] J. H. Wilkinson, The Evaluation of the Zeros of Ill-Conditioned Polynomials, Num. Math. 1, (1959) 150-180.

Trends in the Theory and Practice of Non-Linear Analysis
V. Lakshmikantham (Editor)

THE PARALLEL SUM OF GENERALIZED GRADIENTS

Gregory B. Passty and Ricardo Torrejón

Department of Mathematics and Computer Science
Southwest Texas State University
San Marcos, Texas 78666

Let ∂f and ∂g be impedance operators which are the generalized gradients of locally Lipschitzian functions defined in a Banach space. The parallel sum $\partial f : \partial g$ is defined using the joint resistance formula, and is shown to conform with Maxwell's Principle in this general setting.

1. INTRODUCTION

In an electrical network, consider a circuit element which consists of two resistors joined in parallel. Let the resistances be R_1 and R_2, respectively, and let x be the current flowing through the element. Using Kirchhoff's and Ohm's Laws, we may conclude that this element acts as a single resistor with joint resistance

$$R = (R_1^{-1} + R_2^{-1})^{-1} .$$

We may draw the same conclusion by instead using Maxwell's Principle: the current x will divide itself between the two branches so as to minimize the total power dissipated. This may be rephrased as a variational problem: minimize $R_1 y^2 + R_2 z^2$ subject to the condition $y+z = x$.

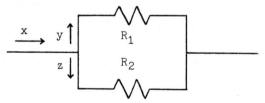

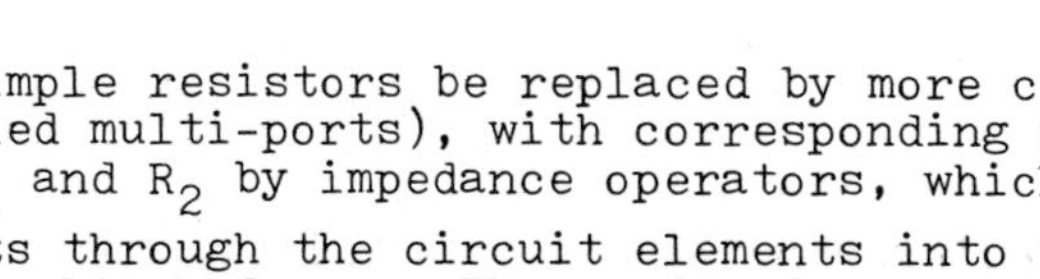

Now let the simple resistors be replaced by more complex circuit elements (called multi-ports), with corresponding replacement of the resistances R_1 and R_2 by impedance operators, which map vector-valued currents through the circuit elements into the corresponding vector-valued voltage drops. The study of parallel sums is concerned with the extension of the joint resistance formula to this more general setting, and, whenever possible, a companion extension of Maxwell's Principle.

In Section 2, we will summarize previous work on the parallel sum, treating impedance operators which are, successively, matrices, linear operators on Hilbert spaces, subdifferentials of convex functions, and monotone operators in Hilbert spaces. In Section 3, we will define the parallel sum of generalized gradients of locally Lipschitzian functions in Banach spaces, and initiate the analysis of Maxwell's Principle in this setting.

2. A BRIEF HISTORY OF PARALLEL SUMS

While the expression $(A^{-1}+B^{-1})^{-1}$ neatly copies the joint resistance formula, one may encounter several difficulties in its interpretation. Questions which need to be considered include: (i) Are A and B invertible? (ii) If A and B are in a certain class of operators, will the expression above give a result in that class? Strangely enough, (i) has been more easily handled in recent work [9] with A and B multivalued and the inverse defined by inverting the graph. On the other hand, the joint impedance of linear operators should be linear, and the expression above is not even defined if A or B is noninvertible.

Anderson and Duffin [1] first defined the parallel sum of Hermitian semidefinite matrices A and B as

$$A:B = A(A+B)^{\dagger} B \ , \tag{1}$$

where $\dagger$ denotes the Moore-Penrose generalized inverse. The class of Hermitian semidefinite matrices is closed under this parallel sum, and A:B is related to the generalized inverses of A and B in Theorem 5 of [1]: If P is the projection onto Range(A) $\cap$ Range(B), then $A:B = (P(A^{\dagger}+B^{\dagger})P)^{\dagger}$. In addition, Maxwell's Principle holds [1, Lemma 18]: for any x, y, z such that y+z = x, we have

$$\langle (A:B)x, x\rangle \le \langle Ay, y\rangle + \langle Bz, z\rangle \quad .$$

If A+B is nonsingular, then (1) can be rewritten as

$$A:B = A(A+B)^{-1}B \quad .$$

If A and B are invertible, then (1) may be further rewritten as

$$A:B = (A^{-1} + B^{-1})^{-1} \quad , \tag{2}$$

and the joint resistance formula is recaptured. (See [1, p. 577].)

Anderson and Trapp [4] defined the parallel sum of positive linear operators on a Hilbert space H, extending notions used in [3] and [6]. The definition given in [4] is stated in terms of shorted operators without reference to the generalized inverse. For positive operators B and C, we write $B \ge C$ if B-C is positive. If A is a positive operator and S is a subspace of H, then the set of positive operators $\{D \mid 0 \le D \le A \ , \ \text{Range}(D) \subset S\}$ has a maximum element, which is defined to be the shorted operator $\mathcal{S}(A)$ of A with respect to S [4, Theorem 1]. To define the parallel sum of positive operators A and B, let S be the subspace $H \oplus \{0\}$ of $H \oplus H$. Then A:B is defined by [4, p. 67]

$$\begin{bmatrix} A:B & 0 \\ 0 & 0 \end{bmatrix} = \mathcal{S}\begin{bmatrix} A & A \\ A & A+B \end{bmatrix} \quad .$$

This definition is shown to be equivalent to that of Fillmore and

Williams [6, p. 277]: For positive operators A and B, let C and D be the uniquely determined operators on H such that

$$A^{\frac{1}{2}} = (A+B)^{\frac{1}{2}}C \ , \quad \ker(C^*) \supset \ker((A+B)^{\frac{1}{2}}) \quad ,$$

$$B^{\frac{1}{2}} = (A+B)^{\frac{1}{2}}D \ , \quad \ker(D^*) \supset \ker((A+B)^{\frac{1}{2}}) \quad .$$

Then $A:B = A^{\frac{1}{2}}C^*DB^{\frac{1}{2}}$.

While Maxwell's Principle was derived as a consequence of the definition of parallel sum in [1] and [4], Morley [8] used it to define the parallel sum of two positive semidefinite linear operators in a complex Hilbert space: let A:B be the unique linear operator such that $\langle (A:B)x, x\rangle = \inf \{ \langle Ay, y\rangle + \langle Bz, z\rangle \mid y+z = x\}$.

Anderson, Morley, and Trapp [2] extended the notion of parallel sum to nonlinear operators through the use of Maxwell's Principle. Their setting is much more general, in that parallel and other hybrid electrical connections are considered as specific examples of a general connection termed a confluence. The impedance functions F and G are subdifferentials of lower semicontinuous proper convex functions f and g, respectively. It is assumed that 0 is an interior point of the domain of each impedance operator and that $0 \in F(0) \cap G(0)$. Generalizing the parallel sum is the router sum. This turns out to be the Lagrange multiplier vector corresponding to Maxwell's Principle, which here seeks the minimum of f(y)+g(z) subject to y+z = x.

Passty [9] approached the parallel sum by considering the natural extension of positive operators: nonlinear, possibly multi-valued monotone operators on a Hilbert space. The natural definition is given by (2) in this case, because of the identification of operators with their graphs. If A and B are monotone, then so is A:B. If A and B are maximal monotone, and if in addition one of them is strongly monotone, then A:B is maximal monotone. While it follows fairly easily from the definition that Range(A:B) coincides with Range(A) $\cap$ Range(B), a characterization of the domain is very difficult because of the implicit nature of the definition. One result in this direction is Theorem 25 of [9]: Let A and B be maximal monotone, with A or B strongly monotone, and assume Domain(A) has nonempty interior. Then

$$\text{Interior}(\text{Domain}(A)) + \text{Domain}(B) \subset \text{Domain}(A:B) \ .$$

In the special case when A and B are subdifferentials of convex functions, Maxwell's Principle is obtained under less restrictive conditions than in [2].

Left open, however, is a curious problem concerning the parallel sum as the limit of approximations. Theorem 8 of [4] states that if A and B are positive operators, and if $\{E_i\}$ and $\{F_j\}$ are monotonically decreasing sequences of positive operators converging strongly to 0, then $(A+E_i):(B+F_j)$ converges strongly to A:B. Kubo [7] actually uses this idea to define the parallel sum of positive linear operators on a Hilbert space:

$$A:B = \text{s-lim } (A+\varepsilon I):(B+\varepsilon I) \text{ as } \varepsilon \searrow 0 \ .$$

Thus we conjecture for nonlinear, possibly multi-valued monotone operators that

$$(A:B)x = \lim_{\varepsilon \searrow 0} ((A+\varepsilon I):(B+\varepsilon I))x$$

for all x in the domain of A:B.

3. EXTENSION TO GENERALIZED GRADIENTS

In this section, $\{E, \|\cdot\|\}$ will denote a Banach space. A real-valued function f defined on an open subset D of E is said to be locally Lipschitzian if for each x in D, there is a neighborhood N of x and a constant K depending on N such that

$$|f(y)-f(z)| \leq K\|y-z\| \text{ for all } y, z \text{ in } N . \qquad (3)$$

Following Clarke [5], for each u in E, the generalized directional derivative of f at x in the direction u is

$$f^{\circ}(x;u) = \limsup_{\substack{c \to x \\ t \searrow 0}} t^{-1}[f(c+tu) - f(c)] . \qquad (4)$$

The generalized gradient of f at x, denoted $\partial f(x)$, is defined by

$$\partial f(x) = \{w \in E^* \mid \langle w, u\rangle \leq f^{\circ}(x;u) \text{ for all } u \text{ in } E\} . \qquad (5)$$

As Clarke points out in [5], the locally Lipschitzian nature of f insures the existence of the lim sup in (4), and consequently the existence of at least one element in $\partial f(x)$.

We now use (2) to define the parallel sum:

DEFINITION. The parallel sum of the generalized gradients ∂f and ∂g is given by

$$\partial f:\partial g = ((\partial f)^{-1} + (\partial g)^{-1})^{-1} ,$$

where we write $p \in (\partial f)^{-1}q$ if and only if $q \in (\partial f)p$, and similarly for $(\partial g)^{-1}$.

LEMMA 1. $w \in (\partial f:\partial g)x$ if and only if there exists y in E such that $w \in \partial f(y) \cap \partial g(x-y)$.

PROOF. The proof of Lemma 2 in [9] can be applied.

To fix notation for the statements of the theorems of this section, let $k(x,y) = f(y) + g(x-y)$, and let $h(x) = \inf_y k(x,y)$.

THEOREM 1. Let f and g be locally Lipschitzian, and let there exist an x_o in E and a neighborhood V of x_o such that for all z in V, h(z) is actually attained. Further, let there be a function γ defined implicitly by $h(z) = k(z,\gamma(z))$ such that γ is continuous at $z = x_o$. Then $\partial h(x_o) \subset (\partial f:\partial g)(x_o)$.

In other words, x_o is in the domain of the parallel sum and elements of $\partial h(x_o)$ are in the parallel sum of ∂f and ∂g at x_o. As the proof will show, the current x_o is split in accordance with Maxwell's Principle.

PROOF OF THEOREM 1. Since g is locally Lipschitzian, h is Lipschitzian in a neighborhood of x_o. Thus $\partial h(x_o)$ is nonempty.

By (4) and (5), if $w \in \partial h(x_o)$, then for all u in E,

$$\langle w, u\rangle \leq h^{o}(x_o;u) = \limsup_{\substack{c \to x_o \\ t \searrow 0}} t^{-1}[h(c+tu) - h(c)] . \qquad (6)$$

An element c which is near enough to x_o is in the neighborhood V of the hypothesis. Thus the right hand side of (6) can be rewritten as

$$\limsup_{\substack{c \to x_o \\ t \searrow 0}} t^{-1}[h(c+tu) - k(c,\gamma(c))]$$

$$\leq \limsup_{\substack{c \to x_o \\ t \searrow 0}} t^{-1}[f(\gamma(c))+g(c+tu-\gamma(c))-f(\gamma(c))-g(c-\gamma(c))]$$

$$\leq \limsup_{\substack{b \to x_o-\gamma(x_o) \\ t \searrow 0}} t^{-1}[g(b+tu) - g(b)] = g^{o}(x_o-\gamma(x_o);u) .$$

Since u is an arbitrary element of E, we conclude from (5) that $w \in \partial g(x_o-\gamma(x_o))$. Starting with (6), we can observe also that

$$\limsup_{\substack{c \to x_o \\ t \searrow 0}} t^{-1}[h(c+tu) - k(c,\gamma(c))]$$

$$\leq \limsup_{\substack{c \to x_o \\ t \searrow 0}} t^{-1}[f(\gamma(c)+tu)+g(c-\gamma(c))-f(\gamma(c))-g(c-\gamma(c))]$$

$$\leq \limsup_{\substack{b \to \gamma(x_o) \\ t \searrow 0}} t^{-1}[f(b+tu) - f(b)] = f^{o}(\gamma(x_o);u) .$$

We may now use (5) to see that $w \in \partial f(\gamma(x_o))$. By Lemma 1, $w \in (\partial f:\partial g)(x_o)$. Q.E.D.

THEOREM 2. Let f be locally Lipschitzian and g be globally Lipschitzian. Assume further that the domain of f is compact and that $h(x) = k(x,z)$ has a unique solution $z = \gamma(x)$ for each x in a neighborhood U of x_o. Then $\partial h(x_o) \subset (\partial f:\partial g)(x_o)$.

PROOF. First we note that γ is continuous at x_o. Indeed, let x approach x_o. Then, by compactness of the domain of f, there is a convergent subnet which we will denote by $\{\gamma(x)\}$ which approaches a limit z_o. For $\gamma(x)$ near enough to z_o,

$$k(x_o,z_o) = k(x_o,z_o)-k(x,z_o)+k(x,z_o)-k(x,\gamma(x))+k(x,\gamma(x))$$

$$\leq L_1\|x_o - x\| + L_2\|z_o - \gamma(x)\| + h(x) - h(x_o) + h(x_o) .$$

Since h is continuous at x_o, we see when x approaches x_o that $k(x_o,z_o) \leq h(x_o)$. Thus $z_o = \gamma(x_o)$. Since this process can be repeated with any subnet, the continuity of γ at x_o is established. Since the domain of f is compact, the infimum h(z) is attained for all z. We are thus in the situation of Theorem 1, and the result follows. Q.E.D.

Note that the hypotheses underlying Theorem 2 imply that E in that theorem is finite-dimensional.

If $\lim_{\|u\|\to\infty} f(u) = \lim_{\|u\|\to\infty} g(u) = +\infty$, with f and g bounded below and weakly lower semi-continuous, then h(z) will actually be attained for each z.

REFERENCES

[1] Anderson, W.N. Jr. and Duffin, R.J., Series and parallel addition of matrices, J. Math. Anal. Appl. 26(1969), 576-594.

[2] Anderson, W.N. Jr., Morley, T.D., and Trapp, G.E., Fenchel duality of nonlinear networks, IEEE Transactions on Circuits and Systems CAS-25(1978), 762-765.

[3] Anderson, W.N. Jr. and Schreiber, M., The infimum of two projections, Acta Sci. Math. (Szeged) 33(1972), 165-168.

[4] Anderson, W.N. Jr. and Trapp, G.E., Shorted operators II, SIAM J. Appl. Math. 28(1975), 60-71.

[5] Clarke, F.H., Optimization and Nonsmooth Analysis, John Wiley & Sons, New York, 1983.

[6] Fillmore, P.A. and Williams, J.P., On operator ranges, Advances in Math. 7(1971), 254-281.

[7] Kubo, F., Conditional expectations and operations derived from network connections, J. Math. Anal. Appl. 80(1981), 477-489.

[8] Morley, T.D., Parallel summation, Maxwell's principle and the infimum of projections, J. Math. Anal. Appl. 70(1979), 33-41.

[9] Passty, G.B., The parallel sum of nonlinear monotone operators, preprint.

This paper is in final form and no version of it will be submitted for publication elsewhere.

Trends in the Theory and Practice of Non-Linear Analysis
V. Lakshmikantham (Editor)

AN EXACT, DIRECT, FORMAL INTEGRAL (DFI) APPROACH TO DIFFERENTIAL EQUATIONS

Fred R. Payne

Aerospace Engineering and Mathematics Departments
The University of Texas at Arlington
Arlington, Texas 76019

A simple iterative procedure is described which converts any differential equation or system into a hierarchy of integral or integro-differential equations easily solved via standard iteration techniques. Analytics and numerics on three classes of computers are presented for six differential systems encompassing sixteen distinct applications in boundary layers and turbulence modelling. Formulations are given for Navier-Stokes, Euler, Burgers, Laplace and Poisson. Advantages over differencing are: 1) improved compatibility with any computing machinery, 2) removal of differencing in at least one coordinate, 3) increased computer speed and accuracy, 4) explicit simulation of inherent physics, 5) ease of coding and transportability, (6) treatment of multi-point boundary conditions by multiple iteration loops. Only parabolic systems have been computed, however, extension to elliptic and hyperbolic systems should pose no new problems.

OVERVIEW AND MOTIVATIONS

Discussion follows an historical sequence; first application of DFI (Payne 1980, 81; Ko 1982) was to Falkner-skan; extension to partial DE (Payne 1982; Mokkapati 1983) solved 2-D boundary flows; inclusion of turbulence effects (Payne 1982) is simple; extensions follow and a summary of DFI structure, physical and mathematical, concludes the paper.

Author readily admits to bias toward integral and away from differencing methods due, in large portion, to years of turbulence studies wherein the influences are inherently global rather than local or limited to neighborhoods. Certain facets of FDM are disturbing, namely, 1) FDM is basically "user unfriendly", i.e., FDM codes are not easily transportable; 2) differencing is incompatible with digital machinery; 3) grid generation is not simple; 4) fluids, as most physics, are global rather than local. FEM obviates many of these objections but its development in fluids, as opposed to that in solid mechanics, has been rather slow due to the lack of extremum principles therein. Elliptic systems are totally global; hyperbolics are "sector-global"; parabolics are "downstream global". Nature, herself, seems to be an integrator in many cases. Integration is far more compatible with digital, or analog, computing machinery than is differencing.

Analytic results (1980) for Blasius' flat-plate flow provide strong heuristic aids for improved understanding of flow mechanisms and computational schemes. Numerical results for Blasius (1981), Falkner-Skan (1982), mixing layer (1982), linear heat conduction (1983), several "psuedo-" airfoils (Howarth flows, 1983), Burgers' (1984), and developing Falkner-Skan similarity (1984) required generally a small fraction of human and computer time via other methods; improved accuracy usually prevails over other methods.

DFI is a two-stage scheme: 1) formally integrate, by parts or partially, in one coordinate direction any ODE or PDE of order N up to N-iterated integrations, e.g., if M is an Nth order, linear or non-linear, differential operator upon f(x,y,...):

$$M(f) = g(x,y),\ldots)$$

and if y is an appropriate direction for DFI:

$$\int_a^y M(f)ds = \int_a^y gds \equiv I_1(x,y,\ldots)$$

$$\int_a^y dz \int_a^z M(f)ds = \int_a^y dz \int_a^z gds \equiv I_2(g) = \int_a^y I_1(g)ds$$

$$\int_a^y dz\ldots M(f)ds = I_k(g),\ k=N, \text{ or } N-1 \text{ (usually)}$$

where I_k is the i-iterated integral of g, the forcing function. During this process all I.C./B.C. at y=a are organically incorporated. The working equation(s) are integral (if ODE) or integro-differential (if PDE).

2) Computation stage requires only "reasonable" initial guess of $f_\alpha(x,y,\ldots)$ where $\alpha=0$ for integral equations and $\alpha=0,1,\ldots,p$ for p=highest order of remaining differential operators; solution is then by any iterative scheme (Picard's, Euler's, ...). For example:

Problem	Value of α	Number of Profiles guessed
Blasius	0	1
F-Skan	0	1
Heat Eq., 1-D	1	2
Prandtl, 2-D	1	2
Burger's	1	2
Parabolized N-Stokes,	2	3
Full N-Stokes, 2-D	2	3
Compress, N-S, 2-D	3	4

Advantages of DFI include:

1. Reduced execution time
2. Minimal Differencing
3. Automatic accuracy control via "smoothing"
4. Enhanced numeric stability
5. Easy to code and "transport" to other users
6. Explicit cause-effect relationships are displayed
7. Simpler modelling is possible (e.g., turbulence)
8. Faithfully and automatically simulates fluid and heat flow physics

Ergo, integral methods, which have been much neglected due to little emphasis even in graduate schools and none in undergraduate schools, deserve careful, serious attention by computational specialists.

FIRST APPLICATIONS OF DFI (FLOSIM)

FLOSIM denotes the FLOw SIMulator application of DFI to fluid flows; indeed, the working equations, shown below, exhibit explicity and clearly all the governing physical mechanisms. The first three DFI applications were to viscous flows which, under similarity assumption, are governed by ODE.

BLASIUS FLOW: This traditional boundary-layer flow, first formulated by Prandtl in 1904 and solved by his student, Blasius, in 1908, was solved, first analytically and then digitally, by Payne (1980, 1981) and by Ko (1982). Physically, the flow over a flat-plate at zero incidence is assumed to have no pressure gradient:

$$f''' + ff'' = 0; \quad f(0) = 0 = f'(0); \quad f'(\infty) = 1 \quad \text{(force equation)} \qquad (1)$$

Integrating once => $\int_0^y dz$ => $f'' = f''_0 - ff' + \int_0^y f'^2\, dz = 0$ (2)

Integrating again => $f' = yf''_0 - 0.5\, f^2 + \int_0^y dz \int_0^y f'^2\, dw$ (3)

which is the velocity equation. Presence of f''_0 means that "shooting" is required for the 2-point BVP; hence, a double iterative procedure is necessary.

SCHEME:

1. Guess f''_0 and hold constant (until step 4)
2. Guess initial profile ("trial solution")
3. Method of Successive Substitutions, or other iterative scheme, until f' converges.
4. Check matching condition, $f'(\infty)$=>1; if not met, then vary f''_0 (secant, Newton, etc.) and GOTO 3
5. If 4 is met, then problem is done.

SOLUTIONS:

1. Analytic (1980) after two iterations:

$$f'(y) \sim y - f''_0 \frac{y^4}{4} + \frac{(f''_0)^2 y^7}{7!} - \frac{(f''_0)^3 y^{10}}{10!}$$

This yielded 22 terms at the next iteration. However, the power series converges slowly.

2. Digital (Payne 1981); Ko 1981-82) used a secant "root finder" to shoot for f''_0. Comparisons (Ko 1982) with 4th order Runge-Kutta solution are:

	f''_0 (converged)	Execute (IBM 4341)	f'^{MAX}	y^{MAX}
FLOSIM	0.469596	1.73 sec	0.99999999	10.0
Runge-K	0.469599	10.30 sec	same	8.93

FALKNER-SKAN SIMILARITY FLOWS (PRANDTL BOUNDARY-LAYERS): The only difference, mathemtically and physically, of these flows from the Blasius "proto-type" boundary-layer is that the pressure gradient is no longer zero but can be favorable (β>0) or adverse (β<0). The force, or linear momentum, equation becomes, in terms of f, the nondimensional stream function:

$$f''' + ff'' + \beta(1-f'^2) = 0 \quad (4)$$

A single, formal integration =>

$$f'' - f''_0 + ff' - \int_0^y f'^2 dz + \beta y - \beta \int_0^y f'^2 dz = 0 \quad (5)$$

Integration once more yields:

$$f' = yf''_0 - 0.5(\beta y^2 + f^2) + (\beta+1) \int_0^y dz \int_0^z dw f'^2 \text{ (velocity Eq)} \quad (6)$$

where β=0 reduces to the Blasius case.

Ko's (1982) results include:

	f'	Time	$f'(\infty)$	y^{MAX}
β=-0.18 (Near separation)				
FLOSIM	0.128634	1.96s	.99999999	10.0
R-Kutta	0.128636	7.36s	.99999999	10.0
β=0.3 (Favorable P)				
FLOSIM	0.77474808	3.36s	.99999999	10.0
R-Kutta	0.77475458	15.51s	.99999999	8.48
β=1.0 (Heimenz plane stagnation flow)				
FLOSIM	1.23257249	2.05s	.99999999	7.0
R-Kutta	1.23258766	6.14s	.99999999	7.0

	f''	Time	$f'(\infty)$	y^{MAX}
β=5 (Very strong, favorable ∇P)				
FLOSIM	2.61568256	32.96	.99999996	4.0
R-Kutta	2.61577944	103.25	.99999997	4.0
β=10				
FLOSIM	3.67498351	7.86	1.00000000	2.0
R-Kutta	3.67523417	26.40	1.00000000	2.0
β=20				
FLOSIM	5.18004105	49.15	1.00000000	2.0
R-Kutta	5.18071784	144.76	.9999915	2.0

Interesting is to compare the ratio of R-Kutta errors to those of DFI/FLOSIM: (Errors computed from analytic expressions (White 1974) for δ_1, δ_2; eight significant digits were used.)

Ratio of error in integral thickness of R-K/FLOSIM:

B=	Displacement (δ_1)	Momentum (δ_2)
- 0.18	∞	∞
0.0	∞	∞
0.3	∞	∞
1.0	∞	∞
5.0	∞	684
10.0	∞	1.8
20.0	∞	52.

where $\delta_1 \equiv \int_0^\infty (1-f')dy$; $\delta_2 \equiv \int_0^\infty (1-f')dy$

MIXING LAYER SIMILARITY FLOW (White, p. 288): This is a 3-point BVP which consequently requires a "double shoot" for the unknown second derivatives at y=0. Here, U_2 and U_1 are the asymptotic velocities at the two edges of the shear layer. A double matching condition exists at y=0 for the first and second derivatives. Mokkapati (1983a) solved the two sides of the layer alternately until both matching conditions were satisfied to the preset error criterion. The force equation, in similarity coordinates, is:

$$f_\alpha''' + f_\alpha f_\alpha'' = 0, \ \alpha = 1,2 \tag{7}$$

and the B.C./M.C. are:

$$f_2'(\infty) = 1 \ ; \ f_1'(-\infty) = U_2/U_1$$

$$f_2(0) = f_1(0) = 0$$

$$f_2'(0) = f_1'(0) \neq 0$$

$$f_2''(0) = f_1''(0)$$

Integrating twice =>

$$f_\alpha' = (f_\alpha')_0 y - 0.5 f_\alpha^2 + \int_0^y dz \int_0^z f_\alpha'^2 \, ds \tag{8}$$

Comparisons of DFI results to R-Kutta were favorable to DFI.

Interim conclusions as of 5/82: DFI/FLOSIM had been demonstrated, for three flow types and eight separate flows, to be 1) simple, both to derive and code/execute, 2) more accurate than 4th order R-Kutta, 3) fast, about 2-3 times Runge-Kutta speeds, 4) "organic" in the sense of naturalness of physical applications, and 5) a simulator of complex physical processes and interactions.

Extension to PDE (Payne 1982, Mokkapati 1983)

The first was 1-D Fourier heat conduction in a bar of length L:

$$\frac{\partial T}{\partial t} = k\,\frac{\partial^2 T}{\partial x^2}\;;\quad \begin{array}{ll} \text{B.C.:} & T(0,t) = T_0,\; T(L,t) = 0 \text{ for } t > 0 \\ \text{I.C.:} & T(x,0) = 0 \text{ for } x > 0 \end{array} \tag{9}$$

which, under DFI, yields

$$kT(x,t) = k\,(\frac{x\partial T_0}{\partial x}) - T_0) + \frac{\partial}{\partial t}\int_0^x dz \int_0^z dwT(w,t) \tag{10}$$

This study was brief but encouraging (Mokkapati 1983a). A curious result is that $\Delta x = \Delta t$ was required for stability; this is probably a function of the initial $T(x,t)$ guess; more work, both numeric and stability analysis, is needed.

EXTENSION TO NON-SIMILAR PRANDTL BOUNDARY-LAYER: The mass conservation and force equations for steady, 2-D, incompressible and laminar flow of a Newtonian fluid are:

$$\nabla \cdot \underline{u} = 0 \tag{11}$$

$$\frac{u\partial u}{\partial x} + \frac{v\partial u}{\partial y} = \frac{-UdU}{dx} + \nu\,\frac{\partial^2 u}{\partial y^2} \tag{12}$$

which, under DFI, yield

$$\nu\,\frac{\partial u}{\partial y} = \frac{t_0}{\rho} + y\,\frac{UdU}{dx} + uv + \frac{\partial}{\partial x}\int_0^y u^2 dz \quad \text{(vorticity)} \tag{13}$$

$$\nu u = \frac{yt_0}{\rho} + \frac{y^2}{2}\,\frac{UdU}{dx} + \int_0^y uvdz + \frac{\partial}{\partial x}\int_0^y dz \int_0^z u^2 dw \quad \text{(velocity)} \tag{14}$$

SCHEME (Mokkapati 1983b):

1. "Bootstrap" initial u-profile via Pohlhausen (1921) method
2. Make any reasonable guess for the second profile at $x + \Delta x$
3. Iterate until the profile converges at current x-station
4. Secant method to update t_0
5. Repeat 3,4 until t_0 convergence
6. Advance to next x-station and repeat 2-5

See Mokkapati's thesis for details of four "psuedo-" airfoils, namely Howarth flows with power index = 2,4,6,8. Comparison with three other and totally differing calculational schemes for these same flows favors DFI. Those methods were: 1) semi-discrete Galerkin FEM (Kim 1983), 2) "Box" FDM (Cebeci and Bradshaw 1977) and the classic Pohlhausen (White 1974).

EXTENSION TO TURBULENCE: The only change to laminar Prandtl (see eqs (11-14)) is the extra or "Reynolds" stress term involving fluctuations in the longitudinal (u') and vertical (v') velocities in (15):

$$\nabla \cdot \underline{u} = 0$$

$$\frac{U\partial U}{\partial x} + \frac{V\partial U}{\partial y} = \frac{-1dP}{\rho dx} + \nu\,\frac{\partial^2 U}{\partial y^2} - \partial\,\frac{\langle u'v'\rangle}{\partial y} \tag{15}$$

Integrating twice yields:

$$\nu U = \frac{t_0}{\rho} y + \frac{1}{\rho}\frac{dP}{dx} y^2/2 + \int_0^y [UV + (u'v')]\, ds + 2\frac{d}{dx}\int_0^y dz \int_0^z U^2\, ds \qquad (16)$$

Where U,V are now time averaged quantities and have a 3rd unknown: $(u'v')$, Reynold's stress which must be modelled.

PROPOSED CALCULATIONAL SCHEME:

1. Chose t_0 and hold constant until step 4.
2. Given pressure gradient from potential solution or from experiment, make initial guesses on u(y) at x, x + Δx.
3. Select turbulence model and iterate until u converges.
4. Update t_0 until matching condition u = Ue at "infinity" is met.
5. Repeat 3,4 until both u and t_0 converge.
6. Increment x and repeat 1-5.

DFI may well provide its greatest benefits in the calculation of turbulent flows due the simplicity with which the turbulent term is generated and the smoothing of any inaccuracy in the modelling by the organic integration from the surface, where turbulence vanishes, to the edge of the boundary layer. A sketch is dramatic:

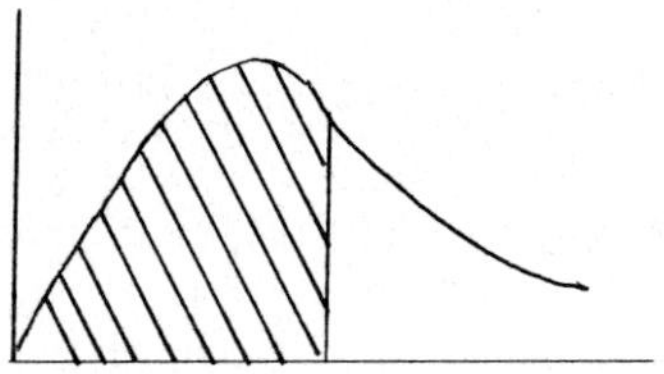

N.B. Usual turbulence approaches model $(u'v')$ at each x-point in a PDE

FURTHER EXTENSIONS (1984)

2-D, incompressible Navier-Stokes (extends easily to compressibility): DFI applied to the usual PDEs yields a triple of coupled PDE,

$$v = -\frac{\partial}{\partial x}\int_0^y u\, ds \qquad (17)$$

$$\nu v = \frac{y t_0}{\rho} + \frac{\partial}{\partial x}\left[\int_0^y dz \int_0^z dw \left\{u^2 + \frac{p}{\rho} - \nu\frac{\partial u}{\partial x}\right\}\right]$$

$$\nu v = \int_0^y \left(\frac{p}{\rho} + v^2\right) dz - y\frac{P_w}{\rho} + \frac{\partial}{\partial x}\int_0^y dz \int_0^z dw \left[u\nu - \nu\frac{\partial v}{\partial x}\right]$$

If "parabolized" then the second x-derivative terms are neglected. One can set P_w=0 with no loss in generality for incompressible flow.

PROPOSED SCHEME (TENTATIVE):

1. Assume two u-profiles at x, x + Δx
2. Obtain v exactly from fluid continuity
3. Estimate t_0 at x + Δx
4. Iterate between the u and v equations until convergence; then ensure the Bernouilli constraint

   $p + 0.5p[u^2 + v^2] = H(\psi)$ {=constant if irrotational external flow}
   is adequately satisfied sufficiently far from the surface.

5. Update t and repeat 3,4 until all three, u, v, t_0, converge.
6. Increment x and complete the "forward sweep"
7. Perform a "backward sweep"; presumably some form of relaxation will be required .
8. Repeat the "outer iteration", steps 6,7, until all converge.

EULER EQUATION for 2-D, steady, and incompressible flows: This is simply Navier-Stokes less the viscous terms; hence, it is practical to integrate only once here rather than twice as before:

$$\rho \, \underline{u} \cdot \nabla \underline{u} = \nabla P \quad , \quad \nabla \cdot \underline{u} = 0 \tag{18}$$

$$vu = -\frac{1}{\rho}\frac{\partial}{\partial x}\int_0^y P ds - \frac{\partial}{\partial x}\int_0^y u^2 ds$$

$$, \quad v = -\frac{\partial}{\partial x}\int_0^y u ds$$

$$v^2 = -\frac{1}{\rho}(P - Pw) - \frac{\partial}{\partial x}\int_0^y uv ds$$

Oddly, the structure here is more complex than Navier-Stokes. The velocities are not now separable. This may well explain the difficulty computerists have with Euler, especially when seeking shocks without the mechanism which causes shocks, namely viscosity. Author recommends little time be spent on Euler for technological application.

LAPLACE and POISSON: DFI, by casting the equations into a new form, can assist in at least, promoting understanding; its calculational advantages may be considerably less than for parabolic problems but this remains to be demonstrated. After the normal (for second order systems) two integrations we have:

$$\nabla^2 G(x,y) = 0 \text{ or } = f\,(x,y)$$

$$G(x,y) = G(x,a) + G'(x,a)y + \frac{\partial^2}{\partial x^2}\int_a^y dw \int_a^w G(x,s)ds \tag{19}$$

$$G(x,y) = G(x,a) + G'(x,a)y + \frac{\partial^2}{\partial x^2}\int_a^y dw \int_a^w G(x,s)\, ds + \int_a^y dw \int_a^w f(x,s)ds \tag{20}$$

SOME EARLY RESULTS AND COMPARISONS

Three different classes of digital equipment have been used: 1) mainframe DEC2060, super-mini IBM 4341, and super-micro SAGE II which is a "desk-top main-frame" (single user) due to the MC68000 8(MHz) CPU. DFI easily accommodates to all three classes of computing machinery.

5.1 Sample numerical experiments included Falkner-Skan upon IBM4341 with execution times of 30-50% for DFI as compared to Runge-Kutta. Blasius was solved on the other equipment with total times including I/O to the CRT of:

SAGE II	62 sec	due, primarily, to 19,200 baud line
DEC2060	92 sec	"low-load" = 4 time-sharing users
DEC2060	300 sec	"high-load" = 50+ users
DEC2060	300 sec	20,000 integration points
DEC2060	3600 sec	1,200,000 points (1200 baud line)

Note: DEC 36-bit word versus 32-bit (IBM, SAGE) reduces the number of iterations to convergence. Double precision (64-, 72-bits) was used. DFI is an efficient DE solver. Formal integration incorporates all I.C. or B.C. at the baseline point. Results prove to be insensitive to initial guesses. Stability is astonishing, at least on DEC; division by an error which became zero to ~22 digits caused overflow but the program recovered in only 2 iterations and quickly converged to the proper value of 0.4695999988. A dynamic range of 39 decades appears adequate. Point-wise as well as global data are available. DFI seems ideally suited to digital machines.

DFI STRUCTURAL SUMMARY

Five classes of fluid flows: All five exhibit the same four mechanisms: linear viscous, quadratic pressure gradient, strongly nonlinear inertia, and interactions. The mathematical forms of Blasius (3), Falkner-Skan (6), Prandtl (14), Navier-Stokes (17), and turbulent Prandtl (14) are all quite similar. Euler (18) has marked dissimilarities as well as several similarities.

Singularity of Prandtl: As distance from the surface increases without bound all four terms in DFI also become unbounded; however, their sum must => 1 (suitably non-dimensionalized). The viscous term will diverge positively and linearly; the pressure term diverges quadratically and either positively or negatively depending upon "adverse" or "favorable" gradient; the Reynolds stress can diverge to either + or - "infinity" as can the interaction term. This, of course, is another reflection of the asymptotic nature of the matching at the "edge" of the layer.

Physical structure of FLOSIM: Conservation of Mass is identical. u-equation terms are:

1. $t_0 y$ - linear viscous term (wall shear) - same for all five
2. Pressure - $(\frac{1dP}{\rho dx})\frac{y^2}{2}$ for Prandtl; $\frac{1}{\rho}\int_0^y dz \int_0^z$ Pds for Navier-Stokes
3. Reynolds stress - same form for all
4. Normal stress - same form for all
5. Streamwise viscous diffusion - Navier-Stokes only

v-equation terms (Navier-Stokes only) are:

1. Average P-gradient
2. Derivative of average Reynolds stress
3. Viscous diffusion = same structure as u-equation

Mathematical structure of DFI/FLOSIM (S,D = no. single, double integrals; "x"=no. derivatives of integrals; T = totals):

Equation	Order	S	Sx	D	Dx	T	Tx
Blasius	3	1		1		2	
Falkner-Skan	3	1		1		2	
Prandtl	3	1	1		1	4	1
turbulent	3	2	1		1	5	1
N-Stokes (parabolized)	4	2	1		2	7	3
N-Stokes	4	2	1		4	11	7
Eulers	3	1	4			5	4

Number of differentials in the equations after 2 formal integrations:

	Originally	FLOSIM
Blasius	5	0
F-Skan	7	0
Prandtl	6	1
N-Stokes	10	3
Prandtl turb:	6	1
Euler	6	4 (only 1 integration is practical)
Total	40	9 (a massive DFI advantage over FDM/FEM)

MODEL EQUATION for TURBULENCE (Payne 1984):

Since the conference author has invented a model equation which exhibits all the behaviors of 2-D turbulent thin-shear-layer flows (Prandtl), i.e., viscosity, pressure, inertia, interactions, and turbulent stresses. u=velocity, ν=viscosity, P=pressure gradient, and Z=Reynolds stress:

$$\nu u'' = u^2 + P + \frac{dZ}{dy} \quad ; \quad Z \equiv - u'v' = \nu e \frac{du}{dy} \tag{21}$$

which, under DFI, becomes

$$\nu u = \nu u_0' y + \frac{Py^2}{2} + \int_0^y dw \int_0^w u\, ds + \int_0^y Z ds \tag{22}$$

Equation (22) exhibits the four major physical mechanisms alluded to above: 1) "viscous sub-layer" which is liner in y; 2) pressure gradient (quadratic); 3) inertia terms in the double integral; and 4) the extra, Reynolds stress (Z-term). Note that the DFI structure of Eq. (22) is identical to that for Falkner-Skan (less Z) and non-linear Prandtl.

Numerical study verifies the above claims and still is in progress. The only turbulence model yet coded is a simple, four-parameter eddy viscosity. Equation (22) is proposed as a vehicle for quantitative study of the multidude of turbulence closure models in a single, ordinary DE or IE. One easy design application is direct comparison of laminar flow with a multitude of turbulence models.

CLOSURE

Sixteen (16) separate physical problems involving 3 ODE and 3 PDE have been coded and solved upon digital equipment with good accuracy and speed results. Seven different physical parameters, four dealing with a turbulent model, have been studied in the past three years by seven persons.

DFI is simple and easy to code or execute. It converts an ODE to a pure IE; PDE is converted to IDE; hence, faster numerics must obtain. DFI removes the necessity for differencing in at least one direction and organically incorporates all I.C. and the B.C. at a single point. It clearly delineates the various physical processes; ease of understanding and data assimilation are thereby enhanced. A major plus is its fundamental capability with computers, digital, analog or hybrid, and consequent improved numeric characteristics over any finite-differencing or finite-element method. DFI offers clearer understanding and better numerics than any other method.

REFERENCES

Cebeci, T. & Bradshaw, P., 1977, Momentum Transfer in Boundary Layers McGHill.
Kim, Sang-Wook, 1983, Ph.D. Dissertation, UTA.
Ko, Fung-Tai, 1982, MSAE Thesis, UTA.
Mokkapati, R., and Payne, F., 1983, Rep. No. CFL/AE 83-01, July.
Mokkapti, R., 1983, MSAE Thesis, UTA.
Payne, F. R., 1980, unpublished lecture notes, AE5305, Oct.
Payne, F. R., 1981, AIAA Symposium, UTA.
Payne, F. R., and Ko, F-T, 1982, Trends in Theory and Practice of Nonlinear Equations p. 467-476, M. Dekker, 1984.
Payne, F. R., 1983, "FLOSIM Extension to Navier-Stokes," CFL/AE 83-03, Aug.
Payne, F. R., 1984, unpublished lecture notes, AE5306, July.
White, F. M., 1974, Viscous Fluid Flow McGHill.

This paper is in final form and no version of it will be submitted for publication elsewhere.

Trends in the Theory and Practice of Non-Linear Analysis
V. Lakshmikantham (Editor)

A QUASI-AUTONOMOUS SECOND-ORDER DIFFERENTIAL INCLUSION

Esteban I. Poffald and Simeon Reich
Department of Mathematics
The University of Southern California
Los Angeles, California 90089

We study a two-point boundary value problem and an incomplete Cauchy problem for a certain nonlinear second-order (elliptic) differential inclusion in Banach spaces. We obtain existence, uniqueness and regularity results for both problems, as well as results on the existence of periodic solutions and the asymptotic behavior of solutions to the incomplete Cauchy problem.

Our purpose in this note is to present several results concerning the boundary value problem

$$\begin{cases} u''(t) \in Au(t) + f(t), & 0 < t < T, \\ u(0) = x, \ u(T) = y \end{cases} \tag{1}$$

and the incomplete Cauchy problem

$$\begin{cases} u''(t) \in Au(t) + f(t), & 0 < t < \infty, \\ u(0) = x \\ \sup \{|u(t)| : t \geq 0\} < \infty, \end{cases} \tag{2}$$

where A is a nonlinear (possibly discontinuous and set-valued) m-accretive operator in a Banach space X and f is a given X-valued function. We refer the reader to [5] for motivation and a detailed treatment of problems (1) and (2) in the autonomous case $f = 0$.

We begin with the approximate problem

$$\begin{cases} u''(t) = A_r u(t) + f(t), & 0 < t < T, \\ u(0) = x, \ u(T) = y, \end{cases} \tag{3}$$

where $r > 0$ and A_r is the Yosida approximation of A.

Proposition 1. Let X be a Banach space and $A \subset X \times X$ an m-accretive operator. Then for each x and y in X and f in $L^2(0,T; X)$ the problem (3) has a unique solution in $W^{2,2}(0,T; X)$.

Proof. Define an operator B in the space $E = L^2(0,T; X)$ by

$$B = \{[u,v] \in E \times E: u \in W^{2,2}(0,T; X),\ u(0) = u(T) = 0 \text{ and } v(t) = -u''(t) \text{ for almost all } t > 0\}.$$

It can be shown (cf. [3]) that this m-accretive operator is also strongly accretive: $\|(I + rB)^{-1}\| \leq T^2/(r + T^2)$. Therefore the result can be established by using Banach's fixed point theorem.

In order to analyze the behavior of the solutions to (3) as $r \to 0+$, we shall need the following differentiation and convergence lemmata (cf. [5]).

Recall that the duality map J from X into the family of nonempty subsets of its dual X^* is always monotone. It is single-valued if and only if X is smooth. In this case we shall say that it is strongly monotone if there is a positive constant M such that

$$(x - y, Jx - Jy) \geq M|x - y|^2 \tag{4}$$

for all x and y in X. It can be shown [5] that a smooth Banach space has a strongly monotone duality map if and only if it is uniformly convex with a modulus of convexity of power type 2. This is the case, for example, when X is one of the Lebesgue spaces L^p, $1 < p \leq 2$.

Lemma 2: Suppose $u: [0,T] \to X$ is continuously differentiable, u' is absolutely continuous, and $u'' \in L^1(0,T; X)$. Let $b: [0,T] \to R^+$ be monotone and absolutely continuous. If $J: X \to X^*$ satisfies (4) and $q: [0,T] \to R$ is defined by $q(t) = (u'(t), Ju(t))$, then

(a) q is differentiable almost everywhere;

(b) $\int_s^t b(t)q'(r)dr \leq b(t)q(t) - b(s)q(s) - \int_s^t b'(r)q(r)dr$

for all $0 \leq s \leq t \leq T$;

(c) $q'(t) \geq (u''(t), Ju(t)) + M|u'(t)|^2$ for almost all t.

Lemma 3. Let X be reflexive and let $\{u_n\}$ be a sequence in $W^{2,2}(a,b; X)$ such that $\{u'_n\}$ converges strongly in $L^2(a,b; X)$, $\{u_n(a)\}$ converges strongly in X, and $\{u''_n\}$ is bounded in $L^2(a,b; X)$. Then there is a function u in $W^{2,2}(a,b; X)$ such that

(a) $u_n \to u$ in $C([a,b]; X)$;

(b) $u'_n \to u'$ in $C([a,b]; X)$;

(c) $u''_n \rightharpoonup u''$ in $L^2(a,b; X)$.

Theorem 4. Let X be a Banach space and $A \subset X \times X$ an m-accretive operator. Assume that X is uniformly smooth and that the duality map $J : X \to X^*$ is strongly monotone. Then for each x and y in D(A) and f in $W^{1,1}(0,T; X)$ the problem (1) has a unique solution in $W^{2,2}(0,T; X)$.

Proof. Let $u = u_r$ be the solution to (3). (It exists by Proposition 1.) We first apply Lemma 2 to $q(t) = (u''(t), Ju'(t))$. Using the strong monotonicity of J we show that $\{u''_r\}$ is bounded in $E = L^2(0,T; X)$. Since E is also uniformly smooth, its duality map is uniformly continuous on bounded sets. Therefore we can apply

Lemma 2 to $a(t) = (u'_p(t) - u'_q(t), J(u_p(t) - u_q(t)))$ and conclude that $\{u'_r\}$ is Cauchy in E. Combining Lemma 3 with the demi-closedness of the m-accretive operator defined by

$$\mathcal{A} = \{[u,v] \in E \times E: u(t) \in D(A) \text{ and } v(t) \in Au(t) \text{ for almost all } 0 < t < T\},$$

we see that $u = \lim_{r \to 0+} u_r$ exists in $C([0,T]; X)$ and is a solution of (1). If v is another solution, then $g(t) = \frac{1}{2}|u(t) - v(t)|^2$ is convex and $g(0) = g(T) = 0$. Therefore $g(t) = 0$ for all $0 \leq t \leq T$ and $u = v$.

In order to extend this result to the case when x and y are in cl(D(A)), we follow [2] and define the norms

$$\|u\|_* = (\int_0^T |u(t)|^2 \beta(t)dt)^{1/2}$$

and

$$\|u\|_{**} = (\int_0^T |u(t)|^2 \beta^3(t)dt)^{1/2},$$

where $\beta(t) = \min\{t, T-t\}$, $0 \leq t \leq T$.

Our next two lemmata are proved with the aid of Lemma 2.

Lemma 5. If u and v are two solutions of (1), then

$$\|u-v\|_\infty \leq \max\{|u(0) - v(0)|, |u(T) - v(T)|\} \tag{5}$$

and

$$\|u'-v'\|_* \leq (|u(0) - v(0)|^2 + |u(T) - v(T)|^2)/(2M). \tag{6}$$

Lemma 6. If f is in $W^{1,2}(0,T; X)$ and u satisfies (1), then

$$\|u''\|_{**} \leq (3/M)\|u'\|_* + (1/3)\|\beta f'\|_{**}. \tag{7}$$

Theorem 7. Let X be a Banach space and $A \subset X \times X$ an m-accretive operator. Assume that X is uniformly smooth and that its duality is strongly monotone. Then for each x and y in cl(D(A)) and f in $W^{1,2}(0,T; X)$ the problem (1) has a unique solution in $C([0,T]; X) \cap W^{2,2}_{loc}(0,T; X)$.

Proof. Let $\{x_n\}$ and $\{y_n\}$ be two sequences in D(A) with $x_n \to x$ and $y_n \to y$. By Theorem 4, for each n there is a solution $u_n \in W^{2,2}(0,T; X)$ of (1) with x replaced by x_n and y by y_n. The estimate (5) shows that $\{u_n\}$ is a Cauchy sequence in $C([0,T]; X)$. Let $u : [0,T] \to X$ be its limit. Moreover, the estimates (6) and (7) show that we can apply Lemma 3 on each compact subinterval [a,b] of (0,T) to conclude that u is also in $W^{2,2}_{loc}(0,T; X)$. Combining Lemma 3 once again with the demi-closedness of the operator $\mathcal{A}$ we see that u is a solution to (1). It is unique by (5).

Consider now the incomplete Cauchy problem (2). In general, this problem has no solution even if $A = 0$ and $f \in W^{1,2}(0,\infty; X)$. It turns out however, that if (2) has a solution for one point x in cl(D(A)), then it has a unique solution for all x in cl(D(A)).

Theorem 8. Let X be a Banach space, $A \subset X \times X$ an m-accretive operator, and $f \in W^{1,2}_{loc}(0,\infty; X)$. Assume that X is uniformly smooth and that its duality map is strongly monotone. If problem (2) has a solution in $C([0,\infty); X) \cap W^{2,2}_{loc}(0,\infty; X)$ for some x in cl(D(A)), then it has a unique solution there for all x in cl(D(A)).

Proof. We first consider the problem with x in D(A). By Theorem 4, for each n there is a solution $u_n \in W^{2,2}(0,n; X)$ of problem (1) with $T = n$ and $x = y$. Fix $n > m > T$ and set

$$p(t) = \frac{1}{2} |u_n(t) - u_m(t)|^2, \quad 0 \leq t \leq m.$$

By Lemma 2 we have

$$M \int_0^m (m-r) \, |u_n'(r) - u_m'(r)|^2 \, dr \leq \int_0^m (m-r) \, p''(r) \, dr \leq$$
$$\leq -mp'(0) + p(m) - p(0) \leq p(m).$$

Hence

$$\int_0^T |u_n'(r) - u_m'(r)|^2 \, dr \leq p(m)/(M(m-T)).$$

Assuming the existence of a solution to (2), we can use (5) to conclude that $\{\|u_n\|_\infty\}$ is bounded. Therefore $\{u_n'\}$ is a Cauchy sequence in $L^2(0,T; X)$. It follows that there exists $u \in C([0,\infty); X)$ such that $u_n \to u$ in $C([0,T]; X)$ for all positive T. We now use (7) and Lemma 3 to conclude that u is in $W^{2,2}_{loc}(0,\infty; X)$ and is a solution of (2). When x is in cl(D(A)), we choose a sequence $\{x_n\} \subset D(A)$ which converges to x, and let u_n be the solution to (2) with x replaced by x_n. Note that if v and w are two solutions of (2), then by Lemma 2 the function $\frac{1}{2} |v(t) - w(t)|^2$ is convex. Since it is also bounded on $[0,\infty)$, it must be non-increasing. This shows that $\{u_n\}$ converges in $C([0,\infty); X)$ to a function u (and proves uniqueness of solutions to (2)). We can now use (6), (7) and Lemma 3 to prove that u belongs to $W^{2,2}_{loc}(0,\infty; X)$ and is a solution of (2). The result is established.

Lemma 2 also leads to the following property of the solutions to (2).

Proposition 9. If u and v are two solutions of (2) with $u(0) = x$ and $v(0) = y$, then

$$\int_0^\infty t|u'(t) - v'(t)|^2 \, dt \leq |x-y|^2/2M. \tag{8}$$

Let γ be a maximal monotone graph in R^1 with $0 \in \gamma(0)$, and let Ω be a bounded domain in R^n with a smooth boundary $\partial\Omega$. Theorem 8 and Proposition 9 can be applied, for example, to the following problem in $L^p(\Omega)$, $1 < p \leq 2$:

$$\begin{cases} u_{tt} \in -\Delta u + \gamma(u) + f(t) & \text{in } (0,\infty) \times \Omega \\ \partial u/\partial n = 0 & \text{on } (0,\infty) \times \partial\Omega \\ u(0,x) = u_0(x) & \text{in } \Omega \\ \sup\{|u(t,\cdot)| : t \geq 0\} < \infty. \end{cases} \tag{9}$$

Our next two results deal with the case when f is periodic. We continue to be in the setting of Theorem 8.

<u>Theorem 10</u>. If problem (2) has a solution and f is periodic of period T, then there is a solution of (2) which is also T-periodic.

<u>Proof</u>. Let u be a solution of (2), and define $u_n : [0,\infty) \to cl(D(A))$, $n = 1,2,\ldots$, by $u_n(t) = u(t + nT)$, $0 \leq t < \infty$. Given x in cl(D(A)) and $s \geq 0$, there is by Theorem 8 a unique solution v of the problem

$$\begin{cases} u''(t) \in Av(t) + f(t), & s < t < \infty, \\ v(s) = x \\ \sup\{|v(t)| : t \geq s\} < \infty . \end{cases} \tag{10}$$

Define $U(t,s) : cl(D(A)) \to cl(D(A))$ by $U(t,s)x = v(t)$, $t \geq s$. U is a nonexpansive evolution system. Since cl(D(A)) is convex and $u_n(t) = U(t + T, t)^n u(t)$ is a bounded sequence of iterates of the nonexpansive mapping $U(t + T,t)$, this mapping must have a fixed point. The result follows by taking $t = 0$.

<u>Theorem 11</u>. If problem (2) has a solution u, f is periodic of period T, and cl(D(A)) is boundedly compact, then there is a T-periodic solution w of (2) such that $u(t) - w(t) \to 0$ and $u'(t) - w'(t) \to 0$ as $t \to \infty$.

<u>Proof</u>. Let the sequence $\{u_n(t)\}$ be defined as in the proof of Theorem 10. The estimate (8) can be used to show that $u_n(t) - u_{n+1}(t) \to 0$ as $n \to \infty$. In the setting of Theorem 10 this implies that $\{u_n(t)\}$ converges weakly to a fixed point w(t) of $U(t + T,t)$ for each $t \geq 0$. Since cl(D(A)) is assumed to be boundedly compact, this convegence is acturally strong. Therefore we are able to identify w as a solution of (2) by combining the demi-closedness of the operator A with Lemma 3. The result now follows by modifying the arguments leading to Theorem 3.2 of [2].

When we assume in addition that the duality map $J : X \to X^*$ is Lipschitzian, then it can be shown that Theorems 4 and 7 remain true when f is merely in $L^2(0,T; X)$, and Theorems 8, 10, and 11 remain true when f belongs to $L^2_{loc}(0,\infty; X)$. The estimate (7) becomes

$$\|u''\|_{**} \leq (1 + L\|f\|_{**})(1 + 3\|u'\|_*)/M, \qquad (11)$$

where L is the Lipschitz constant of J. In this way we obtain complete extensions of the Hilbert space results in [1] and [2] (except that in Theorem 11 we still require that cl(D(A)) be boundedly compact). The duality map of X is indeed Lipschitzian when the modulus of convexity of X^* is of power type 2. Note that spaces for which the moduli of convexity of both X and X^* are of power type 2 need not be isometric to a Hilbert space (cf. [4]).

We conclude this note with another result on the asymptotic behavior of the solutions to (2). Its proof is based on Lemma 2.

Proposition 12. Let X be a uniformly smooth Banach space, $A \subset X \times X$ an m-accretive operator which is also strongly accretive, and $u \in C([0,\infty); X) \cap W^{2,2}_{loc}(0,\infty; X)$ a solution of (2). If $f \in C([0,\infty); X)$ and $\lim_{t\to\infty} f(t) = f_\infty$, then $\lim_{t\to\infty} u(t) = u_\infty$, where u_∞ is the unique solution to the inclusion $0 \in Au_\infty + f_\infty$.

REFERENCES:

[1] Barbu, V., A class of boundary problems for second order abstract differential equations, J. Fac. Sci. Univ. Tokyo 19 (1972), 295-319.

[2] Bruck, R.E., Periodic forcing of solutions of a boundary value problem for a second order differential equation in Hilbert space, J. Math. Anal. Appl. 76 (1980), 159-173.

[3] Da Prato, G., Weak solutions for linear abstract differential equations in Banach spaces, Advances in Math. 5 (1970), 181-245.

[4] Leonard, E. and Sundaresan, K., A note on smooth Banach spaces, J. Math. Anal. Appl. 43 (1973), 450-454.

[5] Poffald, E.I. and Reich, S., An incomplete Cauchy problem, preprint.

This paper is in final form and no version of it will be submitted for publication elsewhere.

Trends in the Theory and Practice of Non-Linear Analysis
V. Lakshmikantham (Editor)

ON SYSTEMS WITH TRANSFER FUNCTIONS RELATED TO THE RIEMANN ZETA FUNCTION

V. M. Popov
University of Florida
Gainesville, Florida 32611

INTRODUCTION.

The qualitative theory of differential equations has been successful in solving many fundamental problems of the theory of control. The development of systems which contain computers leads to new types of problems which require special investigation. Such systems can operate in a succession of nonidentical steps of increasing complexity. Moreover, some critical quantities in the system (e.g. the feedback function) can be determined by testing whether a certain condition is satisfied and by choosing the value of the quantity depending on the result of the test ("logical feedback"). At this level of complexity, new types of behavior may arise at any step and the problem of predicting the future behavior of the systems becomes particularly acute. Sometimes it is not even clear if, or to what extent, the long range behavior is predictable. In any case, although these applications deal with purely deterministic processes executed by machines, the machines themselves are unable to predict their future behavior. It is a task of the theory to study mathematical models of these systems in order to elucidate their behavior.

In this paper, we present three of the simplest mathematical models related to our subject. The first two are simple interpretations, in the control theory framework, of known results of Möbius and von Mangoldt. Both of these systems have transfer functions related to Riemann's zeta function. The qualitative behavior of the systems depends on the location of the zeros of the function zeta, a problem which is still in the conjecture-checking stage. Both models illustrate very clearly the difficulties mentioned above. It is perhaps natural that, in trying to understand the behavior of systems of increasing complexity, one is faced with mathematical problems of extreme difficulty.

Both Möbius and von Mangoldt models are purely algebraic. The third model given in this paper is a nonlinear dynamical system with unbounded delays. This system is studied in more detail, by combining some standard results in the analytic number theory with the frequency domain approach of control theory.

A MATHEMATICAL MODEL FOR A PURE COMPUTER FEEDBACK SYSTEM (MÖBIUS SYSTEM).

Let u be a real function, defined for $T > 1$. We define the Möbius system as

$$x(T) = -\sum_{n=2}^{\infty} x\left(\frac{T}{n}\right) + u(T) \quad , \text{ for } \quad T > 1, \tag{1}$$

$$x(T) = 0 \quad \text{if } 0 < T \leq 1 \tag{2}$$

The "solution" is the unique function x, defined for $T > 1$, which satisfies (1) and (2). These equations can be interpreted as describing a feedback system (fig. 1).

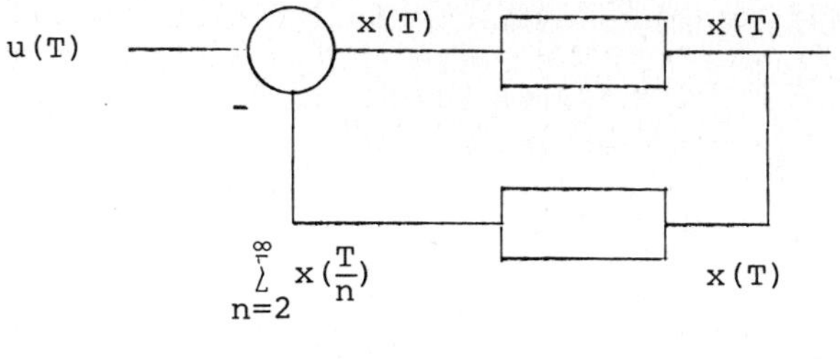

fig. 1

Notice that if u is integer-valued then so also is the solution x. Hence the problem can be handled by computers with absolute accuracy.

From the viewpoint of the qualitative theory of systems, one meets the following problem: Given u (e.g. $u = 1$) find precise estimates of x. But although x can be written explicitly (by Möbius inversion formula) the problem remains unsolved. The theory leads to some (upper and lower) estimates, between which there is an enormous gap. Nor can this problem be solved by running it on a computer (in spite of the fact that it seems to be ideally suited for such an approach). In fact the computer is able to determine precisely the solution on finite intervals, but is absolutely useless for finding long range estimates. Similar comments remain valid for the other systems considered below.

In order to apply traditional methods of control theory, it is convenient to put $t = \log T$ and to consider, instead of (1),(2), the system

$$x(t) = -\sum_{n=2}^{\infty} x(t-\log n) + u(t) \ , \text{ if } t > 0$$

$$x(t) = 0, \quad \text{if} \quad t \leqq 0,$$

the solution x being now defined on the whole real axis R. It is easy to see that the transfer function of this system (from u to x) is equal to

$$1/\sum_{n=1}^{\infty} n^{-s} \ , \text{ for Re } s > 1,$$

i.e. it equals the inverse of Riemann's zeta function, $\zeta(s)$.

The qualitative behavior of this system depends on the location of the zeros of the function zeta - and the elucidation of this difficult problem constitutes the only way for understanding the long range behavior of this system, in spite of its apparent simplicity.

A MATHEMATICAL MODEL FOR SYSTEMS WITH LOGICAL FEEDBACK (VON MANGOLDT SYSTEM).

Let $\Lambda(n)$ be von Mangoldt's function (equal to $\log p$ if n is equal to a prime p or to a power of the prime p and equal to zero otherwise). We define the von Mangoldt's feedback system by

$$\sum_{n=2}^{\infty} \Lambda(n) \, x(t-\log n) = u(t), \text{ if } t > 0 \ ,$$

$$x(t) = 0 \ , \ \text{if } t \leqq 0.$$

This system can again be interpreted as a feedback system, but the coefficients $\Lambda(n)$ must be determined by a test (checking whether n is a power of a prime). Hence this model belongs to the category of systems with logical feedback.

The inverse of the transfer function of this system is $\sum_{n=2}^{\infty} \Lambda(n) n^{-s}$, if Re s > 1. It is known that

$$-\frac{\zeta'(s)}{\zeta(s)} = \sum_{n=2}^{\infty} \Lambda(n) n^{-s} \ , \ \text{if Re } s > 1 \tag{3}$$

[1,(1.1.8) p.4]. Moreover

$$-\frac{\zeta'(s)}{\zeta(s)} = -\sum_{\rho} \frac{1}{s-\rho} + 1 - \sum_{\rho} \frac{1}{\rho} - \frac{\zeta'(0)}{\zeta(0)} + \frac{1}{s-1} + \sum_{n=1}^{\infty} \frac{s}{2n(s+2n)} \tag{4}$$

where ρ runs over all the nonreal zeros of $\zeta(s)$ [2,p.52]. It is known that all the nonreal zeros ρ satisfy the condition Re $\rho \in (0,1)$ (they lie in the "critical strip", symmetrically about the point $(1/2,0)$). The qualitative behavior of the solution of von Mangoldt's system depends on the upper bound of Re ρ - which will be denoted in the following by Θ. But all one knows is that Θ lies somewhere in the interval [1/2,1].

A NONLINEAR DYNAMICAL SYSTEM WITH LOGICAL FEEDBACK.

One can now study a new class of feedback systems, introducing another feedback law, namely assuming that u is related to x according to some suitable feedback relation (this adds a new feedback loop in fig. 1, but the new loop can be combined with the old one to obtain a single feedback loop). The behavior of the solutions will depend, of course, in an essential way on the new feedback relation. It is an interesting problem to find feedback relations for which the growth of the solution is determined by the quantity Θ. The next model satisfies this condition. Moreover the feedback law is nonlinear and contains the derivative of the solution, such that the result is a dynamical system whose form is not very exotic. The system is described by the equations

$$\sum_{n=2}^{\infty} \Lambda(n)\ x(t-\log n) - \int_{-\infty}^{t} e^{t-\tau} x(\tau)\,d\tau - \int_{0}^{t} \sum_{n=1}^{\infty} \frac{1}{2n} e^{-2n(t-\tau)} \dot{x}(\tau)\,d\tau = \dot{x}(t) + f(x(t)), \ \text{if } t > 0 \tag{5}$$

$$x(t) = \phi(t), \ \text{if } t \leqq 0, \tag{6}$$

where ϕ is a C^1 function with compact support, defined for $t \leqq 0$ (relation (6) defines an initial condition). The function f is assumed to be globally Lipschitzian, from R into R. By "solution" of (5),(6) we mean a real C function x, defined for $t \in R$, such that the series in (5) converge and relations (5) and (6) are satisfied.

Since the coefficients $\Lambda(n)$ are determined by a test, as in the von Mangoldt case, the model can be interpreted as a system with logical feedback.

The system falls into the category of the dynamical systems with unbounded delays [3]. A similar system has been considered in [4]. The fact that the kernel in the last integral from (5) is not bounded makes the problem somewhat peculiar, but a contraction-mapping argument (which we omit) proves that, for every C^1 function ϕ with compact support, there exists a unique C solution x of (5), (6).

This model is just a representative of a huge class of systems which can be treated similarly.

In order to estimate the solution x we introduce the notation

$$|\phi| = \sup_{t \leqq 0} (e^{-\theta t}|\phi(t)|) + \int_{-\infty}^{0} e^{-\theta t}|\phi'(t)|dt,$$

where θ is as in the following statement:

Theorem. Suppose that the globally Lipschitzian function f satisfies the condition $f(r)r \geqq 0$, for every real r. Then for every $\theta > \Theta$ there exists a constant k such that, for every C^1 function ϕ with compact support, the solution x of (5),(6) satisfies the inequality

$$|x(t)| \leqq k\, e^{\theta t}|\phi|, \text{ for every } t > 0.$$

The proof uses some standard results of the theory of distribution of prime numbers (see e.g. [5]) and the main ideas of the frequency domain method (see e.g. [6]-[8]).

Fix the initial condition ϕ and the solution x of (5),(6). For each $T > 0$, define the functions y_T, w_T and z_T, from $(0,\infty)$ into R, as follows

$$y_T(t) = \begin{cases} e^{-\theta t}x(t) & \text{if } 0 < t \leqq T \\ 0 & \text{if } t > T \end{cases}$$

$$w_T(t) = \begin{cases} e^{-\theta t}\dot{x}(t) & \text{if } 0 < t \leqq T \\ 0 & \text{if } t > T \end{cases}$$

$$\begin{aligned} z_T(t) = & \sum_{2 \leq n < e^t} \Lambda(n)n^{-\theta}y_T(t-\log n) - \int_0^t e^{(1-\theta)(t-\tau)}y_T(\tau)d\tau \\ & - \sum_{n=1}^{\infty} \frac{1}{2n} \int_0^t e^{-(2n+\theta)(t-\tau)}w_T(\tau)d\tau \\ & - x(0)\sum_{n=1}^{\infty} \frac{1}{2n} e^{-(2n+\theta)t} + e^{-\theta T}x(T)S_T(t), \qquad (7) \end{aligned}$$

where

$$S_T(t) = \begin{cases} 0 \quad \text{if} \quad 0 < t \leqq T \\ \sum_{n=1}^{\infty} \frac{1}{2n} e^{-(2n+\theta)(t-T)} \quad \text{if} \quad t > T. \end{cases}$$

A space-consuming computation shows that z_T tends exponentially to zero as t tends to infinity. (Since the main idea of this computation is contained in the proof of a Lemma below, we omit the details) It follows that the Laplace transform

$$\hat{z}_T(s) = \int_0^\infty e^{-st} z_T(t)\,dt$$

is analytic for $\operatorname{Re} s > -\varepsilon$, where ε is some strictly positive number. Notice also that the Laplace transform $\hat{y}_T(s)$ is entire. Using (5) it is easy to see that, if $0 < t < T$, then

$$z_T(t) = e^{-\theta t}(\dot{x}(t) + f(x(t)) - \sum_{n \geqq e^t} \Lambda(n)\phi(t-\log n)$$

$$- x(0) \sum_{n=1}^{\infty} \frac{1}{2n} e^{-2nt} + \int_{-\infty}^{0} e^{t-\tau}\phi(\tau)\,d\tau), \quad 0<t<T, \qquad (8)$$

since $S_T(t) = 0$ in the interval $0 < t < T$.

From (7) one finds that, if $\operatorname{Re} s + \theta > 1$, then

$$\hat{z}_T(s) = \left(\sum_{n=2}^{\infty} \Lambda(n) n^{-(s+\theta)} - \frac{1}{s+\theta-1} - \sum_{n=1}^{\infty} \frac{s+\theta}{2n(s+\theta+2n)} \right) \hat{y}_T(s).$$

Moreover from (3) and (4) it follows that, if $\operatorname{Re} s + \theta > 1$, then

$$\hat{z}_T(s) = \left(c - \sum_{\rho} \frac{1}{s+\theta-\rho} \right) \hat{y}_T(s), \qquad (9)$$

where

$$c = 1 - \frac{\zeta'(0)}{\zeta(0)} - \sum_{\rho} \frac{1}{\rho}.$$

However, since the functions involved in (9) are analytic for $\operatorname{Re} s > -\varepsilon$, the relation is also true for $\operatorname{Re} s = 0$. From the known relations $\zeta'(0)/\zeta(0) = \log 2\pi$ [2,p.66] and

$$\sum_{\rho} \frac{1}{\rho} = 1 + \gamma/2 - (\log \pi)/2 - \log 2$$

[2, pp. 159,160] - where γ is Euler's constant - it follows that $c < 0$. Moreover, our assumption $\theta > \operatorname{Re} \rho$ implies that

$$\operatorname{Re}\,(1/(i\omega+\theta-\rho)) > 0, \text{ for every real } \omega.$$

These results allow us to obtain, from (9), the relation

$$\hat{z}_T(i\omega) = G(i\omega)\hat{y}_T(i\omega), \text{ for every real } \omega,$$

where $\operatorname{Re} G(i\omega) < 0$, for every real ω. Hence, if one defines

$$J_T = \int_0^\infty y_T(t)\; z_T(t)\; dt$$

and one applies Parseval formula (which is legitimate) one obtains

$$J_T \leqq 0. \qquad (10)$$

On the other hand, from (8), we get

$$J_T = \int_0^T e^{-2\theta t} x(t)(\dot{x}(t) + f(x(t)) - x(0) \sum_{n=1}^{\infty} \frac{1}{2n} e^{-2nt}$$

$$- \sum_{n > e^t} \Lambda(n)\phi(t-\log n) + \int_{-\infty}^{0} e^{t-\tau}\phi(\tau)\,d\tau)\; dt.$$

Part of this expression can be estimated as in the following lemma (whose proof is given in the next section).

Lemma. There exists a constant K (independent of T) such that

$$\left| \int_0^T e^{-2\theta t} x(t) \left(\sum_{n > e^t} \Lambda(n)\phi(t-\log n) - \int_{-\infty}^0 e^{t-\tau}\phi(\tau)d\tau \right) dt \right|$$

$$\leqq K|\phi| \sup_{0<\tau<T} (e^{-\theta\tau}|x(\tau)|), \text{ for every } T > 0.$$

From this lemma and from the condition $f(r)r \geqq 0$ it follows that (10) becomes

$$\int_0^T e^{-2\theta t} x(t)\dot{x}(t)dt \leqq x(0) \int_0^T e^{-2\theta t} x(t) \sum_{n=1}^{\infty} \frac{1}{2n} e^{-2nt} dt$$

$$+ K|\phi| \sup_{0<\tau<T} (e^{-\theta\tau}|x(\tau)|) \qquad (11)$$

The left hand side can be written as

$$\frac{1}{2}\int_0^T e^{-2\theta t} \frac{d}{dt}(x^2(t))dt = \frac{1}{2} e^{-2\theta T} x^2(T) - \frac{1}{2}\phi^2(0)$$

$$+ \theta \int_0^T e^{-2\theta t} x^2(t)dt$$

and the first term in the right hand side of (11) can be estimated as follows: there exists a constant k' such that

$$\int_0^T e^{-2\theta t} x(t) \sum_{n=1}^{\infty} \frac{1}{2n} e^{-2nt} x(0)dt \leqq k'|\phi(0)| \sup_{0<\tau<T} (e^{-\theta\tau}|x(\tau)|)$$

Putting $y(t) = e^{-\theta t}x(t)$, inequality (11) becomes

$$y^2(T) \leqq \phi^2(0) + K'|\phi| \sup_{0<\tau<T} |y(\tau)|,$$

for some new constant K'. Since the condition is valid for every $T > 0$, the quantity $v(T) = \sup_{0<\tau<T} |y(\tau)|$ satisfies the inequality

$$v^2(T) \leqq \phi^2(0) + K'|\phi| v(T).$$

By completing the square one obtains the conclusion of the theorem.

PROOF OF THE LEMMA.

Using the function

$$\psi(x) = \sum_{n \leqq x} \Lambda(n),$$

one can write the Stieltjes integral

$$\sum_{n>e^t} \Lambda(n)\phi(t-\log n) = \int_{e^t}^{\infty} \phi(t-\log x)\, d\psi(x)$$

$$= \int_{e^t}^{\infty} \phi(t-\log x)\, d(\psi(x)-x) + \int_{e^t}^{\infty} \phi(t-\log x)dx.$$

The last integral equals $\int_{-\infty}^{0} e^{t-\tau}\phi(\tau)d\tau$. The first integral in the right hand side is integrated by parts and one uses the fact that there exists a number $r < \theta$ such that

$$\psi(x) - x = O(x^r).$$

This last fact follows from our assumption on θ and from Theorem 30 in [5,p.83]. With these results, the inequality to be proved follows easily from our assumptions on the function ϕ.

CONCLUSIONS.

Problems which have been intensively studied for a long time for their own sake turn out to be relevant for understanding the behavior of some actual systems in control theory. Inherent difficulties make the theory still incomplete, in spite of more than a century of efforts and continuous progress. The fundamental interest of these problems in so many different areas points to the need for further investigations.

REFERENCES:

[1] Titchmarsh, E. C., The Theory of The Riemann Zeta-Function (Oxford University Press, Oxford, 1951)

[2] Edwards, H. M., Riemann's Zeta Function (Academic Press, New York, 1974)

[3] Corduneanu, C. and Lakshmikantham, V., Equations with Unbounded Delay: A Survey, Dept. of Math., University of Texas at Arlington, Arlington, Texas, Technical Report No 113, August 1979.

[4] Kolmanovskii, V. B. and Nosov, V. R., On the stability of first order nonlinear equations of neutral type, J. Appl. Math. Mech. 34 (1971) 560-567.

[5] Ingham, A. E., The Distribution of Prime Numbers (Cambridge Tracts in Mathematics and Mathematical Physics, No. 30, 1932)

[6] Popov, V. M., Hyperstability of Control Systems (Springer Verlag, New York, 1973)

[7] Halanay, A., Differential Equations: Stability, Oscillations, Time Lags (Academic Press, New York, 1966).

[8] Corduneanu, C., Integral Equations and Stability of Feedback Systems (Academic Press, New York, 1973).

This paper is in final form and no version of it will be submitted for publication elsewhere.

Trends in the Theory and Practice of Non-Linear Analysis
V. Lakshmikantham (Editor)

LIMIT SETS IN INFINITE HORIZON OPTIMAL CONTROL SYSTEMS

Emilio O. Roxin

Department of Mathematics
University of Rhode Island
Kingston, Rhode Island 02881
U.S.A.

For an autonomous control system, the reachability relation between points of the state space can be used to define classes of points and a partial order among these. Basic properties of such classes and some examples are given. The importance of these classes for the optimal control problem in the infinite horizon case is briefly indicated.

1. INTRODUCTION. GENERAL ASSUMPTIONS. NOTATION.

Let $X \subset R^n$ be the state space, U compact $\subset R^m$ the control set, $t \in R$ the time, $f(x,u): X \times U \to R^n$ continuous in (x,u), lipschitz in x. "Admissible controls" will be all measurable functions $u(t)$ defined on some appropriate interval $I \subset (-\infty,\infty)$. It is well known that under these conditions the differential equation

(1) $$\frac{dx}{dt} = \dot{x} = f(x,u)$$

defines an autonomous control system, and given an initial condition

(2) $$x(t_0) = x_0$$

and an admissible controller $u(t)$, (1) has a unique local solution. A growth condition of the type

(3) $$\langle x , f(x,u)\rangle \leqq c_1 \|x\|^2 + c_2,$$

where $\|.\|$ is the norm and $\langle .,. \rangle$ the inner product, insures the extendability of the solution to the whole real line $(-\infty,\infty)$.

The set

(4) $$V(x) = f(x,U) = \{f(x,u) \mid u \in U\}$$

is called the "velocity set" at x; according to the above assumptions it will be compact (for each x). We make the additional assumption that $V(x)$ is convex for each x.

The above assumptions insure all the "desired" properties of the solutions of the control system (1). They are, of course, quite stringent and can be relaxed, still maintaining several of the desirable properties. Here, as this paper deals with some basic ideas which are better presented within a framework which avoids more subtle technicalities, no emphasis will be put on generalities. It still would be quite interesting to investigate how the concepts defined below would have to be redefined in the case of a non-autonomous control systems.

In the applications to the optimal control problem, a scalar-valued function $f^0(x,u)$, continuous in (x,u), is used to define the "cost functional"

(5) $$J(T) = J_{u(.)}(T) = \int_{t_0}^{T} f^0(x(t),u(t))\, dt$$

The optimal control problem consists then in minimizing $J_{u(.)}(T)$ over all admissible controllers u(.). The "infinite horizon" problem corresponds to $T = \infty$, in which case it must be specified in which sense this "minimization" should be understood (there is an interesting discussion about the concept of optimality in the infinite horizon case, see for example [4]).

2. REACHABILITY RELATIONS

1. Definition. Let $A(t_1-t_0,x_0) = \{x(t_1)|x(t)$ is the solution of (1),(2) corresponding to some admissible controller $u(.)\}$. This set is usually called the "attainable set at t_1, starting at (t_0,x_0)". It is introduced here mainly for comparison with the other sets to be defined.

2. Definition. The symbol "$\leadsto$" (handwritten "$\nearrow$") will be used to denote the following relation: $x_1 \leadsto x_2$ if x_2 is reachable from x_1 in a finite positive time interval, i.e. if there exist $t_1 < t_2$ and an admissible u(.) such that for the corresponding trajectory x(t) satisfying (1) and $x(t_1) = x_1$, we have $x(t_2) = x_2$. Let us call "$\leadsto$" the "reachability relation".

3. Definition. The relation $x_1 \leftrightsquigarrow x_2$ (handwritten $x_1 \leftrightsquigarrow x_2$) shall mean $x_1 \leadsto x_2$ and $x_2 \leadsto x_1$. Let us call "$\leftrightsquigarrow$" the "reversible reachability relation".

4. Properties. The reachability relation is transitive:

$$\text{if } x_1 \leadsto x_2 \text{ and } x_2 \leadsto x_3 \text{ then } x_1 \leadsto x_3.$$

Therefore the reversible reachability is also transitive:

$$\text{if } x_1 \leftrightsquigarrow x_2 \text{ and } x_2 \leftrightsquigarrow x_3 \text{ then } x_1 \leftrightsquigarrow x_3.$$

The reversible reachability is also symmetric:

$$\text{if } x_1 \leftrightsquigarrow x_2 \text{ then also } x_2 \leftrightsquigarrow x_1.$$

5. Remark. The reachability relation is not reflexive:

$x_1 \leadsto x_1$ is not necessarily true (hence $x_1 \leftrightsquigarrow x_1$ is not necessarily true)

6. Definition. The set

$$H(x_0) = \{x | x \leftrightsquigarrow x_0\}$$

shall be called the "holding set" from x_0. Due to symmetry and transitivity, this set is well determined independently of $x_0 \in H$.

7. Remark. $H(x_0)$ may be empty, but if it is not empty, then $x_0 \in H(x_0)$. Indeed, $H(x_0)$ may consist of a single point (which is then x_0 and a rest-point), but if $H(x_0)$ contains two different points x_0 and x_1, then it contains also all points on the trajectory going from x_0 to x_1 and coming back to x_0.

8. Remark. $H(x_0)$ is connected.

9. Definition. If $x_1 \leftrightsquigarrow x_1$ is false, then x_1 shall be called "transient". The "transient set" is defined as

$$T = \{x | x \text{ is transient}\}.$$

10. A subdivision of the state space X is induced by the above relation in the form:

$$X = T \cup \bigcup_{i \in \Lambda} H_i ,$$

where H_i are the holding sets defined above. Note that T and the H_i's are all disjoint.

11. Relations between the holding sets.

$$\text{If } x_1 \rightsquigarrow y_1,\ x_2 \in H(x_1) \ \text{ and } \ y_2 \in H(y_1)$$

$$\text{then } \ x_2 \rightsquigarrow y_2.$$

Hence it will be written that $H(x_1) \rightsquigarrow H(y_1)$.

12. Reachability sets.

$H^+(x_0) = \{x \mid x_0 \rightsquigarrow x\}$ ("reachable set").

$H^-(x_0) = \{x \mid x \rightsquigarrow x_0\}$ (inverse reachable set).

13. Corollaries.

a) for every $x_0 \in X$: $H(x_0) \subset H^+(x_0)$;

b) if $x_1 \rightsquigarrow x_2$ then $H^+(x_1) \supset H^+(x_2)$;

c) if $x_1 \leftrightsquigarrow x_2$ then $H^+(x_1) = H^+(x_2)$.

14. The converse is not true in general:

$H^+(x_1) \supset H^+(x_2)$ does not imply $x_1 \rightsquigarrow x_2$.

$H^+(x_1) = H^+(x_2)$ does not imply $x_1 \leftrightsquigarrow x_2$.

Counterexample:

$$x = \begin{pmatrix} x^1 \\ x^2 \end{pmatrix}, \quad \begin{pmatrix} \dot{x}^1 \\ \dot{x}^2 \end{pmatrix} = \begin{pmatrix} 1 \\ u \end{pmatrix}, \quad u \geqq 0 \ ; \text{ then } H^+\left(\begin{pmatrix} a \\ b \end{pmatrix}\right) = \left\{\begin{pmatrix} x^1 \\ x^2 \end{pmatrix} \;\middle|\; x^1 > a\right\}.$$

Hence $x_1 = \begin{pmatrix} a \\ b \end{pmatrix}$ and $x_2 = \begin{pmatrix} a \\ b' \end{pmatrix}$, with $b \neq b'$ have the same reachable set H^+, but are not reachable from each other (see fig.1). Note that in this example, the assumption $u(t) \in U$ compact is violated; indeed, for this type of counterexample to work, $H^+(x_0)$ cannot be closed (cannot contain x_0).

The proof of all statements above is reasonably easy and can be left to the reader.

3. EXAMPLES

1. If the control system becomes an ordinary differential system $f(x,u) = f(x)$ and if solutions of the initial value problem are unique, then

a) the only H^+ sets are half-trajectories,

b) the only H sets are periodic orbits and critical points.

2. Let $x(t) \in R^2$. Consider the system

$$\dot{x}^1 = -x^1 + u, \qquad \dot{x}^2 = -x^2 + u, \qquad |u| \leqq 1.$$

The "rest-point set" $= \{x \mid 0 \in f(x,U)\}$ is $\left\{\begin{pmatrix} x^1 \\ x^2 \end{pmatrix} = \begin{pmatrix} u \\ u \end{pmatrix}, \ |u| \leqq 1\right\}$,

hence a line segment $[x_+, x_-]$ (see fig.2). $H^+(x_0)$ is the triangle $[x_+, x_-, x_0]$ without the vertex x_0 and without the side $[x_+, x_-]$. $H(x_0)$ is empty except if x_0 belongs to $[x_+, x_-]$; if x_0 is x_+ or x_-, then $H(x_0) = \{x_0\}$, and if $x_0 \in (x_+, x_-)$, then $H(x_0) = (x_+, x_-)$. The transient set T is all R^2 minus the segment $[x_+, x_-]$.

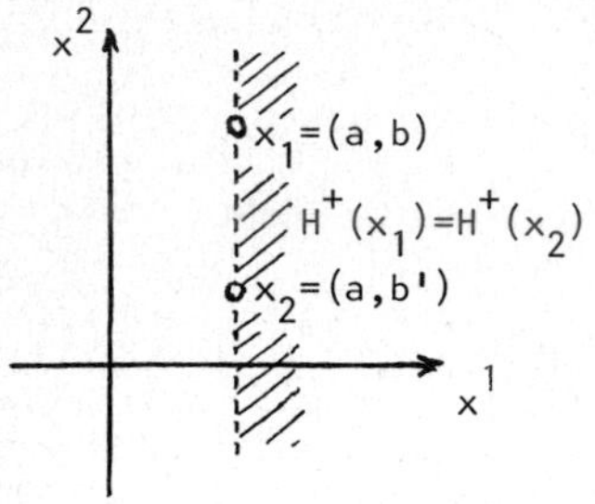

Figure 1

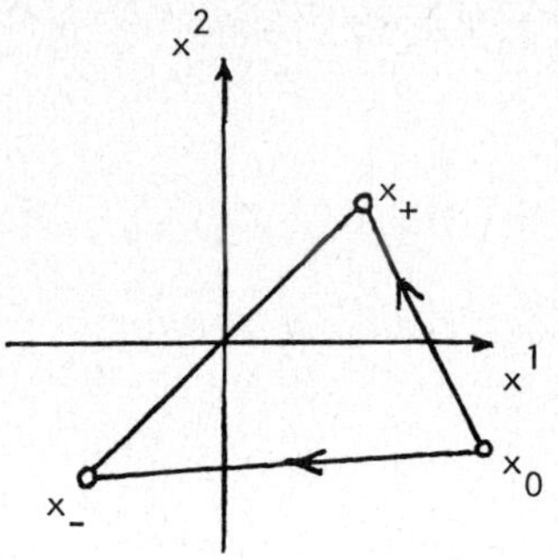

Figure 2

3. Let $x(t) \in R^2$. Consider the system

$$\dot{x}^1 = -x^1 + u \qquad |u(t)| \leqq 1.$$
$$\dot{x}^2 = -2x^2 + 2u$$

The rest-point set is again the line segment $[x_+, x_-]$. $H(x)$ is the interior of the area between the parabolas (see fig.3). The transient set is the exterior of this area plus the two sides (parabolas) but except the vertices (x_+ and x_-).

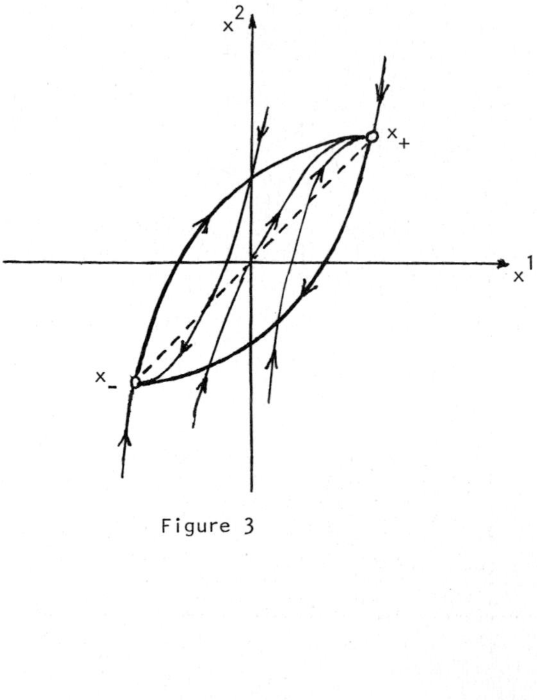

Figure 3

4. This example shows that there can be infinitely many H_i sets. Again $x \in R^2$.

$$\dot{x}^1 = 2 + \cos x^1 + \cos x^2 + u^1 \qquad |u^i| \leqq 1, \; i = 1,2.$$
$$\dot{x}^2 = 2 + \cos x^1 + \cos x^2 + u^2$$

Note that for $u^1 = u^2 = 0$, we have $\dot{x}^1 = \dot{x}^2 \geqq 0$.

Figure 4a shows possible velocity sets $V(x) = f(x,U)$. If $\cos x^1 + \cos x^2 \leqq -1$, then x is a rest-point (for some u, $\dot{x} = 0$). These rest-point sets are indicated in figure 4b (they look like circles). Within these sets, $\dot{x}^1$ and $\dot{x}^2$ can be either positive or negative, hence we can maneuvre the point $x(t)$ forth and back, hence we are within a set H_i. All these sets H_i are different, as one can see that it is impossible to come back from, say, $x^1 = 3\pi$ to $x^1 = \pi$ (because at $x^1 = 2\pi$ the velocity $\dot{x}^1$ is always positive).

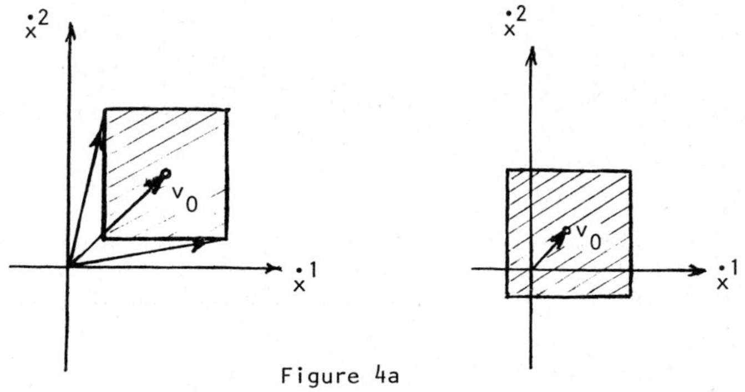

Figure 4a
Velocity sets
v_0 = velocity for $u = 0$

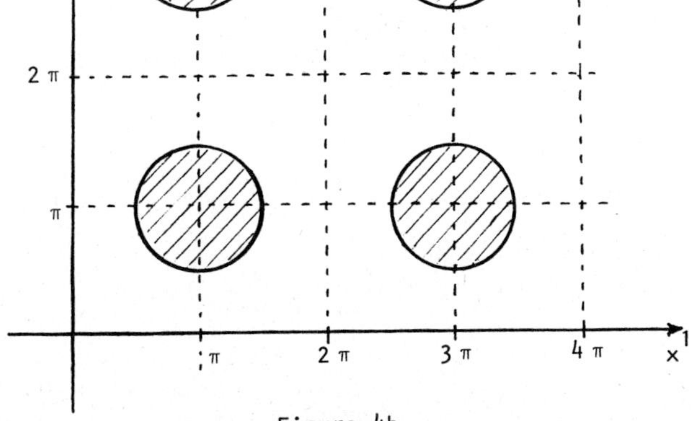

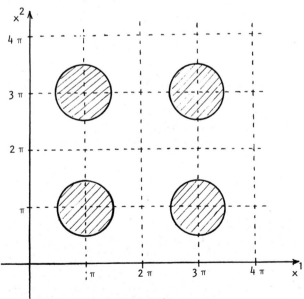

Figure 4b

4. LIMIT SETS

The theory of limit sets is well known for dynamical systems (classical references are [2] and [3]). The ω-limit set of a trajectory $x(t)$ is the set of points z having the following property: given any $\varepsilon > 0$ and any T_0, there is a $T > T_0$ such that $\|x(T) - z\| < \varepsilon$. For a control system as given by equation (1), the situation is more complicated, because given any initial point x_0, there is a whole family of trajectories starting at x_0.

1. Definition. Given a control system (1), and an initial point x_0, let the ω-limit set from x_0 be defined as

$$\Omega(x_0) = \{z \mid \text{there is a trajectory } x(t) \text{ satisfying (1) and } x(0) = x_0, \text{ such that } z \text{ is an } \omega\text{-limit point of } x(t)\}.$$

2. Definition. For a set $A \subset X$, define

$$\Omega(A) = \bigcup_{x_0 \in A} \Omega(x_0).$$

3. Definition. The ω-limit set of the control system S (given by (1)) is

$$\Omega_S = \Omega(X),$$

where X is the whole state space.

In the case X is compact (and this can be achieved by compactification, for example adding the "point at infinity" to R^n), the ω-limit sets are necessarily non-empty. This can be useful to describe the asymptotic behavior of the control system.

4. Relation with the holding sets.

Theorem: $H(x_0) \subset \Omega(x_0)$.

The proof lies in the fact that, if $H(x_0)$ is non-empty, starting from x_0 it is possible to return to x_0 after a finite time $\tau > 0$, and this can then be repeated indefinitely, giving a periodic solution (this happens because the control system is autonomous).

5. Corollary. For every $x \in X$,

$$H(x) \subset \Omega_S.$$

6. Structure of Ω_S. Given the control system by equation (1), the ω-limit set Ω_S contains, according to the above corollary, all the holding sets $H(x)$. It may also contain some of the transient states. For example, if the control system degenerates into a dynamical system with uniqueness (of the solution of the initial value problem)(this means no control, or U is a single point), then a possible ω-limit set is a "path polygon" (see, for example, Lefschetz [2]). All noncritical points of such a path polygon are transient, hence Ω_S includes then such transient states.

5. APPLICATIONS TO THE INFINITE HORIZON OPTIMAL CONTROL PROBLEM

The intention of this section is to indicate very briefly, how the above relations may play an important role within the infinite horizon optimal control problem. No specific results will be given, hence no proofs will be attempted.

As stated in section 1, the optimal control problem with infinite horizon consists in minimizing the cost functional given by (5), with $T \to \infty$. There are several ways to define minimization in the infinite time interval, as was discussed by L. Stern in her thesis [4].

Some basic results concerning existence, uniqueness and related questions of this problem were given by Brock and Haurie in [1]. In that paper, very strong assumptions are made concerning compactness of the state space X, convexity of the functions defining the control system and reachability of specific points. Under such restrictive assumptions, it is then shown that for each initial point x_0, there is an optimal control $u^*(t)$ and a corresponding trajectory $x^*(t)$, and that this optimal trajectory necessarily tends asymptotically to the "optimal rest point" (i.e. the pair $(\bar{x},\bar{u})$ with $f(\bar{x},\bar{u}) = 0$ and $f^0(\bar{x},\bar{u})$ minimal). The optimal trajectory may, indeed, reach the optimal rest point in finite time and stay there forever after. The assumption has to be made that this optimal rest point $\bar{x}$ is reachable from the initial state x_0.

Several problems open up if one wants to relax the assumptions of Brock and Haurie. What happens if the state space is not compact? What if the system is not convex? What if the optimal rest point is not reachable from x_0? Of course, the straight answer to these questions is that in such cases the result is not valid anymore. Counterexamples can be construct to show this. But the question arises: are there weaker conditions such that some of the conclusions are still preserved?

For example, a conjecture will be that if the reachable set from x_0 is bounded and contains the optimal rest point $\bar{x}$, then there is an optimal trajectory $x^*(t)$ starting at x_0, and then necessarily $\lim_{t\to\infty} x^*(t) = \bar{x}$. Another conjecture is that if the optimal rest point $\bar{x}$ is not reachable from x_0, then the similar property will be true with the rest point $\bar{x}_1$ which is optimal (i.e. $f^0(\bar{x}_1,\bar{u}_1)$ minimal) within the reachable set $H^+(x_0)$.

Connecting these problems with the sets defined in section 2, several relations become evident. The optimal rest point $\bar{x}$ (and actually any rest point) must necessarily belong to some holding set H_i. Any periodic solution must also stay within a certain holding set H_i. Any trajectory (possibly optimal) will eventually stay in one holding set, or in the transient set T, or alternate indefinitely between the transient set and (always different) holding sets.

In conclusion, it becomes clear that in the infinite horizon optimal control problem there are still some good questions to investigate, and that the concepts given here can be useful in the related discussions.

REFERENCES

1. Brock, W. A. and Haurie, A., "On Existence of Overtaking Optimal Trajectories over an Infinite Time Horizon", Math. Oper.Res. 1 (1976), 337-346.

2. Lefschetz, S., Differential Equations, Geometric Theory. Wiley-Interscience, New York, 1957.

3. Nemytskii, V. V. and Stepanov, V. V., Qualitative Theory of Differential Equations. Princeton University Press, Princeton, 1960.

4. Stern, L. E., The Infinite Horizon Optimal Control Problem, University of Rhode Island Ph.D. Dissertation, 1980. (Parts of it are going to be published elsewhere.)

This paper is in final form and no version of it will be submitted for publication elsewhere.

Trends in the Theory and Practice of Non-Linear Analysis
V. Lakshmikantham (Editor)

EXCHANGE OF STABILITY AND BIFURCATION FOR PERIODIC DIFFERENTIAL SYSTEMS

L. Salvadori
Dipartimento di Matematica
Universita di Trento
38050 Povo (Trento)
ITALY

INTRODUCTION

Consider a one parameter family of periodic differential equations, $(1)_\mu$ $\dot{u} = f(t,u,\mu)$, $f \in C^1[\mathbb{R} \times \mathbb{R}^n \times \mathbb{R}, \mathbb{R}^n]$. We call $(1)_0$ the unperturbed equation. In case $(1)_\mu$ is autonomous there has been much work concerned with the appearance of compact invariant asymptotically stable sets, M_μ, bifurcating from an invariant compact set M_0 of the unperturbed equation (see for instance [1], [2], [3]). According to the structure of M_0 and M_μ, one may list many types of bifurcation phenomena, for example bifurcation from an equilibrium to equilibrium, or to periodic orbits (Hopf bifurcation), or to periodic orbits and to invariant tori [4], or from periodic orbits to periodic orbits (see for instance [1],[5]). In most significant cases bifurcation occurs as a consequence of a drastic change in the stability properties of M_0 as μ crosses the critical value $\mu = 0$. If, for instance, M_0 is an invariant set of $(1)_\mu$, $\mu \geq 0$, and is asymptotically stable for $\mu = 0$ and completely unstable for $\mu > 0$, then for each small $\mu > 0$ there exists a compact invariant asymptotically stable set, M_μ, of $(1)_\mu$ disjoint from M_0, contained in a fixed neighborhood of M_0 and tending to M_0 as $\mu \to 0$ [6].

The role of an exchange of stability properties in determining bifurcation phenomena was analyzed for periodic systems under particular hypotheses on the Floquet exponents of $D_u f(t,0,\mu)$ (see for instance [7],[8],[9]). The present paper concerns general periodic systems. In contrast with the autonomous case the natural setting in which to analyze the existence and the stability properties of the bifurcating sets is $\mathbb{R} \times \mathbb{R}^n$, rather than $\mathbb{R}^n$. The results of [6] are thus extended to periodic systems. For simplicity we assume $M_0 = \{(t,u),\ t \in \mathbb{R}\ ,\ u = 0\}$. Each bifurcating set M_μ lie in a neighborhood H of M_0 in $\mathbb{R} \times \mathbb{R}^n$ and is an invariant asymptotically stable set of $(1)_\mu$, is disjoint from M_0 and tends to M_0 as $\mu \to 0$. The sections $M_\mu(t) = \{u \in \mathbb{R}\ ,\ (t,u) \in M_\mu\}$ are compact and contained in a fixed neighborhood of the origin of $\mathbb{R}^n$.

The main technique used to determine the existence and the stability properties of the bifurcating sets, M_μ, is to consider a family of autonomous discrete dynamical systems $\pi_\mu^{t_0}(i,u) = u(t_0+iT,t_0,\mu,\mu)$. Because of the autonomous character of $\pi_\mu^{t_0}$, arguments similar to those in [6] carry over. Namely we have for each t_0 the existence of asymptotically stable, compact invariant sets in $\mathbb{R}^n$, disjoint from the origin, under the flow $\pi_\mu^{t_0}$. These sets, $M_\mu(t_0)$, are shown to be the sections of an invariant and asymptotically stable set of $(1)_\mu$ lying in $\mathbb{R}\times\mathbb{R}^n$. In another result which is very related to our previous discussion we consider the case in which there exists a family of periodic (in t) invariant manifolds, S_μ, lying in $\mathbb{R}\times\mathbb{R}^n$. In this instance the bifurcating sets lie on S_μ.

The achievement of these results may be regarded as a first step into the systematic analysis of the existence and stability of periodic orbits or invariant tori considered now as invariant subsets of the bifurcating sets M_μ.

PRELIMINARIES

Let I be the set $\mathbb{R}$ of the real numbers or the set $\mathbb{Z}$ of the integers. Let $\|\cdot\|$ be a norm in $\mathbb{R}^n$ and ρ the induced distance. We denote by U a T-periodic dynamical system, $T\in I$, defined by a continuous map $u\colon I\times I\times\mathbb{R}^n\to\mathbb{R}^n$ satisfying for all t, t_0, $t_1\in I$ and $u_0\in\mathbb{R}^n$ the conditions:

$$u(t,t_1,u(t_1,t_0,u)) = u(t,t_0,u_0)$$
$$u(t+T,t_0+T,u_0) = u(t,t_0,u_0)$$
$$u(t_0,t_0,u_0) = u_0$$

If $I=\mathbb{Z}$, then U is called a periodic discrete dynamical system. We say that U is autonomous or t-independent if it is λ-periodic for any $\lambda\in I$. In this last case we set $\pi(t,u) = u(t,0,u)$ and it is well known that a continuous map $\pi\colon I\times\mathbb{R}^n\to\mathbb{R}^n$ defines an autonomous dynamical system satisfying $u(t,0,u) = \pi(t,u)$ if and only if $\pi(t_2,\pi(t_1,u)) = \pi(t_1+t_2,u)$ and $\pi(0,u) = u$.

Let M be a set in $I\times\mathbb{R}^n$ such that for every $t\in I$ the set $M(t) = \{u\in\mathbb{R}^n\colon (t,u)\in M\}$ is not empty. If there exists a compact set Q in $\mathbb{R}^n$ such that $M(t)\subset Q$ for all $t\in I$, then M is said to be s-bounded. If in addition each $M(t)$ is compact, then we say that M is s-compact. When the mapping $t\to M(t)$ is λ-periodic for some $\lambda\in I$ or in particular t-independent, we will say that M is λ-periodic or t-independent respectively. Throughout the paper we will adopt some stability concepts concerning M, where M is s-compact and periodic with the same period of U. These concepts are essentially those discussed in [10] for dynamical systems generated by general non-autonomous differential equations in $\mathbb{R}^n$. Because of periodicity of U, M, and the s-compactness of M, we need not distinguish now between stability and uniform stability, and between asymptotic stability and uniform asymptotic stability.

Definition 2.1: Let M be T-periodic and s-compact. Then M is said to be:

(a) an invariant (resp. positively invariant, negatively invariant) set of the dynamical system U if $(t_0,u_0)\in M$ inplies $u(t,t_0,u_0)\in M$ for all $t\in I$ (resp. $t\geq t_0$, $t\leq t_0$);

(b) a stable set of U if for any $\varepsilon>0$ there exists a $\delta(\varepsilon)>0$ such that $(u_0,M(0))<\delta(\varepsilon)$ implies $\rho(u(t,0,u_0),M(t))<\varepsilon$ for all $t\geq 0$;

(c) an attracting set of U, or simply an attractor, if there exists a $\sigma > 0$ such that $\rho(u_0, M(0)) < \sigma$ implies $\rho(u(t,0,u_0), M(t)) \to 0$ as $t \to +\infty$;

(d) an uniformly attracting set of U, or a uniform attractor, if there exists a $\sigma > 0$ with the property that for each $\nu > 0$ there exists $\tau(\nu) > 0$ such that $\rho(u_0, M(0)) < \sigma$ implies $\rho(u(t,0,u_0), M(t)) < \nu$ for all $t \geq \tau(\nu)$;

(e) an asymptotically set of U if it is stable and attracting.

By using continuity arguments it is easy to see that M is asymptotically stable if and only if it is positively invariant and a uniform attractor. When M is t-independent with $M(t) \equiv N$, it is customary to replace in Definition 2.1 M by N and thus look at (a) - (e) as properties of a set in $\mathbb{R}^n$.

Consider now an autonomous dynamical system π and let M be a compact subset of $\mathbb{R}^n$. We shall say that $u \in \mathbb{R}^n$ is uniformly attracted by M under π if for any $\nu > 0$ there exist $a > 0$, $\tau > 0$ such that $\rho(u,v) < a$ implies $\rho(\pi(t,v), M) < \nu$ for all $t \geq \tau$. The set of all points which are uniformly attracted by M, say A(M), is called the region of uniform attraction of M. Clearly A(M) is an invariant set of π and M is a uniform attractor if and only if A(M) is a neighborhood of M. In this case A(M) is open. The uniform attractivity property of M may be expressed by the mapping $u \to J^+(u)$, where

$$J^+(u) = \{v \in \mathbb{R}^n: \text{ there exist a sequence } (t_n),\ t_n \to +\infty, \text{ and a sequence } (u_n),\ u_n \to u, \text{ such that } \pi(t_n,u_n) \to v\}.$$

The set $J^+(u)$ is called the positive prolongational limit set of u. Similarly we may define $J^-(u)$, the negative prolongational limit set of u. The sets $J^+(u)$, $J^-(u)$ are closed and invariant. A point in $\mathbb{R}^n$ is uniformly attracted by M if and only if $J^+(u) \neq \phi$ and $J^+(u) \subset M$. These properties are well known and when $I = \mathbb{R}$ the proofs may be found for instance in [11]. Their extension to the case of a discrete autonomous dynamical system may be easily obtained by some slight modifications in the proofs.

Corresponding stability concepts for the dynamical system as $t \to -\infty$ will be referred to as stability in the past. In particular the asymptotic stability in the past will be called complete instability.

RESULTS

Consider a one-parameter family of differential equations in $\mathbb{R}^n$

$$(3.1)_\mu \qquad \dot{u} = f(t,u,\mu),$$

where $f \in C^1[\mathbb{R} \times \mathbb{R}^n \times \mathbb{R}, \mathbb{R}^n]$ is T-periodic in t, $T > 0$. We assume that for each $\mu \in \mathbb{R}$, $t_0 \in \mathbb{R}$ and $u_0 \in \mathbb{R}^n$, the solution of $(3.1)_\mu$ through (t_0,u_0), denoted by $u(t,t_0,u_0,\mu)$, exists for all $t \in \mathbb{R}$. Moreover we suppose that $f(t,0,\mu) \equiv 0$ so that $u(t) \equiv 0$ is a solution for every μ. We call $(3.1)_0$ the unperturbed equation. We will denote by $B^n(a)$, $a > 0$, the open ball $\{u \in \mathbb{R}^n: \|u\| < a\}$ and by M_0 the set in $\mathbb{R} \times \mathbb{R}^n$ $\{(t,u):\ t \in \mathbb{R},\ u = 0\}$.

Definition 3.1: Let $\bar{\mu} > 0$. We say that $\mu = 0$ is a bifurcation value on the right for the family $(3.1)_\mu$ if there exist a $\bar{\mu} > 0$ and a family (M_μ) $\mu \in (0,\bar{\mu})$ of s-compact subsets of $(\mathbb{R} \times \mathbb{R}^n)/M_0$ having the following properties:

(a) for each $\mu \in (0,\bar{\mu})$ M_μ is a periodic invariant set of $(3.1)_\mu$ and $M_\mu(t) \neq \phi$ for all $t \in \mathbb{R}$;

(b) $M_\mu(t) \to \{0\}$ when $\mu \to 0$ uniformly in t.

Theorem 3.1. Suppose that $u(t) \equiv 0$ is an asymptotically stable solution of $(3.1)_0$ and a completely unstable solution of $(3.1)_\mu$ for $\mu > 0$ small. Then $\mu = 0$ is a bifurcation value on the right for (3.1). Precisely there exist a $\bar{\mu} > 0$ and an s-compact neighborhood H of M_0 such that for each $\mu \in (0,\bar{\mu})$ the largest s-compact invariant set of $(3.1)_\mu$ contained in H/M_0, say M_μ, is nonempty, T-periodic and the family (M_μ) satisfies (b) in Definition 3.1. Moreover each M_μ is an asymptotically stable set of $(3.1)_\mu$.

Proof. The asymptotic stability of the origin of $(3.1)_0$ implies the existence of a number $\gamma > 0$ and a function $V \in C^1[\mathbb{R} \times \mathbb{R}^n, \mathbb{R}]$, T-periodic in t such that

$$a(\|u\|) \leq V(t,u) \leq b(\|u\|) \tag{3.2}$$

$$\dot{V}_{(3.1)_0}(t,u) \leq -c(\|u\|), \tag{3.3}$$

for all $t \in \mathbb{R}$ and $u \in B^n(\gamma)$. Here a,b,c are continuous strictly increasing functions from $\mathbb{R}_+$ into $\mathbb{R}_+$ with $a(0) = b(0) = c(0) = 0$, and the left hand side of (3.3) is the derivative of V along the solutions of the unperturbed equation. By a known procedure [10] we determine now a $\bar{\mu} > 0$ and a s-compact neighborhood H of M_0 such that H is an asymptotically stable set of $(3.1)_\mu$ for all $\mu \in (0,\bar{\mu})$ and invariant only in the future. Precisely we choose a number $\lambda \in (0,a(\gamma))$ and consider the subset of $\mathbb{R} \times \mathbb{R}^n$

$$H = \{(t,u): \|u\| \leq \gamma,\ V(t,u) \leq \lambda\}.$$

Clearly $(t,u) \in H$ implies $\|u\| < \gamma$. Moreover we see that H contains the points for which $t \in \mathbb{R}$ and $\|u\| \leq b^{-1}(\lambda)$. Thus each section $H(t)$ is a compact neighborhood of $u = 0$ and is contained in the open ball $B^n(\gamma)$. By (3.3) and continuity arguments we may select $\bar{\mu} > 0$ so that

$$\dot{V}_{(3.1)_\mu}(t,u) \leq -c(b^{-1}(\lambda))/2$$

for all $\mu \in (0,\bar{\mu})$, $t \in \mathbb{R}$, and $u \in B^n(\gamma)/B^n(b^{-1}(\lambda))$. We see that H has all the properties we have required above and in addition the region of attraction of H under $(3.1)_\mu$ contains a fixed neighborhood $\bar{H}$ of H.

For any fixed $t_0 \in \mathbb{R}$ and $\mu \in (0,\bar{\mu})$ consider the map $\pi = \pi_\mu^{t_0}: \mathbb{Z} \times \mathbb{R}^n \to \mathbb{R}^n$ defined by $\pi(i,u) = u(t_0+iT,t_0,u,\mu)$. Clearly we have $\pi(i_1+i_2,u) = \pi(i_1,\pi(i_2,u))$ for every $i_1,i_2 \in \mathbb{Z}$ and $u \in \mathbb{R}^n$. Hence π defines an autonomous discrete dynamical system depending on t_0, μ. For $u \in \mathbb{R}^n$ we denote by $J^+(u)$, $J^-(u)$ the positive and negative prolongational limit set of u under π. The set $H(t_0)$ is a uniform attractor under π and the region of uniform attraction contains $\bar{H}(t_0)$. Therefore $u \in \bar{H}(t_0)$ implies $J^+(u) \neq \phi$ and $J^+(u) \subset H(t_0)$. Let $F_\mu(t_0)$ be the largest invariant set of π contained in $H(t_0)$. We show that

$F_\mu(t_0) \supset A^-(0)$, where $A^-(0)$ is the region of negative uniform attraction of the origin of $\mathbb{R}^n$ and under our assumptions is a neighborhood of $u = 0$. Since $A^-(0)$ is invariant under π, it is sufficient to prove that $A^-(0) \subset H(t_0)$. For, $u \in A^-(0)$ implies the existence of $i \in \mathbb{Z}^-$ such that $\pi(i,u) \in H(t_0)$ and then $\pi(-i,\pi(i,u)) \in H(t_0)$ by virtue of the positive invariance of $H(t_0)$. We set now $M_\mu(t_0) = F_\mu(t_0)/A^-(0)$ and prove that $M_\mu(t_0)$ is a uniform attractor under π and the region of uniform attractivity contains $\bar{H}(t_0)/\{0\}$. For assume $u \in \bar{H}(t_0)/\{0\}$. We have seen that $J^+(u) \neq \phi$ and $J^+(u) \subset H(t_0)$. Since $J^+(u)$ is invariant under π, we have $J^+(u) \subset F_\mu(t_0)$. Hence it remains to prove that $v \in J^+(u)$ implies $v \notin A^-(0)$. Indeed we have $u \in J^-(v)$. Therefore if $v \in A^-(0)$, then $J^-(v) = \{0\}$ and consequently $u = 0$. This is a contradiction. Let $M_\mu = \{(t,u):\ t \in \mathbb{R},\ u \in M(t)\}$. Since $H(t_0) = H(t_0+T)$ and $\pi_\mu^{t_0} = \pi_\mu^{t_0+T}$, we have $M_\mu(t_0) = M_\mu(t_0+T)$. Then M_μ is a periodic set in $\mathbb{R} \times \mathbb{R}^n$. It is immediate to recognize that M_μ is s-compact and that $M_\mu \to \{0\}$ as $\mu \to 0$, uniformly in t. We now prove that M_μ is an invariant set of $(3.1)_\mu$. Clearly it is sufficient to show that $M_\mu(t_0)$ is the image of $M_\mu(0)$ under the flow generated by $(3.1)_\mu$. Let $G = u(t_0,0,M_\mu(0),\mu)$. For any $i \in \mathbb{Z}$ we have $u(t_0+iT,t_0,G,\mu) = u(t_0+iT,t_0,u(t_0,0,M_\mu(0),\mu),\mu) = u(t_0+iT,0,M_\mu(0),\mu) \subset G$. Indeed $u(iT,0,M_\mu(0),\mu) \subset M_\mu(0)$, since $M_\mu(0)$ is invariant under π_μ^0. Thus G is a compact invariant set of $\pi_\mu^{t_0}$ not containing the origin. Therefore $G \subset M_\mu(t_0)$. By using the same argument we find that $M_\mu(0) \supset W$ where $W = u(0,t_0,M_\mu(t_0),\mu)$. Since $u(t_0,0,W,\mu) \subset G$ and $u(t_0,0,W,\mu) = M_\mu(t_0)$, we have $M_\mu(t_0) \subset G$. Hence $G = M_\mu(t_0)$, and this completes the proof of the invariance of M_μ. It is also clear now that M_μ is the largest s-compact invariant set of $(3.1)_\mu$ contained in H/M_0.

Finally we prove that M_μ is an asymptotically stable set of $(3.1)_\mu$ and for this it is sufficient to prove that M_μ is an uniform attractor under $(3.1)_\mu$. Let σ denote any positive number such that the set $\{u:\ \rho(u,M_\mu(0)) < \sigma\}$ is contained in $\bar{H}(t_0)/\{0\}$. Since $M_\mu(0)$ is a uniform attractor under π_μ^0 we have that for any $\eta > 0$ there exists an integer $r(\eta) > 0$ such that $\rho(u_0,M_\mu(0)) < \sigma$ implies $\rho(\pi_\mu^0(i,u_0),M_\mu(0)) < \eta$ for $i \in \mathbb{Z}$ and $i \geq r(\eta)$. On the other hand given any $\nu > 0$ we can find a number $\delta(\nu) > 0$ such that $\rho(u_0,M_\mu(t_0)) < \delta(\nu)$ implies $\rho(u(t,t_0,u_0,\mu),M_\mu(t)) < \nu$ for all $t_0 \in \mathbb{Z}$ and $t \in [t_0,t_0+T]$. Indeed since $M_\mu(t_0)$ is compact, there exists $v \in M_\mu(t_0)$ for which $\rho(u,M_\mu(t_0)) = \rho(u,v)$. Since M_μ is invariant under $(3.1)_\mu$ we have $u(t,t_0,v,\mu) \in M_\mu(t)$ for all $t \in \mathbb{R}$ and then we have $\rho(u(t,t_0,u,\mu),M_\mu(t)) \leq \rho(u(t,t_0,u,\mu),u(t,t_0,v,\mu))$. The existence of the number $\delta(\nu)$ then follows by continuity arguments and the T-periodicity of $(3.1)_\mu$. In conclusion setting $h(\nu) = r(\delta(\nu))$ we see that $\rho(u_0,M_\mu(0)) < \sigma$ implies $\rho(u(t,0,u_0,\mu),M_\mu(t)) < \nu$ for all real $t \geq h(\nu)$. The proof of Theorem (3.1) is now complete.

We assume now that $(3.1)_\mu$ has the form

$$(3.4)_\mu \qquad \dot{u} = C(\mu)u + w(t,u,\mu),$$

with $w\colon \mathbb{R} \times \mathbb{R}^n \times \mathbb{R} \to \mathbb{R}$ of class C^1, $w(t,0,\mu) \equiv 0$, $(\partial w/\partial u)(t,0,\mu) \equiv 0$,

$$C(\mu) = \begin{bmatrix} A(\mu) & 0 \\ 0 & B(\mu) \end{bmatrix},$$

where $A(\mu)$ and $B(\mu)$ are $m \times m$ and $(n-m) \times (n-m)$ matrices respectively. Moreover we suppose that: (1) the eigenvalues of $A(\mu)$ have zero real part for $\mu = 0$ and positive real part for $\mu > 0$; (2) the eigenvalues of $B(\mu)$ have negative real part. As we are concerned with local problems in which only neighborhoods of $u = 0$ and $\mu = 0$ are involved, given any $\sigma > 0$ we may assume without loss of generality that the norm of w in the C^1 topology satisfies $|w| < \sigma$. If $u = (p,q)$, $w = (P,Q)$, $p, P \in \mathbb{R}^m$, $q, Q \in \mathbb{R}^{n-m}$, then (3.4) is the same as

$$(3.5)_\mu \qquad \begin{aligned} \dot{p} &= A(\mu)p + P(t,p,q,\mu) \\ \dot{q} &= B(\mu)q + Q(t,p,q,\mu). \end{aligned}$$

If $\mu^* > 0$ and σ are sufficiently small, then there exists a function $g\colon \mathbb{R} \times \mathbb{R}^m \times [0,\mu^*] \to \mathbb{R}^{n-m}$, $g \in C^1$, $g(t,0,\mu) \equiv 0$, which is T-periodic in t such that for every $\mu \in [0,\mu^*]$ the set

$$S_\mu = \{(p,q,t)\colon\ q = g(t,p,\mu)\}$$

is an invariant manifold of $(3.5)_\mu$, and

$$(3.6) \qquad \|q(t_0,p_0,q_0,\mu) - g(t,p(t,t_0,p_0,q_0,\mu),\mu)\| \le Le^{-\beta(t-t_0)}, \quad t \ge t_0,$$

where $\beta > 0$, $L > 0$ are constants with L depending continuously on p_0,q_0 ([12]; see also [3]). We have the following result.

<u>Theorem 3.2</u>: Suppose that $p(t) \equiv 0$, $q(t) \equiv 0$ is an asymptotically stable solution of $(3.5)_0$. Then the conclusion of Theorem 3.1 holds and each bifurcating set M_μ lies on the invariant manifold S_μ.

<u>Proof</u>. Since the proof is very similar to that of Theorem 3.1, we only indicate the necessary changes. Define again H, $\bar{\mu}$ and $\pi = \pi_\mu^{t_0}$ and denote by $\tilde{\pi}$ the restriction of π to S_μ. The origin of $\mathbb{R}^n$ is asymptotically stable under $\tilde{\pi}_0^{t_0}$ and is completely unstable under $\tilde{\pi}_\mu^{t_0}$ if $\mu \in (0,\bar{\mu})$, $\bar{\mu}$ small. Therefore we may define $M_\mu(t_0)$ to be the largest invariant compact set of $\tilde{\pi}_\mu^{t_0}$ contained in $S_\mu(t_0) \cap H(t_0)/M_0$. Moreover $M_\mu(t_0)$ is a uniform attractor under $\tilde{\pi}$. We prove that $M_\mu(t_0)$ is also a uniform attractor under π. For if $u = (p,q)$ is close to $M_\mu(t_0)$, then the properties of the matrix A imply that $0 \notin J^+(u)$, where $J^+(u)$ is again the positive prolongational limit set of u under π. Since $J^+(u)$ is a compact invariant set of π and (because of (3.6)) it lies on

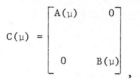

$S_\mu(t_0) \cap H(t_0)$, we have then $J^+(u) \subset M_\mu(t_0)$. This shows that u is uniformly attracted by $M_\mu(t_0)$ under π. Defining again $M_\mu = \{(t,p,q)\colon\ t \in \mathbb{R},\ (p,q) \in M_\mu(t)\}$, we may proceed as before to prove that the family (M_μ) has all the required properties.

REFERENCES

[1] Andronov, A.A., Leontovich, E.A., Gordon, I.I., and Maier, A.G., Theory of Bifurcation of Dynamical Systems in the Plane, Israel Program of Scientific Translations, Jerusalem, 1973.

[2] Marsden, J. and McCraken, M.F., The Hopf Bifurcation and its Applications, Springer-Verlag, New York, 1976.

[3] Chow, S.N. and Hale, J.K., Methods of Bifurcation Theory, Springer-Verlag, New York, 1982.

[4] Samoilenko, A.M. and Polesya, I.V., Genesis of Invariant Sets in the Neighborhood of an Equilibrium Position, Differentsial'nye Uravneniya, 11, 8, 1409-1415 (1975).

[5] Moauro, V., Bifurcation of Closed Orbits from a Limit Cycle in $\mathbb{R}^2$, Rend. Sem. Mat. Univ. Padova 65, 277-291 (1981).

[6] Marchetti, F., Negrini, P., Salvadori, L. and Scalia, M., Liapunov Direct Method in Approaching Bifurcation Problems, Ann. Mat. Pure & Appl. 108, 211-225 (1976).

[7] De Oliveira, J.C. and Hale, J.K., Dynamic Behavior from Bifurcation Equations, Tohoku Math. J., 32, 189-199 (1980).

[8] Salvadori, L. and Visentin, F., Sul Problema della Biforcazione Generalizzata di Hopf per Sistemi Periodici, Rend. Sem. Mat., Univ. Padova, 68, 129-147 (1982).

[9] Bernfeld, S.R., Salvadori, L. and Visentin, F., Hopf Bifurcation and Related Stability Problems for Periodic Differential Systems, to appear.

[10] Yoshizawa, T., Stability Theory by Liapunov's Second Method, The Math. Soc. of Japan, Tokyo, 1966.

[11] Bhatia, N.P. and Szego, G.P., Dynamical System: Stability, Theory and Applications, Springer-Verlag, New York, 1967.

[12] Kelley, A., "The Stable, Center-Stable, Center-Unstable and Unstable Manifolds," J. Diff. Equations, 3, 546-570 (1967).

The final version of this paper will be submitted for publication elsewhere.

Trends in the Theory and Practice of Non-Linear Analysis
V. Lakshmikantham (Editor)

ON INVARIANT TRANSFORMATIONS OF A CLASS OF EVOLUTION EQUATIONS

Károly Seitz

Department of Mathematics
Technical University of Budapest
Hungary

In this work the transformation properties of a class of second order nonlinear evolution equations in a given Hilbert space are investigated by using group theoretical tools.

INTRODUCTION

Consider the following evolution equation

$$\ddot{u}(t) + F\,u = 0, \qquad (-\infty < t < \infty), \tag{1}$$

in a given real Hilbert space $\mathcal{H}$, where F is a nonlinear operator of $\mathcal{H}$. Denote by $\mathcal{T}(\mathcal{H})$ the set of all invertible linear transformation of $\mathcal{H}$ which map $\mathcal{H}$ onto itself. In this work we shall investigate on a very simple way the structure of the $G(\mathcal{H},F)$ set of all elements of $\mathcal{T}(\mathcal{H})$ which map (1) to itself. The reading of this paper will be easier in knowing the works [1],[2],[3].

MAIN RESULTS

If $T \in G(\mathcal{H},F)$, then

$$(T\,u(t))^{\cdot\cdot} + F\,T\,u(t) = 0 \tag{2}$$

has the form

$$\ddot{u}(t) + F\,u(t) = 0 \tag{3}$$

Since T does not depend from t, we have that

$$\ddot{u}(t) + (T^{-1}\,F\,T)\,u(t) = 0. \tag{4}$$

From (3) and (4)

$$T^{-1}\,F\,T = F \tag{5}$$

follows. If

$$S^{-1}\,F\,S = F, \qquad (S \in G(\mathcal{H},F)) \tag{6}$$

then

(7) $$(T\,S)^{-1}\,F\,(TS) = F\ ,$$

and

(8) $$(S^{-1})^{-1}\,F\,S^{-1} = F.$$

Therefore we have proved the following theorem.

Theorem 1. $G(\mathcal{H}, F)$ is a group and $T \in G(\mathcal{H}, F)$ iff (5) holds.

Next we consider the equation,

(9) $$\ddot{u}(t) + (u,\ Au)\ u = 0,$$

where A is a linear operator of $\mathcal{H}$ which maps $\mathcal{H}$ into itself.

In this case

(10) $$F\,u = (u,\ Au\,)\,u\ ,$$

and

(11) $$(T^{-1}\,F\,T\,)\,u = F\,u$$

has the form

(12) $$T^{-1}\,(Tu,\ A\,Tu)Tu = (u,\ Au)\ u$$

from which

(13) $$(Tu,\ A\,Tu) = (u,\ Au)$$

follows.

If T is an isometry of $\mathcal{H}$ and $AT = TA$, then (13) holds.

Denote by $J(A)$ the set of all T isometries of $\mathcal{H}$ for which $AT = TA$. If $T, S \in J(A)$ then $TS, T^{-1} \in J(A)$.

Therefore we have proved the following theorem:

Theorem 2. If $Fu = (u, Au)u$, $u \in \mathcal{H}$, where A is a linear operator of $\mathcal{H}$ which maps $\mathcal{H}$ onto itself, then $J(A)$ is a subgroup of $G(\mathcal{H}, F)$.

Now consider the equation

(14) $$\ddot{u}(t) + \frac{1}{(u\ ,\ u)^{\frac{3}{2}}}\ A\,u(t) = 0,$$

where A is a linear operator of $\mathcal{H}$ which maps $\mathcal{H}$ onto itself.

In this case

(15) $$F u = \frac{1}{(u,u)^{\frac{3}{2}}} Au,$$

and

(16) $$(T^{-1} F T) u = F u,$$

has the form

(17) $$T^{-1} \frac{1}{(Tu,Tu)^{\frac{3}{2}}} A T u = \frac{1}{(u,u)^{\frac{3}{2}}} Au,$$

from which

(18) $$\frac{1}{(Tu, Tu)^{\frac{3}{2}}} A T u = \frac{1}{(u,u)^{\frac{3}{2}}} T A u,$$

follows.

If $T \in J(A)$ then (18) holds.

Therefore we have proved the following theorem.

Theorem 3. If $F u = \frac{1}{(u,u)^{\frac{3}{2}}} A u$, $u \in \mathcal{H}$, where A is a linear operator of $\mathcal{H}$ which maps $\mathcal{H}$ onto itself then $J(A)$ is a subgroup of $G(\mathcal{H},F)$.

REFERENCES

[1] Riesz,F. and Sz.Nagy,B., Lecons d'Analyse Fonctionelle (Akadémiai Kiadó, Budapest, 1953).

[2] Halmos, P.R., A Hilbert space problem book (D.Van Nostrand Company, Inc., Princeton-Toronto-London, 1967).

[3] Yosida, K., Functional analysis (Springer Verlag,Berlin,1965).

This paper is in final form and no version of it will be submitted for publication elsewhere.

Trends in the Theory and Practice of Non-Linear Analysis
V. Lakshmikantham (Editor)

CAUCHY PROBLEM FOR HYPER-PARABOLIC PARTIAL DIFFERENTIAL EQUATIONS

R.E. Showalter

Department of Mathematics, RLM 8.100
University of Texas at Austin
Austin, Texas
U.S.A.

INTRODUCTION

We shall consider initial-boundary-value problems for partial differential equations of the form

(1) $$u_t = u_{xx} - u_{yy}.$$

Such equations arise in a discussion of classification. Specifically, every linear second-order constant-coefficient partial differential equation in n variables can be reduced to the form

$$\sum_{i=1}^{k} u_{x_i x_i} - \sum_{i=1}^{j} u_{y_i y_i} + au_t + bu = f(x,y,t,\ldots).$$

The general parabolic case is $k+j+1 = n$; we are interested in the non-normal case of $kj \neq 0$ which we call <u>hyper-parabolic</u>.

Equations of this type have arisen in diverse non-standard applications [1,4,7-11, 15], most of which require a solution subject to classical initial and boundary conditions on a space-time cylinder. However such a problem is not well posed but is "hyper-sensitive" to variations in the data [12]. It is clear that neither the initial-value nor the final-value problem is well posed for (1) and, moreover, the boundary-value problem for stationary solutions is ill posed.

Our plan is to develop some elementary notions of generalized solution of an abstract model of (1) as an evolution equation in Hilbert space. The Cauchy problem can be approximated by a <u>quasi-reversibility method</u> [5]. Then we present some well posed problems for this equation, and these results suggest a more natural method of approximating solutions of the ill posed Cauchy problem. This new approximation scheme we call the <u>quasi-boundary-value method</u>.

INITIAL-VALUE PROBLEM

Hereafter A and B denote self-adjoint non-negative operators on a Hilbert space H, and we assume their resolvents commute. Thus $-A$ generates a (holomorphic) semigroup of contractions $\{\exp(-At) : t \geq 0\}$ on H; their inverses are unbounded operators $\exp(At)$ which could also be obtained by the spectral theorem. If $u \in C^1$ is a solution of the evolution equation

$$u'(t) + Au(t) = 0, \quad \tau < t$$

then $\frac{d}{ds} \exp(-A(t-s))u(s) = 0$, hence, $\xi(t) \equiv \exp(-A(t-s))u(s)$, $\tau \leq s \leq t$, is independent of s. Thus one obtains the semi-group representation $u(t) = \exp(-A(t-s))u(s)$, $s \leq t$; the operators $\exp(-A(t-s))$ are the <u>propagators</u> for the evolution equation and their continuity implies the initial-value problem is well posed.

Proceeding similarly for the hyper-parabolic equation

$$u'(t)+Au(t)-Bu(t)=0, \tag{2}$$

we see that if $u \in C^1$ is a solution on the interval $[\tau,t]$, then $\frac{d}{ds}\exp(-A(t-s))\exp(-B(s-\tau))u(s)=0$, $\tau<s<t$, so

$$\xi(t,\tau)=\exp(-A(t-s))\exp(-B(s-\tau))u(s), \quad \tau\le s\le t \tag{3}$$

is independent of s. Thus, we are led to define a weak solution of (2) on $[\tau,t]$ as a continuous H-valued function for which (3) holds, i.e., the right side of (3) is independent of $s\in[\tau,t]$.

Lemma. If u is a weak solution on $[\tau,t]$ then it is a weak solution on each $[\tau_1,t_1]\subset[\tau,t]$ and then $\xi(t,\tau)=\exp(-A(t-t_1))\exp(-B(\tau_1-\tau))\xi(t_1,\tau_1)$, and $u(t)=\xi(t,t^-)$.

If u is a continuous H-valued function on $[0,1]$, then u is a weak solution iff $\exp(-Bt)u(t)=\exp(-At)u(0)$, $0\le t\le 1$. Thus the initial-value problem of finding a weak solution of (2) on $[0,1]$ with $u(0)=f$ given in H is equivalent to

$$u\in C^0([0,1],H) \text{ with } \exp(-Bt)u(t)=\exp(-At)f, \quad 0\le t\le 1.$$

Since each $\exp(-Bt)$ is one-to-one, there is at most one solution of the initial-value problem. Also, the representation via unbounded operators as $u(t) = \exp(Bt)\exp(-At)f$ shows the initial-value problem is not well posed. Considering existence, we see that if $f\in Rg\{\exp(-B)\}=dom\{\exp(B)\}$, then

$$u(t)\equiv\exp(-At)\cdot\exp(-B(1-t))\cdot\exp(B)f$$

defines a strong solution (C^∞) of the initial-value problem. More generally we have the following

Proposition 1. There exists a weak solution of the initial-value problem if and only if $\exp(-A)f=\exp(-B)g$ for some $g\in H$, i.e., $f\in dom\{\exp(B-A)\}$.

QUASI-REVERSIBILITY METHOD

Since the lack of well-posedness of the initial-value problem for (2) is due to the unboundedness of B, we use a $Q-R$ method [3,5,13,14] to obtain an approximate solution. First replace B by its bounded Yosida approximation $B_\varepsilon\equiv B(I+\varepsilon B)^{-1}$, $\varepsilon>0$, and solve the equation (2) for $\exp((B_\varepsilon-A)t)f$, $0\le t\le 1$. The final-value $\exp(B_\varepsilon-A)f$ belongs to $Rg\{\exp(-A)\}$ so we obtain a (strong) solution of (2) backward from here,

$$\begin{aligned}u_\varepsilon(t)&=\exp((A-B)(1-t))\exp(B_\varepsilon-A)f\\&=\exp(-At)\exp(-B(1-t))\exp(B_\varepsilon)f, \quad 0\le t\le 1,\end{aligned}$$

and it satisfies $u_\varepsilon(0)=\exp(B_\varepsilon-B)f$. Using results from [14] we obtain the following

Theorem 1. For any $f\in H$, $\lim_{\varepsilon\to 0} u_\varepsilon(0)=f$ and $\|u_\varepsilon(0)\|\le\|f\|$. There exists a (weak) solution u of the initial-value problem for (2) on $[0,1]$ if and only if

$\lim_{\varepsilon\to 0}\{u_\varepsilon(t)\}$ exists in H for all $t \in [0,1]$, and then $\lim_{\varepsilon\to 0} u_\varepsilon(t) = u(t)$.

This QR-method is theoretically incisive: there exists a solution if and only if it converges. It is slightly more subtle than the usual case $(A = 0)$ in [14] since the method must give a $u_\varepsilon(1) \in Rg\{\exp(-A)\}$ in order that the backward problem have a strong solution. The method is always stable and convergent at $t = 0$, but only at $t = 0$. For example, $\|u_\varepsilon(1)\|$ may grow like $\mathcal{O}(\exp(1/\varepsilon))$, so as a numerical method it is essentially worthless. Even if one uses log-convexity estimates to stabilize the method [2,6], the use of initial-value or final-value problems in the procedure is not natural for (2).

BOUNDARY-VALUE PROBLEM

Suppose there is a weak solution of (2) on $[0,1]$. From the Lemma above it follows that $\xi(0,1)$ and hence the solution u will depend on both $u(0)$ and $u(1)$. One need only determine the domain of influence of u on ξ through the formulae of the Lemma. This suggests that a boundary-value problem on the interval $[0,1]$ is more appropriate than an initial-value problem.

We can substantiate this observation as follows. First let C be a self-adjoint operator whose spectrum is unbounded in both positive and negative real numbers, thus $C = A - B$ as above where A and B are the positive and negative parts of C, respectively. We seek a representation of a solution of

$$u'(t) + Cu(t) = 0, \quad 0 < t < 1, \tag{4}$$

in the form $u(t) = \int_\Gamma \exp(Cz)U(t,z)dz$. In order to choose the contour Γ in $\mathbb{C}$ so $\{\exp(Cz) : z \in \Gamma\}$ is bounded, we take $z = i\tau$, $\tau \in \mathbb{R}$, so we have

$$u(t) = \int_{\mathbb{R}} \exp(i\tau C)U(t,\tau)d\tau . \tag{5}$$

Substitution of (5) into (4) yields

$$\int_{\mathbb{R}} \exp(i\tau C)(U_t + iU_\tau)d\tau + (1/i)\exp(i\tau C)U(t,\tau)\Big|_{\tau=-\infty}^{\tau=\infty},$$

so we need require the kernel $U(t,\tau)$ to satisfy the Cauchy-Riemann equation in the slab $0 < t < 1$ and to vanish at $\tau \to \pm\infty$. Thus U will be determined by its remaining boundary-values, $U(0,\tau)$, $U(1,\tau)$, $-\infty < \tau < +\infty$. These in turn are determined by $u(0)$ and $u(1)$ through (5). These formal calculations can (and will) be made precise elsewhere but they already suggest that the equation (1) is elliptic and that the following problem is well posed. The boundary-value problem is to find a weak solution u of (2) on $[0,1]$ for which $au(0) + bu(1) = f$. Here $f \in H$ and $a,b \in \mathbb{R}$ are given.

Proposition 2. If u is a solution of the boundary-value problem then

$$(a\exp(-B) + b\exp(-A))u(t) = \exp(-At)\exp(-B(1-t))f, \quad 0 \le t \le 1.$$

If also $a,b \ge 0$ and not both are zero, then there is at most one solution. If both a and b are strictly positive then there exists a solution u for each $f \in H$ and it satisfies

$$u(t) \le \|f\|/a^{1-t}b^t, \quad 0 \le t \le 1.$$

QUASI-BOUNDARY-VALUE METHOD

Consider again the initial-value problem for (2) on the interval $[0,1]$ with

$u(0) = f$. The quasi-reversibility method of approximation was to regularize the problem by perturbing the equation, i.e., replace B by B_ε. The method suggested by Proposition 2 is to regularize the problem by perturbing the initial condition, i.e., replace it by the boundary condition

$$u(0) + \varepsilon u(1) = f . \tag{6}$$

Thus for each $\varepsilon > 0$ we let u_ε be the solution of the boundary-value problem (2), (6).

<u>Theorem 2</u>. For any $f \in H$, $\lim_{\varepsilon \to 0} u_\varepsilon(0) = f$. There exists a solution u of the initial-value problem for (2) on $[0,1]$ if and only if $\lim_{\varepsilon \to 0} \{u_\varepsilon(t)\}$ exists in H for all $t \in [0,1]$, and then $\lim_{\varepsilon \to 0} u_\varepsilon(t) = u(t)$. The solutions u_ε of (2), (6) satisfy the estimates

$$\|u_\varepsilon(t)\| \leq \|f\|/\varepsilon^t , \quad 0 \leq t \leq 1 , \quad \varepsilon > 0 . \tag{7}$$

The regularization procedure of Theorem 2, the QB-method, and the QR-method of Theorem 1 both approximate with a well posed problem for each $\varepsilon > 0$. Moreover, the estimate (7) is $\mathcal{O}(1/\varepsilon)$ at $t = 1$ in contrast to $\mathcal{O}(\exp(1/\varepsilon))$ in the QR-method, so the QB-method is reasonable for numerical implementation. However the regularized problems in the QB-method are global in t, so marching methods and their resultant sparse matrices and reduced storage requirements are not directly available in numerical work. Our preceding remarks on the "elliptic" nature of these equations suggest that such difficulties may be implicit in the problem, not just this method.

There is a fundamental deficiency in the use of Theorem 2 to actually find a solution u from data f; namely, the data is never measured exactly. This measurement error can be handled if we stabilize the problem by considering only those solutions which satisfy a prescribed global bound. Whereas Theorem 2 merely guarantees a good approximation at the initial time $t = 0$, we shall get a global approximation on $t \in [0,1]$.

<u>Theorem 3</u>. Let u be a weak solution of (2) on $[0,1]$. Let $M \geq 1$, $\delta > 0$, and $f \in H$ be given such that $\|u(0) - f\| < \delta$ and $\overline{u(1)} \leq M$. Choose $\varepsilon \equiv M/\delta$ and let u_ε be the solution of the boundary-value problem (2), (6). Then we have the estimate

$$\|u(t) - u_\varepsilon(t)\| \leq 2\delta^{1-t}M^t , \quad 0 \leq t \leq 1 ,$$

for the error.

The procedure above is the <u>stabilized quasi-boundary-value method</u>. It is appropriate in applied problems where one knows from physical considerations there is a solution with a bound but the data $u(0)$ is not known exactly.

REFERENCES

[1] Bammann, D.J. and Aifantis, E.C., On a proposal for a continuum with microstructure, Acta Mechanica 45 (1982) 91-121.

[2] Ewing, R.E., The approximation of certain parabolic equations backward in time by Sobolev equations, SIAM J. Math. Anal. 6 (1975) 283-294.

[3] Lagnese, J., The Final Value Problem for Sobolev Equations, Proc. Amer. Math. Soc. (1976).

[4] Lambropoulis, P., Solution of the Differential Equation $P_{xy} + axP_x + byPy + cxyP + P_t = 0$, J. Math. Phys. 8 (1967) 2167-2169.

[5] Lattes, R. and Lions, J.L., Methode de Quasi-Reversibility et Applications (Dunod, Paris, 1967). (English trans., R. Bellman, Elsevier, New York, 1969.)

[6] Miller, K., "Stabilized quasireversibility and other nearly best possible methods for non-well-posed problems," Symposium on Non-Well-Posed Problems and Logarithmic Convexity, Lecture Notes in Mathematics, Vol. 316 (Springer-Verlag, Berlin, 1973) 161-176.

[7] Miller, M. and Steinberg, S., The Solution of Moment Equations Associated with a Partial Differential Equation with Polynomial Coefficients, J. Math. Phys. 14 (1973) 337-339.

[8] Multhei, H.N., Initial-Value Problem for the Equation $u_t + au_x + bu_y + cu + du_{xy} = f$ in the complex domain, J. Math. Phys. 11 (1970) 1977-1980.

[9] Multhei, H.N. and Neunzert, H., Pseudoparabolische Differentialgleichungen mit Charakteristischen Vorgaben im Komplexen Gebeit, Math. Z. 113 (1970) 24-32.

[10] Neuringer, J.L., Closed-form solution of the differential equation $P_{xy} + axP_x + byPy + cxyP + P_t = 0$, J. Math. Phys. 10 (1969) 250-251.

[11] Neunzert, H., Z. Angew. Math. Mech. 48 (1968) 222.

[12] Payne, L.E., "Some general remarks on improperly posed problems for partial differential equations," Symposium on Non-Well-Posed Problems and Logarithmic Convexity, Lecture Notes in Mathematics, Vol. 316 (Springer-Verlag, Berlin, 1973) 1-30.

[13] Showalter, R.E., Initial and Final-Value Problems for Degenerate Parabolic Evolution Systems, Indiana Univer. Math. J. (1979) 883-893.

[14] Showalter, R.E., The Final Value Problem for Evolution Equations, J. Math. Anal. Appl. 47 (1974) 563-572.

[15] Steinberg, S. and Treves, S., Pseudo-Fokker Planck Equations and Hyperdifferential Operators, J. Diff. Eq. 8 (1970) 333-366.

The final (detailed) version of this paper will be submitted for publication elsewhere.

Trends in the Theory and Practice of Non-Linear Analysis
V. Lakshmikantham (Editor)

EXISTENCE, UNIQUENESS, AND GALERKIN APPROXIMATIONS FOR SEMILINEAR PERIODICALLY FORCED WAVE EQUATIONS AT RESONANCE

Michael W. Smiley

Department of Mathematics
Iowa State University
Ames, Iowa 50011
U.S.A.

The abstract boundary value problem $Lu + Gu = f$, $u \in \mathrm{dom}(L) \subset H$, is considered. Here H is used to denote a real separable Hilbert space, L a closed symmetric linear operator, and G a continuous nonlinear operator assumed to be monotone and sublinear. A recent result on the existence of unique solutions, depending continuously on f, for this abstract boundary value problem is stated and then applied to the problem of the periodically forced nonlinear vibrating membrane. Galerkin approximations are also considered and error estimates are given.

Let us consider first the example of the semilinear periodically forced vibrating string, in which one seeks to find a function $u = u(t,x)$ describing the displacement of a string and satisfying

(1) $$u_{tt} - u_{xx} + g(u) = f, \quad (t,x) \in \mathbb{R}\times(0,a),$$

(2) $$u(t,0) = u(t,a) = 0, \quad t \in \mathbb{R} = \{\text{real numbers}\},$$

(3) $$u(t+T,x) = u(t,x), \quad (t,x) \in \mathbb{R}\times(0,a),$$

where $f \in L^2((0,T)\times(0,a))$ is assumed to satisfy $f(t+T,x) = f(t,x)$, (a.e.)$(t,x) \in \mathbb{R}\times(0,a)$, and $g : \mathbb{R} \to \mathbb{R}$ is continuous. When $g(u) = 0$ for all $u \in \mathbb{R}$, we obtain an ill-posed linear problem in which a Fredholm alternative may apply. For instance when $a = \pi$, $T = 2\pi$ we find that each of the functions $u_k(t,x) = \sin kt \sin kx$, $k \geq 1$, satisfy $u_{tt} - u_{xx} = 0$ and the boundary conditions (2)-(3). Hence the corresponding linear homogeneous problem has an infinite dimensional subspace of solutions. This is true whenever the ratio T/a is rational; and in this case solutions of the linear problem exist if and only if f is orthogonal to the null space of the associated linear operator (cf. Smiley [4]).

In the past two decades, an extensive literature has developed regarding (1)-(3). Usually $a = \pi$, $T = 2\pi$ is assumed for convenience. Most approaches to the problem require g to be monotone. For example we have the following result due to Brezis and Nirenberg [2].

<u>Theorem</u> Let $a = \pi$, $T = 2\pi$. If $g(u)$ is a continuous nondecreasing function satisfying $g(0) = 0$, $\lim_{u \to \pm\infty} g(u) = \pm\infty$, and

$$|g(u)| \leqslant \gamma|u| + C, \qquad \forall u \in \mathbb{R},$$

for some constants γ, C with $\gamma < 3$, then (1)-(3) has a weak solution for every $f \in L^{\infty}((0,2\pi) \times (0,\pi))$.

Observe that $g(u)$ is also assumed to be sublinear with constant $\gamma < 3$. The significance of the number 3 stems from the fact that $\lambda_{-1} = -3$ is the first (largest) negative eigenvalue for the associated linear operator ($\Box$ plus the boundary conditions). One can show that the solution is unique if $g(u)$ is strictly increasing and satisfies $|g(u) - g(v)| \leqslant \gamma|u-v|$ for all $u,v \in \mathbb{R}$, where $\gamma < 3$. We refer the interested reader to the recent article of Brezis [1] for other results and further references.

Our purpose here is to briefly describe a constructive approach, due to the author [5], for establishing the existence of solutions for problem (1)-(3) as well as higher dimensional analogs. We work within the following framework. Let H denote a real separable Hilbert space, and L denote a closed symmetric linear operator from $\mathrm{dom}(L) \subset H$ into H. With $G: H \to H$ denoting a continuous nonlinear mapping we consider the abstract boundary value problem

$$Lu + Gu = f, \quad u \in \mathrm{dom}(L), \tag{4}$$

where $f \in H$. Of course we must allow $\ker(L) = \{u \in H : Lu = 0\}$ to be a nontrivial, possibly infinite dimensional subspace of H. The following assumptions on L and G will be enforced. We assume that L has a collection of eigenvalues $\{\lambda_i : i \in I\} \subset \mathbb{R}$ and eigenfunctions $\{\varphi_i : i \in I\} \subset H$, where I is a countable index set. Furthermore we require the set of eigenfunctions to be a complete orthonormal system in H. Let $\|\cdot\|$ and $(\cdot,\cdot)$ denote the norm and inner product in H. Of $G : H \to H$ we require the properties

(5) $$\|Gu - Gv\| \leq \beta_1 \|u - v\|, \quad \forall\, u,v \in H,$$

(6) $$(Gu - Gv, u - v) \geq \beta_0 \|u - v\|^2, \quad \forall\, u,v \in H,$$

where $0 < \beta_0 \leq \beta_1 < +\infty$. There is no loss of generality in assuming $G(0) = 0$.

We make some notational definitions. Let $I_0 = \{i \in I : \lambda_i = 0\}$, $I_1 = \{i \in I : \lambda_i \neq 0\}$, and $I_{1-} = \{i \in I : \lambda_i < 0\}$. When $I_1 \neq \emptyset$ we define $\delta = \inf\{|\lambda_i| : i \in I_1\}$, and when $I_{1-} \neq \emptyset$ we define $\kappa = \inf\{|\lambda_i| : i \in I_{1-}\}$. Clearly we have $0 \leq \delta \leq \kappa$. We point out that if $I_0 \neq \emptyset$ then the problem is at resonance; that is $\dim(\ker(L)) \geq 1$.

The following theorem guarantees existence, uniqueness and continuous dependence of solutions on the data for problem (4). In proving this theorem (cf. [5]) the solution u is shown to be the unique fixed point of a contraction mapping. Because of this we are able to define an algorithm for constructing approximate solutions; and in this setting the Galerkin approximations are natural to consider.

Theorem I Assume that $I_0 \neq \emptyset$, $I_{1-} \neq \emptyset$, and $\delta > 0$. If $\beta_1^2 < \kappa\beta_0$ then there is a unique solution of the abstract boundary value problem (4) for every $f \in H$. Moreover if $v \in \mathrm{dom}(L)$ satisfies $Lv + Gv = h$, for some $h \in H$, then $\|u - v\| \leq c(\kappa\beta_0 - \beta_1^2)^{-1}\|f - g\|$, where c is a positive constant.

We comment briefly on the assumptions above. As was mentioned, $I_0 \neq \emptyset$ implies that the problem is at resonance, which is the case of interest here. The assumption that I_{1-} is not empty results from considering the case in which $L + G$ is an indefinite operator. This is typically the situation for wave equations subject to boundary conditions in both space and time. Finally we assume $\delta > 0$ so that we are assured that the associated linear problem admits a duality structure analogous to that guaranteed by the Lax-Milgram theory for elliptic equations (cf. [4]).

The proof of Theorem I uses the alternative method (cf. [3]) in conjunction with the contraction mapping principle and some properties of monotone operators. Necessarily one must consider a subspace $\mathcal{H} \subset H$, which is itself a Hilbert space, with norm denoted by $\|\cdot\|_{\mathcal{H}}$ and such that $\|u\| \leq c\|u\|_{\mathcal{H}}$ for all $u \in \mathcal{H}$. It is the existence of this space and its dual $\mathcal{H}^*$ that the hypothesis $\delta > 0$ insures.

Remarks 1) Under the hypotheses of Theorem I, one can also establish the stronger estimate:

$$\|u - v\|_{\mathcal{H}} \leq c(\kappa\beta_0 - \beta_1^2)^{-1}\|f - g\|.$$

2) We have assumed G to be monotone increasing, but it is only the monotonicity that is needed. If G satisfies (5) and

$$(6') \qquad (Gu - Gv, u - v) \leqslant -\beta_o \|u - v\|^2, \qquad \forall\, u,v \in H,$$

then we multiply (4) by -1 and redefine $\kappa = \inf\{\lambda_i : \lambda_i > 0\}$, under the assumption that $I_{1+} = \{\lambda_i : \lambda_i > 0\}$ is nonempty.

3) Consider the problem $Lu + \alpha Gu = f$, $u \in \mathrm{dom}(L)$. If both I_o and I_{1-} are nonempty and $\delta > 0$, then this problem has a unique solution for every $f \in H$ and α in the interval $(0, \kappa\beta_o\beta_1^{-2})$. Moreover, the correspondence $\alpha, f \to u$ is continuous.

As an application of Theorem I we consider the problem of nonlinear oscillations of a rectangular membrane. Let $\Omega = (0,\alpha_1\pi) \times (0,\alpha_2\pi)$ be a rectangle in the plane with boundary $\partial\Omega$. We consider the problem

$$(7) \qquad u_{tt} - u_{xx} - u_{yy} + g(u) = f, \quad (t,x,y) \in \mathbb{R} \times \Omega$$

$$(8) \qquad u(t,x,y) = 0, \quad (t,x,y) \in \mathbb{R} \times \partial\Omega,$$

$$(9) \qquad u(t + 2\pi, x, y) = u(t,x,y), \quad (t,x,y) \in \mathbb{R} \times \Omega,$$

where $f(t,x,y)$ is square integrable on $(0,2\pi) \times \Omega$, and satisfies $f(t + 2\pi, x, y) = f(t,x,y)$, (a.e.)$(t,x,y) \in \mathbb{R} \times \Omega$. The continuous function $g : \mathbb{R} \to \mathbb{R}$ will be assumed to satisfy $g(0) = 0$ and $0 < \beta_o \leqslant [g(u) - g(v)](u - v)^{-1} \leqslant \beta_1 < +\infty$ whenever $u \neq v$.

__Theorem II__ Let α_1, α_2 be integers. If $\beta_1^2 < \pi^4\beta_o/A^2$, where A denotes the area of Ω then there is a unique weak solution of (7)-(9) for every $f \in L^2((0,2\pi) \times \Omega)$.

Proof: We derive this result from Theorem I. First we notice that the eigenvalues for the associated linear operator are $\lambda_{k\ell} = (\ell_1/\alpha_1)^2 + (\ell_2/\alpha_2)^2 - k^2$, with $\ell_1, \ell_2 \geqslant 1$ and $k \geqslant 0$. Hence $I_{1-} \neq 0$ and one can show that I_o is an infinite set. In addition we know that either $\lambda_{k\ell} = 0$ or $|\lambda_{k\ell}| \geqslant (\alpha_1\alpha_2)^{-2}$. Therefore $\delta > 0$ and $\pi^4/A^2 = (\alpha_1\alpha_2)^{-2} \leqslant \kappa$.

__Remark:__ Theorem I also applies to the n-dimensional wave equation. In fact the rectangle Ω can be replaced by an n-dimensional rectangle $(0,\alpha_1\pi) \times \ldots \times (0,\alpha_n\pi) \subset \mathbb{R}^n$, $n \geqslant 1$, where the numbers α_i are all rational (cf. [5]). This is the only result for the $n \geqslant 2$ case known to the author.

In the context of the preceding framework it is natural to consider Galerkin approximations. Given any finite index subset $I_m \subset I$, chosen so that the linear span of the set $\{\varphi_i : i \in I_m\}$ is finite dimensional, we may define a Galerkin approximation U_m with respect to I_m according to the equations

$$U_m = \sum_{i \in I_m} u_i \varphi_i , \tag{10}$$

$$(LU_m + GU_m - f, \varphi_i) = 0, \qquad \forall i \in I_m. \tag{11}$$

That U_m is well-defined is one of the consequences of the next result. We again need to introduce some notation. Assuming $I_{1-} \neq \emptyset$, we define $\kappa = \inf\{|\lambda_i| : i \in I_{1-}\}$ as before. Let $J_o = \{i \in I : 0 \leq \lambda_i < \kappa\}$ and $J_1 = \{i \in I : \lambda_i < 0 \text{ or } \lambda_i \geq \kappa\}$. Let $I \setminus I_m$ denote the complement of I_m in I and set $\kappa_m = \inf\{|\lambda_i| : i \in J_1 \cap (I \setminus I_m)\}$. We use PP_m, $(I-P_m)P$ and $(I-P)P_m$ to denote projection operators defined by

$$PP_m u = \sum_{i \in J_o \cap I_m} u_i \varphi_i ,$$

$$(I-P_m)Pu = \sum_{i \in J_o \cap (I \setminus I_m)} u_i \varphi_i$$

$$(I-P)P_m u = \sum_{i \in J_1 \cap I_m} u_i \varphi_i ,$$

where $u_i = (u, \varphi_i)$ and $u \in H$. We note that the range of $(I-P_m)P$ may not be finite dimensional and in fact must be infinite dimensional whenever the null space of L is infinite dimensional. Finally, we use K to denote the partial inverse of L, which is known to exist as a continuous linear operator according to the hypothesis that $\delta > 0$.

<u>Theorem III</u> Assume that $I_o \neq \emptyset$, $I_{1-} \neq \emptyset$, $\delta > 0$ and $\beta_1^2 < \kappa\beta_o$. There is a unique Galerkin approximation U_m defined by (10) and (11), which can be shown to be the limit of a sequence of successive approximations $\{U_m^k\}$ defined as follows. We choose the set of real numbers $\{u_i^o : i \in I_m \cap J_o\}$ arbitrarily and define for $k = 1,2,3,\ldots$ the successive approximation $U_m^k = \psi_o^k + \psi_1^k$ according to the equations

(12) $\psi_o^k = \sum_{i \in J_o \cap I_m} u_i^k \varphi_i, \qquad \psi_1^k = \sum_{i \in J_1 \cap I_m} u_i^k \varphi_i,$

(13) $\psi_1^k = K(I-P)P_m[f - G(\psi_o^{k-1} + \psi_1^k)],$

(14) $0 = PP_m[L(\psi_o^k + \psi_1^k) + G(\psi_o^k + \psi_1^k) - f].$

In addition it can be shown that

(15) $\|U_m - U_m^k\| \leq c\,\alpha^{k-1}\|\psi_o - \psi_o^0\|$, and

(16) $\|u - U_m\| \leq c\left(\|(I-P_m)Pu\| + \kappa_m^{-1}\|f\|\right),$

where c denotes a positive constant and $\alpha = \beta_1(\beta_1-\beta_o)[\beta_o(\kappa-\beta_1)]^{-1} < 1$.

Remarks 1) We see that $U_m^k \to U_m$ at a guaranteed rate but that $U_m \to u$ at an indeterminate rate whenever $(I-P_m)P \neq 0$. This is certainly the case if $\dim(\ker(L)) = +\infty$.

2) Estimates similar to (15)-(16) can also be given in terms of the stronger norm $\|\cdot\|_{\mathcal{H}}$.

3) The algorithm (12)-(14) is a two stage method, in which $U_m^k = \psi_o^k + \psi_1^k$ is computed by first using (13) to obtain ψ_1^k and then using (14) to obtain ψ_o^k. To implement (13) a fixed point subroutine can be used. To compute ψ_o^k satisfying (14) one can use minimization techniques. In particular, one considers a functional $\mathfrak{F}(\psi_o)$ with the property that $\mathfrak{F}'(\psi_o) = PP_m[L(\psi_o+\psi_1^k) + G(\psi_o+\psi_1^k) - f]$. Then ψ_o^k can be determined as the unique function minimizing $\mathfrak{F}(\psi_o)$. The author, using this approach, has carried out some numerical experiments on problems (1)-(3) with $a = \pi$, $T = 2\pi$. Rapid convergence of the successive approximations has been observed. Further details and some results are reported in [5].

References

[1] H. Brezis, Periodic solutions of nonlinear vibrating strings and duality principles, Bull. Amer. Math. Soc., 8(1983) no. 3, pp. 409-426.

[2] H. Brezis and L. Nirenberg, Forced vibrations for a nonlinear wave equation, Comm. Pure Appl. Math. 31(1978), pp. 1-30.

[3] L. Cesari, Functional analysis, nonlinear differential equations, and the alternative method, Nonlinear Functional Analysis and Differential Equations, Cesari-Kannan-Schuur, eds., Marcel Dekker, New York 1976, pp. 1-197.

[4] M. Smiley, Hyperbolic boundary value problems - a Lax-Milgram approach and the vibrating string, to appear in Boll. Unione Mat. Ital.

[5] M. Smiley, Eigenfunction methods and nonlinear hyperbolic boundary value problems at resonance, preprint to appear.

The final (detailed) version of this paper will be submitted for publication elsewhere.

Trends in the Theory and Practice of Non-Linear Analysis
V. Lakshmikantham (Editor)
Elsevier Science Publishers B.V. (North-Holland), 1985

HOPF BIFURCATION FOR PERIODIC SYSTEMS (*)

Francesca Visentin

Dipartimento di Matematica e Applicazioni
Università di Napoli
80134 Napoli
ITALY

This paper concerns with the problem of Hopf bifurcation from an equilibrium position to periodic solutions, in the case of n dimensional periodic differential systems. Results about existence and uniqueness of bifurcating periodic solutions are obtained.

1. INTRODUCTION

We are concerned with the problem of Hopf bifurcation for periodic ordinary differential systems. Let m be the number of Floquet multipliers (of the unperturbed system) lying on the unite circle. In the case m = 1 the problem of existence,uniqueness and stability of bifurcating periodic solutions has been analized in [3 - 6]. For m>1 this problem presents more difficult aspects. Here we will examine the case m = 2. Precisely, we consider a family of T-periodic systems depending on a parameter μ, and suppose that the linear part of the unperturbed system ($\mu = 0$) admits a couple of purely imaginary characteristic exponents $\pm i$. If the period T is different from a multiple of 2π, then (locally) T-periodic bifurcating solutions do not occur. Suppose the period T is equal to a multiple of 2π. In this case, we find that we need to have another parameter, say ε, in order to obtain Hopf bifurcation. This additional parameter controls the angular velocity of the solutions and allows us to obtain periodic solutions of a period exactly equal to T. The case m=2 was analized in [2] for a two dimensional family of 2π-periodic systems of the kind

$$(1.1)\qquad \begin{aligned} \dot{x} &= \alpha(\mu,\varepsilon)x - \beta(\mu,\varepsilon)y + X(t,x,y,\mu,\varepsilon) \\ \dot{y} &= \alpha(\mu,\varepsilon)y + \beta(\mu,\varepsilon)x + Y(t,x,y,\mu,\varepsilon) \end{aligned} ,$$

where the parameters μ and ε are close to 0 and 1 respectively. Moreover $\alpha(0,\varepsilon) \equiv 0$, $\beta(0,1) = 1$, $\alpha'_{\mu}(0,1) \neq 0$, $\beta'_{\varepsilon}(0,1) \neq 0$, and X, Y are C^{∞} functions 2π-periodic in t and of order $\geqslant 2$ in x, y when $(x,y) \to (0,0)$. For such systems it was proved in [2] the existence of bifurcating solutions. Furthermore, results about uniqueness and stability were obtained under some additional requirements concerning the systems for which $\mu = 0$. Precisely, it was required that for $\mu=0$ and each ε near 1 system (1.1) is autonomous and its zero solution is h-asymptotically stable or h-completely unstable. By h-asymptotic stability it is

(*) This work was partially supported by C.N.R. (Italian Council of Research) , and by M.P.I. (Italian Ministry of Education).

meant that the origin is asymptotically stable and this property is recognizable on the terms up to order h in the Taylor expansions of X and Y. Similarly we define h-complete instability. In the above mentioned hypotheses in particular in [2] was proved that for each μ sufficiently small there exists an annulus M_μ wrapping around the origin in the (x,y) plane which contains all 2π-periodic orbits of (1.1) lying near the origin and corresponding to this μ. We notice that, in contrast with the autonomous case, the periodic solutions of (1.1) may not have asymptotic stability behavior. However, for practical purposes, the annulus M_μ takes on the role of the single periodic orbits. This is particularly substantiated by the fact that the width of the asymptotically stable annulus M_μ is much smaller than the distance it is away from the origin. So it is the annulus we are able to observe rather than the single periodic orbits.

In this paper we extend the above results given in [2] to 2π-periodic systems in R^n, $n>2$, satisfying a suitable non resonance condition. The extension is obtained by means of a procedure already used in [1],[7]. In particular, this procedure includes the construction of two dimensional quasi invariant manifolds. We essentially analize the problem of existence and uniqueness of bifurcating 2π-periodic solutions. However, we observe that if the remaining (n-2) Floquet multipliers are all inside the unite circle, an application of the center manifold theorem allows us to prove the same stability results obtained in [2].

2. BIFURCATION FROM THE ORIGIN IN PERIODIC SOLUTIONS FOR PERIODIC EQUATIONS

Consider the n dimensional family of systems

$$\dot{u} = f(t,u,\mu,\varepsilon), \tag{2.1}$$

where $f \in C^\infty(R \times R^n \times R \times R,\ R^n)$, $f(t,0,\mu,\varepsilon) \equiv 0$, and f is T-periodic in t, $T>0$. We may assume that the Jacobian matrix $D_u f(t,0,\mu,\varepsilon)$ is independent of t. Denote its eigenvalues by $\alpha(\mu,\varepsilon) \pm \beta(\mu,\varepsilon)$, $\lambda_1(\mu,\varepsilon), \ldots, \lambda_{n-2}(\mu,\varepsilon)$. We suppose that $\alpha(\mu,\varepsilon)$, $\beta(\mu,\varepsilon)$ satisfy the conditions

$$\alpha(0,\varepsilon) \equiv 0 \qquad \beta(0,1) = 1$$

$$\alpha'_\mu(0,1) \neq 0 \qquad \beta'_\varepsilon(0,1) \neq 0,$$

while the remaining eigenvalues $\lambda_j(0,\varepsilon)$, $j = 1,\ldots,n-2$ satisfy a non resonance condition. By means of a linear transformation on u independent of t and involving μ,ε we may write (2.1) in the form

$$\begin{aligned} \dot{x} &= \alpha(\mu,\varepsilon)x - \beta(\mu,\varepsilon)y + X(t,x,y,z,\mu,\varepsilon) \\ \dot{y} &= \alpha(\mu,\varepsilon)y + \beta(\mu,\varepsilon)x + Y(t,x,y,z,\mu,\varepsilon) \\ \dot{z} &= A(\mu,\varepsilon)z + Z(t,x,y,z,\mu,\varepsilon)\ , \end{aligned} \tag{2.2}$$

where $x,y \in R$, $z \in R^{n-2}$ and $A(\mu,\varepsilon)$ is an $(n-2)\times(n-2)$ constant matrix whose eigenvalues λ_j, $j = 1,\ldots,n-2$ for $\mu = 0$ satisfy the non resonance condition $\lambda_j(0,\varepsilon) \neq m\beta(0,\varepsilon)i$, $m \in \mathbb{Z}$. The functions X, Y, Z are C^∞ in all their arguments and begin with terms of at least degree 2 on x, y, and z in their Taylor expansions. We first observe that it is possible to adapt to the periodic case the two following propositions proved in [7] for autonomous systems.

Proposition 2.1 there exist three positive numbers $\bar{a}$, $\bar{b}$, $\bar{c}$ and a function $\sigma \in C^\infty(R \times [-\bar{c},\bar{c}]^2 \times [-\bar{b},\bar{b}] \times [1-\bar{a},1+\bar{a}],\ R^{n-2})$, $(t_0,x_0,y_0,\mu,\varepsilon) \to \sigma(t_0,x_0,y_0,\mu,\varepsilon)$, $\sigma(t_0,0,0,\mu,\varepsilon) \equiv 0$, such that for any $t_0 \in R$, $x_0,y_0 \in [-\bar{c},\bar{c}]$, $|\mu| \in [0,\bar{b}]$ and

$|\varepsilon - 1| \in [0,\bar{a}]$ the equation $z(t_0+2\pi,t_0,x_0,y_0,z_0,\mu,\varepsilon) = z_0$ is satisfied if and only if $z_0 = \sigma(t_0,x_0,y_0,\mu,\varepsilon)$.

The proof of this proposition easily follows by applying the variation of constants formula to the z-equation of (2.2) and the implicit function theorem. Now a transformation in polar coordinates of the first two equations of (2.2) allows us to prove the following proposition.

Proposition 2.2 The numbers $\bar{a}$, $\bar{b}$, $\bar{c}$ can be chosen such that if $t_0 \in R$, $|\mu| \in [0,\bar{b}]$, $|\varepsilon - 1| \in [0,\bar{a}]$ and γ is a nontrivial 2π-periodic solution of (2.2) lying in a suitable neighborhood of the origin then: (i) for any $(x,y,z) \in \gamma$ we have $(x,y) \neq (0,0)$; (ii) the projection of γ on the (x,y) plane intersects any straight line through the origin in exactly two points.

Because of Proposition 2.2, in order to find 2π-periodic solutions of (2.2) we may restrict ourselves to consider the solutions of (2.2) for which at $t=t_0$, $x_0=c \in [0,\bar{c}]$, $y_0=0$, $z_0=\sigma(t_0,c,0,\mu,\varepsilon)$. Denote those solutions by $(x(t,t_0,c,\mu,\varepsilon)$, $y(t,t_0,c,\mu,\varepsilon),z(t,t_0,c,\mu,\varepsilon))$. Now we are in position to give the following result about existence of 2π-periodic solutions of (2.2).

Theorem 2.3 Suppose that $\bar{a}$, $\bar{b}$, $\bar{c}$ are sufficiently small. There exist two functions $\mu^* \in C^\infty(R \times [0,\bar{c}], [-\bar{b},\bar{b}])$, $\varepsilon^* \in C^\infty(R \times [0,\bar{c}], [1-\bar{a},1+\bar{a}])$ such that if $t_0 \in R$, $c \in [0,\bar{c}]$, $|\mu| \in [0,\bar{b}]$, $|\varepsilon - 1| \in [0,\bar{a}]$, then the solution $(x(t,t_0,c,\mu,\varepsilon)$, $y(t,t_0,c,\mu,\varepsilon),z(t,t_0,c,\mu,\varepsilon))$ of (2.2) is 2π-periodic in t if and only if $\mu = \mu^*(t_0,c)$, $\varepsilon = \varepsilon^*(t_0,c)$.

Outline of the proof. From Propositions 2.1 and 2.2 it follows that $z(t_0+2\pi,t_0,c,\mu,\varepsilon) = 0$. Then, in order to find 2π-periodic solutions we have to inspect only the two first equations of (2.2). For those equations, first we insist that $y(t_0+2\pi,t_0,c,\mu,\varepsilon) = 0$. As in Proposition 2.1 an application of the implicit function theorem provides us with the following $\varepsilon = \bar{\varepsilon}(t_0,c,\mu)$. Successively, requiring that

$$V(t_0,c,\mu) = x(t_0+2\pi,t_0,c,\mu,\bar{\varepsilon}(t_0,c,\mu)) - c = 0,$$

we determine $\mu = \mu^*(t_0,c)$. By using the function μ^* we see that $\bar{\varepsilon}$ can be considered as a function $\varepsilon^*(t_0,c)$. This completes the proof.

Remark 2.4 The function $V \in C^\infty(R \times [0,\bar{c}] \times [-\bar{b},\bar{b}], R)$ defined in the proof of Theorem 2.3 will be called the displacement function relative to system (2.2). Clearly, as seen in the above proof, the zeros of V correspond to the 2π-periodic solutions of (2.2). This function will be fruitfully used in the next section in order to obtain a uniqueness result for the bifurcating periodic solutions.

3. UNIQUENESS PROPERTY OF BIFURCATING PERIODIC SOLUTIONS

In this section we obtain a uniqueness property. Toward this end we assume that (2.2) is independent of t at $\mu = 0$. Then system (2.2) may be written as

$$\begin{aligned} \dot{x} &= \alpha(\mu,\varepsilon)x - \beta(\mu,\varepsilon)y + \tilde{X}(x,y,z,\varepsilon) + \mu X^*(t,x,y,z,\mu,\varepsilon) \\ \dot{y} &= \alpha(\mu,\varepsilon)y + \beta(\mu,\varepsilon)x + \tilde{Y}(x,y,z,\varepsilon) + \mu Y^*(t,x,y,z,\mu,\varepsilon) \qquad (3.1) \\ \dot{z} &= A(\mu,\varepsilon)z + \tilde{Z}(x,y,z,\varepsilon) + \mu Z^*(t,x,y,z,\mu,\varepsilon) \ . \end{aligned}$$

For $\mu = 0$ we obtain

$$(3.2)\qquad \begin{aligned} \dot{x} &= -\beta(0,\varepsilon)y + \tilde{X}(x,y,z,\varepsilon) \\ \dot{y} &= \beta(0,\varepsilon)x + \tilde{Y}(x,y,z,\varepsilon) \\ \dot{z} &= A(0,\varepsilon)z + \tilde{Z}(x,y,z,\varepsilon)\ . \end{aligned}$$

For every $h>0$ it is possible to determine (see for instance [1]) an $(n-2)$ dimensional polinomial $\Phi^{(h)}(x,y,\varepsilon) = \phi_1(x,y,\varepsilon) + \ldots + \phi_h(x,y,\varepsilon)$ where $\phi_j(x,y,\varepsilon)$ is homogeneous of degree j in (x,y), such that along the solutions of (3.2)

$$\frac{d}{dt}\left[z - \Phi^{(h)}(x,y,\varepsilon)\right]_{z=\Phi^{(h)}(x,y,\varepsilon)} = o((x^2+y^2)^{h/2}).$$

The two dimensional surface $z = \Phi^{(h)}(x,y,\varepsilon)$ is called a quasi invariant manifold of order h. It is tangent at the origin to the eigenspace corresponding to the eigenvalues $\pm i$. By using the transformation $w = z - \Phi^{(h)}(x,y,\varepsilon)$ we may write (3.2) in the form

$$(3.2)'\qquad \begin{aligned} \dot{x} &= -\beta(0,\varepsilon)y + \tilde{X}^{(h)}(x,y,w,\varepsilon) \\ \dot{y} &= \beta(0,\varepsilon)x + \tilde{Y}^{(h)}(x,y,w,\varepsilon) \\ \dot{w} &= A(0,\varepsilon)w + \tilde{W}^{(h)}(x,y,w,\varepsilon)\ , \end{aligned}$$

where $\tilde{W}^{(h)}(x,y,0,\varepsilon)$ is of order $>h$ in (x,y). In connection with (3.2)' consider the two dimensional system

$$(S_h)\qquad \begin{aligned} \dot{x} &= -\beta(0,\varepsilon)y + \tilde{X}^{(h)}(x,y,0,\varepsilon) \\ \dot{y} &= \beta(0,\varepsilon)x + \tilde{Y}^{(h)}(x,y,0,\varepsilon) \end{aligned}$$

We find that the following result holds.

Theorem 3.1 Suppose there exists an odd integer $h\geq 3$ such that the origin of (S_h) is h-asymptotically stable for every $\varepsilon \in [1-\bar{a},1+\bar{a}]$. Then if $\alpha'_\mu(0,1)>0$ (resp. $\alpha'_\mu(0,1)<0$) the bifurcating 2π-periodic solutions of (3.1) occur for $\mu>0$ (resp. for $\mu<0$). Moreover the positive numbers $\bar{a}$, $\bar{b}$, $\bar{c}$ of Theorem 2.3 can be chosen such that for any $t_0 \in R$ and $\mu \in [0,\bar{b}]$ (resp. $\mu \in [-\bar{b},0]$) there exists one and only one $c \in [0,\bar{c}]$ such that $\mu = \mu^*(t_0,c)$. A similar result holds when the above assumption of h-asymptotic stability is replaced by that of h-complete instability.

Outline of the proof. First we note that in this case the functions $\bar{\varepsilon}$ and V defined in the proof of Theorem 2.3 are independent of t_0. We may set $\tilde{\varepsilon}(c) = \bar{\varepsilon}(0,c,0)$ and $\tilde{V}(c) = V(0,c,0)$. Furthermore, setting $x(t,c) = x(t,0,c,0,\tilde{\varepsilon}(c))$, $y(t,c) = y(t,0,c,0,\tilde{\varepsilon}(c))$, $w(t,c) = w(t,0,c,0,\tilde{\varepsilon}(c))$ we may write

$$\begin{aligned} x(t,c) &= (\cos t)c + u_2(t)c^2 + \ldots + u_h(t)c^h + o(c^h), \\ y(t,c) &= (\sin t)c + v_2(t)c^2 + \ldots + v_h(t)c^h + o(c^h), \\ w(t,c) &= w_1(t)c + w_2(t)c^2 + \ldots + w_h(t)c^h + o(c^h), \\ \tilde{\varepsilon}(c) &= 1 + \varepsilon_1 c + \varepsilon_2 c^2 + \ldots + \varepsilon_h c^h + o(c^h). \end{aligned}$$

From the hypotheses on $A(0,\varepsilon)$ and $\tilde{W}^{(h)}(x,y,0,\varepsilon)$ it follows that $w_j(t) = 0$ for $j = 1,\ldots,h-1$. Consequently, in order to compute $u_j(t)$ and $v_j(t)$, $j = 1,\ldots,h$, we may set $w = 0$ in the first two equations of (3.2)'. Thus, if $(\tilde{x}(t,c),\tilde{y}(t,c))$ denotes the solution of (S_h) corresponding to $x_0 = 0$, $y_0 = 0$, we have

$$(3.3)\qquad x(t,c) - \tilde{x}(t,c) = o(c^h) \quad \text{and} \quad y(t,c) - \tilde{y}(t,c) = o(c^h).$$

In this way, our original problem can be reduced to the two dimensional system (S_h). Taking into account that the origin of (S_h) is h-asymptotically stable, a classical procedure due to Poincaré (see for instance [8]) tells us there exists a polynomial of degree h+1 having the form

$$F(x,y,\varepsilon) = x^2 + y^2 + f_3(x,y,\varepsilon) + \dots + f_{h+1}(x,y,\varepsilon)$$

(f_j is homogeneous of degree j in (x,y)) such that along the solutions of (S_h)

$$\dot{F}(x,y,\varepsilon) = -G(\varepsilon)(x^2+y^2)^{(h+1)/2} + o((x^2+y^2)^{(h+1)/2}). \tag{3.4}$$

Here G is of class C^∞ and $G(\varepsilon)>0$. Integrating (3.4) over $[0,2\pi]$ and equating the coefficients of the same degree in c, we have $u_j(2\pi) = 0$ for $j = 2,\dots,h-1$ and $u_h(2\pi) = -\pi G(1)<0$. Then, taking into account (3.3) we have

$$(\partial^j \tilde{V}/\partial c^j)(0) = 0 \quad j = 1,\dots,h-1 \qquad (\partial^h \tilde{V}/\partial c^h)(0) = h!u_h(2\pi)<0.$$

Now, proceeding as in [2], from the identity $V(t_0,c,\mu^*(t_0,c)) \equiv 0$ by successive derivations with respect to c it follows

$$(\partial^{s+1}V/\partial c^{s+1})(t_0,0,0) = -(s+1)[(\partial^2 V/\partial\mu\partial c)(t_0,0,0)][(\partial^s\mu^*/\partial c^s)(t_0,0)]$$

for every s such that $(\partial^j\mu^*/\partial c^j)(t_0,0) = 0$ for $j = 1,\dots,s-1$. Since $\tilde{V}(c) = V(0,c,0)$ and

$$(\partial^j V/\partial c^j)(t_0,0,0) = (\partial^j \tilde{V}/\partial c^j)(0) \; j = 1,\dots,h-1 \text{ and } (\partial^2 V/\partial\mu\partial c)(t_0,0,0)=2\pi\alpha'_\mu(0,1),$$

we can conclude that $(\partial^{h-1}\mu^*/\partial c^{h-1})(t_0,0) \neq 0$ and has the same sign of $\alpha'_\mu(0,1)$. Thus the function $\mu^*(t_0,\cdot)$ is invertible in the interval $[0,\bar{c}]$. Furthermore, this function is respectively strictly increasing or strictly decreasing in $[0,\bar{c}]$ according to $\alpha'_\mu(0,1)>0$ or $\alpha'_\mu(0,1)<0$. Since $\mu^*(t_0,0) = 0$, we immediately recognize that the bifurcating periodic solutions arise for $\mu>0$ in the first case, and for $\mu<0$ in the second case. Then, the proof of Theorem 3.1 is completed.

REFERENCES

1. Bernfeld S.R., Negrini P. and Salvadori L., Quasi-Invariant Manifolds, Stability, and Generalized Hopf Bifurcation, Ann. Mat. Pura Appl. (IV) CXXX (1982), pp. 105-119.

2. Bernfeld S.R., Salvadori L., and Visentin F., Hopf Bifurcation and Related Stability Problems for Periodic Differential Systems, (to appear).

3. Chow S.N. and Hale J.K., Methods of Bifurcation Theory (Springer-Verlag , New York, Heidelberg, Berlin, 1982).

4. De Oliveira J.C. and Hale J.K., Dynamic Behavior from Bifurcation Equations, Tohoku Math. J. 32 (1980), pp. 189-199.

5. Salvadori L. and Visentin F., Sul problema della biforcazione generalizzata di Hopf per sistemi periodici, Rend. Sem. Mat. Univ. Padova, 68 (1982), pp. 1-19.

6. Salvadori L., Methods of Qualitative Analysis in Bifurcation Problems,

in Proceedings of the IUTAM-ISIMM Symposium on "Modern Developments in Analytical Mechanics", (Atti dell'Accademia delle Scienze di Torino, 1983).

7. Salvadori L., An approach to Bifurcation Via Stability Theory, in Proceedings of the Conference "Recent Advances in Nonlinear Analysis and Differential Equations", Madras, India, 1981 (to appear).

8. Sansone G. and Conti R., Nonlinear Equations (MacMillan, New York, 1964).

Trends in the Theory and Practice of Non-Linear Analysis
V. Lakshmikantham (Editor)

REMARKS ON SOME STRONGLY NONLINEAR DEGENERATE STURM-LIOUVILLE EIGENVALUE PROBLEMS

Pierre A. Vuillermot

Department of Mathematics
University of Texas at Arlington
Arlington, TX 76019
U.S.A.

INTRODUCTION AND OUTLINE

This paper is devoted to discussing a new theorem concerning the existence and the $C^{1,1}$-regularity of countably many eigensolutions for a class (SL) of strongly nonlinear degenerate Sturm-Liouville eigenvalue problems with Young function nonlinearities in their principal part. The basic idea inherent in the proof of the theorem, given elsewhere [1], is threefold: for each $\varepsilon \in (0,1)$, we first replace problem (SL) by a class $(SL)_\varepsilon$ of regularized, nondegenerate Sturm-Liouville eigenvalue problems with strictly convex nonlinearities; we then invoke the monotonicity and compactness arguments of [2] and [3] to infer that problem $(SL)_\varepsilon$ possesses at least countably many $C^{(2)}$- eigensolutions; we finally get the existence and the $C^{1,1}$ - regularity of countably many eigensolutions to problem (SL) by taking the $\varepsilon\downarrow 0$-limit in problem $(SL)_\varepsilon$. The necessary bounds uniform in ε required to do so are obtained through new convexity inequalities which characterize the shape of the given nonlinearities. Two examples and a counter-example are given which illustrate the role of the various hypotheses, and which allow us to show that $C^{1,1}$-regularity is a nearly optimal result. For proofs and complete details, we refer the reader to [1]; for an extension of the method to elliptic boundary-value problems on bounded regions of $\mathbb{R}^n$, $2 \leq n \in \mathbb{N}^+$, see [4].

$C^{1,1}$-EIGENSOLUTIONS TO A CLASS OF STRONGLY NONLINEAR DEGENERATE STURM-LIOUVILLE BOUNDARY-VALUE PROBLEMS

Let $Y_{1,2}$ be two $C^{(2)}$ - Young functions in the sense of [5]; we consider the real-valued Sturm-Liouville eigenvalue problem

$$\left\{\begin{array}{l} \{Y_1'(z'(x))\}' + \lambda Y_2'(z(x)) = F(x;z(x)) \text{ in } \Omega \\ \gamma Y_1'(z(a)) - \tilde{\gamma} Y_1'(z'(a)) = 0 \\ \delta Y_1'(z(b)) + \tilde{\delta} Y_1'(z'(b)) = 0 \end{array}\right\} \quad \text{(SL)}$$

In (SL), we have defined $\Omega = (a,b) \subset \mathbb{R}$ and we shall write $\overline{\Omega} = [a,b]$; moreover, $\gamma, \tilde{\gamma}, \delta, \tilde{\delta} \in \mathbb{R}$ with $\gamma^2 + \tilde{\gamma}^2 = \delta^2 + \tilde{\delta}^2 = 1$ and we shall assume throughout that $\tilde{\gamma}, \tilde{\delta} \neq 0$ with $\mathrm{sgn}(\gamma) = \mathrm{sgn}(\tilde{\gamma})$ if $\gamma \neq 0$ and/or $\mathrm{sgn}(\delta) = \mathrm{sgn}(\tilde{\delta})$ if $\delta \neq 0$. The case with either $\tilde{\delta} = 0$ or $\tilde{\gamma} = 0$ can be treated separately using similar methods; finally, $F\colon \Omega \times \mathbb{R} \to \mathbb{R}$ is a Carathéodory function odd in its second argument. Our main result concerning problem (SL) is the following

Theorem II.1. Consider problem (SL) and pick $\mu \in (0;+\infty)$; assume moreover that the following hypotheses hold:

(H_1) The $C^{(2)}$ - Young functions $Y_{1,2}$ are convex in t^2 in the sense of [5].

(H_2) There exist $\nu > 2$ and $t \geq 0$ such that $Y_1(2t) \leq \nu Y_1(t)$ for each $t \in \mathbb{R}$ with $|t| \geq t_0$.

(H_3) There exist $f \in L^\infty(\Omega)$ and $\eta > 0$ such that

$$|F(x;\tau)| \leq f(x) + \eta \tilde{Y}_2^{-1}(Y_2(\tau))$$

where $\tilde{Y}_2^{-1}$ denotes the reciprocal inverse on $\mathbb{R}^+$ of $\tilde{Y}_2(s) = \max_{t \geq 0} \{|s|t - Y_2(t)\}$, $s \in \mathbb{R}$ (The Legendre transform of Y_2); moreover, $z \in C(\bar{\Omega};\mathbb{R})$ implies $x \to F(x;z(x))$ continuous on $\bar{\Omega}$.

Then problem (SL) possesses at least a countable infinity $\{\pm z_\mu^{(\alpha)}\}_{\alpha \in \mathbb{N}^+}$ of one-parameter antipodal pairs of $C^{1,1}(\bar{\Omega})$ - eigensolutions, which satisfy (SL) almost everywhere on Ω with $\int_\Omega dx\, Y_2(z_\mu^{(\alpha)}(x)) = \mu$.

Remark. With the additional hypothesis of strict convexity of Y_1, one can improve the $C^{1,1}(\bar{\Omega})$ - regularity of Theorem II.1 to a $C^2(\bar{\Omega})$ - regularity; this follows from the methods of [2].

For instance, we have the following

Example II.1. Consider the Sturm-Liouville eigenvalue problem

$$\left\{ \begin{array}{l} z''(x) + \lambda Y'(z(x)) = F(x;z(x)) \text{ in } \Omega \\ \\ \gamma z(a) - \tilde{\gamma} z'(a) = 0 \\ \delta z(b) + \tilde{\delta} z'(b) = 0 \end{array} \right\} \quad \text{(II.1)}$$

where $Y = Y_2$ and F satisfy (H_1) and (H_3), respectively. Problem (II.1) is of the form (SL) with the strictly convex Young function $Y_1(t) = t^2/2$; hence problem (II.1) has countably many $C^{(2)}(\bar{\Omega})$ - eigensolutions.

The above result contrasts sharply with the following

Example II.2. Consider the Sturm-Liouville eigenvalue problem

$$\left\{ \begin{array}{l} 3(z'(x))^2 z''(x) + \lambda Y'(z(x)) = F(x;z(x)) \text{ in } \Omega \\ \\ \gamma z^3(a) - \tilde{\gamma}(z')^3(a) = 0 \\ \\ \delta z^3(b) + \tilde{\delta}(z')^3(b) = 0 \end{array} \right\} \quad \text{(II.2)}$$

where $Y \equiv Y_2$ and F again satisfy (H_1) and (H_3), respectively. Problem (II.2) is still of the form (SL), but with $Y_1(t) = t^4/4$: Y_1 is not strictly convex since $Y''(0) = 0$; we infer from Theorem (II.1) that problem (II.2) has countably many eigensolutions with only $C^{1,1}$ - regularity since the problem is degenerate.

Remark. For nonlinear eigenvalue problems of the form (SL), $C^{1,1}$ - regularity is a nearly optimal result since a boundary-value problem like (II.2) does not have any C^2 - eigensolution in general. Indeed, we have the following

Example II.3. Consider the Sturm-Liouville problem

$$\left.\begin{cases} 3(z'(x))^2\, z''(x) + \lambda Y'(z(x)) = F(x;z(x)) \text{ in } \Omega \\ z(a) = z(b) = 0 \end{cases}\right\} \quad \text{(II.3)}$$

where Y and F satisfy (H_1) and (H_3) respectively and are chosen such that $F(x;\tau) - \lambda Y'(\tau) > 0$. Then problem (II.3) has no classical C^2 - eigensolutions; for if this were the case, there should be $x_0 \in (a,b)$ with $z'(x_0) = 0$, a contradiction since $(z')^2(x_0)\, z''(x_0) > 0$ from the first relation of (II.3).

The proof of Theorem (II.1) essentially follows from the $\varepsilon\downarrow 0$-limit taken in the regularization of problem (SL) defined in the next section.

$C^{(2)}(\bar{\Omega})$ - EIGENSOLUTIONS TO A REGULARIZATION OF PROBLEM (SL)

For $\varepsilon \in (0;1)$ and $i \in \{1,2\}$, define the Young functions $Y_{i,\varepsilon}(t) = Y_i(t) + \varepsilon t^2/2$; consider the regularized Sturm-Liouville eigenvalue problems

$$\begin{cases} Y_{1,\varepsilon}'(z_\varepsilon'(x))' + \lambda_\varepsilon Y_{2,\varepsilon}'(z_\varepsilon(x)) = F(x;z_\varepsilon(x)) \text{ in } \Omega \\ \gamma Y_{1,\varepsilon}'(z_\varepsilon(a)) - \tilde{\gamma} Y_{1,\varepsilon}'(z'_\varepsilon(a)) = 0 \\ \delta Y_{1,\varepsilon}'(z_\varepsilon(b)) + \tilde{\delta} Y_{1,\varepsilon}'(z'_\varepsilon(b)) = 0 \end{cases}$$

Since the Young functions $Y_{i,\varepsilon}$ are strictly convex, the methods of [2] and [3] give the following result:

Theorem III.1. Consider the regularized problem $(SL)_\varepsilon$ and pick $\mu \in (0;+\infty)$; assume moreover that hypotheses (H_1), (H_2) and (H_3) hold; then problem $(SL)_\varepsilon$ possesses at least countably many one-parameter antipodal pairs $\{\pm z^{\alpha}_{\mu,\varepsilon}\}_{\alpha \in \mathbb{N}^+}$ of $C^{(2)}(\bar{\Omega})$ - eigensolutions which satisfy $\int_\Omega dx\, Y_{2,\varepsilon}(\pm z^{(\alpha)}_{\mu,\varepsilon}(x)) = \mu$.

The proof of Theorem (II.1) follows upon taking the limit $\varepsilon\downarrow 0$ in Theorem (III.1); for complete details regarding the transition mechanism from $C^{(2)}$- regularity to $C^{1,1}$ - regularity as $\varepsilon\downarrow 0$, see [1]. For an extension of the method regarding the transition from C^2 - regularity to $C^{1,\alpha}_{loc}$ - regularity, $\alpha \in (0;1)$, in elliptic boundary-value problems over bounded regions on $\mathbb{R}^n$, $2 \leq n \in \mathbb{N}^+$, see [4], which goes beyond the recent results of [6], [7], [8] and [9].

REFERENCES

[1] Vuillermot, P. A., An optimal regularity result for a class of strongly nonlinear degenerate Sturm-Liouville eigenvalue problems, to be submitted to the Journal of Differential Equations (1984).

[2] Vuillermot, P. A., A class of Sturm-Liouville eigenvalue problems with polynomial and exponential nonlinearities, Nonlinear Analysis, Theory, Methods and Applications, 8 (1984) 775-796.

[3] Vuillermot, P. A., A class of elliptic partial differential equations with exponential nonlinearities, Mathematische Annnalen, 268 (1984) 497-518.

[4] Vuillermot, P. A., Regularity theory for strongly nonlinear degenerate Dirichlet boundary-value problems on Orlicz-Sobolev spaces, to be submitted to Inventiones Mathematicae (1985).

[5] Vuillermot, P. A., A class of Orlicz-Sobolev spaces with applications to variational problems involving nonlinear Hill's equations, J. Math. Anal. Appl. 89 (1982) 327-349.

[6] Evans, L. C., A new proof of $C^{1,\alpha}$- regularity of solutions of certain degenerate elliptic partial differential equations, J. Diff. Equations, 45 (1982) 356-373.

[7] Tolksdorf, P., Regularity of a more general class of quasilinear elliptic equations, J. Diff. Equations, 51 (1984) 126-150.

[8] Wardi-Lamrini, S., Régularité höldérienne de la solution d'une équation aux dérivées partielles fortement nonlinéare, Thèse de 3ème cycle, Publication Number 2868, Université Paul Sabatier, Toulouse, France (1983).

[9] Vuillermot, P. A., $C^{2,\alpha}(\Omega) \cap C^{0,\beta}(\bar{\Omega})$-Regularity for the solutions of strongly nonlinear eigenvalue problems on Orlicz-Sobolev spaces, Proceedings of the American Mathematical Society (1985).

The final (detailed) version of this paper will be submitted for publication elsewhere.

Trends in the Theory and Practice of Non-Linear Analysis
V. Lakshmikantham (Editor)

POINT DATA BOUNDARY VALUE PROBLEMS FOR FUNCTIONAL DIFFERENTIAL EQUATIONS*

Joseph Wiener and A. R. Aftabizadeh

Department of Mathematics and Computer Science
Pan American University
Edinburg, Texas
U.S.A.

Differential equations of mixed type with argument deviations having intervals of constancy are studied. They have the structure of continuous dynamical systems within intervals of unit length. Continuity of a solution at a point joining any two consecutive intervals implies recursion relations for the values of the solution at such points. The equations are thus similar in structure to those found in certain "sequential-continuous" models of disease dynamics.

1. INTRODUCTION

In [1] and [2] analytic solutions to differential equations with linear transformations of the argument have been studied. The initial values are given at the fixed point of the argument deviation. The author of [3] introduced a special class of functional differential equations (differential equations with involutions) and investigated point data initial value problems for them. This is possible because such equations can be reduced to ordinary differential equations. The discussion in [4] concerns two-point boundary value problems for functional differential equations with reflection of the argument. The method is based on the transformation of the functional differential equation to a higher-order ordinary differential equation. Differential equations with reflection found important applications in the study of stability of linear systems of differential-difference equations [5].

Biological models also often lead to functional differential equations which are reducible to systems of ordinary differential equations [6]. In the present article we consider functional differential equations with arguments having intervals of constancy. These equations are similar in structure to those found in certain "sequential-continuous" models of disease dynamics as treated in [7]. Our equations are directly related to difference equations of a discrete argument, the theory of which has been very intensively developed in numerous works. Bordering on difference equations are also impulse functional differential equations with impacts and switching and loaded equations (that is, those including values of the unknown solution for given constant values of the argument). Here we study differential equations with arguments $[t]$, $[t-n]$, and $[t+n]$, where $[t]$ denotes the greatest-integer function and n is a natural number. Connections are established between differential equations with piecewise constant deviations and difference equations of an integer-valued argument. Impulse and loaded equations may be included in our scheme too. Indeed, consider the equation

$$x'(t) = ax(t) + a_{-1}x([t-1]) + a_0x([t]) + a_1x([t+1]) \tag{1.1}$$

and write it as

---

*Research partially supported by U.S. Army Research Grant No. DAAG29-84-G-0034

$$x'(t) = ax(t) + \sum_{i=-\infty}^{\infty} (a_{-1}x(i-1) + a_0x(i) + a_1x(i+1))(H(t-i) - H(t-i-1)),$$

where $H(t) = 1$ for $t > 0$ and $H(t) = 0$ for $t < 0$. If we admit distributional derivatives, then differentiating the latter relation gives

$$x''(t) = ax'(t) + \sum_{i=-\infty}^{\infty} (a_{-1}x(i-1) + a_0x(i) + a_1x(i+1))(\delta(t-i) - \delta(t-i-1)),$$

where δ is the delta functional. This impulse equation contains the values of the unknown solution for the integral values of t. Differential equations of retarded and advanced types with piecewise constant argument deviations have been studied in [8] and [9], respectively. Therefore, in this paper we concentrate on equations of neutral type. In the second section, Eq. (1.1) is considered. The boundary value problem is posed at $t=-1$ and $t=0$, and the solution is sought for $t>0$. The existence and uniqueness of solution is proved and sufficient conditions of asymptotic stability of the trivial solution are determined. Then, the foregoing results are generalized for equations with many deviations. We show that these equations are intrinsically closer to difference rather than to differential equations. The main feature of equations with piecewise constant deviations is that it is natural to pose boundary value problems for them not on intervals, but at a number of individual points. In the last section, linear equations with variable coefficients are studied. The existence and uniqueness of solution on $[0, \infty)$ is proved for linear systems with continuous coefficients. A simple algorithm of computing the solution by means of continued fractions is indicated for a class of scalar equations. An existence criterion of periodic solutions to linear equations with periodic coefficients is established. Some nonlinear equations are also tackled.

2. EQUATIONS WITH CONSTANT COEFFICIENTS

Consider the scalar boundary value problem

$$x'(t) = ax(t) + a_{-1}x([t-1]) + a_0x([t]) + a_1x([t+1]), \tag{2.1}$$

$$x(-1) = c_{-1}, \quad x(0) = c_0$$

with constant coefficients. Here $[t]$ designates the greatest-integer function. We introduce the following

DEFINITION. A solution of Eq. (2.1) on $[0, \infty)$ is a function $x(t)$ that satisfies the conditions:

(i) $x(t)$ is continuous on $[0, \infty)$.

(ii) The derivative $x'(t)$ exists at each point $t \in [0, \infty)$, with the possible exception of the points $[t] \in [0, \infty)$ where one-sided derivatives exist.

(iii) Eq. (2.1) is satisfied on each interval $[n, n+1) \subset [0, \infty)$ with integral endpoints.

Denote

$$b_0(t) = e^{at} + a^{-1}a_0(e^{at}-1), \quad b_i(t) = a^{-1}a_i(e^{at}-1), \quad i = \pm 1 \tag{2.2}$$

and let λ_1 and λ_2 be the roots of the equation

$$(b_1(1) - 1)\lambda^2 + b_0(1)\lambda + b_{-1}(1) = 0. \tag{2.3}$$

THEOREM 2.1. Problem (2.1) with $a_1 \neq (e^a-1)/a$ and $a_{-1} \neq 0$ has on $[0, \infty)$ a unique solution

$$x(t) = b_{-1}(\{t\})c_{[t-1]} + b_0(\{t\})c_{[t]} + b_1(\{t\})c_{[t+1]}, \tag{2.4}$$

where $\{t\}$ is the fractional part of t and

$$c_{[t]} = (\lambda_1^{[t+1]}(c_0-\lambda_2 c_{-1}) + (\lambda_1 c_{-1}-c_0)\lambda_2^{[t+1]})/(\lambda_1-\lambda_2), \tag{2.5}$$

and this solution cannot grow to infinity faster than exponentially.

Proof. For $n \leq t < n+1$, Eq. (2.1) takes the form

$$x(t) = ax(t) + a_{-1}x(n-1) + a_0x(n) + a_1x(n+1)$$

with the general solution

$$x(t) = e^{a(t-n)}c - a^{-1}(a_{-1}x(n-1) + a_0x(n) + a_1x(n+1)),$$

where c is an arbitrary constant. Hence, a solution of Eq. (2.1) on the given interval satisfying the conditions

$$x(n+i) = c_{n+i}, \quad i = -1, 0, 1$$

is

$$x_n(t) = e^{a(t-n)}c - a^{-1}(a_{-1}c_{n-1} + a_0c_n + a_1c_{n+1}).$$

To determine the value of c, put $t = n$, then

$$c = a^{-1}a_{-1}c_{n-1} + (1+a^{-1}a_0)c_n + a^{-1}a_1c_{n+1},$$

and

$$x_n(t) = b_{-1}(t-n)c_{n-1} + b_0(t-n)c_n + b_1(t-n)c_{n+1}. \tag{2.6}$$

By virtue of the relation

$$x_n(n+1) = x_{n+1}(n+1) = c_{n+1},$$

we have

$$c_{n+1} = b_{-1}(1)c_{n-1} + b_0(1)c_n + b_1(1)c_{n+1},$$

whence

$$(b_1(1)-1)c_{n+1} + b_0(1)c_n+b_{-1}(1)c_{n-1} = 0, \ n \geq 0. \tag{2.7}$$

We look for a nontrivial solution of this difference equation in the form $c_n = \lambda^n$. Then

$$(b_1(1)-1)\lambda^{n+1} + b_0(1)\lambda^n + b_{-1}(1)\lambda^{n-1} = 0,$$

and λ satisfies (2.3). The hypotheses of the theorem imply $b_1(1) \neq 1$, $b_{-1}(1) \neq 0$. If the roots λ_1 and λ_2 of (2.3) are different, the general solution of (2.7) is

$$c_n = k_1\lambda_1^n + k_2\lambda_2^n,$$

with arbitrary constants k_1 and k_2. In fact, it satisfies (2.7) for all integral n. In particular, for n=-1 and n=0 this formula gives

$$\lambda_1^{-1}k_1 + \lambda_2^{-1}k_2 = c_{-1}, \quad k_1 + k_2 = c_0,$$

and

$$k_1 = \lambda_1(c_0-\lambda_2 c_{-1})/(\lambda_1-\lambda_2), \quad k_2 = \lambda_2(\lambda_1 c_{-1}-c_0)/(\lambda_1-\lambda_2).$$

These results, together with (2.6), establish (2.4) and (2.5). If $\lambda_1 = \lambda_2 = \lambda$, then

$$c_n = \lambda^n(c_0(n+1) - \lambda c_{-1}n),$$

which is the limiting case of (2.5) as $\lambda_1 \to \lambda_2$. Formula (2.4) was obtained with the implicit assumption $a \neq 0$. If $a=0$, then

$$x(t) = c_{[t]} + (a_{-1}c_{[t-1]} + a_0 c_{[t]} + a_1 c_{[t+1]})\{t\},$$

which is the limiting case of (2.4) as $a \to 0$. The uniqueness of solution (2.4) on $[0, \infty)$ follows from its continuity and from the uniqueness of the coefficients c_n for each $n \geq 0$. The conclusion about the solution growth is an implication of the estimates for expressions (2.5).

From (2.4) and (2.5) it follows that the solution $x(t) = 0$ of Eq. (2.1) is asymptotically stable as $t \to +\infty$ if and only if the roots λ_i of Eq. (2.3) satisfy the inequalities $|\lambda_i| < 1$.

THEOREM 2.2. The solution $x = 0$ of Eq. (2.1) is asymptotically stable if

$$\left(a_1 - \frac{a}{e^a-1}\right)a_{-1} < 0 \tag{2.8}$$

and either of the following hypotheses is satisfied:

(i) $(a+a_0+a_1+a_{-1})\left(a_0+\dfrac{ae^a}{e^a-1}\right) < 0,$

(ii) $\left(a_0+\dfrac{ae^a}{e^a-1}\right)\left(a_0-a_1-a_{-1}+\dfrac{a(e^a+1)}{e^a-1}\right) < 0,$

where the first factors in (i) and (ii) retain the sign of $a_1-a(e^a-1)^{-1}$.

Proof. With the notations $b_j = b_j(1)$, it follows from (2.8) that

$$D^2 = b_0^2 - 4(b_1-1)b_{-1} > 0,$$

hence, the roots

$$\lambda_{1,2} = (-b_0 \pm D)/2(b_1-1)$$

of (2.3) are real. If $b_1 > 1$, the inequalities $|\lambda_i| < 1$ are equivalent to

$$-b_0 + D < 2(b_1-1) \tag{2.9}$$

and

$$b_0 + D < 2(b_1-1). \tag{2.10}$$

For $b_0 < 0$, we take into account only (2.9) and obtain

$$b_0 + b_1 + b_{-1} > 1,$$

which in terms of the coefficients a_j coincides with hypothesis (i). If $b_0 > 0$, we consider only (2.10) which yields the result

$$b_1 - b_0 + b_{-1} < 1,$$

equivalent to hypothesis (ii). The case $b_1 < 1$ is treated similarly.

Let $x_n(t)$ be a solution of the equation

$$x'(t) = ax(t) + \sum_{i=-N}^{N} a_i x([t+i]), \quad N \geq 2 \tag{2.11}$$

with constant coefficients on the interval $[n, n+1)$. If the initial conditions for (2.11) are

$$x(n+i) = c_{n+i}, \quad -N \leq i \leq N$$

then we have the equation

$$x_n'(t) = ax_n(t) + \sum_{i=-N}^{N} a_i c_{n+i}$$

the general solution of which is

$$x_n(t) = e^{a(t-n)} c - \sum_{i=-N}^{N} a^{-1} a_i c_{n+i}, \quad a \neq 0.$$

For $t=n$ this gives

$$c_n = c - \sum_{i=-N}^{N} a^{-1} a_i c_{n+i}$$

and

$$x_n(t) = e^{a(t-n)} c_n + (e^{a(t-n)} - 1) \sum_{i=-N}^{N} a^{-1} a_i c_{n+i}. \tag{2.12}$$

Taking into account that $x_n(n+1) = x_{n+1}(n+1) = c_{n+1}$, we obtain

$$c_{n+1} = e^a c_n + \sum_{i=-N}^{N} (e^a-1) a^{-1} a_i c_{n+i}.$$

With the notations

$$b_0 = e^a + a^{-1} a_0 (e^a-1), \quad b_1 = a^{-1} a_1 (e^a-1) - 1,$$

$$b_i = a^{-1} a_i (e^a-1), \quad i = -1, \pm 2, \ldots, \pm N$$

this equation takes the form

$$\sum_{i=-N}^{N} b_i c_{n+i} = 0. \tag{2.13}$$

Its particular solution is sought as $c_n = \lambda^n$; then

$$\sum_{i=-N}^{N} b_i \lambda^{N+i} = 0. \tag{2.14}$$

If all roots $\lambda_1, \ldots, \lambda_{2N}$ of (2.14) are simple, the general solution of (2.13) is given by

$$c_n = \sum_{i=1}^{2N} k_i \lambda_i^n , \tag{2.15}$$

with arbitrary constant coefficients. The boundary value problem for (2.11) may be posed at any 2N consecutive points. Thus we consider the existence and uniqueness of solution to (2.11) for $t \geq 0$ satisfying the conditions

$$x(i) = c_i, \quad -N \leq i \leq N - 1 \tag{2.16}$$

Then letting $n = -N, \ldots, N - 1$ and $c_n = x(n)$ in (2.15) we get a system of equations with Vandermond's determinant $\det(\lambda_j^i)$ which is different from zero. Hence, the unknowns k_j are uniquely determined by (2.16). If some roots of (2.14) are multiple, the general solution of (2.13) contains products of exponential functions by polynomials of certain degree. The limiting case of (2.12) as $a \to 0$ gives the solution of (2.11) when $a = 0$. We proved

THEOREM 2.3. Problem (2.11) - (2.16) has a unique solution on $[0, \infty)$ if $a_{\pm N} \neq 0$. This solution cannot grow to infinity faster than exponentially.

Remark. The condition $a_{-N} \neq 0$ is nonessential. If $a_{-N} = 0$, solution (2.12) on $[0, \infty)$ depends only on $2N - 1$ boundary conditions $x(i) = c_i$, $-(N-1) \leq i \leq N - 1$ and does not contain c_{-N}.

3. EQUATIONS WITH VARIABLE COEFFICIENTS

Along with the equation

$$x'(t) = A(t)x(t) + \sum_{i=-N}^{N} A_i(t)x([t+i]), \; N \geq 2, \; (0 \leq t < \infty) \tag{3.1}$$

$$x(i) = c_i, \quad -N \leq i \leq N - 1$$

the coefficients of which are rxr- matrices and x is an r- vector, we consider

$$x'(t) = A(t)x(t). \tag{3.2}$$

If $A(t)$ is continuous, the problem $x(0) = c_0$ for (3.2) has a unique solution $x(t) = U(t)c_0$, where $U(t)$ is the solution of the matrix equation

$$U'(t) = A(t)U(t), \; U(0) = I. \tag{3.3}$$

The solution of the problem $x(s) = c_0$, $s \in [0, \infty)$ for (3.2) is represented in the form

$$x(t) = U(t)U^{-1}(s)c_0.$$

Let

$$B_{0n}(t) = U(t)\left(U^{-1}(n) + \int_n^t U^{-1}(s)A_0(s)ds\right),$$

$$B_{in}(t) = U(t)\int_n^t U^{-1}(s)A_i(s)ds, \; i = \pm 1, \ldots, \pm N. \tag{3.4}$$

THEOREM 3.1. Problem (3.1) has a unique solution on $0 \leq t < \infty$ if $A(t)$ and $A_i(t) \in [0, \infty)$, and the matrices $B_{Nn}(n+1)$ are nonsingular for $n = 0, 1, \ldots$.

Proof. On the interval $n \le t < n+1$ we have the equation

$$x'(t) = A(t)x(t) + \sum_{i=-N}^{N} A_i(t)c_{n+i}, \quad c_{n+i} = x(n+i).$$

Its solution $x_n(t)$ satisfying the condition $x(n) = c_n$ is given by the expression

$$x_n(t) = U(t)(U^{-1}(n)c_n + \int_n^t U^{-1}(s) \sum_{i=-N}^{N} A_i(s)c_{n+i}ds). \tag{3.5}$$

Hence, the relation $x_n(n+1) = x_{n+1}(n+1) = c_{n+1}$ implies

$$c_{n+1} = U(n+1)(U^{-1}(n)c_n + \int_n^{n+1} U^{-1}(s) \sum_{i=-N}^{N} A_i(s)c_{n+i}ds).$$

With the notations (3.4), this difference equation takes the form

$$B_{Nn}(n+1)c_{n+N} + \ldots + B_{2n}(n+1)c_{n+2} + (B_{1n}(n+1) - I)c_{n+1} +$$

$$+ B_{0n}(n+1)c_n + B_{-1n}(n+1)c_{n-1} + \ldots + B_{-Nn}(n+1)c_{n-N} = 0, \quad (n \ge 0).$$

Since the matrices $B_{Nn}(n+1)$ are nonsingular for all $n \ge 0$, there exists a unique solution $c_n (n \ge N)$ provided that the values $c_{-N}, \ldots, c_{N-1}$ are prescribed. Substituting the vectors c_n in (3.5) yields the solution of (3.1). For the scalar equation

$$x'(t) = a(t)x(t) + a_{-1}(t)x([t-1]) + a_0(t)x([t]) + a_1(t)x([t+1]),$$

$$x(-1) = c_{-1}, \quad x(0) = c_0$$

with coefficients continuous on $[0, \infty)$ we can indicate a simple algorithm of computing the solution. According to (3.4) and (3.5), we have

$$x_n(t) = B_{-1n}(t)c_{n-1} + B_{0n}(t)c_n + B_{1n}(t)c_{n+1}, \tag{3.6}$$

with

$$U(t) = \exp(\int_0^t a(s)ds)$$

The coefficients c_n satisfy the equation

$$(B_{1n}(n+1) - 1)c_{n+1} + B_{0n}(n+1)c_n + B_{-1n}(n+1)c_{n-1} = 0, \quad n \ge 0.$$

Denote

$$d_{-1}(n+1) = B_{-1n}(n+1)/(1-B_{1n}(n+1)), \quad d_0(n+1) = B_{0n}(n+1)/(1-B_{1n}(n+1)),$$

$$r_n = c_{n+1}/c_n.$$

Then from the relation

$$c_{n+1} = d_0(n+1)c_n + d_{-1}(n+1)c_{n-1}$$

it follows that

$$r_n = d_0(n+1) + \frac{d_{-1}(n+1)}{r_{n-1}},$$

which yields

$$r_0 = d_0(1) + \frac{d_{-1}(1)}{r_{-1}},$$

$$r_1 = d_0(2) + \frac{d_{-1}(2)}{r_0} = d_0(2) + \cfrac{d_{-1}(2)}{d_0(1) + \cfrac{d_{-1}(1)}{r_{-1}}},$$

and continuing this procedure leads to the continued-fraction expansion

$$r_n = d_0(n+1) + \frac{d_{-1}(n+1)}{d_0(n) +} \cdots \frac{d_{-1}(2)}{d_0(1) +} \frac{d_{-1}(1)}{c_0/c_{-1}}$$

and to the formula

$$c_n = r_{-1} r_0 \cdots r_{n-1} c_{-1}, \quad n \geq 1$$

for the coefficients of solution (3.6).

THEOREM 3.2. Assume that the coefficients in (3.1) are periodic matrices of period 1. If there exists a periodic solution $x(t)$ of period 1 to Eq. (3.1), then $\lambda = 1$ is an eigenvalue of the matrix

$$S = \sum_{i=-N}^{N} B_{i0}(1),$$

where $B_{i0}(t)$ are given in (3.4) and $x(0)$ is a corresponding eigenvector. Conversely, if $\lambda = 1$ is an eigenvalue of S and c_0 is a corresponding eigenvector and if the matrices $B_{\pm N0}(1)$ are nonsingular, then Eq. (3.1) with the conditions

$$x(i) = c_0, \quad -N \leq i \leq N - 1$$

has a unique solution on $-\infty < t < \infty$, and it is 1-periodic.

Along with the scalar equation

$$x'(t) = f(x(t), \{x([t+i])\}_{i=-N}^{N}), \quad (0 \leq t < \infty) \tag{3.7}$$

$$x(i) = c_i, \quad -N \leq i \leq N - 1,$$

where f is continuous in the space of its variables, we consider the ordinary differential equation with parameters

$$x' = f(x, \{\lambda_i\}_{-N}^{N}). \tag{3.8}$$

THEOREM 3.3. If the solutions of (3.8) can be extended over $[0, \infty)$ and the equation

$$\int_{\lambda_0}^{\lambda_1} \frac{dx}{f(x, \{\lambda_i\}_{-N}^{N})} = 1$$

has a unique solution with respect to λ_N, then on $[0, \infty)$ there exists a unique solution of problem (3.7).

REFERENCES

[1] Shah, S. M. and Wiener, J., Distributional and entire solutions of ordinary differential and functional differential equations, Internat. Jrnl. Math &

Math. Sci. 6(2) (1983), 243-270.

2 Cooke, K. L. and Wiener, J., Distributional and analytic solutions of functional differential equations, Jrnl. Math. Anal. and Appl. 98(1) (1984), 111-129.

3 Wiener, J., Differential equations with involutions, Differencial'nye Uravnenija 5(6) (1969), 1131-1137.

4 Wiener, J. and Aftabizadeh, A. R., Boundary value problems for differential equations with reflection of the argument, Internat. Jrnl. Math. & Math. Sci. (to appear).

5 Castelan, W. B. and Infante, E. F., On a functional equation arising in the stability theory of difference-differential equations, Quart. Appl. Math. 35(1977), 311-319.

6 Busenberg, S. N. and Travis, C. C., On the use of reducible-functional differential equations in biological models, Jrnl. Math. Anal. and Appl. 89(1) (1982), 46-66.

7 Busenberg, S. N. and Cooke, K. L., Models of vertically transmitted diseases with sequential-continuous dynamics, in "Nonlinear Phenomena in Mathematical Sciences" (V. Lakshmikantham, Ed.), pp. 179-187 (Academic Press, New York, 1982).

8 Cooke, K. L. and Wiener, J., Retarded differential equations with piecewise constant delays, Jrnl. Math. Anal. and Appl. 99(1) (1984), 265-297.

9 Shah, S. M. and Wiener, J., Advanced differential equations with piecewise constant argument deviations, Internat. Jrnl. Math. & Math. Sci. 6(4) (1983), 671-703.

Trends in the Theory and Practice of Non-Linear Analysis
V. Lakshmikantham (Editor)

ON THE USE OF ITERATIVE METHODS WITH SUPERCOMPUTERS FOR SOLVING PARTIAL DIFFERENTIAL EQUATIONS*

David M. Young and David R. Kincaid

Center for Numerical Analysis
The University of Texas
Austin, Texas
U.S.A.

The paper describes research in the Center for Numerical Analysis of The University of Texas on the numerical solution of elliptic partial differential equations by descretization methods. The emphasis is on iterative algorithms and software for solving large sparse systems of linear equations. Particular attention is paid to the treatment of nonsymmetric systems, which correspond to non-self-adjoint problems, and to the use of supercomputers.

INTRODUCTION

In this paper we describe research which is being done in the Center for Numerical Analysis of The University of Texas at Austin on numerical methods for solving partial differential equations. A key subproblem in the numerical solution of partial differential equations is that of solving a system of linear algebraic equation where the number of equations is very large and where the coefficient matrix is very sparse. Our research is primarily concerned with the use of iterative methods rather than direct methods for solving these systems.

The ITPACK 2C software package has been developed as a research tool for studying the behavior of various iterative algorithms for solving large sparse linear systems. This package provides for a limited number of alternative algorithms and is designed primarily to handle symmetric, or nearly symmetric, systems using a conventional, or scalar, computer. Our current work involves the expansion of ITPACK to allow for a greater number of alternative algorithms, to handle nonsymmetric systems, and to run efficiently on vector and parallel computers.

In Section 2 we describe briefly how linear systems arise in the solution of partial differential equations. In Section 3 we give a brief discussion of the relative merits of direct and iterative methods. Sections 4 and 5 contain a description of ITPACK 2C and the underlying iterative algorithms. Sections 6 and 7 provide a description of an expanded package, ITPACK 3A, which provides for a greater variety of iterative algorithms including algorithms designed to handle nonsymmetric systems. Our work on the use of supercomputers is summarized in Section 8.

---

* This work was supported in part by the National Science Foundation through Grant MCS-8214731, by the Department of Energy through Grant DE-AS05-81ER10954, by the Control Data Corporation through Grant 81T01 and PACER Fellowship 84PCR54B, and by the North Atlantic Treaty Organization through Grant 648/83 with The University of Texas at Austin.

PARTIAL DIFFERENTIAL EQUATIONS

Let us consider the solution of the elliptic partial differential equation

$$(2.1) \qquad L[u] = (Au_x)_x + (Cu_y)_y + Du_x + Eu_y + Fu = G$$

where A, C, D, E, F, and G depend on x and y and where $A > 0$, $C > 0$ and $F \leq 0$ in a bounded plane region R. The solution u is required to satisfy Dirchlet, Neumann or mixed boundary conditions on the boundary, S, of R. By the use of finite difference methods or finite element methods the above problem can be reduced to the solution of a linear system of the form

$$(2.2) \qquad Au = b$$

where A is a given square matrix and b is a given column vector. Normally, in order to obtain an accurate solution to (2.1) it is necessary to choose a fine mesh (or to choose small elements) and in such cases the number of equations is very large. However, in most cases the matrix A will be sparse. Thus, if a five-point finite difference equation is used, there will be at most five nonzero elements in any given row of A.

We are also interested in time-dependent problems of the form

$$(2.3) \qquad \frac{\partial u}{\partial t} = L[u] - G$$

Here, for each t, boundary conditions are given on S and for an initial value, t_0, of t values of u are assumed to be given in R. The usual numerical procedure is to obtain values of u in the region R successively for $t = t_1 = t_0 + (\Delta t)_0$, $t_2 = t_1 + (\Delta t)_1$, etc. Explicit methods such as the forward difference method can be used. However, because of stability considerations, the allowable time step size must usually be extremely small. In order to be able to use a larger time step one must usually use a fully implicit scheme, such as the backward difference method, or use a fully implicit scheme, such as the backward difference method, or a semi-implicit scheme, such as the Crank-Nicolson method. Such methods require the solution of a linear system of the form (2.2) at each time step.

ITERATIVE METHODS AND DIRECT METHODS

For the solution of the linear system (2.2) one can use an iterative method or a direct method such as Gaussian elimination. A number of software packages based on the use of direct methods are available: including LINPACK, the Yale sparse matrix package (YSMP), and SPARSPAK to mention only a few. An excellent survey of such methods is given in the paper by Duff [1984].

In general, however, for almost any "family" of problems of increasing size, the computer time and storage required using direct methods will increase faster than that required using iterative methods. As the problems get larger there will be a "crossover point", i.e., a problem such that for larger problems of the family the amount of storage and machine time required for iterative methods will be less than for direct methods.

Problems where iterative methods appear to have definite advantages over direct methods include problems involving three space dimensions, time-dependent problems involving two space dimensions (unless linear with constant coefficients), steady state problems involving two space dimensions which are nonlinear and/or where the mesh is very fine.

There are several reasons why many people who are not familiar with iterative methods have often tended to avoid them and to use direct methods instead. The first is the difficulty in choosing an iterative method appropriate for a given problem. Second, for many iterative methods it is necessary to choose certain iteration parameters such as the relaxation factor for the SOR method. In many cases, these parameters must be chosen very accurately or else the convergence will be slow. Third, it is often difficult to determine when the approximate solution obtained by an iterative method is sufficiently accurate so that the procedure can be terminated.

One of the objectives of the ITPACK project has been to try to contribute to improving the "image" of iterative methods among non-experts. A package of research-oriented programs has been provided to demonstrate that, for many problems at least, iterative algorithms can be successfully applied by non-experts.

ITERATIVE ALGORITHMS

Each of the iterative algorithms used in ITPACK consists of the following four components: a basic iterative method; an acceleration procedure; an adaptive procedure for choosing any necessary iteration parameters; a stopping procedure to decide when to terminate the iteration process. In this section we will give a brief description of each component. Further details can be found in Hageman and Young [1981] and in Grimes, Kincaid and Young [1979].

Given the system (2.2), a very simple basic iterative method is the Richardson method defined by

$$u^{(n+1)} = u^{(n)} + (b - Au^{(n)}) = Gu^{(n)} + k \tag{4.1}$$

where

$$G = I - A, \qquad k = b . \tag{4.2}$$

A more general basic iterative method can be defined by first choosing a non-singular matrix Q and applying Richardson's method to the <u>preconditioned system</u>

$$Q^{-1}Au = Q^{-1}b \tag{4.3}$$

This yields the method (4.1) with

$$G = I - Q^{-1}A, \qquad k = Q^{-1}b \tag{4.4}$$

Another approach is to choose a nonsingular "splitting" matrix Q and to represent A in the form of $A = Q - (Q - A)$. If we write (2.2) in the form

$$Qu = (Q - A)u + b \tag{4.5}$$

and then determine $u^{(n+1)}$ from $u^{(n)}$ by

$$Qu^{(u+1)} = (Q - A)u^{(n)} + b \tag{4.6}$$

we get (4.1) and (4.4) after solving for $u^{(n+1)}$.

The splitting matrix Q must be chosen to be such that for any given vector y one can easily solve the system $Qx = y$ for x. Examples of suitable splitting matrices are diagonal matrices, tridiagonal matrices, triangular matrices or products of triangular matrices. The choices of Q for the standard basic iterative methods are:

Richardson's method	$Q = I$
Jacobi method	$Q = D$
SOR method	$Q = \omega^{-1}D - C_L$
SSOR method (symmetric SOR)	$Q = (2 - \omega)^{-1}(\omega^{-1}D - C_U)D^{-1}(\omega^{-1}D - C_L)$
incomplete Cholesky (ICC)	$Q = LU$

Here $A = D - C_L - C_U$ where D is a diagonal matrix, C_L is a strictly lower triangular matrix and C_U is a strictly upper triangular matrix. The number ω, which lies between 0 and 2, is the relaxation factor. Also L is a lower triangular matrix and U is an upper triangular matrix such that if $a_{ij} = 0$ then $\ell_{i,j} = u_{i,j} = 0$.

Another basic iterative method, known as the RS method can be defined if the matrix A of (2.2) is a red-black matrix of the form

$$A = \begin{pmatrix} D_R & H \\ K & D_B \end{pmatrix} \tag{4.7}$$

where D_R and D_B are diagonal matrices. In this case we can write u in the block form $u = (u_R^T, u_B^T)^T$ and eliminate u_R thus obtaining the reduced system

$$u_B = (D_B^{-1} K D_R^{-1} H)u_B + D_B^{-1}(K D_R^{-1} b_R + b_B) \tag{4.8}$$

where $b = (b_R^T, b_B^T)^T$. Applying Richardson's method to (4.8) yields the RS method.

A basic iterative method of the form (4.1) is symmetrizable if $I - G$ is similar to a symmetric and positive definite (SPD) matrix. If A and Q are SPD then the method is symmetrizable. If A is SPD then the Jacobi, SSOR and ICC methods are symmetrizable, but the SOR method is not, in general. The convergence of a symmetrizable basic iterative method can be greatly accelerated by using Chebyshev acceleration or conjugate gradient acceleration. In either case the formula can be written in the form

$$u^{(n+1)} = \rho_{n+1}\{\gamma_{n+1}(Gu^{(u)} + k) + (1 - \gamma_{n+1})u^{(n)}\} + (1 - \rho_{n+1})u^{(n-1)} \tag{4.9}$$

For Chebyshev acceleration the parameters ρ_{n+1} and γ_{n+1} involve estimates of m(G) and M(G), the smallest and largest eigenvalues of G, respectively. (Note that since the method is symmetrizable the eigenvalues of G are real and less than one.) For conjugate gradient acceleration the formulas for ρ_{n+1} and γ_{n+1} involve certain inner products of vectors. To apply conjugate gradient acceleration one must choose an auxiliary matrix Z such that Z and $Z(I - G)$ are SPD. Such a matrix Z exists if the method is symmetrizable; if A and Q are SPD one can choose $Z = A$ or $Z = Q$. While conjugate gradient acceleration requires more work per iteration than Chebyshev acceleration, it has two important advantages. First, no iteration parameters need be estimated, unless, as in the case of the SSOR method, the basic method itself involves an iteration parameter. Second, the convergence is at least as fast as that of Chebyshev acceleration and is in many uses very much faster.

The use of an acceleration procedure for a symmetrizable iteration method can result in an order-of-magnitude improvement in convergence. Thus if the condition number $K(I - G)$, which is the ratio of the largest eigenvalue of $I - G$ to the smallest eigenvalue, is large, the number of iterations required with the accelerated scheme is on the order of $\sqrt{K(I - G)}$ as compared to $K(I - G)$ for the unaccelerated scheme. In the case of a linear system arising from a 5-point discretization of Laplace's equation with mesh size h, the accelerated Jacobi method requires on the order of h^{-1} iterations as compared with h^{-2} for the unaccelerated Jacobi method. Here h is the mesh size.

In order to carry out an iterative procedure involving a basic iterative method and an acceleration procedure, one or more iteration parameters may be required. Thus, for Chebyshev acceleration one needs estimates for $m(G)$ and $M(G)$. Also, for the SSOR method the parameter ω is required both for Chebyshev and for conjugate gradient acceleration. However, no parameters are required with conjugate gradient acceleration with the Jacobi or RS methods.

The sensitivity of the iterative procedures to the choice of iteration parameters is usually so great as to make it impractical, in general, to estimate them a priori. Fortunately, adaptive procedures are available to determine the parameters automatically. With these procedures one estimates initial values of the iteration parameters, which may be very crude, and begins the iteration process. If, at any stage, the observed convergence rate is appreciably less than the anticipated convergence rate then the iteration parameters are modified.

An often neglected problem with iterative algorithms is that of deciding when to terminate the iteration process. In other words, when is the vector $u^{(n)}$ a sufficiently accurate approximation to the true solution $\bar{u} = A^{-1}b$? Ideally, we would like to stop the iterations when

$$||u^{(n)} - \bar{u}||_\alpha / ||u||_\alpha \leqq \zeta \tag{4.10}$$

where $||\cdot||_\alpha$ is a suitable norm and ζ is a stopping number in the range, say, $10^{-3} \leqq \zeta \leqq 10^{-8}$. It can be shown that for certain norms, α, (4.10) is satisfied if the condition

$$(1 - M(G))^{-1} ||\delta^{(n)}||_\alpha / ||\bar{u}||_\alpha \leqq \zeta \tag{4.11}$$

is satisfied where $\delta^{(n)} = Gu^{(n)} + k - u^{(n)}$. Instead of (4.11) we use the test defined by

$$(1 - M_E)^{-1} ||\delta^{(n)}||_\alpha / ||u^{(n)}||_\alpha \leqq \zeta \tag{4.12}$$

For adaptive Chebyshev acceleration M_E is the latest estimate for $M(G)$. For conjugate gradient acceleration M_E is the largest eigenvalue of a certain tri-diagonal matrix of order n whose elements involve the $\{\rho_i\}$ and the $\{\gamma_i\}$. The test (4.12) has been found to be satisfactory in a wide class of cases.

ITPACK 2C

A package of programs, known as ITPACK 2C, has been prepared which includes seven alternative iterative algorithms. These algorithms include the Jacobi, SSOR and RS methods with Chebyshev acceleration (the J-SI, SSOR-SI and RS-SI methods); the same methods with conjugate gradient acceleration (the J-CG, SSOR-CG, and RS-CG methods); and the SOR method.

Iteration parameters are determined adaptively where appropriate. For the J-SI method it is assumed that a number $\underline{m}$ is available such that $\underline{m} \leq m(G)$. In this case $M(G)$ is determined adaptively. For the SSOR and RS methods, since it can be shown that $m(G) \geqq 0$, only $M(G)$ and ω are to be determined adaptively for the SSOR-SI method and only $M(G)$ is to be determined adaptively for the RS-SI method. No iteration parameters are needed for the J-CG and RS-CG methods. The relaxation factor ω is determined adaptively for the SSOR-CG and SOR methods.

As described by Kincaid, Respess, Young and Grimes [1982], the use of ITPACK 2C involves calling any one of the seven subroutines. It is necessary to have the matrix A stored in the format used in the Yale Sparse Matrix Package, described by Eisenstat et al. [1977], and to provide a number of input quantities to control the iteration process; otherwise default values are automatically inserted.

As described by Kincaid and Young [1984], the programs of ITPACK 2C have been incorporated into the ELLPACK software package. The ELLPACK package, see, e.g., Rice [1981] and Rice and Boisvert [1984], is a collection of routines for solving a class of partial differential equations by various procedures. The user provides information, using a special language, to specify such things as the differential equation, domain, boundary conditions, mesh discretization, solution of the algebraic system, etc. A preprocessor constructs a FORTRAN program using various modules. Included among the solution modules for ELLPACK are seven modules based on the seven ITPACK 2C routines. These routines were modified so as to use the ELLPACK data structure.

The ITPACK 2C package has been made available to the scientific community through the distribution services of the International Mathematical and Statistical Libraries (IMSL) and the Transactions on Mathematical Software. The package has been used on a variety of computers throughout the world.

THE NONSYMMETRIZABLE CASE

In this section we consider the <u>nonsymmetrizable case</u> where the basic iterative method (4.1) is not symmetrizable, as defined in Section 4. We remark that in the <u>nonsymmetric case</u>, when the matrix A of the system (2.2) is not SPD, most standard basic iterative methods, such as the Jacobi, SSOR, and RS methods, used in ITPACK 2C are not symmetrizable. Even if A is SPD the basic iterative may be nonsymmetrizable, as in the case of the SOR method.

The nonsymmetrizable case is much more complicated than the symmetrizable case. For example, in the nonsymmetrizable case the eigenvalues of G need not be real and less than unity as in the symmetrizable case. Quite the contrary, they may be complex and they may even have real parts greater than unity. In the second place, the Jordan canonical form of G need not be diagonal. Both of these complications must be considered in choosing an effective acceleration procedure.

In this section we discuss very briefly several ways of dealing with nonsymmetrizable iterative methods. These include: generalized normal equations; Chebyshev acceleration; generalized conjugate gradient acceleration; Lanczos methods; and the "GCW method" of Concus and Golub [1976] and of Widlund [1978] where $Q = \frac{1}{2}(A + A^T)$.

Nonsymmetric systems frequently arise in the solution of partial differential equations. An example is the convection-diffusion equation

$$u_{xx} + u_{yy} + Du_x = f(x,y) \tag{6.1}$$

If one uses ordinary central difference (finite difference) representations of the derivatives, one will obtain a nonsymmetric system. Various techniques have been proposed to modify the difference equations to improve the convergence of various iterative methods; see, e.g., Axelsson and Gustaffson [1977], and Eisenstat et al. [1979].

The programs of ITPACK 2C, though not designed to handle nonsymmetric systems have been found to work in some nonsymmetric, but nearly symmetric problems. Thus for example, some of the programs have been found to work for some problems involving (6.1) where D is small. For larger values of D the procedures have failed.

Generalized Normal Equations

Given a linear system (2.2) with a matrix which is nonsingular but not necessarily SPD we can construct the normal equation $A^TAu = A^Tb$ by multiplying both sides of (2.2) by A^T. The resulting system has a matrix, A^TA, which is SPD. However, the condition number of A^TA may be much greater than that of A. However, Dongarra et al. [1981] and Elman [1982] considered the use of the generalized normal equations which are based on the preconditioned system (4.3). Other, slightly more general normal equations are considered by Elman [1982] and by Young, Jea, and Kincaid [1984].

Chebyshev Iteration

If all of the eigenvalues of the iteration matrix G have real parts less than one then one can speed up the convergence of a basic iterative method (4.1) by using Chebyshev acceleration. We again use (4.9). The optimum values of the iteration parameters can be determined if all of the eigenvalues of G are known. Programs for doing so have been developed by Manteuffel [1978] and by Huang [1983]. Procedures for estimating key eigenvalues of G have been developed by Manteuffel [1978] and by Elman [1984]. We are also developing such procedures based on the use of Lanczos acceleration.

Generalized Conjugate Gradient Acceleration

The form of the conjugate gradient acceleration used in ITPACK, see (4.9), is only one of several possible forms which can be used in the symmetrizable case. We refer to this form as ORTHORES. Other forms include a more familiar form, which we refer to as ORTHOMIN, and another form which we refer to as ORTHODIR. In order to find $u^{(n+1)}$ it is only necessary to have information available from the n-th iteration and the (n-1)st iteration for ORTHORES and ORTHOMIN. ORTHODIR also requires information from the (n-2)nd iteration.

The three forms of conjugate gradient acceleration can be generalized to non-symmetrizable iterative methods, see, e.g., Young and Jea [1980], [1981]. One chooses an auxiliary matrix Z; usually Z or Z(I - G) is SPD. The procedures converge, in theory, to the true solution in at most N iterations, where N is the order of the system. However, in the general case, in order to compute $u^{(n+1)}$ it is necessary to have information from all previous iterations. In most cases, the amount of computer time and storage required for these methods is prohibitively large. One procedure for reducing computer time and storage is to modify the formulas to choose an integer $s \geq 1$ and to discard all information except that involving $u^{(n)}, u^{(n-1)}, \ldots, u^{(n-s)}$. Thus we obtain ORTHODIR(s), ORTHOMIN(s), and ORTHORES(s). In the symmetrizable case, with the proper choice of Z,

ORTHODIR, ORTHOMIN and ORTHORES automatically reduce to ORTHODIR(2), ORTHOMIN(1), and ORTHORES(1), respectively. Unfortunately, in the general case many of the theoretical properties of the methods are lost. However, some results are known; see, for example, Eisenstat, Elman and Schultz [1983], Elman [1982] and Young and Jea [1980], [1981]. Thus, for example, Elman [1982] has shown that if $A + A^T$ is positive real then ORTHOMIN(s) applied to Richardson's method, with $Z = I$, converges.

As shown by Jea [1982] and by Jea and Young [1983] it is always possible to simplify the formulas for the three forms of the generalized conjugate gradient method by suitably choosing the auxiliary matrix Z. Such a simplification occurs if for some s ORTHODIR, ORTHOMIN and ORTHORES are equivalent to ORTHODIR(s+1), ORTHOMIN(s), and ORTHORES(s), respectively. However, it is often not practical to find such a Z and even when such a Z can be found the resulting method may not converge. However, if $I - G$ is similar to a symmetric matrix or to a normal matrix then, for suitable Z and s, ORTHODIR(s) converges and is equivalent to ORTHODIR. It also follows from results of Faber and Manteuffel [1984] that if such simplification occurs for some SPD matrix Z then, except for some very special cases, $I - G$ is similar to a normal matrix.

Lanczos Methods

As an alternative to generalized conjugate gradient acceleration, we can use any one of three versions of the Lanczos method. These three versions are referred to by Jea and Young [1983] as Lanczos/ORTHODIR, Lanczos/ORTHOMIN, and Lanczos/ORTHORES, respectively. Lanczos/ORTHOMIN is equivalent to the biconjugate gradient method considered by Fletcher [1975]. These three methods require about twice as much work per iteration as ORTHODIR(2), ORTHOMIN(1) and ORTHORES(1), respectively. They have the property that they converge to the true solution in at most N steps, provided that they do not break down.

GCW Method

Concus and Golub [1976] and Widlund [1978] considered the use of the splitting matrix $Q = \frac{1}{2}(A + A^T)$ for the case where the matrix A is PR. We refer to the resulting method as the GCW method. In order for the GCW method to be feasible it is important that one be able to solve conveniently systems of the form $Qx = y$ for x, given any y. This can be done for some problems arising from partial differential equations using fast direct methods. Otherwise one may have to use an "inner iteration process" to solve $Qx = y$. In any case, ORTHORES = ORTHORES(1) for the GCW method with $Z = Q$ (see, e.g., Young, Jea and Kincaid [1984]). Numerical experiments indicate that the use of ORTHORES(1) with the GCW method is effective in many cases, especially when A is nearly symmetric.

ITPACK 3

As we have seen, there are a number of promising methods for solving nonsymmetric systems. On the other hand there are relatively few theoretical results concerning the convergence and rate of convergence of these methods. While continuing to search for theoretical results, we have emphasized the experimental approach. To facilitate this approach a package of programs, known as ITPACK 3 is being developed by T.S. Mai. This package will allow the user to test anyone of a large variety of iterative algorithms on various test problems.

A subset of ITPACK 3, known as "ITPACK 3A", is nearly completed and is now being tested. ITPACK 3A is designed for the non-expert user and can be used in a manner similar to ITPACK 2C. The user of ITPACK 3A can select one of several basic methods, acceleration procedures, and auxiliary matrices as well as certain other quantities. He can, if he wishes, construct his own splitting matrix Q and his own auxiliary matrix Z. The available basic methods include those provided in ITPACK 2C as well as the GCW method, certain incomplete factorization methods, and the alternating direction implicit method. Acceleration procedures include Chebyshev acceleration, and three versions of generalized conjugate gradient acceleration and of Lanczos acceleration. Provision has also been made for the use of the generalized normal equations.

For the more sophisticated user, facilities are being developed to make it possible to construct new algorithms from the set of subroutines of ITPACK 3. As an example one might want to test the GCW method with ORTHORES(1), perhaps using ITPACK 3A to carry out the inner iteration procedure. Alternatively, the user might wish to use a hybrid method to estimate the eigenvalues of G using the Lanczos method and then compute the optimum Chebyshev parameters prior to shifting over to Chebyshev acceleration.

THE USE OF SUPERCOMPUTERS

An important part of our research on iterative algorithms relates to the use of supercomputers. While we are interested in both vector and parallel machines, our work to date has been primarily concerned with vector machines such as the Control Data CYBER 205 and the CRAY-1.

With supercomputers there is a large potential increase in speed as compared with conventional, or <u>scalar</u> computers. In most cases, however, considerable effort is required on the part of the user to realize this potential. This is because the greater speed is achieved when a large stream of operations are being performed which are independent of the others. As an example, the vector addition operation $c_i = a_i + b_i$, $i = 1,2,\dots,N$ can be carried out very rapidly <u>per addition</u> as long as no a_i or b_i depends on any previously computed c_i. We refer to the organization of a program into operations of this type as <u>vectorization</u>. Unfortunately, solving a linear system $Lx = y$ for x, where L is a sparse lower triangular matrix, a very common operation with iterative algorithms, is very difficult to vectorize although it is very suitable for a scalar computer.

Several levels of effort can be expended on vectorization with varying degrees of improvement. A certain amount of improvement can often be made by changing the program and the storage format but without changing the basic algorithms used. We refer to such modifications as "short range modifications". Modifications involving major changes in the basic algorithms used are referred to as "long range modifications".

We have already completed a short-range modification of ITPACK 2C to obtain a new package ITPACKV 2C; see Kincaid et al. [1984]. This new package is based on certain programming modifications as well as a change in the storage format. The storage format used in ITPACK 2C is that used in the Yale Sparse Matrix Package. This storage format, which stores the matrix elements and column numbers in one-dimensional arrays, is very general but for linear systems arising from the discretization of elliptic partial differential equations is not as efficient on a vector computer as the storage format used in the ELLPACK package. This latter storage format involves storing the elements of the matrix and the column numbers in two-dimensional arrays. On typical problems the change in storage format has resulted in increases in speed of factors of approximately 10; see Kincaid, Oppe, and Young [1984].

Our work is continuing on long range modifications to ITPACK 2C, and eventually to ITPACK 3, which involves changes in some of the basic algorithms. We mention a few of the highlights. The algorithms based on the use of the Jacobi and the RS methods are already vectorizable. The SOR method and the SSOR method are not vectorizable, in general, but if A has Property A, one can change the ordering of the equations so that the SOR method becomes vectorizable. The most serious difficulty for vectorization among the algorithms used in ITPACK 2C is with the SSOR method. This method loses its effectiveness, in terms of convergence rate, if the red-black ordering is used. On the other hand, with the so-called "natural" ordering, the method has a good convergence rate but poor vectorization. A similar difficulty arises for some incomplete factorization methods. We are now seeking ways to overcome these difficulties; see Kincaid, Oppe, and Young [1984] and Axelsson [1984].

ACKNOWLEDGEMENTS

The work described in this paper represents the joint efforts of a number of people. Particular acknowledgement should be given to Roger Grimes, Tom Oppe and John Respess for their work on ITPACK 2C, to T. S. Mai for his work on ITPACK 3A, and to Tom Oppe for his work on ITPACKV 2C.

REFERENCES

[1] Axelsson, O., Incomplete block matrix factorization preconditioning methods: the ultimate answer?, Report CNA-195, Center for Numerical Analysis, The University of Texas, Austin, Texas (July 1984).

[2] Axelsson, O., On the B-convergence of the θ-method over infinite time for time stepping for evolution equations, Report CNA-194, Center for Numerical Analysis, The University of Texas, Austin, Texas (June 1984).

[3] Axelsson, O. and Gustafsson, I., A modified upwind scheme for convective transport equations and the use of a conjugate gradient method for the solution of nonsymmetric systems of equations, Department of Computer Sciences, Chalmers University of Technology (1977).

[4] Concus, P. and Golubs, G.H., A generalized conjugate gradient method for nonsymmetric systems of linear equations, in: Glowinski, R. and Llions, J.L. (eds.), Lecture Notes in Economics and Mathematical Systems, Vol. 34 (Springer-Verlag, Berlin, 1976), pp. 56-65.

[5] Dongarra, J.J., Leaf, G.K., and Minkoff, M., A preconditioned conjugate gradient method for solving a class of non-symmetric linear systems, Argonne Nat. Lab. Rep. ANL-81-71, Argonne National Laboratory, Argonne, Illinois (1981).

[6] Duff, I.S., A survey of sparse matrix software, chapter 8, in: Cowell, W.R. (ed.), Sources and Development of Mathematical Software (Prentice Hall, Englewood Cliffs, N.J., 1984).

[7] Eisenstat, S.E., Elman, H.C., and Schultz, M.H., Variational iterative methods for nonsymmetric systems of linear equations, SIAM J. Numer. Anal. 20 (1983) 345-357.

[8] Eisenstat, S., Elman, H., Schultz, M., and Sherman, A.H., Solving approximations to the convective diffusion equation, Symposium on Reservoir Simulation, SPE (1979).

[9] Eisenstat, S.C., Gursky, M.C., Schultz, M.H., and Sherman, A.H., Yale sparse matrix package, I, the symmetric codes, Research Report No. 112, Department of Computer Science, Yale University, New Haven, Conn. (1977).

[10] Eisenstat, S.C., Gursky, M.C., Schultz, M.H., and Sherman, A.H., Yale sparse

matrix package, II, the nonsymmetric codes, Research Report No. 114, Department of Computer Science, Yale University, New Haven, Conn. (1977a).

[11] Elman, H.C., Iterative methods for large, sparse, nonsymmetric systems of linear equations, Res. Rep. 229, Department of Computer Science, Yale University, New Haven, Conn. (1982).

[12] Elman, H.C., Iterative methods for non-self-adjoint elliptic problems, in: Birkhoff, G. and Schoenstadt, A. (eds.), Elliptic Problem Solvers II (Academic Press, 1984), pp. 271-283.

[13] Faber, V. and Manteuffel, T., Necessary and sufficient conditions for the existence of a conjugate gradient method, SIAM Jour. of Numer. Anal. 21 (1984) 352-362.

[14] Fletcher, R., Conjugate gradient methods for indefinite systems, in: Watson, G.S. (ed.), Proc. Dundee Biennial Conf. on Numer. Anal. (Springer-Verlag, Berlin and New York, 1975), p. 73.

[15] Grimes, R.G., Kincaid, D.R., and Young, D.M., ITPACK 2.0, user's guide, Report CNA-150, Center for Numerical Analysis, The University of Texas, Austin, Texas (July 1979).

[16] Hageman, L.A. and Young, D.M., Applied Iterative Methods (Academic Press, New York, 1981).

[17] Huang, R., On the determination of iteration parameters for complex SOR and Chebyshev methods, Master's Thesis, Department of Mathematics; also Report CNA-187, Center for Numerical Analysis, The University of Texas, Austin, Texas (September 1983).

[18] Jea, K.C., Generalized conjugate gradient acceleration of iterative methods, Ph.D. Thesis; also Report CNA-176, Center for Numerical Analysis, The University of Texas, Austin, Texas (February 1982).

[19] Jea, K.C. and Young, D.M., On the simplification of generalized conjugate-gradient methods for nonsymmetrizable linear systems, Linear Algebra and Its Applications 52/53 (1983) 399-417.

[20] Kincaid, D.R., Oppe, T.C., Respess, J.R., and Young, D.M., ITPACKV 2C user's guide, Report CNA-191, Center for Numerical Analysis, The University of Texas, Austin, Texas (February 1984).

[21] Kincaid, D.R., Oppe, T.C., and Young, D.M., Vector computations for sparse linear systems, Report CNA-189, Center for Numerical Analysis, The University of Texas, Austin, Texas (February 1984).

[22] Kincaid, D.R., Respess, J.R., Young, D.M., and Grimes, R.G., Algorithm 586, ITPACK 2C: a FORTRAN package for solving large sparse linear systems by adaptive accelerated iterative methods, ACM Transactions on Mathematical Software 8 (1982) 302-322.

[23] Kincaid, D.R. and Young, D.M., The ITPACK project: past, present, and future, in: Birkhoff, G. and Schoenstadt, A. (eds.), Elliptic Problem Solvers II (Academic Press, 1984), pp. 53-63.

[24] Manteuffel, T.A., The Tchebychev iteration for nonsymmetric linear systems, Numer. Math. 31 (1978) 183.

[25] Rice, J.R., ELLPACK: progress and plans, in: Schultz, M. (ed.), Elliptic

Problem Solvers (Academic Press, 1981), pp. 135-162.

[26] Rice, J.R. and Boisvert, R.F., Solving Elliptic Problems With ELLPACK, (Springer-Verlag, 1984).

[27] Widlund, O., A Lanczos method for a class of non-symmetric systems of linear equations, SIAM J. Numer. Anal. 15 (1978) 801-812.

[28] Young, D.M. and Jea, K.C., Generalized conjugate gradient acceleration of iterative methods. Part II: The nonsymmetrizable case, Report CNA-163, Center for Numerical Analysis, The University of Texas, Austin, Texas (September 1981).

[29] Young, D.M. and Jea, K.C., Generalized conjugate gradient acceleration of nonsymmetrizable iterative methods, J. of L.A.A. 34 (1980) 159.

[30] Young, D.M., Jea, K.C. and Kincaid, D.R., Accelerating nonsymmetrizable iterative methods, in: Birkhoff, G. and Schoenstadt, A. (eds.), Elliptic Problem Solvers II (Academic Press, 1984), pp. 323-342.

The final (detailed) version of this paper will be submitted for publication elsewhere.

Trends in the Theory and Practice of Non-Linear Analysis
V. Lakshmikantham (Editor)

A NON-LINEAR RESULT ABOUT ALMOST-PERIODIC SOLUTIONS OF ABSTRACT DIFFERENTIAL EQUATIONS

S. Zaidman

Department of Mathematics and Statistics
Universite de Montreal
Montreal, Quebec
CANADA

We obtained an almost-periodicity theorem for an abstract differential equation of the form

$$u'(t) = A\,u(t) + f(u(t),t)$$

-a quasi-linear problem- in the Banach space X.

Here A is the infinitesimal generator of a C_0-semigroup $S(t)$ which has an exponential decay at $+\infty$; $f = f(x,t)$ from $X \times R$ into X verifies a Lipschitz condition with respect to $x \in X$ (uniformly in $t \in R$) and is almost-periodic in t, uniformly for x in compact subsets of X.

We prove existence and uniqueness of a mild almost-periodic solution on the whole real line which means -in the sense of Foïas, - Zaidman, Ann. Sc. Norm. Sup. Pisa, 1962- that an abstract functional equation

$$u(t) = S(t-a)u(a) + \int_a^t S(t-v)f(u(v),v)dv$$

is satisfied, for all $a \in R$ and for all $t \geq a$.

In the proof one has first to study almost-periodic solutions for linear non-homogeneous equations of the form

$$u'(t) = A\,u(t) + g(t)$$

where g is almost-periodic, $R \to X$, and then to consider the general case by means of the contraction mapping principle, when we assume that the Lipschitz constant is sufficiently small.

The final (detailed) version of this paper has been submitted for publication elsewhere.

Trends in the Theory and Practice of Non-Linear Analysis
V. Lakshmikantham (Editor)

A MATHEMATICAL MODEL FOR THE STUDY OF THE MOTION OF A MIXTURE OF TWO VISCOUS INCOMPRESSIBLE FLUIDS

Anna Zaretti

Dipartimento di Matematica
Politecnico di Milano
Milano, Italy

We consider the problem of the motion of two viscous incompressible fluids in a closed basin, taking into account the molecular diffusion process. We prove the existence of a unique solution for the 3-dimensional motion provided the model is physically consistent.

INTRODUCTION

A mathematical model of a physical problem consists generally of:

a) a system of equations (generally partial differential equations): constitutive equations of the model.

b) initial and boundary conditions (associated to equations a)).

c) physical hypotheses under which the model a), b) mantains its validity, i.e., it is physically consistent: consistency conditions.

When we are studying a mathematical model from the classic point of view, conditions c) are generally overlooked and the model is associated only to a) and b); consequently the eventual solutions of a), b) may not have physical meaning or, as we shall say, the model may not be physically consistent. Taking into account conditions c), it is in many cases possible to prove that the model is well posed whenever the solution is physically significant. In other words it is possible in this way to prove an existence, uniqueness and continuous dependence theorem for the solution of a), b) in the time interval $0 \leq t \leq \bar{t}$ where $\bar{t}$ is the infimum of the values of t for which c) does not hold.

Now it must be noted that, in general, conditions c) consist in imposing to the solutions to belong to appropriate convex sets; therefore, as it is well known, the fact of taking into account the consistency conditions leads us to substitute to the equations a):

a') a system of inequalities: constitutive inequalities of the model.

The theory of variational inequalities (see for instance [1]) grants us then that if the solutions of the constitutive inequalities satisfy the consistency conditions in a time interval $(0\ \bar{t})$, they are also solutions of the constitutive equations. (For a complete interpretation of the substitution of the constitutive inequalities to the constitutive equations in the study of a mathematical model we refer to [2]).

On the other hand it is often possible to prove a global existence theorem of a unique solution for the inequalities even when only local existence has been proved for the solutions of the equations. The study of inequalities enables us therefore to have further information on the model: the time interval $(0\ \bar{t})$ is the largest one in which the solutions of the equations correspond to the physical problem. It must be noted that if a local existence and uniqueness theorem for the equations does not hold, we do not obtain in this way any new information on the model. Indeed it might happen that there exists no interval $(0\ \bar{t})$ of positive measure, in which the consistency conditions are verified: the model would then not be physically consistent for any value of t. It is then obvious that if we know a global existence, uniqueness, continuous dependence and a local regularity theorem for a), b), the study of the associated inequalities is useless.

THE NAVIER STOKES DIFFUSION MODEL

As an example of what has been said above, we shall now consider the motion of a mixture consisting of two viscous incompressible fluids (say pure water and a salt solution) in a closed basin with a molecular diffusion effect obeying the Fick's law. This study is of particular interest, for example, in the analysis of problems connected with pollution and can be effected by means of various mathematical models deduced, under more or less stringent hypoteses from the general equations which govern the motion of a mixture (see for example [2] , [3]).

It appears reasonable to assume that the simplest mathematical model that describes correctly the phenomenon is given by the following equations that correspond to what we shall from now on call the Navier - Stokes diffusion model (N.S.D. model):

$$(1)\quad \begin{cases} \rho\left(\dfrac{\partial \vec{u}}{\partial t} + (\vec{u}\cdot\nabla)\vec{u} - \vec{f}\right) = -\nabla p + \mu\Delta\vec{u} \\ \dfrac{\partial \rho}{\partial t} + \vec{u}\cdot\nabla\rho = \lambda\Delta\rho \\ \nabla\cdot\vec{u} = 0 \end{cases}$$

where:
$\vec{u}$ is the mean-volume velocity of the mixture
ρ is the density
p is the pressure
$\vec{f}$ is the external mass force
μ = const. is the viscosity coefficient
λ = const. >0 is the molecular diffusion coefficient.

Assuming that Ω is the open bounded set in $\mathbb{R}^3 = \{x_1,x_2,x_3\}$ in which the motion takes place and that the boundary Γ of Ω is constituted by a solid fixed surface, it is natural to associate to system (1) the usual initial conditions:

$$(2)\quad \begin{cases} \vec{u}(x_1,x_2,x_3,0) = \vec{u}_o(x_1,x_2,x_3) \\ \rho(x_1,x_2,x_3,0) = \rho_o(x_1,x_2,x_3) \end{cases} \qquad ((x_1,x_2,x_3)\in\Omega)$$

and the classic boundary conditions:

$$(3)\quad \begin{cases} \vec{u}(x_1,x_2,x_3,t) = 0 \\ \dfrac{\partial\rho}{\partial\nu}(x_1,x_2,x_3,t) = 0 \end{cases} \qquad ((x_1,x_2,x_3)\in\Gamma,\ 0\le t\le T)$$

($\vec{\nu}$ unit outward normal to Γ).

The conditions (3) interpret the fact that there is no flux through the boundary of Ω.

It must be noted that the model (1) (2) (3) is well posed at least locally: we recall indeed the results of Kazhikov-Smagulov [4] and Beirão da Veiga [5] that hold also for a more general model.

ASSOCIATED VARIATIONAL INEQUALITIES

If we refer to the N.S.D. model it seems reasonable to assume that the consistency conditions are the following:
1) the velocity $|\vec{u}|$ must be bounded (the model is not relativistic);
2) the pressure must be bounded;
3) the density of the mixture must be strictly positive and bounded;
4) the internal stresses must be bounded, hence:

$$|\nabla\rho|, \quad |\Delta\rho|, \quad |\frac{\partial\rho}{\partial t}|$$

must be bounded (as it can easily be deduced from the equations (1) recalling the Fick's law).

It must be noted that the conditions 2),3) are automatically satisfied in the functional class chosen for the solution.

In order to take into account the consistency conditions 1),4), we shall require the solution to belong to appropriate convex sets; that leads us to substitute to (1) the following system of <u>constitutive inequalities</u>: (1)

$$(4)\begin{cases} \int_0^t\int_\Omega \tilde{\rho}(\frac{\partial\vec{u}}{\partial t} + (\vec{u}.\nabla)\vec{u}-\vec{f}+\nabla p-\mu\Delta\vec{u})\ (\vec{u}-\vec{\phi})\,d\Omega d\eta\leq 0 \\ \int_0^t\int_\Omega (\frac{\partial\rho}{\partial t} + \vec{u}.\nabla\rho-\lambda\Delta\rho)\ (\frac{\partial\rho}{\partial t} - \frac{\partial\psi}{\partial t})\,d\Omega d\eta\leq 0 \\ \nabla.\vec{u} = 0 \end{cases}$$

where $\vec{\phi}$ and ψ are appropriate test functions.

We have proved a global existence theorem of a unique solution of the inequalities (4) with conditions (2) (3) (as we will see later). According to what has been said in § 1, the result of global existence for a unique solution of the problem (4) (2) (3) and the result of local existence for the solution of the problem (1)(2)(3), enable us to say that the N.S.D. model is well posed whenever it can be expected that the solution is physically consistent. That means that there exists in any case a time interval $(0\ \bar{t})$ of positive measure in which our model is physically consistent.

---

(1) Setting:
$$\tilde{\rho} = \begin{cases} \rho \text{ where } \alpha = \operatorname{Inf}_{x\in\Omega} \rho_o(x) \leq \rho \leq \operatorname{Sup}_{x\in\Omega} \rho(x) = \beta \\ \alpha \text{ where } \rho < \alpha \\ \beta \text{ where } \rho > \beta \end{cases}$$

the first of (1) can be replaced by the equivalent equation:

$$\tilde{\rho}(\frac{\partial\vec{u}}{\partial t} + (\vec{u}.\nabla)\vec{u} - \vec{f}) = -\nabla p + \mu\Delta\vec{u}.$$

WEAK FORMULATION OF THE CONSTITUTIVE INEQUALITIES AND THEOREM.

In order to give the existence and uniqueness theorem we need to introduce some basic definitions and notations.

Let Ω be an open bounded set $\subset \mathbb{R}^3$, with boundary Γ and $\vec{v}(x) = \{v_1(x), v_2(x), v_3(x)\}$ a vector defined on Ω. Denoting by $L^2 = L^2(\Omega)$, $H^s = H^s(\Omega)$, $H^s_o = H^s_o(\Omega)$ the usual Sobolev spaces, let us introduce the following notations:

$N = \{\vec{v}(x);\ v_j \in \mathcal{D}(\Omega),\ \nabla.\vec{v} = 0\}$

N^s= closure of N in H^s, with $(\vec{u},\vec{v})_{N^s} = (\vec{u},\vec{v})_{H^s_o}$ $(s \geqslant 0)$

$(N^s)' = N^{-s}$ dual space of N^s, with $(N^o)' = N^o$

$b(\vec{u},\vec{v},\vec{w}) = ((\vec{u}.\nabla)\vec{v},\vec{w})_{L^2}$

Let us introduce also the following closed convex sets:

$K_1 = \{\vec{v} \in L^2:\ |\vec{v}| \leqslant M_1 \text{ a.e.}\}$

$K_2 = \{g \in L^2:\ |\nabla g| \leqslant M_2,\ |\Delta g| \leqslant M_3 \text{ a.e.}\}$

$K_3 = \{g \in L^2:\ |g| \leqslant M_4 \text{ a.e.}\}$

The consistency conditions given will then be imposed by assuming that:

$$\vec{u}(t) \in \mathring{K}_1,\quad \rho(t) \in \mathring{K}_2,\quad \rho'(t) \in \mathring{K}_3 \text{ a.e. in } (0\ T).$$

Observe now that, if $\vec{\phi}(t) \in L^2(0,T;N^1)$, $\psi(t) \in H^1(0,T;H^1)$ and assuming that $\vec{u}(t)$ takes its values in N^o, (4) can be written in the form:

$$(5)\begin{cases} \int_0^t \{(\tilde{\rho}\vec{u}' - \mu\Delta\vec{u} - \tilde{\rho}\vec{f},\ \vec{u} - \vec{\phi})_{L^2} + b(\tilde{\rho}\vec{u},\vec{u},\vec{u}-\vec{\phi})\}\, d\eta \leqslant 0 \\ \int_0^t \{(\rho' + (\vec{u}.\nabla)\rho, \rho'-\psi')_{L^2} - \lambda(\rho,\psi')_{H^1}\}d\eta + \frac{\lambda}{2}\|\rho(t)\|^2_{H^1} - \\ \quad - \frac{\lambda}{2}\|\rho_o\|^2_{H^1} \leqslant 0 \\ \nabla.\vec{u} = 0 \end{cases}$$

We shall then call $\{\vec{u},\rho\}$ <u>a solution of (4) in (0,T) satisfying (2), (3) and the consistency conditions</u> if:

i) $\vec{u} \in L^\infty(0,T;N^1 \cap K_1) \cap L^2(0,T;H^2) \cap H^1(0,T;N^1) \cap H^{1,\infty}(0,T;N^o)$

$\rho(t) \in L^\infty(0,T;H^1 \cap K_2)$, $\rho'(t) \in L^\infty(0,T;K_3)$

ii) $\vec{u},\rho$ satisfy (5) a.e. in $(0\ T)$ $\forall \vec{\phi}(t) \in L^2(0,T;K_1)$,

$\psi(t) \in L^\infty(0,T;K_2)$, $\psi'(t) \in L^\infty(0,T;H^1 \cap K_3)$.

We have proved the following Theorem (global existence and uniqueness):

Let $\vec{u}_o \in N^1 \cap H^2 \cap K_1; \rho_o \in H^1 \cap K_2; \vec{f}(t) \in H^1(0,T;N^o)$ (Γ of class C^2).

Then there exists a unique solution $\{\vec{u},\rho\}$ of the problem (4), (2), (3) in (0,T).

For the proof and further details we refer to [2] .

REMARK

This problem has been studied from the numerical point of view by P. Bulgarelli and R.M. Spitaleri [6] . They solve the problem by means of a finite difference technique and the method yields numerical solutions that are in complete agreement with physical expectations and, therefore, with the choice of the N.S.D. model to describe the phenomenon. In particular they propose a computer example that proves that the model describes properly a diffusion phenomenon when we want to take into account the molecular diffusion and not only the transport type diffusion.

REFERENCES

[1] Lions J.L., Quelques méthodes de resolution des problemès aux limites non lineaires, (Dunod, Paris, 1969).

[2] Prouse G., Zaretti A., On the inequalities associated to a model of Graffi for the motion of a mixture of two viscous incompressible fluids, Archive for Rat. Mech. and Anal., (to appear).

[3] Frank D.A., Kamenetskii V.I., Diffusion and heat tranfer in chemical Kinetics, (Plenum Press, 1969).

[4] Kazhikov A.V., Smagulov S.H., The correctness of boundary value problem in a diffusion model of an inhomogeneous fluid, Sov. Phys. Dokl., 22 (5), (1977).

[5] Beirão da Veiga H., Diffusion on viscous fluids, existence and asymptotic properties of solutions, Annali Scuola Normale Pisa, (to appear).

[6] Bulgarelli U., Spitaleri R.M., A numerical method for the motion of a mixture of two viscous incompressible fluids, Preprint I.A.C., Roma, (1983).

The detailed version of this paper has been submitted for publication elsewhere.

Trends in the Theory and Practice of Non-Linear Analysis
V. Lakshmikantham (Editor)

A SURVEY OF THE OSCILLATION OF SOLUTIONS TO FIRST ORDER DIFFERENTIAL EQUATIONS WITH DEVIATING ARGUMENTS

B. G. Zhang
Department of Mathematics
Shandong College of Oceanography
CHINA

I. INTRODUCTION

The oscillation theory of differential equations with deviating arguments is a relatively new and rapidly developing branch of the theory of ODE. Numerous research papers have been devoted to this theory. Some monographs [20,61,63,47] are also published in the literature. recently, attempts have been made by many mathematicians to develop the oscillation theory of first order differential equations with deviating arguments. There are two reasons to pursue this research. First, theoretically, it is well known that the ODE $y'(t) + p(t)y(t) = 0$, $p(t) \in C(R_+)$ has no oscillatory solution, but the equation $y'(t) + y(t-\frac{\pi}{2}) = 0$, has the oscillatory solution $y = \sin t$. Therefore the oscillations of this type of first order equation is caused by deviating arguments. Second, this problem arises in many industrial and scientific problems; see, for example, [9,39,47,69]. In this paper, we attempt to present a recent development and to show some unsolved problems in this subject.

As usual, in this paper, we restrict our discussion to those solutions $y(t)$ of the equation which exist on some ray $[T_y, +\infty)$ and satisfy $\sup\{|y(t)| : t \geq T\} > 0$ for every $T \geq T_y$, in other words, $|y(t)| \not\equiv 0$ on any infinite interval $[T,\infty)$. Such a solution is said to be oscillatory if it has arbitrarily large zeros on $t \geq t_0$. Otherwise, $y(t)$ is said to be nonoscillatory. Alternative definitions of oscillations are given in [61]. Such definitions will be stated when we use them.

II. FIRST ORDER DIFFERENTIAL EQUATIONS WITH SINGLE DEVIATING ARGUMENT

We consider the first order delay equation

$$y'(t) + p(t)y(\tau(t)) = 0. \quad t \geq t_0 \tag{1}$$

where $0 \leq p(t) \in C(R_+)$, $(t) = t - \Delta(t)$, $0 \leq \Delta(t) \in C(R_+)$, $\lim_{t\to\infty} \Delta(t) = +\infty$.

<u>Theorem 1.</u>[48] If

$$\overline{\lim_{t\to\infty}}\, \Delta(t) < \infty, \quad \underline{\lim_{t\to\infty}}\, \Delta(t) \quad \underline{\lim_{t\to\infty}}\, p(t) > \frac{1}{e} \tag{2}$$

then all solutions of (1) are oscillatory.

In [30] an integral criteria that includes (2) is first discussed. For related results see [39-41,19,26,29,68]. A typical result is the following theorem.

<u>Theorem 2.</u> If

$$\underline{\lim_{t\to\infty}} \int_{\tau(t)}^{t} p(s)ds > \frac{1}{e} \tag{3}$$

or

$$\overline{\lim_{t\to\infty}} \int_{\tau(t)}^{t} p(s)ds > 1, \quad \tau'(t) \geq 0, \quad p(t) > 0 \tag{4}$$

then all solutions of (1) oscillate.

In [30] it is shown that the condition $\tau pe > 1$ is a necessary and sufficient condition for oscillation when $p(t) \equiv p > 0$, $\tau(t) \equiv \tau > 0$ in (1), so (3) can not be improved. When $\lim_{t\to\infty} \int_{\tau(t)}^{t} p(s)ds$ does not exist, (3) and (4) is intersecting. The existence of nonoscillatory solutions of (1) was given in [39].

Theorem 3 [39]. If

$$\overline{\lim_{t\to\infty}} \int_{\tau(t)}^{t} p(s)ds < \frac{1}{e} \tag{5}$$

then (1) has a nonoscillatory solution. For related work see [49].

The result in [30] was extended to first order delay differential inequality [26,36]

$$y'(t)\operatorname{sgn} y(t) + p(t)|y(\tau(t))| \leq 0, \; p(t) \geq 0, \; t \geq 0 \tag{6}$$

i.e. under condition (3), every solution of (6) oscillates. In [75,76,28,33], it was pointed out that the results of Theorems 2 and 3 are true for advanced type equations: $p(t) \leq 0$, $\tau(t) > t$ in (1).

If $p(t)$ has oscillatory behavior in (1), some oscillation criteria is obtained in [35]. Following the result in Theorem 1, we can improve Theorem 2.2 in [35] a little. For asymptotic behavior of solutions of (1) see [11,34]. Several papers [55,64,78] attempted to extend the above results to the nonlinear case. We consider the equation

$$y'(t) + p(t)f(y(\tau(t))) = 0 \tag{7}$$

Theorem 4. Assume that

(i) $t > \tau(t) \in C(R_+)$ and increasing, $\lim_{t\to\infty} \tau(t) = \infty$;

(ii) $f \in C(R)$ and nondecreasing, $yf(y) > 0$ as $y \neq 0$, and

$$\lim_{y\to\infty} \frac{y}{f(y)} = M < \infty; \tag{8}$$

(iii) $p(t) \geq 0$ is locally integrable;

(iv) [78]

$$\lim_{t\to\infty} \int_{\tau(t)}^{t} p(s)ds > \frac{M}{e} \tag{9}$$

or

(iv)' [64]

$$\overline{\lim_{t\to\infty}} \int_{\tau(t)}^{t} p(s)ds > M \tag{10}$$

then every solution of (7) oscillates.

In [64], one shows that the similar result of (10) does not hold for superlinear case. For related work of [64] see [55]. In superlinear case of (7), a sufficient condition of existence of monotone solution is obtained in [25]. For other results, see [45].

The asymptotic behavior of solutions of (7) is discussed in [7,10,11,16]. Some advanced type equations are discussed in [17,21,43].

III. THE EQUATIONS WITH SEVERAL DEVIATING ARGUMENTS

We consider the equations with several delays

(11) $$y'(t) + \sum_{i=1}^{m} p_i(t)y(t-\tau_i(t)) = 0, \quad t \geq 0$$

where $p_i \leq p_i(t) \in C(R_+)$, $\tau_i \leq \tau_i(t) \in C(R_+)$, $\lim_{t\to\infty}(t-\tau_i(t)) = \infty$, p_i and τ_i are nonnegative and do not equal zero altogether, $i \in I_m = \{1,2,\ldots,m\}$.

Theorem 5 [72]. If

(12) $$f(\lambda) = \lambda + \sum_{i=1}^{m} p_i e^{-\tau_i\lambda} > 0, \quad \text{for any real } \lambda$$

then all solutions of (11) oscillate. (A solutions always changes signs after any large number).

Obviously, condition (12) can not be improved, i.e. if $f(\bar\lambda) = 0$, $\bar\lambda$ is real, then $e^{\bar\lambda t}$ is a nonoscillatory solution when $p_i(t) \simeq p_i$, $\tau_i(t) \equiv \tau_i$ in (11). For related work see [37]. But it is difficult to check (12). Therefore, researchers try to find effective conditions for oscillation of (11). In [33], four independent sufficient conditions are obtained. In Theorems 6 - 9, we assume that $p_i(t) \equiv p_i > 0$, $\tau_i(t) \equiv \tau_i > 0$, $i \in I_m$ in (11).

Theorem 6 [33]. If anyone of the following conditions holds

(i) $p_i\tau_i > \frac{1}{e}$, some $i \in [1,\ldots,m]$; (ii) $\tau_{min}(\sum_{i=1}^{m} p_1) > \frac{1}{e}$;

(iii) $\frac{1}{m}(\prod_{i=1}^{m} p_i)^{1/m}(\sum_{i=1}^{m}\tau_i) > \frac{1}{e}$; (iv) $\frac{1}{m}(\sum_{i=1}^{m}(p_i\tau_i)^{1/2})^2 > \frac{1}{e}$

then all solutions of (11) oscillate.

Some researchers attempt to find a better sufficient condition that will include the above conditions (i)-(iv). In this direction we have the following results.

Theorem 7. If $f(\lambda_0) > 0$, $f(\lambda)$ is defined by (12), where λ_0 satisfies the equation

(13) $$\sum_{i=1}^{m} p_i\tau_i e^{-\lambda_0\tau_i} = 1$$

then all solutions of (11) oscillate.

Theorem 7 is better than Theorem 6, since (13) can be solved by the numerical method. From Theorem 7 we obtain the following results.

Theorem 8. If there exists $N_i > 0$, $\sum_{i=1}^{m} N_i = 1$ such that

(14) $$\sum_{i=1}^{m} \frac{N_i}{\tau_i}\left(1-\ln\frac{N_i}{p_i\tau_i}\right) > 0$$

then all solutions of (1) oscillate.

Theorem 8 shows that there exist many sufficient conditions for oscillation of (11) depending on the choice of N_i. The following sufficient conditions are obtained by using different N_i.

Theorem 9. If any one of the following conditions hold

(a) $\sum_{i=1}^{m} p_i\tau_i > \frac{1}{e}$, $\left(N_i = \frac{p_i\tau_i}{\sum_{i=1}^{m} p_i\tau_i}\right)$

(b) $\left(\prod_{i=1}^{m} p_i\right)^{1/m}\left(\sum_{i=1}^{m}\tau_i\right) > \frac{1}{e}$, $\left(N_i = \frac{\tau_i}{\sum_{i=1}^{m}\tau_i}\right)$

(c) there exists some $j \in [1,\dots,m]$ such that

$$\sum_{i\neq j} p_i\tau_i + p_j\tau_j e > \exp\left(-\frac{\sum_{i\neq j} p_i}{\sum_{i\neq j} p_i + p_j e}\right), \left(N_j = \frac{p_j\tau_j e}{\sum_{i\neq j} p_i\tau_i + p_j\tau_j e}, N_i = \frac{p_i\tau_i}{\sum_{i\neq j} p_i\tau_i + p_j\tau_j e}\right)$$

then all solutions of (11) oscillate.

Obviously condition (a) includes conditions (i), (ii) and (iv) of Theorem 6. Condition (a) was obtained in [18] [2] by another method.

For the variable coefficient case of (11), the following results are known.

Theorem 10. If any one of the following conditions holds

(1)[33] $\underline{\lim}_{t\to\infty} \int_{t-\tau_i(t)}^{t} p_i(s)ds > \frac{1}{e}$, for some $i \in [1,\dots,m]$

(2)[18] $\underline{\lim}_{t\to\infty} \sum_{i=1}^{m} p_i(t)\tau_i(t) > \frac{1}{e}$, $\tau_i(t)$ have uniform upper bound

(3)[41] $\underline{\lim}_{t\to\infty} \int_{t-\tau_{min}(t)}^{t} \sum_{i=1}^{m} p_i(s)ds > \frac{1}{e}$, $\tau_{min}(t) = \min(\tau_1(t),\dots,\tau_m(t))$

then all solutions of (11) oscillate. Furthermore, if

$$\overline{\lim}_{t\to\infty} \int_{t-\tau_{max}(t)}^{t} \sum_{i=1}^{m} p_i(s)ds < \frac{1}{e} \tag{15}$$

then (11) has a nonoscillatory solution, where $\tau_{max}(t) = \max(\tau_1(t),\dots,\tau_m(t))$. When $\tau_i(t) \equiv \tau_i > 0$, in [2], a uniform and more simple method is used to obtain all results of [33].

In [75,33], they pointed out that for equations with several advanced arguments similar results with Theorem 6 - 10 hold. Using a simple transformation we can reduce the equation

$$y'(t) + p(t)y(t) + \sum_{i=1}^{m} p_i(t)y(t-\tau_i(t))) = 0 \tag{16}$$

so all results with respect to (11) can be applied to (16). The forced equation to (11) is studied in [51,69]. In the papers [1,22,23,26] the nonlinear equations

$$y'(t) + \delta\sum_{i=1}^{m} p_i(t)f_i(y(\tau_i(t))) = 0,\ p_i(t) \geq 0,\ i \in I_m \tag{17}$$

and

(18) $$y'(t) + \delta f(t,y(\tau_1(t)),\ldots,y(\tau_m(t))) = 0$$

where $\delta = 1$ as $\tau_i(t) < t$ and $\delta = -1$ as $\tau_i(t) > t$, $i = 1,2,\ldots,m$, are discussed, one found that $\delta = 1$, f is sublinear and $\delta = -1$, f is superlinear. These two have the same oscillation properties.

The following theorem is a typical result.

Theorem 11 [22]. Suppose that each f_i, $1 \leq i \leq m$ satisfies

(19) $$\int_0^m \frac{du}{f_i(u)} < \infty \quad \text{and} \quad \int_0^{-m} \frac{du}{f_i(u)} < \infty, \quad \text{for any } m > 0,$$

$\tau_i(t) < t$, $1 \leq i \leq m$, then

(20) $$\sum_{i=1}^{m} \int^{\infty} p_i(t)dt = \infty$$

is a necessary and sufficient condition for all solutions of (17) to be oscillatory.

If $\tau_i(t) > t$, $1 \leq i \leq m$, and (19) is replaced by

(21) $$\int_M^{\infty} \frac{du}{f_i(u)} < \infty, \quad \int_{-M}^{-\infty} \frac{du}{f_i(u)} < \infty, \quad M > 0$$

then the conclusion of Theorem 11 is valid.

The mixed type case was also discussed in [22]. The delay (and advanced) differential inequality

(22) $$y'(t) + a(t)y(t) + p(t)f(y(t-\tau_1),\ldots,y(t-\tau_m)) \lesseqgtr 0$$

was discussed in [50,66,1*]. Nonlinear equation with distributed type deviating argument was discussed in [77]. The more general functional differential equation was discussed in [6,45]. In the above papers, the some former results were extended to more general equations.

IV. UNSTABLE TYPE EQUATION WITH DELAY

We consider the following equation

(23) $$y'(t) = p(t)y(t-\tau), \; p(t) \geq 0, \; \tau > 0, \; t \geq t_0.$$

It is easy to see that (23) always has unbounded nonoscillatory solutions when $p(t) \equiv p > 0$. What condition can guarantee that (23) has nonoscillatory solutions in the variable coefficient case? We have not seen any result to solve this problem yet. The following result is new.

Theorem 12. If

(24) $$p(t) \geq \frac{1}{t \,\ln t \,\ln \ln t \ldots \ln \ldots \ln t}$$

then (23) has an unbounded oscillatory solution. Theorem 12 is proved by the comparison theorem.

In [3,4,5,11,21,42-27,49,52] the property of oscillatory solution or asymptotic behavior of nonoscillatory solution of (23) or more general unstable type equations with deviating arguments was discussed.

V. OSCILLATION ON BOTH SIDES

Assume that equation (1) is defined on $-\infty < t < +\infty$, and $\tau(t) \to -(+\infty)$, as $t \to +\infty(-\infty)$ in (1).

Theorem 13. Assume that there exists $T \geq 0$ such that $p(t)$ and $\tau(t)$ are continuous for $|t| \geq T$, and $p(t)$ is either nonpositive or nonnegative, and

$$(25) \qquad \int_T^{\infty} p(t)dt = \infty.\mathrm{sgn}\, p, \qquad \int_{-\infty}^{-T} p(t)dt = \infty.\mathrm{sgn}\, p$$

then every solution of (1) is oscillatory on both sides, i.e. there exists two sequences of $t: t_1 < t_2 < \ldots \to +\infty$, and $\bar{t}_1 > \bar{t}_2 > \ldots > -\infty$ such that $y(t_i) = 0$, $i = 1,2,\ldots$ and $y(\bar{t}_i) = 0$, $i = 1,2,\ldots$.

For example, $y'(t) + y(\frac{\pi}{2} - t) = 0$ satisfies the conditions of Theorem 13, as expected, it has oscillatory solution on both sides of $y = \cos t$. There exists an example to show that (1) has a nonoscillatory solution on both sides even one of them converge in (25) [58].

In brief, the nonoscillation is a character of first order ODE, in general, first order ODE with deviating arguments has contrary character, i.e., it usually has oscillatory solutions. These oscillatory phenomena are caused by deviating arguments.

VI. SOME PROBLEMS

1. There extists a gap between conditions (3) and (5), also for (4) and (5) when $\lim_{t\to\infty} \int_{\tau(t)}^{t} p(s)ds$ does not exist. How do we fill this gap?
2. Does there exist a necessary and sufficient condition similar with condition $p\tau e > 1$ for $m = 1$ for oscillation of (11) with $p_i(t) \equiv p_i > 0$, $\tau_i(t) \equiv \tau_i > 0$, $i \in I_m$?
3. Can we establish a better sufficient condition that will include both conditions (a) and (b) in Theorem 9? How do we further apply Theorem 9?
4. Can we extend condition (2) of Theorem 10 to an integral form? For example,
$$\underline{\lim}_{t\to\infty} \sum_{i=1}^{m} \int_{\tau_i(t)}^{t} p_i(s)ds > \frac{1}{e} .$$
5. Can we improve the condition (15)?
6. Provide conditions for the existence of oscillatory solutions or nonoscillatory solutions of unstable type equations (23).
7. Establish conditions for the existence of oscillatory solutions to superlinear delay differential equation (7) $(p(t) \geq 0)$ and sublinear advanced differential equation (7) $(p(t) \leq 0)$.
8. Provide conditions for the existence of nonoscillatory solutions in both sides for (1), especially, consider the case that one of them in (25) converges.

REFERENCES

[1] Anderson, C. H. J. Math. Anal. Appl. 24 (1968) 430-439.

[2] Arino, O. and Gyori, I. and Jawhari, A., J. Diff. Equations 53,1 (1984)115-123

[3] Birkhoff, G. and Kotin, K. J. Math. Anal. Appl. 13 (1966) 8-18.

[4] Birkhoff, G. and Kotin, K. J. Diff. Equations 2 (1966) 320-327.

[5] Buchanan, J. SIAM J. Appl. Math. 27 (1974) 539-543.

[6] Burkowski, F. J. and Ponzo, P. J. Canad. Math. Bull. 17 (2) (1974) 185-188.

[7] Burton, T. A. and Haddock, J. R. J. Math. Anal. Appl. 54 (1976) 37-48.

[8] Bykov, Y. V. and Matekaev, A. I. (Russian) Studies in integro-differential equations, pp. 20-28, 2560257. "Ilim" Frunze, 1979.

[9] Cooke, K. L. and Yorke, J. Math. Biosci. 16 (1973) 75-101.

[10] Cooke, K. L. J. Math. Anal. Appl. 19 (1967) 160-173.

[11] Driver, R. D., Sasser, D. W. and Slater, M. L. Amer. Math. Monthly 80 (1973) 990-995.

[12] Driver, R. D. in Delay and Dunctional Differential Equations and Their Application (New York, 1972) 103-119.

[13] Elbert, A. Studia Sci. Math. Hungar. 11 (1976) 259-267.

[14] Fite, W. B. Trans. Amer. Math. Soc. 22 (1921) 311-319.

[15] Fox, L., Mayers, D. F., Ockendon, J. R. and Taylor, A. B. J. Inst. Math. Appl. 8 (1971) 271-307.

[16] Haddock, J. R. SIAM J. Math. Anal. 5 (1974) 569-573.

[17] Heard, M. L. J. Math. Anal. Appl. 44 (1973) 745-757.

[18] Hunet, B. R. and Yorke, J. A. J. Diff. Equations 53, 2 (1984).

[19] Ivanov, A. F. and Shevelo, V. N. Ukrain. Mat. Zh. 33 (1981) 745-751, 859.

[20] Kartsatos, A. G. in: Graef, J. R. (ed.), Stability of Dynamical Systems Theory and Applications 28 (1977) Chapter IV.

[21] Kato, T. and McLeod, J. B. Bull. Amer. Math. Soc. 77 (1971) 891-937.

[22] Kitamura, Y. and Kusano, T. Proc. Amer. Math. Soc. 78 (1980) 64-67.

[23] Koplatadze, R. G. Coob Akad Nauk G.S.S.R. 70 (1973) 17-20.

[24] Koplatadze, R. G. Differentcial'nye Uravnenija 10 (1978) 1400-1405.

[25] Koplatadza, R. G. (Russian) Tbilisi. Gos, Univ. Inst. Priki. Mat. Trudy 8 (1980) 24-28, 163.

[26] Koplatadze, R. G. and Canturija, T. A. Differentcial'nye Uravnenija 18 (1982) 1463-1465, 1472.

[27] Kusano, T. and Onose, H. J. Diff. Equations 15 (1974) 269-277.

[28] Kusano, T. J. Diff. Equations 45 (1982) 75-84.

[29] Ladas, G., Lakshmikantham, V. and Papadakis, J. S. in: Delay and Functional Differential Equations and Their Application (New York, London, 1972) 219-231.

[30] Ladas, G. Appl. Anal. 9 (1979) 95-98.

[31] Ladas, G. J. Math. Anal. Appl. 69, 2 (1977) 410-41 .

[32] Ladas, G. and Stavroulakis, I. P. Proceedings of the 18th International Conference on Nonlinear Oscillations, held in Kiev, USSR from August 30 to September 6, 1981.

[33] Ladas, G. and Stavroulakis, I. P. J. Diff. Equations 44 (1982) 135-152.

[34] Ladas, G., Sficas, Y. G. and Stavroulakis, I. P. Proc. Amer. Math. Soc. 88 (1983) 247-25 .

[35] Ladas, G. Sficas, Y. G. and Stavroulakis, I. P. in: Nonlinear Analysis, Lakshmikantham, V. (ed.) 1983, 277-284.

[36] Ladas, G. and Stavroulakis, I. P. Funkcialaj Ekvac. 25 (1982).

[37] Ladas, G. and Stavroulakis, I. P. Amer. Math. Monthly (1983).

[28] Ladas, G. and Stavroulakis, I. P. Canad. Math. Bull. 25 (1982) 348-354.

[39] Ladde, G. S. Int. J. System Sci. 10 (1979) 621-637.

[40] Ladde, G. S. Nonlinear Analysis 21 (1978) 259-261.

[41] Ladde, G. S. Atti. Acad. Naz. Lincei. Rend. Cl. Sci. Fis. Mat. Natur. 63 (1977) 351-359.

[42] Liilo, J. C. J. Diff. Equations 6 (1969) 1-35.

[43] Lim, Eng-Bin. J. Math. Anal. Appl. 55 (1976) 794-806.

[44] Lim, Eng-Bin. SIAM J. Math. Anal. 9 (1978) 915-920.

[45] Limanskii, V. G. Izvestiia Akademica Nauk. S.S.S.R. Ser Matematika 34 (1970) 156-174.

[46] McCalla, C. SIAM J. Math. Anal. 9 (1978) 843-847.

[47] Myskis, A. D. Moscow Nauka (1972) (Russian).

[48] Myskis, A. D. Uspekhi Matem. Nauk 5 (1950) 160-162.

[49] Myskis, A. D. Doklady Akademia Mauk S.S.S.R. 70 (1950) 953-956.

[50] Onose, H. Nonlinear Analysis 8 (1984) 171-180.

[51] Onose, H. J. Austral. Math. Soc. Ser. A 26 (1978) 323-329.

[52] Pandoti, L. J. Math. Anal. Appl. 67 (1979) 483-4 .

[53] Pesin, I. B. Differentcial'nye Uravnenija 101 (1974) 1025-1036.

[54] Romanenko, E. I. and Sharkovskii, A. V. in: Asymptotic Behavior of Solutions of Functional Differential Equations, Kiev Institut. Matematiki. Akad. Nauk. Ukr., S.S.R. (1978) 6-41.

[55] Sficas, Y. G. and Staikos, V. A. Proc. Amer. Math. Soc. 46, 2 (1974) 259-264.

[56] Sficas, Y. G. and Staikos, V. A. Funkcialaj Ekvac. 19 (1976) 35-43.

[57] Sficas, Y. G. U.I.G., Tech. Report #103 (1977) 1-14.

[58] Sharkovskii, A. V. and Shevelo, V. N. in: Problems of Asymptotic Theory of Nonlinear Oscillation, Kiev Nauk. Dunka (1977) 257-263.

[59] Sharkovskii, A. V. and Shevelo, V. N. in: Theoretical and Applied Mechanics, 3rd Congress of Speeches, V. I. Sofia (1977) 49-53.

[60] Sharkovskii, A. V., Shevelo, V. N. and Ivanov, A. F. in: Functional Differential Systems and Related Topics II, Proceedings of the Second International Conference held in Blazenjko, May 3-10, 1981. Kisielewicz, M. (ed.) Zielona Gora (1981).

[61] Shevelo, V. N. (Russian) Kiev Nauk. Damka (1978).

[62] Shevelo, V. N. and Ivanov, A. F. in: Asymptotic Behavior of Solutions of Functional Differential Equations, Kiev Institut. Matem. Akad. Mauk. Uk. S.S.R. (1978) 78-86.

[63] Shevelo, V. N. and Varekh, N. V. Proc. All. Union Conf. (1977) 247-278.

[64] Shreve, W. E. Proc. Amer. Math. Soc. 41 (1973) 565-568.

[65] Smith, H. L. J. Math. Anal. Appl. 56 (1976) 223-23 .

[66] Stavroulakis, I. P. Nonlinear Analysis 6 (1982) 389-396.

[67] Tomaras, A. Bull. Austral. Math. Soc. 17 (1977) 91-95.

[68] Tomaras, A. J. Austral. Math. Soc. 9 (1978) 183-190.

[69] Tomaras, A. Bull. Austral. Math. Soc. 13 (1975) 255-260.

[70] Tomaras, A. Bull. Austral. Math. Soc. 12 (1975) 425-451.

[71] Tramov, M. I. (Russian) Comput. Appl. Math. (1971) N.S. 31-36.

[72] Tramov, M. I. Izvestuya Vysshikh Uchebnykh Zavedenii Matematicas 19 (1975) 92-96.

[73] Winston, E. J. Math. Anal. Appl. 29 (1970) 455-463.

[74] Yorke, J. A. Lecture Notes in Math. 243 (1971) 16-28.

[75] Zhang, B. G. Science Exploration 3 (1982).

[76] Zhang, B. G. and Ding, Y. D. A Monthly Journal of Science 27, 11 (1982).

[77] Zhang, B. G. Oscillation behavior of solutions of the first order functional differential equations, Funkcialaj Ekvac. (to appear).

[78] Zhang, B. G., Ding., Y. D., Feng, R. L., Wu, D. and Wang, Q. S. J. Shandong College of Oceanology 12, 3 (1982).

[1]* Gyorim I. Nonlinear Analysis 8 (1984) 429-439.

The final (detailed) version of this paper will be submitted for publication elsewhere.

AUTHOR ADDRESS LIST

Editor

V. Lakshmikantham

Department of Mathematics
University of Texas at Arlington
Arlington, Texas 76019
USA

Authors Address List

A. R. Aftabizadeh & J. Wiener
Department of Mathematics
Pan American University
Edinburg, Texas 78539

Ravi P. Agarwal
Department of Mathematics
National University of Singapore
Kent Ridge, Singapore 0511

M. Altman
Department of Mathematics
Louisiana State University
Baton Rouge, Louisiana 70803

Ovide Arino and
Departement de Mathematiques
Universite de Pau
Avenue Louis Sallenave
64000 Pau, France

Marek Kimmel
Memorial Sloan-Kettering Cancer Center
Department of Pathology
1275 York Avenue
New York , New York 10021

O. Axelsson*
Department of Mathematics
Catholic University
Nijmegen, The Netherlands

*(from June on use this address)
C/O D. M. Young
CNA - RLM 13.150
The University of Texas at Austin, TX 78712

Prem N. Bajaj
Department of Mathematics & Statistics
Wichita State University
Wichita, Kansas 67208

Peter W. Bates
Department of Mathematics
292 Talmage Math/Computer Bldg.
Brigham Young University
Provo, Utah 84602

L. P. Belluce and
Department of Mathematics
University of British Columbia
Vancouver, B.C. V6T 1W5
Canada

W. A. Kirk
Department of Mathematics
The University of Iowa
Iowa City, Iowa 52242

S. R. Bernfeld
Department of Mathematics
University of Texas at Arlington
Arlington, Texas 76019

S. R. Bernfeld and
Department of Mathematics
University of Texas at Arlington
Arlington, Texas 76019

M. Pandian
Department of Mathematics
University of Alabama
Tuscaloosa, Alabama 35486

John A. Burns, Terry L. Herdman and Janos Turi
Department of Mathematics
Virginia Polytechnic Institute and State University
Blacksburg, Virginia

T. A. Burton
Department of Mathematics
Southern Illinois University
Carbondale, Illinois 62901

V. Capasso and L. Maddalena
Dipartimento di Matematica
Universitá di Bari
70100 Bari, Italy

Herminio Cassago, Jr.
Inst. de Ciências Matemáticas de
São Carlos - U.S.P.
Departamento de Matemática
Ar, Dr. Carlos Botelho-1465
São Carlos 13560, Brazil

(temporary address)
Department of Mathematics
University of Texas at Arlington
Arlington, Texas 76019

Alfonso Castro and R. Shivaji
Department of Mathematics
Southwest Texas State University
San Marcos, Texas 78666

Jagdish Chandra and
US Army Research Office
Box 1221
Research Triangle Park, N.C. 27709

Paul Davis
Mathematical Sciences Department
Worcester Polytechnic Institute
Worcester, MA 01609

C. Corduneanu and H. Poorkarimi
Department of Mathematics
University of Texas at Arlington
Arlington, Texas 76019

M. Crandall
Department of Mathematics
University of Wisconsin
Madison, Wisconsin 53706

G. Da Prato
Scuola Normale Superiore
Science Department
P.zza dei Cavalieri, 7
56100 Pisa, Italy

Hung Dinh and Graham F. Carey
Aerospace Engineering/Engineering Mechanics Department
University of Texas at Austin
Austin, Texas 78712

Lance D. Drager and
Department of Mathematics
Texas Tech University
Lubbock, Texas 79409

William Layton
School of Mathematics
Georgia Institute of Technology
Atlanta, Georgia 30332

Lance D. Drager, William Layton and

M. M. Mattheij
Mathematics Institute
Catholic University
Nijmegen, The Netherlands

S. Elaydi and
Department of Mathematics
University of Colorado
Colorado Springs, Colorado 80933

O. Hajek
Department of Mathematics
Case Western Researve University
Cleveland, Ohio 44106

Saber Elaydi and

Saroop K. Kaul
Department of Mathematics & Statistics
University of Regina
Regina, Saskatchewan, S4S 0A2, Canada

Alexander Eydeland
Mathematics Research Center
University of Wisconsin-Madison
Madison, Wisconsin 53705

W. E. Fitzgibbon
Department of Mathematics
University of Houston
Houston, Texas 77004

John R. Haddock
Department of Mathematical Sciences
Memphis State University
Memphis, Tennessee 38152

Charles J. Holland and
Office of Naval Research
Code 411
Math. Division
800 N. Quincy St.
Arlington, Virginia 22217

James G. Berryman
Lawrence Livermore National Laboratory
P. O. Box 808, L-200
Livermore, CA 94550

F. A. Howes
Department of Mathematics
University of California-Davis
Davis, California 95616

W. J. Hrusa and
Department of Mathematics
Carnegie-Mellon University
Pittsburgh, Pennsylvania 15213

M. Renardy
Mathematics Research Center
University of Wisconsin
Madison, Wisconsin 53706

Barbara Kaškosz
Department of Mathematics
University of Rhode Island
Kingston, Rhode Island 02881

Dorothea A. Klip
Physiology and Biophysics
Department of Computer and Information Sciences
University of Alabama in Birmingham
Birmingham, Alabama 35294

Ronald A. Knight
Mathematics Division
Northeast Missouri State University
Kirksville, Missouri 67208

Kazuo Kobayasi
Department of Mathematics
Sagami Inst. of Technology
1-1-25 Tsujido-Nishikaigan
Fujisawa 251, Japan

G. S. Ladde and
Department of Mathematics
University of Texas at Arlington
Arlington, Texas 76019

M. Sambandham
Department of Mathematics
and Computer Science
Atlanta University
Atlanta, GA 30314

G. S. Ladde and O. Sirisaengtaksin
Department of Mathematics
University of Texas at Arlington
Arlington, Texas 76019

G. S. Ladde and
Department of Mathematics
University of Texas at Arlington
Arlington, Texas 76019

A. S. Vatsala
Department of Mathematics
University of Southwestern Louisiana
Lafayette, Louisiana 70504

John E. Lavery
Department of Mathematics & Statistics
Case Western Reserve University
Cleveland, Ohio 44106

Daniel S. Levine
Department of Mathematics
University of Texas at Arlington
Arlington, Texas 76019

Howard A. Levine
Department of Mathematics
Iowa State University
Ames, Iowa 50011

Allesandra Lunardi
Universitá di Pisa
Dipartimento di Matematica
Via Buonarroti 2
Pisa, Italy

C. D. Luning and
Department of Mathematics
Sam Houston State University
Huntsville, Texas 77340

W. L. Perry
Department of Mathematics
Texas A&M University
College Station, Texas 77843

Carla Maderna and Sandro Salsa and
Dipartimento di Matematica
"F.Enriques"
Via C.Saldini 50
Milano, Italy

Carlo D. Pagani
Departimento di Matematica
Politednico di Milano
Pazzo L. da Vincei, 32
20133, Milano, Italy

Toru Maruyama
Department of Economics
Keio University
2-15-45 Mita, Minato-ku
Tokyo, Japan

Vinicio Moauro
Dipartimento di Matematica
Università di Trento
Povo, Trento, Italy

Olavi Nevanlinna
Institute of Mathematics
Helsinki University of Technology
SF-02150 Espoo 15
Finland

B. Nicolaenko
Los Alamos National Laboratory
UCB, Group T-7, Mail Stop B28
P.O.B. 1663
Los Alamos, New Mexico 87544

Juan J. Nieto
Departamento de Teoria de Funciones
Facultad de Matematicas
Universidad de Santiago
Spain

M. N. Oguztoreli, T. M. Caelli and G. Steil
Departments of Mathematics
and Psychology
University of Alberta
Edmonton, Alberta, Canada T6G 2G1

M. C. Pandian
Department of Mathematics
University of Alabama
Tuscaloosa, Alabama 35486

L. Pasquini and
Dipartimento di Metodi e Modelli Matematici
per le Scienze Applicate, 1ª Università, via
A. Scarpa 10, 00161 Roma, Italy

D. Trigiante
Dipartimento di Matematica
Università di Bari
via Nicolai 2, 70121-Bari, Italy

Gregory B. Passty and Ricardo Torrejon
Department of Mathematics and Computer Science
Southwest Texas State University
San Marcos, Texas 78666

Fred R. Payne
Department of Mathematics and Aerospace Engineering
University of Texas at Arlington
Arlington, Texas 76019

Estaban I. Poffald and Simeon Reich
Department of Mathematics
DRB 306, University Park
University of Southern California
Los Angeles, California 90089-1113

V. M Popov
Department of Mathematics
University of Florida
Gainesville, Florida 32611

Emilio O. Roxin
Department of Mathematics
University of Rhode Island
Kingston, Rhode Island 02881

L. Salvadori
Dipartimento di Matematica
Universita di Trento
38050 Povo (Trento) Italy

Károly Seitz
Department of Mathematics
Technical University of Budapest
Budapest, Hungary

R. E. Showalter
Department of Mathematics, RLM 8.100
University of Texas at Austin
Austin, Texas 78712

Michael W. Smiley
Department of Mathematics
Iowa State University
Ames, Iowa 50011

Francesca Visentin
Dipartimento di Matematica e Applicazioni
Università di Napoli
80134 Napoli, Italy

Pierre A. Vuillermot
Department of Mathematics
University of Texas at Arlington
Arlington, Texas 76019

David M. Young and David R. Kincaid
Center for Numerical Analysis
University of Texas at Austin
Austin, Texas

S. Zaidman
Department of Mathematics and Statistics
Universite de Montreal
Montreal, Quebec, Canada

B. G. Zhang
Department of Mathematics
Shandong College of Oceanography
Shandong, PEOPLE'S REPUBLIC OF CHINA

Anna Zaretti
Dipartimento di Matematica
Politecnico di Milano
Piazza L. da Vinci,32
20133, Milano, Italy